Deltas of the World

Series Editor

Abdelazim Negm, Zagazig, Egypt

Worldwide, over 600 million people live in deltas. Extensive human activities in deltas, their supplying rivers, and related maritime currents increase the pressure on deltas' natural resources and habitat equilibrium, leading to degradation of the available resources and jeopardizing inhabitants' livelihoods – and in some cases even their lives.

This book series is intended to promote a sustainable future for deltas around the globe and the communities that live in them. Accordingly, we publish studies focusing on various aspects of deltas, including hydrogeology, soil pollution, water pollution, environmental hydrology, environmental hydraulics, groundwater engineering and management, morphology, anthropology, socioeconomics, wastewater management, water resources engineering and management, agriculture, greenhouse agriculture, irrigation and drainage engineering, hydraulic structures, geophysics and geology, sedimentology, mineralogy, remote sensing and GIS studies, climate change variabilities and impacts, marine-delta interactions and sustainability, and solid waste management. The findings presented here support the SDGs and Agenda 2030 with regard to helping countries with deltas and related environments to preserve their natural resources for future generations.

The editors invite scientists, researchers and scientific communities to contribute to the series by submitting delta atlases, edited books, monographs, and high-quality conference proceedings focused on deltas around the world.

Shalini Yadav · Ravi Galkate ·
Ram Narayan Yadava
Editors

Water Resources in Delta Region in India

Water Management, Quality Assessment and Modeling

Springer

Editors
Shalini Yadav
Department of Civil Engineering
Rabindranath Tagore University
Bhopal, Madhya Pradesh, India

Ravi Galkate
National Institute of Hydrology
Central India Hydrology Regional Center
Bhopal, Madhya Pradesh, India

Ram Narayan Yadava
Research and International Affairs
Mansarovar Global University
Bhopal, Madhya Pradesh, India

ISSN 2731-832X ISSN 2731-8338 (electronic)
Deltas of the World
ISBN 978-3-032-06499-8 ISBN 978-3-032-06500-1 (eBook)
https://doi.org/10.1007/978-3-032-06500-1

This Springer imprint is published by the registered company Springer Nature Switzerland AG
The registered company address is: Gewerbestrasse 11, 6330 Cham, Switzerland

Preface

River deltas are among the most dynamic and productive landscapes on Earth, yet they remain acutely vulnerable to climate change, sea-level rise, subsidence, groundwater depletion, salinity intrusion, and anthropogenic pressures. India's major delta systems, the Ganga–Brahmaputra, Mahanadi, Godavari, Krishna, and Cauvery-constitute vital socio-ecological interfaces that sustain tens of millions of people, support extensive agricultural production, and harbour globally significant ecosystems. The edited volume "**Water Resources in Delta Region in India: Water Management, Quality Assessment and Modeling**", published under the Book Series: Deltas of the World (DELTAS), brings together scientific insights, methodological advances, and interdisciplinary analyses aimed at addressing the complex water challenges confronting India's deltaic regions.

The volume titled "**Water Resources in Delta Region in India: Water Management, Quality Assessment and Modelling**" contains 15 chapters covering various technologies and methodologies on water management, and quality assessment & modelling. The chapters are divided into two parts. Part I deals with the Water Management to the Indian context consisting of 9 chapters, and Part II presents the Quality Assessment and Modelling consisting of 6 chapters. The next para will briefly introduce the 15 chapters of the book to the audiences.

Chapter 1 authored by Mr. Kaushlandra Bairwa entitled "**An Overview of Water Resources Protection and Integrated Management in Delta Regions**" emphasizes delta regions, where water bodies are subject to severe natural and human stresses. It explores important ideas, obstacles, and tactics for efficient water resource protection. The goal is to summarize current water protection research, regulations, and technical developments thoroughly. This chapter examines the variables influencing water quality and availability, such as urbanization, agricultural runoff, industrial waste, and climatic impacts, drawing on existing. research. Using an interdisciplinary approach, the chapter examines approaches including water quality evaluations, hydrological models, and Geographic Information System (GIS) mapping to determine how well they understand and manage water resources. This chapter also examines new strategies, evaluating the possible advantages and disadvantages of community-based water management, nature-based remedies, and sophisticated pollution control technologies. Further the chapter concludes by highlighting the necessity of flexible and integrated management approaches while providing readers with a comprehensive understanding of the issues and best practices in water resource preservation. Aligned with the United Nations Sustainable Development Goals (SDGs) particularly SDG 6 (Clean Water and Sanitation), SDG 11 (Sustainable Cities and Communities), SDG 13 (Climate Action), and SDG 15 (Life on Land). This study is a useful resource for scholars, decision-makers, and practitioners pursuing sustainable water management in complex and sensitive contexts because it connects research findings with technology and policy.

Chapter 2 authored by Dr. Ravi Galkate entitled "**Delta Regions of India, Their Role and Significance in Water and Land Resources**" evaluates major Indian deltas, their ecological significance, and vulnerabilities. The Ganga–Brahmaputra Delta spans nearly 105,000 km^2. The Godavari Delta covers 7,800 km^2, the Krishna Delta 5,000 km^2, the Mahanadi Delta 9,500 km^2, and the Cauvery Delta 7,000 km^2. Collectively, they support more than 30 million people and irrigate over 6 million hectares of farmland. Irrigation efficiency ranges from 66.6 to 80 %, highlighting their importance for national food security. Ecologically, deltas such as the Sundarbans, spanning 10,000 km^2 and harbouring 693 faunal species, provide critical services. They buffer cyclones, which strike the east coast five to six times annually. However, deltas face growing threats. Salinity intrusion in the Cauvery Delta exceeds 4 dS/m, while subsidence reaches 4 cm/year in the Krishna–Godavari Delta. Sea-level rise and reduced sediment inflow could cause the loss of over 2,000 km^2 of deltaic land by 2050. Water use is largely agriculture-driven, with more than 80 % of available water consumed in many regions. Reservoirs like Nagarjuna Sagar (11.6 BCM) and Polavaram (8.0 BCM) support irrigation but require climate-sensitive management. Policy interventions such as Namami Gange, the Integrated Coastal Zone Management Project, and the National Adaptation Fund for Climate Change (covering 30+ projects in 25 states) play a vital role. Delta management directly links with United Nations Sustainable Development Goals (UNSDG) as SDG 2 on zero hunger, SDG 6 on clean water and sanitation, SDG 13 on climate action, SDG 14 on life below water, and SDG 15 on life on land. Further the chapter highlights the need for integrated and climate-resilient strategies. Water governance, ecological restoration, and legislative protection are crucial to maintain agriculture, biodiversity, and livelihoods in Indian deltas.

Chapter 3 authored by Dr. Nitin Dutt Chaturvedi entitled "**Long-Term Water Allocation and Assessment Strategies for the Delta Region**" argues that proactive, collaborative management is essential for delta resilience. It explores novel technologies and scientific frameworks for optimising water allocation and restoration, aligning resource use with ecological limits. Examining governance reforms and promoting alliances for knowledge sharing, the chapter outlines a pathway toward integrated water management plans. Moreover, the chapter briefs the discussion on long-term water management that fosters knowledge collaboration for strategies to mitigate cascading environmental and socioeconomic risks, thereby directly supporting SDGs related to water security (SDG 6), sustainable communities (SDG 11), climate action (SDG 13), and life on land (SDG 15).

Chapter 4 authored by Mr. Vinay Arya entitled "**Assessment of Hydroclimatic and Groundwater Conditions in the Middle Ganga Basin**" highlights that less than 33% of the basin in Uttar Pradesh is suitable for groundwater recharge, primarily in the Terai region, influenced by higher rainfall, flat terrain, and lineaments. However, high drainage density and clayey soils in the south impede infiltration. Artificial recharge potential is significant, with 92 identified sites on third-order streams for recharge structure construction. Isotopic investigations reveal varied groundwater recharge sources, including rainfall and surface water. Deeper aquifers are generally older than shallow ones, with some shallow aquifers showing surprisingly higher ages due to terrain features. Recharge mapping and isotopic data emphasize the need for localized interventions. The Middle Ganga Basin, a vital ecological and socio-economic region in India, faces critical challenges in water resource management due to rising demand, climate variability, and pollution. This review synthesizes current knowledge on surface and groundwater availability, sectoral demand patterns, and water quality issues. It maps the spatial distribution and seasonal dynamics of water resources, identifies contamination sources, and assesses implications for resource sustainability.

Chapter 5 authored by Dr. H. Ramesh entitled "**Stream Flow Impacts of Inter-Basin Water Transfer: A Case Study of the Nethravathi River**" focuses on most effective and popular methods for hydrological research is basin-scale distributed modelling. The "**inter-basin water transfer**" may be necessary to balance regions with excess rainfall against those with deficit rainfall through inter-basin water transfer (IBWT). It can be applied to the upstream of the river and the deltas downstream to provide a sustainable water supply for domestic, agricultural, and environmental needs. The present study is linked to the United Nations sustainable development goals SDG 6 (Clean Water and Sanitation), which addresses water scarcity receiving basin. However, they also have a significant impact on other SDGs, such as SDG 1 (No Poverty) and SDG 2 (Zero Hunger) through increased agricultural productivity, and SDG 12 (Responsible Consumption and Production) through their influence on water resource management of the donor basin. However, this could change the donor basin's hydrological and environmental regimes. Inter-basin water transfer can significantly impact the streamflow regime of rivers, as illustrated by studies on the Nethravathi River in Southwest India in the state of Karnataka, which originates in the Western Ghats. The river basin receives an average annual rainfall between 3000 and 5000 mm. However, the eastern parts of Karnataka state, including Kolar, Chikkaballapur, Ramanagar, and Tumkur, receive around 600–800 mm of rainfall. 24 TMC is diverted from this basin in an easterly direction, whereas the flow in Nethravathi is westward. The implications of rerouting 24 TMC of water from the Nethravathi River into the Cauvery basin's beneficiary regions were simulated using the SWAT model. The objective of the study was to examine the diversion of 24 TMC of water from the upper watersheds of the Nethravathi basin on downstream flows of the donor basin. Merely 20 km upstream from the estuary, at the Bantwal gauging station, the ArcSWAT model was verified and calibrated. The coefficient of determination (R2) and the Nash-Sutcliffe (N-S) coefficient were used to evaluate the model's performance between observed and simulated flows. The N-S values for calibration and validation were found to be 0.88 and 0.891, respectively, with coefficient determination values of 0.91 and 0.9. Using the Flow Duration Curve (FDC), the impact of diverting 24 TMC of water on downstream flows was carefully examined. FDC deviation was analyzed following the predicted diversion. Based on the model findings, the project impacts the higher altitude sub-basins (up to second-order streams) of the Nethravathi basin. The Nethravathi river has not been affected much overall since both the simulated and actual flows were almost identical to those at the Bantwal gauging station. This reason was due to the higher contribution of baseflow into the downstream sub-basins. The results, therefore, indicate that the extreme outflow of the basin is not considerably affected by the inter-basin water transfer of 24 TMC from the Nethravathi river.

Chapter 6 authored by Dr. Akshar Tripathi entitled "**Surface Displacement Analysis of Krishna Godavari Basin Delta Plain Using Synthetic Aperture RADAR (SAR) & GRACE Satellite Sensor Interoperability**" shows substantial surface subsidence in the region which varies from season and time to time. This subsidence is a combined effect of both petroleum exploitation and groundwater tapping from underground aquifers, which is attributed to a combination of climate change-induced sea-level rise, petroleum exploitation, and groundwater extraction. The DInSAR-based surface subsidence measurements are attributed to highly localized phenomena, but the inherent causes are often spread over long time spans and large areas. The established interoperability of the SAR data with GRACE further concretes the findings in terms of the seasonal groundwater storage changes and petroleum exploitation in the region. The sensors' interoperability establishes the versatility and reliability of the results, which show that the groundwater storage changes and petroleum exploitation is one of the many causes of the gravimetric anomalies and associated surface subsidence variations. It is a cost-effective study, as it utilizes the freely available satellite data from Sentinel-1 and GRACE, and reduces the ground truth validation. Ultimately, the study establishes a technological foundation for evidence-based policy framing, that is essential for the advancement of key SDGs related to climate action, water security and life on land.

Chapter 7 authored by Dr. Abul Amir Khan entitled "**Water, Inequality, and Development in India's Delta Regions: A Socio-Ecological Analysis**" explores the complex relationships among water resource management, social inequality, and economic development across four major Indian deltas: the Kaveri, Krishna-Godavari, Mahanadi, and Sundarbans. Analyzing secondary data from government reports, water resource studies, and scholarly research reveals how these ecologically vital and agriculturally productive landscapes face mounting pressures from climate change, environmental degradation, and socio-political marginalization. The research demonstrates that water-related vulnerabilities in these regions stem not merely from physical scarcity, but from structural inequities embedded in governance frameworks that disproportionately impact lower-caste communities, women, and gender minorities. The chapter systematically examines how sectoral water use across agriculture, aquaculture, industry, and domestic domains shapes economic outcomes and creates differentiated vulnerabilities. Each delta faces distinct challenges: the Kaveri suffers from declining freshwater availability and inter-state conflicts; the Krishna-Godavari from fractured governance and rising coastal salinity; the Mahanadi struggles from industrial competition for water and frequent flooding; while the Sundarbans endures accelerated sea-level rise and livelihood disruptions. Further, the study emphasizes how intersections of caste and gender aggravate water access disparities, leaving underprivileged groups to bear the greatest environmental hazards and resource constraints. Finally, the chapter advocates for measures to support fair and sustainable water management, advancing environmental justice, integrating multiple knowledge systems, and promoting participatory decision-making. Although such approaches face considerable challenges, they offer pathways to reduce entrenched inequalities and enhance the ecological and economic resilience of India's delta ecosystems.

Chapter 8 authored by Dr. Bhabagrahi Sahoo entitled "**Rapid Coastal Flood Extent Mapping with Sentinel-1 SAR Polarization Combinations: Thresholding and Change Detection Methods**" highlights the potential of polarization combinations such as squared addition, squared multiplication, simple addition, and multiplication of VH and VV polarizations in improving the accuracy of flood mapping using thresholding and change detection methods. The study further improves classification accuracy by eliminating misclassified pixels in areas like roads and pavements, which often share similar backscatter properties with water. This methodology reduces overestimation in flood maps, ensuring that only areas with significant water depth are classified as flooded. The integration of Google Earth Engine (GEE) provides a robust computational platform for processing large-scale Sentinel-1 SAR datasets efficiently, offering scalability and ease of implementation for emergency response and flood management. Additionally, the integration of social media data and real-time validation techniques could enhance the reliability of flood maps, supporting timely and effective decision-making in disaster management and mitigation strategies. The VV×VH polarization combination demonstrates the highest effectiveness, achieving an average accuracy of 85% with the change detection method in deriving reliable flood inundation maps using Sentinel-1 SAR data. This is followed by the VH+VV combination with an average accuracy of 82%, highlighting their potential for reliable flood mapping in disaster management. These findings underline the significance of SAR data for operational flood mapping and emergency response, particularly in regions with limited access to optical data due to cloud cover or other constraints. This study reinforces the value of Sentinel-1 SAR data in addressing global flood challenges and advancing urban resilience. Overall, the workflow advances SDG 11 (Sustainable Cities and Communities) and SDG 13 (Climate Action) by providing scalable, near-real-time flood mapping that helps cities reduce disaster impacts and build climate resilience.

Chapter 9 authored by Dr. Subhadeep Biswas entitled "**Recent Studies on the North-east Indian Rivers: A Crucial Water Resource of the Region**" reviews recent research on major river systems in Northeast India, including the Brahmaputra, Barak, Lohit, and Dibang rivers. Water obtained from these rivers is utilized for drinking as well as for other important activities for ages. However, with the due course of time, rapid industrialization as well as climate change have severely impacted water quality and ecosystem health. The focus of this chapter is to highlight the recent research works that have been carried out for determining the suitability of the water obtained from these rivers against specific purpose. Water quality index as well as different physical, chemical, biological parameters of the river water from this region will be compared. Hydrogeological and hydrochemical characteristics of several river basins in the recent times in this area are focused. This chapter also highlights the use of hydrological and remote sensing models for discharge prediction and climate impact assessment. Lastly, the review offers recommendations for sustainable water management and policymaking.

Chapter 10 authored by Dr. M. Mansoor Ahammed entitled "**Greywater Recycling and Reuse: Opportunities and Challenges**" discusses the sources and characteristics of greywater reported from around the world. It also covers various reuse applications, along with the relevant standards or guidelines from different countries. Options for its treatment are then outlined along with their suitability for different reuse. The chapter also addresses the problems and challenges involved in greywater recycling and reuse. These challenges are particularly pronounced in deltas, where high population densities, climate risks, and fragile hydrological balance demand robust, locally appropriate systems. Finally, different strategies for the successful implementation of greywater reuse are reported. This study supports the Sustainable Development Goals (SDG), particularly SDG 6, SDG 11 and SDG 12, through sustainable greywater treatment and reuse.

Chapter 11 authored by Dr. Subhadeep Biswas entitled "**Recent Development in Wastewater Recycling and Future Perspective**" focuses on an importance of the reuse of treated wastewater. Various techniques, including adsorption, SPE, photocatalysis, MFC, green wall, etc., are explored by researchers across the globe for reclamation purposes. Various nations, including India, the USA, Turkey, Italy, Australia, Saudi Arabia, etc., are practicing wastewater recycling extensively. Several benefits are highlighted by the researchers. Recent developments in this field and the status of reuse practice across various countries have been elaborated in the present article. Although reuse and recycling have become an integral part of present-day waste management policies, there are several existing challenges associated with them. In the near future, more thorough analysis of these issues and obtaining prominent solutions may be obtained from the upcoming works.

Chapter 12 authored by Dr. Manoj Kumar Yadav entitled "**Water Quality Assessment and Management in Rural India: Challenges, Contaminants, Health Risks, and**

Sustainable Solutions" contributes to achieve Sustainable Development Goals (SDGs) directly related to water, sanitation, agriculture, ecosystem, and health by promoting sustainable water management and universal access to safe drinking water. The chapter throws light on emerging contaminants and other long-known toxics, associated health risks, primarily affecting economically and socially disadvantaged populations. This is further discussed in state level case studies such as West Bengal, Bihar, and Rajasthan, demonstrating the regional water quality variability and the water safety challenges in rural India. To identify and sort water related challenges, the chapter further discusses sustainable management field-scale techniques, such as affordable water purification technologies, water conservation and community-based management models, particularly relevant to rural settings. The chapter completes itself by showing the major involvement and key roles of government policy frameworks, capacity building, and community engagement to ensure the long-term sustainability of water quality interventions in rural India.

Chapter 13 authored by Dr. Leelambar Singh entitled "**Techniques in Water Quality Modeling**" presents a brief discussion of the important techniques in water quality modeling, specifically addressing their applications and approaches. These techniques are employed to model and evaluate water quality parameters, including pollutants, dissolved oxygen, nutrient biology, and effects of land use change and climate change on the aquatic ecosystems. Methodologies used include deterministic models where the mathematically determined formulae are used to ascertain water quality, from the predictable processes, and stochastic models that challenge the regularity of processes involved in a water system with variation in the results obtained. Empirical models designed with historical data as the determinant provide basic strategies for data-rich scenarios, while mechanistic or process-based models give precise physical and chemical occurrence trends of the water basin. Other techniques, more complex, include hybrid models, artificial intelligence (AI), machine learning (ML), and professional Geographic Information Systems (GIS) that assimilate various data and have high precision in the determination of water quality parameters. In the evaluation, we compare and contrast these models and specify unique contexts in which each model works best. The conclusion has revealed that the deterministic and mechanistic models are effective mainly for system understanding, and the AI, GIS-based, and hybrid models appear to have potential in water quality real-time monitoring/forecasting. The results underscore the need for further research and the application of complex modeling methodologies for efficient future water resource management. Moreover, the strategies presented in this paper are closely related to the United Nations Sustainable Development Goals (SDGs), and SDG 6 (Clean Water and Sanitation) by enhancing water quality management practices. These modeling methods also address climate-related and anthropogenic effects, supporting SDG 13 (Climate Action) and indirectly supporting SDG 14 (Life Below Water) and SDG 15 (Life on Land), as they enable the sustainable management of aquatic and terrestrial ecosystems.

Chapter 14 authored by Dr. Leelambar Singh entitled "**Water Quality Assessment and Management**" focuses on the Chambal River Basin, aiming to identify methodologies for monitoring water quality and evaluating approaches for maintaining water quality standards. The primary goal is to assess the effectiveness of monitoring techniques and management practices for protecting various ecosystems. A detailed analysis of key physicochemical parameters such as pH, EC, hardness, HCO3, NO3, SO4, Ca, Mg, Cl, K, and F was carried out through systematic field sampling and laboratory investigations. Further, Geographic Information Systems (GIS) and remote sensing tools were integrated to perform spatial assessments of pollution sources and visualize water quality trends across the basin. The results were compared with World Health Organization (WHO) standards to identify non-compliant zones and high-risk areas. Findings reveal that while progress has been made in controlling point-source pollution, non-point sources like agricultural runoff, wastewater discharge, and urban stormwater remain significant threats. Climate change further exacerbates water quality degradation by intensifying extreme rainfall events that enhance pollutant transport and groundwater vulnerability. The study emphasizes the necessity of integrated water management strategies combining advanced monitoring technologies, spatial tools, regulatory frameworks, and community participation to secure water resources for the future. Importantly, the work contributes to the achievement of Sustainable Development Goals (SDGs), particularly SDG 6 (Clean Water and Sanitation), SDG 3 (Good Health and Well-Being), SDG 13 (Climate Action), and SDG 15 (Life on Land), by promoting safe access to water, protecting human health, fostering climate resilience, and conserving ecosystems.

Chapter 15 authored by Dr. Shanti Gopal Patra entitled "**Water Quality Assessment and Management of Various Sources in Barak Valley, South Assam, India**" describes the water quality of various sources in different parts of Barak Valley, Assam, India. Key findings of the recently conducted studies are presented in brief. Firstly, after the introduction section, a description of where the investigations are performed is provided. Traditional laboratory-based techniques and new analytical methods for measuring parameters are discussed. Apart from experimental methods, different statistical techniques are also practiced. The Water Quality Index of samples from different locations is highlighted as reported by different groups. Compilation of such reports will help identify which water from which location will be suitable for which purpose. Further, it will help the scientific community decide the treatment needed for purification purposes. It aims to contribute significantly to developing policies for overall water quality, management, and distribution in these parts of South Assam.

This volume has been edited by Dr. Shalini Yadav, Associate Professor, Rabindranath Tagore University, Bhopal, Dr. Ravi Galkate, Scientist-G & Head, National Institute of Hydrology, Central India Hydrology Regional Centre, Bhopal. and Dr. Ram Narayan Yadava, Advisor of International Affairs and Research, Mansarovar Global University, Bhopal, Madhya Pradesh, India. The editors acknowledge the dedicated efforts of all contributing authors and reviewers whose scientific rigour and cooperation made this volume possible. Together, the chapters in this volume present an integrated, interdisciplinary synthesis of contemporary knowledge on water resources in India's delta regions. I hope that this book will serve as a valuable reference for researchers, policymakers, practitioners, and students, and will inspire further research, innovation, and action toward sustainable and equitable management of these critical landscapes.

The volume editor would be happy to receive any comments to further improve future editions. Comments, feedback, suggestions for further improvement, or proposal for new chapters for next editions are welcome and should be sent directly to the volume editors.

Bhopal, India
December 2025

Shalini Yadav
Ravi Galkate
Ram Narayan Yadava

Contents

About the Editors

Prof. Shalini Yadav Professor in Civil Engineering and Head of the Centre for Advanced Water and Environmental Research at Rabindranath Tagore University, India. Her expertise includes solid and hazardous waste management, environmental quality, and water resources. She has overseen numerous R&D projects, guided 35 M.Tech. and 15 Ph.D. students, and authored over 90 publications. Prof. Yadav has contributed to several Springer Nature book series on climate change, water management, and environmental topics and serves on various conference committees and as a reviewer for international journals.

Dr. Ravi Galkate Senior Scientist and Head, Central India Hydrology Regional Centre, National Institute of Hydrology, Bhopal, India. His research focuses on water resource management, hydrological modeling, drought management, and climate change. Dr. Galkate has led national and international projects, supervised over 20 M.Tech. and 3 Ph.D. students, and authored more than 100 research papers and 10 technical reports. He has also contributed to multiple book chapters.

Prof. Ram Narayan Yadava Advisor for Research & International Affairs at Mansarovar Global University, Bhopal, and founding Vice Chancellor of AISECT University, India. Also, a founding member of CSIR-AMPRI, he specializes in solid mechanics, environmental quality, and water resources. Prof. Yadava has executed diverse projects, supervised numerous M.Tech. and Ph.D. students, and delivered invited talks globally. With over 100 publications, four textbooks, and 16 edited books, he has been recognized for his leadership in environmental sciences and hydrology. He also serves as Vice President of the International. Association of Water, Environment, Energy, and Society.

Part I

Water Management

1 An Overview of Water Resources Protection and Integrated Management in Delta Regions

Kaushlandra Bairwa, Ajay Singh Jethoo, and Leelambar Singh

Abstract

Delta regions worldwide are among the most critical yet vulnerable landscapes, supporting millions of people with fertile land, freshwater resources, and rich biodiversity. At the same time, they face severe natural and human-induced stresses, including climate change, sea-level rise, urbanization, and unsustainable water use, which threaten water quality, availability, and ecosystem stability. Protecting and managing water resources in these regions has therefore become a global priority for ensuring long-term sustainability and resilience. This chapter emphasizes delta regions, where water bodies are subject to severe natural and human stresses. It explores important ideas, obstacles, and tactics for efficient water resource protection. The goal is to summarize current water protection research, regulations, and technical developments thoroughly. This chapter examines the variables influencing water quality and availability, such as urbanization, agricultural runoff, industrial waste, and climatic impacts, drawing on existing research. Using an interdisciplinary approach, the chapter examines approaches including water quality evaluations, hydrological models, and Geographic Information System (GIS) mapping to determine how well they understand and manage water resources. This chapter also examines new strategies, evaluating the possible advantages and disadvantages of community-based water management, nature-based

K. Bairwa (✉) · A. S. Jethoo · L. Singh
Department of Civil Engineering, Malaviya National Institute of Technology, Jaipur, India
e-mail: 2023rce9315@mnit.ac.in

A. S. Jethoo
e-mail: asjethoo.ce@mnit.ac.in

L. Singh
e-mail: leelambar.ce@mnit.ac.in

S. Yadav et al. (eds.), *Water Resources in Delta Region in India*, Deltas of the World,
https://doi.org/10.1007/978-3-032-06500-1_1

remedies, and sophisticated pollution control technologies. The chapter concludes by highlighting the necessity of flexible and integrated management approaches while providing readers with a comprehensive understanding of the issues and best practices in water resource preservation. Aligned with the United Nations Sustainable Development Goals (SDGs) particularly SDG 6 (Clean Water and Sanitation), SDG 11 (Sustainable Cities and Communities), SDG 13 (Climate Action), and SDG 15 (Life on Land). This study is a useful resource for scholars, decision-makers, and practitioners pursuing sustainable water management in complex and sensitive contexts because it connects research findings with technology and policy.

1.1 Introduction

Delta regions, situated at the intersection of rivers and seas, represent some of the most dynamic and productive landscapes on the planet. Known for their fertile soils, rich biodiversity, and economic significance, these regions serve as vital hubs for agriculture, fisheries, and transportation. However, their critical role is increasingly jeopardized by challenges such as rising sea levels, saltwater intrusion, pollution, and the impacts of climate change (Hernández-Delgado 2024). Effectively addressing these issues necessitates a comprehensive framework for water resource protection and integrated management, leveraging scientific advancements, innovative technologies, and informed policymaking (Santos et al. 2023; Nwokediegwu et al. 2024).

Even though deltas cover less than 1% of the earth's land surface, they provide more goods and services for societies and ecosystems (Reader et al. 2022). These geographical features formed from sedimentary deposits at the mouth of a river contain millions of human residents and shelter some of the earth's most crucial bioregions (Boulon 2005). Traditionally, deltas have favored humman development in the region by supporting agriculture, commerce, and other aspects of culture (Tran et al. 2025). However, it is evident that their roles cut across local domains of operation and are central to food security, trade, and ecosystem conservation (Chen & Yu 2024). For example, the Ganges–Brahmaputra-Meghna Delta is the largest in the world and supports more than 140 million people and fresh water, agricultural land, and biodiversity resources (Ghosh et al. 2023; Kabir et al. 2024).

Deltas are always important and unique biological environments along coastlines, and they have many problems (Nantavisai et al. 2024). The various activities that human beings have embarked on, like constructing dams, cutting down trees, or otherwise engaging in habitual farming for years, have interfered with natural occurrences like deposition and erosion (Mazumder et al. 2024). These present challenges are, however, compounded by climate change effects such as rising sea levels, more frequent and severe natural disasters, and intrusion of salt water (Richardson et al. 2024; Daraz et al. 2024; Sharma et al.

2025). Such changes pose great risks to the future of such populations and the many other ecosystem services that deltas across the world provide (Yi et al., 2025).

Deltas face many challenges resulting from competing demands on water in the delta region. Central and Western Europe also face similar issues with the delta region, with a high percentage of the population relying on the deltas (Nantavisai et al. 2024; Teng et al. 2025). Population growth and the development of successive industrialization widen pollution waves, decrease the water table, and bring instability to river rhythms (Olson et al. 2024; Gamble et al. 2025). Also, the governance and management systems are most of the time silo-based, therefore not yielding capacities to address the multifaceted problems of water resource conservation holistically (Mazzetto 2024). Because deltas involve complicated interactions between environments, societies, and economies, they form sensitive areas that require a combination of diversified and flexible approaches (Cannatella 2023; Jia et al. 2024).

It is not only the environmental priority to preserve water resources in the delta regions for social and economic stability as well (Fouad et al. 2022; Scown et al. 2023). Lack of water and increasingly poor quality of water impact agriculture, fishery, and public health, in turn affecting the economy and livelihood (Gupta et al. 2021). For example, in Mekong Delta, the construction of dams upstream has led to depletion of fresh water, which has led to low rice production, a situation that has threatened food security and income source for millions of people (Yuen et al. 2021; Phuong et al. 2024). Similarly, the Nile Delta is vulnerable to increased flooding and the loss of potentially arable land from a combination of sea level rise and a decline in sedimentation, which has been causing increased land sinking (Abd-Elaty et al. 2024, Abulnaga et al. 2024; Abd-Elhamid et al. 2025).

Integrated Water Resources Management (IWRM) (Dass et al. 2014) provides a framework for improving water management due to the increased water stress, population pressure, and, last but not least, climate change (Ali et al. 2024). He claimed that stakeholders' cooperation, intersectoral integration, and flexibility are some factors that IWRM positively supports as the strategy of containing and managing water sources (Katusiime et al., 2020). IWRM is based on the concept of engaging stakeholders, meaning that they have to get involved in the process and include the government agencies, the public, industries, NGOs, and academics (Shukla et al. 2024). This process is not only a consultative model to seek information from stakeholders; it is an engagement tool for co-decision-making (Pagliero et al. 2024). The use of multiple voices, First Nations wisdom, and fostered insights bridging to universal and effective strategies makes it successful (Mehltretter et al. 2024). Public awareness activities, such as capacity building through participatory activities that include carrying out workshops and launching forums, build this engagement, with special focus on minorities and indigenous populaces whose input is immensely crucial in the sustainable management of water (Datta et al. 2024).

Ministerial collaboration is another component of IWRM where one sector, for example, agriculture or industry, does not harm water in the other sector (Talpur et al. 2024). For instance, poor collaboration between the water sector and the agricultural sector leads to the adoption of proper, efficient usage of water in irrigation that has no negative impact on the yield (Bwambale et al. 2022). The removal of intersectoral barriers and establishment of better cooperation within intersectoral committees, task forces and research collaboration guarantee a decision-making process that is coherent and correspondingly synchronized among the institutions and sectors. Also, the inclusion of water into generic national and regional plans for other sectors, including agriculture, energy, land, and water use and environmental conservation, enhances coordination and avoids conflict, in line with water resources management for sustainable development (Shams et al. 2022; De Luque-Villa et al. 2024).

Another of the foundations of IWRM is named adaptive governance, which is convenient if there are changes in the conditions, such as climate change, urbanization, and changing population. Management decisions require the integration of new information and recommendations to handle changing issues. Data accuracy and timeliness are the key factors in this method. It becomes easier to monitor water resource information when applying Geographic Information Systems (GIS) technologies and remote sensing, whereby decisions can be made from actual coverage information (Darkwah et al. 2021). As important as the provision of infrastructure and technologies is the development of human and institutional resources that are needed for sustainable water resource management practices at the local institutional, community, and government levels. Education and training, strategic application, learning institutions, and reinforcement involve the different stakeholders actively in water management to tackle future challenges (Alaerts et al. 2022; Adom & Simatele 2024).

This chapter provides the user with a clear and concise outline of water resources protection and an overview of integrated delta management. First, we define deltas and consider their basic features and significance. In the second step, we analyze the threats that deltas are facing. The chapter then focuses on the general concepts of integrated water resource management (IWRM), concerning stakeholders and the possibilities brought by advancing technology. It is our intention, through case studies and future recommendations, to highlight the importance of a coordinated and evolving strategy to protect these fragile habitats for ourselves and generations to come.

1.2 Delta Regions: Characteristics and Importance

1.2.1 Ecologically Significant

Delta regions are a source of the richness of the biosphere, assuming unique ecosystems: mangrove forests, swamps, and estuaries. Such ecosystems are essential in matters such as the purification of water, absorption of carbon, as well as in holding off floods (Neog et al. 2025). For instance, mangroves protect people from storms and storm surges, which are cyclone-related floods on coastlines. Deltas are also important in terms of conservation and sustaining the balance of populations of delta migratory birds and other amphibious species, which breed in the wetlands of those areas. The Sundarbans mangroves are home to an iconic species such as the Bengal tiger while the Mekong Delta supports animals like Irrawaddy dolphin (Biswas et al. 2023). Besides, delta ecosystems participate in carbon sequestration, hence provide a great help to climate change mitigation. Nevertheless, these offered ecological benefits are gradually diminished by habitat destruction, pollution, and alteration caused by climate change which further validates the call for preventive conservation practices.

1.2.2 Economically Vital

Delta regions indeed are economic contributing prosperity hubs to the local, national and international economy. Because of their structures, they are key areas for farming basic foods like rice, wheat and sugar cane. For instance, the Mekong Delta is a source of more than half of rice production in Vietnam and a significant share of its food imports and export earnings fisheries and aquaculture are other economic support structures of the delta regions (Tran et al. 2024). Seas and shores and estuaries offer the fish and shellfish resources that feed millions around the world. The Ganges–Brahmaputra-Meghna Delta, for instance, is one of the major producers of fish in fresh and brackish water (Uddin et al. 2024). Furthermore, numerous deltas contain significant ports and centres of transportation, which support the processes of buying and selling. Cities in the deltas include Cairo in Nile Delta and Kolkata in the Ganges Delta and are economically productive regions featuring industrial and service sectors of the economy; even then, they have issues like resource degradation, environmental pollution, and imbalanced growth. Economically efficient strategies combined with policy recommendations are necessary for achieving the main objective of the paradigm: achieving sustainable development and protecting the ecology of deltas and the communities dependent on them.

1.2.3 Culturally Important

Deltas have always been habited area due to the availability of good soils for cultivation and water sources, which people have been using since prehistoric periods. These species are valued for their cultural and religious importance in people's customs, celebrations and heritage. The Nile Delta, for example, is almost equated with the ancient Egyptians; the Ganges Delta, with the Hindu epic, continues to encourage millions on their pilgrimage of faith; cultural multiplicity prevails in the delta forming zones which offers innovative possibilities through the interacting deltaic environment. Several of the events, folklores, and pieces of artworks in these regions tend to tap the rivers within the regions. The European-style floating markets of the Mekong Delta where people buy fresh produce to traders aboard boats and the Bonbibi festival in the Sundarbans are true deltas of the specific culture and lifestyle (Srinivas et al. 2025). Nevertheless, the cultural demographics of the deltas are at risk due to urbanization and developmental processes, environmental degradation, etc. Members of culture need to feel that it is safe to maintain their cultural heritage and follow sustainable development principles. Imagination can attempt to achieve and promote the countercultural to help people become more aware of deltas and their value to protect.

1.3 Variables Affecting Water Resources in Delta Regions

Delta areas are rather diverse environments, the tendencies of which depend on numerous natural and man-made factors. These variables affect water quality and quantity and the general health of the ecosystems in various ways. The next subsections provide a detailed overview.

1.3.1 Climate Change and Weather Variability

Climate change and weather Variability in coastal delta regions is due to the weather is highly volatile and is likely to affect water supply in the regions. Global warming escalates the process of evaporation, resulting in less water in rivers and underlying aquifers (Singh et al. 2020). As well, changes in weather, such as long dry seasons or short periods of intense rainfall will also affect the flow of the rivers and the distribution of the water resources. Another primary issue that coastal deltas are likely to experience is the rising sea levels, which will intrude into the delta and pollute freshwater sources. The identified climate trends therefore imply that delta-dependent regions need to implement robust water management strategies to address these fluctuating and frequently volatile weather.

1.3.2 Human Activities and Urbanization

It can be stressed that various human impacts affecting water resources have a significant impact on the delta areas. Population increases and or growth puts much demand on water and this means that both surface waters and groundwater sources are overstretched. The formation of towns, black-topped roads, and other junctures interfering with water systems reduces the capacity of ecosystems to purify, store, and distribute water. Development also replaces natural water regulators such as wetlands, forests, and other natural water sources with infrastructure, a situation that worsens the water situation (Adeoba et al. 2025). Moreover, excessive exploitation of water from the groundwater sources for domestic, industrial, and agricultural purposes results in a dangerous phenomenon that indicates diminishing water tables and creates problems for that water supplies cannot be sustained in the future. Another negative result of increased urbanization is the management of water pollution since wastes and other pollutants end up in a river or a lake.

1.3.3 Pollution and Contamination

Nevertheless, water pollution continues to pose a major challenge to delta areas. World deltas are densely populated and industrialized; thus, a great portion of waste and contaminants from these places reach water bodies. Sewage without treatment, industrial effluents, and chemicals from agriculture contribute to rivers, hence lowering water quality. It also impacts the local population's health, threatens the life of aquatic organisms, and decreases the water supply for irrigation and human consumption (Mishra et al. 2023). More often, there is an absence of waste disposal facilities that would accommodate the Byzantine waste, coupled with unhygienic water usage. When pollution levels threaten the supply and quality of water, it becomes problematic to deltas in that human health and productivity in the economy will be affected.

1.3.4 Geological and Hydrological Characteristics

The geological features of delta regions significantly influence how water moves and is stored. Deltas are formed by the deposition of sediment carried by rivers, and their soils are often loose and highly permeable (Mostern 2021). This can make groundwater recharge highly effective in some areas, while in others, impermeable layers beneath the surface can restrict the flow of water. The natural water systems in delta regions, rivers, lakes, and groundwater depend on both the geological structure and the hydrological dynamics of the area (Singh et al. 2021). Human interventions, such as the construction of dams or changes in river channels, can disrupt these natural processes and affect how water is distributed and stored, potentially leading to flooding, water shortages, or changes in water quality.

1.3.5 River Flow and Sedimentation Patterns

The movement of water in delta regions is heavily influenced by the flow of rivers and the patterns of sedimentation they create. Rivers that feed into deltas carry both water and sediment, which help build and maintain the land (Singh et al. 2020). When river flow is altered by human activity, such as the construction of dams, water diversion projects, or industrial use sedimentation patterns can change. In some cases, sediment build-up can prevent the natural replenishment of wetlands and coastal areas, leading to land loss and increased vulnerability to flooding. Conversely, a reduction in sediment flow can cause coastal erosion and affect water quality by limiting the natural filtration process. Managing the balance between water flow and sediment deposition is crucial for maintaining the stability and health of delta ecosystems.

1.3.6 Saltwater Intrusion and Coastal Erosion

In coastal delta regions, saltwater intrusion is a growing concern. Rising sea levels caused by climate change are pushing saltwater into freshwater aquifers, contaminating the water supply. This poses a significant threat to agricultural lands, as saltwater renders soil infertile and damages crops. Coastal erosion, often exacerbated by human activities like construction and deforestation, further amplifies the effects of saltwater intrusion. When coastal buffers, such as mangroves and wetlands, are lost, the coastline becomes more exposed to erosion and the inland spread of saltwater. Managing saltwater intrusion requires proactive measures, such as constructing barriers, restoring coastal ecosystems, and implementing sustainable water extraction practices to protect vital freshwater resources (Tran et al. 2024).

1.3.7 Land Use and Agricultural Practices

Farming is a major activity practiced in most delta areas and the use of the land to practice this activity can greatly affect water resources. Most of the irrigation works involved huge quantities of water that when supplied can exert pressure on rivers and groundwater. In most delta areas, poor irrigation practices result in the utilization of water and increase pressure on scarce water resources (Haldar et al. 2022).' Further, direct discharges from agriculture produce pollutants such as fertilizers, pesticides, and herbicides which in one way or another affect the quality of water. Sustainable farming practices also affect water resources because unsustainable farming turned the productivity of the land into a desert and made the land to have poor water retention capacity. Introducing enduring strategies in water irrigation, Organic farming, and appropriate agriculture are the most crucial ways that can help future generations have the same source of water as the present generation.

1.3.8 Ecosystem Degradation and Biodiversity Loss

Ecosystems such as wetlands, mangroves, and floodplains are critical to maintaining water quality and regulating water flow in delta regions. Unfortunately, human activities such as deforestation, land reclamation, and industrial development have caused significant degradation of these ecosystems. Wetlands, for instance, serve as natural filters, improving water quality and providing habitat for many species of fish and wildlife. The destruction of these natural buffers not only harms biodiversity but also leaves delta regions more vulnerable to flooding, erosion, and water contamination. Restoring and conserving these ecosystems is essential for maintaining a sustainable balance in water management, as they help regulate water flow, prevent saltwater intrusion, and improve overall water quality (Bayabil et al. 2021).

1.3.9 Governance and Institutional Capacity

Effective governance is key to managing water resources in delta regions. Poor governance, lack of coordination, and inadequate institutional capacity can lead to inefficient management of water resources. In many cases, water resources in delta areas are shared by multiple stakeholders, including governments, local communities, industries, and agriculture, which makes it crucial to have a transparent and inclusive water management framework. Local communities must be involved in the decision-making process to ensure that their needs are met and that water management strategies are effective. Strengthening the institutions responsible for water management, promoting cooperation between different levels of government, and improving policy implementation are all necessary steps to ensure that water resources are managed sustainably in delta regions.

1.4 Approaches to Water Resource Management

A combination of water quality evaluations, hydrological models, and Geographic Information System (GIS) mapping is widely used to manage water resources and ensure water quality protection (Saravani et al. 2024). These approaches help researchers, policymakers, and water resource managers understand the current state of water resources, predict future scenarios, and devise appropriate management strategies. In the next subsections, we present how each of these approaches contributes to water resource management:

1.4.1 Water Quality Evaluations

Water quality evaluations involve assessing the chemical, physical, and biological conditions of water bodies to determine whether they meet specific standards for use (drinking, agriculture, recreation, etc.) (Das 2025).

Components of Water Quality Evaluations

A. *Monitoring Parameters*: They include

a. *Physical Parameters:* Air Temperature, Water Turbidity, Water pH, DO (the level of dissolved oxygen in the water), and conductivity.
b. *Chemical Parameters:* Nitrogen, phosphorus, heavy metals, pesticides, organic matter, and salinity.
c. *Biological Parameters:* Such as the presence of microbial content (E. coli), algae, BOD levels, presence of invasive species, and others.

B. *Sampling and Analysis:* The number of samples can be measured over some time from different sources of water and tested in laboratories for pollutants, toxins, and other products (Singh et al. 2024).
C. *Water Quality Indices:* Under the concerned index, WQI is focused on presenting a single numerical value to generalize the status related to water quality from the spokesperson of different indicators.
D. *Risk Assessment:* Risk assessments related to the contamination of drinkable water to humans and the effects of poor water quality on ecosystems (Singh et al. 2024).

Applications of Water Quality Evaluations

A. *Pollution Source Tracking:* Identifying point and non-point sources of pollution (e.g., industrial discharges, agricultural runoff, urban stormwater).
B. *Compliance Monitoring:* Ensuring that water bodies meet regulatory standards for designated uses.
C. *Early Warning Systems:* Using water quality data to forecast potential contamination events or public health risks (e.g., harmful algal blooms).
D. *Restoration and Mitigation:* Guiding water quality improvement efforts, such as wastewater treatment enhancements, riparian buffer restoration, or nutrient control.

1.4.2 Hydrological Models

Hydrological models are mathematical representations of the water cycle that simulate how water moves through the environment (Perera et al. 2025). These models are essential for understanding the distribution, movement, and availability of water within a specific area or watershed.

A. *Surface Water Models:* These models are used for the representation of the water behavior on the land surface and include the river discharge, the stormwater runoff, and the flooding plain behavior.

Examples *HEC-HMS (Hydrologic Engineering Centre's Hydrologic Modelling System):* Employed in the analysis of rainfall-runoff process, flood prediction, and watershed hydraulics.

SWMM (Storm Water Management Model): It is one of the commonly used tools for modeling urban stormwater systems, including drainage networks, drainage, and water hazards such as erosion, flooding, and water pollution. It simulates the flow of water over land surfaces, including river discharge, stormwater runoff, and floodplain dynamics.

B. *Groundwater Models:* These models mimic the motion of water beneath the surface and the relationship of groundwater with surface water structures.

Examples *MODFLOW:* A groundwater modeling system that would help in modeling the groundwater flow and evaluating the outcome of management strategies involving the utilization of the water resource.

Hydro-Geosphere: Surf-GW 1D–A comprehensive coupled model of surface water and groundwater hydrology operating in one model environment.

Integrated Hydrological Models: Using a combination of one-dimensional models focusing on surface water, and the two-dimensional ones that consider groundwater as an important element of the watershed, for instance, MIKE SHE or WASIM.

Applications of Hydrological Models

A. *Flood Prediction and Management:* Estimation of flood occurrence using rainfall values, types of land use, and drainage network systems. Automated models are used in the construction of various structures used in the mitigation of floods, such as dams, levees, and drainage channels.
B. *Water Resource Allocation:* Determining how much water is in a particular supply for different purposes (farming, domestic use, etc.) and how best to distribute this water during low water availability.
C. *Climate Change Impact Assessment:* Making future water availability assessments based on climate change projections, ranging from rainfall changes to temperature changes (Singh et al., 2020).
D. *Water Quality Modelling:* Combining linkages between hydrological models with water quality models, such as QUAL2K, to predict the transportation and levels of pollutants in rivers and lakes.

1.4.3 Geographic Information System (GIS) Mapping

GIS is an excellent application for mapping, analyzing, and displaying Geographical information about water resources (Abijith et al. 2020). Due to the availability of satellite imagery, field surveys, and hydrological models, GIS provides water managers the opportunity to analyze the water resource data spatially and temporally.

1.4.3.1 Components of GIS Mapping for Water Resources

A. *Data Collection:* Some of the data sources include topographical maps, land use, hydrological data, soil type, rainfall, and water quality measurements.
B. *Spatial Analysis:* Some GIS applications include water resources mapping of geographic distribution, accurate identification of high pollution areas, flood-prone regions, and transitional changes in a given region, for example, in a given period of years (Yaseen 2024).
C. *Overlay Analysis:* The process of superimposition of different thematic layers such as land use, water quality, and hydrological features to determine the relation, pattern, or area requiring attention.
D. *Model Integration:* Hydrological modeling results can be inputted and displayed by the GIS so that landscape users can simplify complicated information and make appropriate decisions.
E. *Remote Sensing Integration:* One of the GISP is the use of satellite or drone images that have high spatial and temporal data of changes in water quality, for example, land cover, or water access.

Applications of GIS in Water Resource Management

A. *Watershed Management:* GIS can capture watersheds, demarcate drainage basins, and concomitantly analyze the effects of land use on water quality and quantity.
B. *Flood Risk Mapping:* Flood risks are simulated using GIS, and areas susceptible to floods, as well as approaches to averting floods, are determined.
C. *Water Distribution Network Analysis:* GIS plays a vital role in the management and overall improvement measures applied to water systems within cities. This involves surveys on physical structures, consumption patterns in pipe networks, and leak detection (Farah et al. 2024).
D. *Water Resource Planning:* GIS can be used for the analysis of surface water supplies and groundwater, and the potential allocation of new construction projects like reservoirs and water treatment plants.
E. *Ecosystem Management:* GIS to analyze the effects of various human activities on water bodies and identify areas of conservation or rehabilitation.

1.5 Innovative Strategies for Water Resource Protection

These are new strategies for water resource management, including community-based approaches, nature-based solutions, and advanced pollution control technologies, which can provide valuable insights into improving water quality and availability (Banso et al.

2023). Each strategy offers unique advantages but also poses specific challenges and limitations. The next subsections presents an evaluation of these strategies:

1.5.1 Community-Based Water Management

Community-based water management (CBWM) involves local stakeholders (such as residents, community groups, local governments, and non-governmental organizations) in the management, conservation, and equitable distribution of water resources. This approach emphasizes local knowledge, participation, and collaborative decision-making (Tantoh et al. 2021) (Table 1.1).

Example The Participatory Irrigation Management (PIM) program in India, where local farmer groups manage irrigation systems, has helped improve water use efficiency and reduce conflicts over water distribution.

Table 1.1 Advantages and disadvantages of CBWM

Advantages	Disadvantages
Empowerment and Ownership: By involving local communities in water management, CBWM fosters a sense of ownership and responsibility, leading to more sustainable and effective management practices	*Limited Resources:* Communities may lack the financial, technical, and human resources required to implement complex water management projects, especially in areas with limited capacity
Cost-Effectiveness: Community-driven management can reduce the need for expensive top-down interventions, and the approach often focuses on low-cost, locally appropriate solutions	*Coordination Challenges:* Ensuring effective coordination among diverse local stakeholders can be difficult, especially when interests and priorities conflict
Local Knowledge Integration: Communities possess valuable traditional knowledge and a deep understanding of local water systems, which can lead to more context-sensitive and relevant solutions	*Vulnerability to External Influences:* Local initiatives may be undermined by external factors such as political instability, corruption, or large-scale development projects that prioritize economic growth over sustainability
Increased Awareness and Education: Community involvement enhances environmental education and awareness, which can lead to better water conservation practices and behavior changes	*Sustainability Issues:* Over time, community-based initiatives may struggle with long-term sustainability, especially if they rely heavily on external funding or if there is insufficient institutional support

Table 1.2 Advantages and disadvantages of NBS

Advantages	Disadvantages
Environmental Co-Benefits: NBS often provides multiple environmental benefits, such as enhancing biodiversity, improving soil health, increasing carbon sequestration, and creating recreational spaces *Cost-Effectiveness:* In many cases, nature-based solutions can be more cost-effective than engineered infrastructure. For example, wetlands act as natural filters for pollutants, and restoring floodplains can reduce the need for costly flood defenses *Resilience to Climate Change:* NBS helps improve ecosystem resilience, providing a buffer against climate change impacts, such as increased flooding, droughts, and sea-level rise *Sustainable and Long-Term Solutions:* Unlike traditional infrastructure, nature-based solutions evolve and can provide long-term solutions as ecosystems regenerate and adapt	*Uncertainty of Effectiveness:* The effectiveness of NBS can vary significantly depending on local conditions (e.g., climate, soil types, and species composition), and their long-term performance may be unpredictable *Land Use Conflicts:* Implementing nature-based solutions in urban areas or agricultural lands may face resistance due to land use conflicts or concerns about competing economic interests *High Initial Investment:* Although nature-based solutions can be cost-effective in the long term, they may require significant upfront investment in restoration or infrastructure development *Maintenance and Monitoring:* Nature-based solutions require ongoing monitoring and maintenance to ensure they continue to function as intended. This can be resource-intensive, particularly in large-scale projects

1.5.2 Nature-Based Solutions (NBS)

Nature-based solutions involve using natural ecosystems and processes to address water-related challenges. These solutions can include wetland restoration, reforestation, urban green spaces, and floodplain restoration, all of which aim to improve water quality, reduce flood risk, and enhance water availability (O'Hogain et al. 2018) (Table 1.2).

Example The restoration of the Huangpu River wetland in Shanghai, China, to reduce urban flooding and improve water quality is an example of NBS being used to address water challenges while providing recreational and biodiversity benefits.

1.5.3 Sophisticated Pollution Control Technologies (SPCT)

Advanced pollution control technologies focus on reducing contaminants in water bodies through engineered solutions, such as wastewater treatment plants, advanced filtration systems, chemical treatment, and nutrient removal technologies (Shah et al. 2020). These technologies are often deployed in urban areas or industrial zones where pollution is a major concern (Table 1.3).

Table 1.3 Advantages and disadvantages of SPCT

Advantages	Disadvantages
High Efficiency: Advanced technologies, such as membrane filtration, reverse osmosis, and electrochemical treatment, can effectively remove a wide range of pollutants, including heavy metals, organic compounds, and pathogens *Scalability:* Many pollution control technologies can be scaled to meet the demands of different regions, ranging from small-scale decentralized systems to large, centralized treatment plants *Precision and Control:* Sophisticated systems provide precise control over water quality, enabling compliance with stringent environmental standards and ensuring safe water for human consumption, agriculture, and industrial use *Immediate Impact:* These technologies can quickly address pollution issues, providing an immediate improvement in water quality, especially in heavily contaminated areas	*High Capital and Operating Costs:* Sophisticated pollution control technologies require significant initial investment and ongoing operational costs, making them financially challenging for developing regions or communities with limited resources *Energy Intensity*: Many advanced treatment processes, particularly desalination or reverse osmosis, are energy-intensive, which can contribute to high operating costs and increase the environmental footprint Technological Complexity: The operation and maintenance of advanced treatment systems require specialized knowledge and training, which can be a barrier in regions with limited technical expertise *Residual Waste:* Some technologies, such as reverse osmosis, generate waste by-products (e.g., brine, sludge), which require proper disposal and management to avoid creating secondary environmental problems

Example The Membrane Bioreactor (MBR) systems used in wastewater treatment in cities like Singapore and New York are advanced technologies that remove contaminants effectively while minimizing space requirements (Table 1.4).

Table 1.4 Comparative evaluation of strategies

Strategies	Advantages	Disadvantages	Best use cases
Community-based water management	Empowerment, cost-effective, and integration of local knowledge	Limited resources, coordination challenges, vulnerability to external pressures	Small communities, rural areas, and Indigenous systems
Nature-based solutions (NBS)	Environmental co-benefits, resilience to climate change, cost-effective	Uncertainty in effectiveness, land use conflicts, maintenance needs	Coastal areas, urban flood mitigation, and wetland restoration
Sophisticated pollution control technologies	High efficiency, precision, and scalability	High costs, energy-intensive, complex maintenance	Urban areas, industrial zones, high-pollution areas

The combination of community-based water management, nature-based solutions, and advanced pollution control technologies offers a holistic approach to addressing water challenges. Each strategy has its own set of advantages and limitations, and its effectiveness depends on local conditions, available resources, and the nature of the water-related issues being addressed.

Community-based approaches are particularly effective in small-scale, local contexts where community engagement and low-cost solutions are essential.

Nature-based solutions offer sustainable, resilient, and multi-benefit approaches, making them particularly valuable in areas vulnerable to climate change and natural disasters.

Sophisticated pollution control technologies provide high-efficiency solutions for heavily polluted environments but come with significant financial and operational challenges.

The key to successful water management lies in integrating these strategies, tailoring them to local needs, and ensuring long-term sustainability through collaboration, innovation, and adaptive management practices.

Integrated water management is to be considered as the key need and strategy when dealing with water issues and addressing the problems of water conservation (Kolahi et al. 2024). These approaches are used to achieve goals of economic growth, environmental conservation, and societal welfare through learning approaches such as factor conditions changes like climate variation, population growth, and land use change (Arevalo 2024). These are flexibilities, systems approach, stakeholder engagement, ecological protection, climate adaptation, and integrated water management, at multiple scales, affordability, monitoring, and policies. The positive development strategies are; the protection and efficient use of water, efficiency in the prevention of pollution, enhancement of water infrastructure, restoration of natural systems, and efficiency in the provision of information to the public (Sunny 2024). For this reason, implementation is at a local, regional, and global level, which involves the constant use of data and policies to foster adaptive and sustainable management of water resources. As a result, anyone wishing to achieve long-term goals in the community must seek to innovate and draw on the support of different stakeholders to ensure sustainable water practices that can continue to supply water in the future.

1.6 Discussion

On the basis of the analysis of the current and finished studies, it is possible to note that this study is a collection of knowledge about how to protect and manage water resources in the countries of the delta and explains how the complex relationship of issues arising due to urbanization, industrialization, and climate change is intertwined. These results prompt the need to embrace the idea of integrated water resources management as a comprehensive strategic policy in managing water, based on the extensive involvement of stakeholders, representing all the layers of society, and engineering interventions, including hydrological modeling and Geographic information system (GIS) map. The notion and theories such as integrated water resources management, decentralization of water, natural capital and advanced water pollution control are tested as more appropriate methods of water management.

Notably, the issues and plans presented in this chapter are closely related to the United Nations Sustainable Development Goals (SDGs). Specifically, the priorities of access to safe and equitable water resources address SDG 6 (Clean Water and Sanitation), whereas priorities to control urban pressures and industrialization relate directly to SDG 11 (Sustainable Cities and Communities). The paper on climate-related stresses and adaptive management requirement addresses SDG 13 (Climate Action) and the awareness of natural capital and ecosystem-based management systems addresses SDG 15 (Life on Land). The chapter presents the global applicability of delta governance by putting water resource protection in the context of SDGs and granting opportunities of sustainable, inclusive, and resilient water governance. Nonetheless, some research shortcomings should be mentioned. The disaggregated data was applied when classifying delta countries and it might not reflect particular situations in a given delta region. Other constraints refer to the aspect of availability of data and the difficulty of having all the stakeholders participating optimally. The institutional, financial, and technical obstacles to the implementation of integrated management strategies further exert pressure on such issues. Lastly despite the purpose of the study being to identify the objectives of the approach of adaptive management, it is less specific on the long-term trajectories of supporting water management practices. Addressing these constraints is crucial to a successful implementation and maintenance of the water solutions in the delta areas, especially when considering the SDG targets.

1.7 Conclusion

This chapter has explored the crucial role of delta regions and the pressing challenges they face in managing water resources sustainably. Deltas, rich in biodiversity and economic importance, are increasingly under threat from climate change, pollution, urban growth, and resource overuse. These challenges highlight the need to shift from traditional, fragmented management to more holistic and adaptive strategies. One of the key takeaways is the power of community involvement. By engaging local communities in water management, we can ensure that practices are not only sustainable but also tailored to the unique needs of each area. Nature-based solutions, like restoring wetlands and forests, offer a natural, cost-effective way to combat climate impacts while preserving ecosystems and contribute to achieving the SDGs. On the other hand, advanced pollution control technologies help tackle severe pollution issues, though they require significant investment and expertise. The chapter emphasizes the importance of integrating various scientific, technological, and policy-driven approaches, to create a flexible and comprehensive water management system. Tools like hydrological models and GIS mapping help us better understand water resources and make informed decisions that balance environmental health with economic and social needs. Looking ahead, the future of our delta regions depends on collaboration. Governments, communities, and the private sector must work together to develop inclusive, adaptable policies that protect these vital areas for generations to come. This chapter serves as a guide for anyone looking to understand and contribute to the sustainable management of our precious delta ecosystems.

1.8 Recommendations

Adopt Integrated Water Resources Management (IWRM): Implement IWRM frameworks that bring together agriculture, industry, and urban sectors to ensure balanced and sustainable water use, directly supporting SDG 6 (Clean Water and Sanitation).

Promote Nature-Based Solutions (NBS): Restore wetlands, mangroves, and other ecosystems to enhance flood regulation, biodiversity, and climate resilience, advancing SDG 13 (Climate Action) and SDG 15 (Life on Land).

Invest in Advanced Monitoring and Technology: Expand the use of hydrological models, GIS, and remote sensing tools for real-time water quality monitoring, early warning systems, and adaptive management in delta regions.

Strengthen Community Participation: Empower local communities and marginalized groups in decision-making to foster inclusive governance and sustainable practices, aligning with SDG 11 (Sustainable Cities and Communities).

Ensure SDG Integration and Policy Alignment: Explicitly link water protection and management strategies to SDGs 6, 11, 13, and 15, ensuring that local and regional efforts contribute meaningfully to global sustainability targets.

Acknowledgements The authors want to thank the Ministry of Education of India and MNIT Jaipur for their support.

State whether you used AI or not in preparing the chapter.

No AI was used in the preparation of this chapter.

Author Contribution Kaushlandra Bairwa: Data collection, investigation, formal analysis and supervision; Ajay Singh Jethoo: Review and drafting; Leelambar Singh: Review and supervision.

Funding No.

Data Availability Not applicable.

Declarations **Ethics approval and consent to participate:** Not applicable.

Consent for publication Not applicable.

Competing Interests The authors declare no competing interests.

References

Abd-Elaty I, Kuriqi A, Ramadan EM, Ahmed AA (2024) Hazards of sea level rise and dams built on the River Nile on water budget and salinity of the Nile Delta aquifer. J Hydrol: RegNal Stud 51:101600. https://doi.org/10.1016/j.ejrh.2023.101600

Abd-Elhamid HF, Abdelfattah M, Zeleňáková M, Kabeel AE, Barańczuk J, Gharbia SS, Mahdy M (2025) Monitoring coastal changes in Port Said, Egypt using multi-temporal satellite imagery and GIS-DSAS. Model Earth Syst Environ 11(1):56. https://doi.org/10.1007/s40808-024-02266-y

Abijith D, Saravanan S, Singh L, Jennifer JJ, Saranya T, Parthasarathy KSS (2020) GIS-based multi-criteria analysis for identification of potential groundwater recharge zones-a case study from Ponnaniyaru watershed, Tamil Nadu, India. HydroResearch 3:1–14. https://doi.org/10.1016/j.hydres.2020.02.002

Abulnaga BE, Naga RHA (2024) Sediment management for the Nile River in times of global warming. In: Integration of core sustainable development goals in rural areas: current practices of water, energy, food, climate change, and ecosystems. Cham: Springer Nature Switzerland, pp 307–346. https://doi.org/10.1007/978-3-031-60149-1_13

Adeoba MI, Odjegba EE, Pandelani T (2025) Nature-based solutions: opportunities and challenges for water treatment. Smart Nanomater Environ Appl 575–596. https://doi.org/10.1016/B978-0-443-21794-4.00011-9

Adom RK, Simatele MD (2024) Overcoming systemic and institutional challenges in policy implementation in South Africa's water sector. Sustain Water Resour Manag 10(2):69. https://doi.org/10.1007/s40899-024-01040-3

Alaerts GJ, Kaspersma JM (2022) Facing global transitions in water management: advances in knowledge and capacity development and towards adaptive approaches. Water Policy 24(5):685–707. https://doi.org/10.2166/wp.2022.301

Ali MA, Kamraju M (2024) The role of community participation in sustainable integrated water resources management: challenges, opportunities, and current perspectives. Integr Manag Water Resour India: Comput Approach: Optim Sustain Plan 325–344. https://doi.org/10.1007/978-3-031-62079-9_18

Arevalo JAT (2024) Climate change versus economic growth: case of greenhouse apply a study of European Union Countries and England from 2010 to 2019 using linear regression and neural networks. Sustainability 16(5):1884. https://doi.org/10.3390/su16051884

Banso AA, Olurin JO, Okem ES, Ogunjobi OA (2023) Integrated water resource management in South West Nigeria: a comprehensive review of strategies and outcomes. Int J Appl Res Soc Sci 5(8):330–351. https://www.academia.edu/download/110357030/748.pdf

Bayabil HK, Li Y, Tong Z, Gao B (2021) Potential management practices of saltwater intrusion impacts on soil health and water quality: a review. J Water Clim Chang 12(5):1327–1343. https://doi.org/10.2166/wcc.2020.013

Biswas H (2023) Wetlands and Ramsar Sites in India with special reference to West Bengal: From Crisis to Cure. CONSCIENTIA 8. https://www.krishnagargovtcollege.ac.in/FileNotice/Conscientia%20Dec%202023.pdf#page=9

Boulon R (2005) Atlantic Acropora status review. https://repository.library.noaa.gov/view/noaa/16200/noaa_16200_DS1.pdf

Bwambale E, Abagale FK, Anornu GK (2022) Smart irrigation monitoring and control strategies for improving water use efficiency in precision agriculture: a review. Agric Water Manag 260:107324. https://doi.org/10.1016/j.agwat.2021.107324

Cannatella D (2023) Industrial transformation and sustainable urban planning in the Pearl River Delta: a landscape-based approach. In: Adaptive urban transformation: urban landscape dynamics, regional design and territorial governance in the Pearl River delta, China. Cham: Springer International Publishing, pp 209–232. https://library.oapen.org/bitstream/handle/20.500.12657/63959/1/978-3-030-89828-1.pdf#page=212

Chen X, Yu G (2024) The Impact of urban-rural integration on food security: evidence from provincial panel data in China. Sustainability 16(9):3815. https://doi.org/10.3390/su16093815

Daraz U, Bojnec Š, Khan Y (2024) The impact of climate change on migration patterns in coastal communities. Climate 12(11):180. https://doi.org/10.3390/cli12110180

Darkwah SO, Scoville MD, Wang LK (2021) Geographic information systems and remote sensing applications in environmental and water resources. Integr Nat Resour Manag 197–236. https://doi.org/10.1007/978-3-030-55172-8_5

Das A (2025) Surface water potential zones delineation and spatiotemporal variation characteristics of water pollution and the cause of pollution formation in Brahmani River Basin, Odisha. HydroResearch 8:99–112. https://doi.org/10.1016/j.hydres.2024.10.002

Dass A, Jethoo AS, Poonia MP (2014) Integrated wastewater management in Jaipur City, Rajasthan, India. Nat Environ Pollut Technol 13(3):637. http://www.neptjournal.com/upload-images/NL-49-34-(32)B-3038Com.pdf

Datta R, Kairy B (2024) Decolonizing climate change adaptations from Indigenous perspectives: Learning reflections from Munda Indigenous communities, coastal areas in Bangladesh. Sustainability 16(2):769. https://doi.org/10.3390/su16020769

De Luque-Villa MA, González-Méndez M (2024) Water management as a social field: a method for engineering solutions. Water 16(19):2842. https://doi.org/10.3390/w16192842

Farah E, Shahrour I (2024) Water leak detection: a comprehensive review of methods, challenges, and future directions. Water 16(20):2975. https://doi.org/10.3390/w16202975

Fouad SS, Heggy E, Abotalib AZ, Ramah M, Jomaa S, Weilacher U (2022) Landscape-based regeneration of the Nile Delta's waterways in support of water conservation and environmental protection. Ecol Ind 145:109660. https://doi.org/10.1016/j.ecolind.2022.109660

Gamble R, Tan GG, Xu H, Beavis S, Maurer P, Pittock J, Wasson RJ (2025) Rivers of the Asian highlands: From deep time to the climate Crisis. https://library.oapen.org/handle/20.500.12657/93993

Ghosh S, Roy S, Islam A, Shit PK, Datta DK, Islam MS, Das BC (2023) Floods of Ganga-Brahmaputra-Meghna Delta in context. In: floods in the Ganga–Brahmaputra–Meghna Delta. Cham: Springer International Publishing, pp 1–17. https://doi.org/10.1007/978-3-031-21086-0_1

Gupta P, Singh J, Verma S, Chandel AS, Bhatla R (2021) Impact of climate change and water quality degradation on food security and agriculture. In: Water conservation in the era of global climate change. Elsevier, pp 1–22. https://doi.org/10.1016/B978-0-12-820200-5.00006-3

Haldar K, Kujawa-Roeleveld K, Acharjee TK, Datta DK, Rijnaarts H (2022) Urban water as an alternative freshwater resource for matching irrigation demand in the Bengal delta. Sci Total Environ 835:155475. https://doi.org/10.1016/j.scitotenv.2022.155475

Hernández-Delgado EA (2024) Coastal restoration challenges and strategies for small island developing states in the face of sea level rise and climate change. Coasts 4(2):235–286. https://doi.org/10.3390/coasts4020014

Jia K, Huang X, Qiao W, Zhong S (2024) Unpacking divergent rural-urban land use dynamics in county urbanization: A comparative socio-spatial analytics approach. Cities 154:105343. https://doi.org/10.1016/j.cities.2024.105343

Kabir A, Hasan MM, Rokonuzzaman S, Abdullah HM, Hossain MS (2024) A safe operating space for the major rivers in the Bangladesh Delta. Environ Res Lett 19(11):114002. https://doi.org/10.1088/1748-9326/ad796d/pdf

Katusiime J, Schütt B (2020) Integrated water resources management approaches to improve water resources governance. Water 12(12):3424. https://doi.org/10.3390/w12123424

Kolahi M, Davary K, Omranian Khorasani H (2024) Integrated approach to water resource management in Mashhad Plain, Iran: actor analysis, cognitive mapping, and roadmap development. Sci Rep 14(1):162. https://doi.org/10.1038/s41598-023-50697-x

Kumar P, Choudhury D (2024) Innovative technologies for effective water resources management. Water Cris Sustain Manag Glob South 555–594. https://doi.org/10.1007/978-981-97-4966-9_18

Mazumder R, Shaw R (eds) (2024) Surface environments and human interactions: reflections from Asia. Springer Nature. https://www.springerprofessional.de/en/surface-environments-and-human-interactions/26849520

Mazzetto S (2024) A review of urban digital twins integration, challenges, and future directions in smart city development. Sustainability 16(19):8337. https://doi.org/10.3390/su16198337

Mehltretter S, Bradford A, Longboat S, Luby B (2024) In a good way: braiding indigenous and western knowledge systems to understand and restore freshwater systems. Water 16(7):934. https://doi.org/10.3390/w16070934

Mishra RK (2023) Fresh water availability and its global challenge. Br J Multidiscip Adv Stud 4(3):1–78. https://www.academia.edu/download/105758199/Fresh_Water_9_.pdf

Mostern R (2021) The Yellow River: a natural and unnatural history. Yale University Press

Nantavisai M, Hashimoto S, Meraj G (2024) Unraveling changes in deltas: Exploring drivers and consequences through the Lens of the IPBES conceptual framework. Sci Total Environ 175865. https://books.google.co.in/books?hl=en&lr=&id=fCQ7EAAAQBAJ&oi=fnd&pg=PP1&dq=Mostern,+R.+(2021).+The+Yellow+River:+A+natural+and+unnatural+history.+Yale+University+Press.+Nantavisai,+M.,+Hashimoto,+S.,+%26+Meraj,+G.+(2024).+Unraveling+changes+in+deltas:+Exploring+drivers+and+consequences+through+the+Lens+of+the+IPBES+conceptual+framework.+Science+of+The+Total+Environment,+175865.&ots=yQdMsYmYQ3&sig=oPxN0-eAlXDea_v6gOouTlYts2U&redir_esc=y#v=onepage&q&f=false

Neog N, Deka K, Goswami R (2025) Flood-resilient water supply and sanitation systems: global case studies. In: Water sustainability and hydrological extremes. Elsevier, pp 219–236. https://doi.org/10.1016/B978-0-443-21499-8.00011-8

Nwokediegwu ZQS, Ugwuanyi ED, Dada MA, Majemite MT, Obaigbena A (2024) Urban water management: a review of sustainable practices in the USA. Eng Sci Technol J 5(2):517–530. https://doi.org/10.51594/estj.v5i2.829

O'Hogain S, McCarton L, O'Hogain S, McCarton L (2018) Nature-based solutions a technology Portfolio of nature based solutions: innovations in water management 1–9. https://doi.org/10.1007/978-3-319-73281-7_1

Olson KR, Frenelus W (2024) Environmental and human impacts of Lancang-Mekong Mainstem and Tributary Dams on China, Laos, Thailand, Myanmar, Cambodia, and Vietnam. Open J Soil Sci 14(10):555–605. https://www.researchgate.net/publication/384863169_Environmental_and_Human_Impacts_of_LancangMekong_Mainstem_and_Tributary_Dams_on_China_Laos_Thailand_Myanmar_Cambodia_and_Vietnam

Pagliero L, McIntyre N, Aitken D, Bolz P, Jamett N, Pérez-Murillo G, Cisternas LA (2024) Sustainable integration of desalinated seawater into regional water supply networks using a participatory modelling framework. Environ Sci Policy 155:103714. https://doi.org/10.1016/j.envsci.2024.103714

Perera MD, Gomes PI (2025) Feasibility of hydrological modelling for intermittent streams using HEC-HMS: a process evaluation. Hydrol Sci J 1–15. https://doi.org/10.1080/02626667.2024.2440012

Phuong TT, Vien TD, Son CT, Thuy DT, Greiving S (2024) Impact of climate change on agricultural production and Food Security: A case study in the Mekong River Delta of Vietnam. Sustainability 16(17):7776. https://doi.org/10.3390/su16177776

Reader MO, Eppinga MB, de Boer HJ, Damm A, Petchey OL, Santos MJ (2022) The relationship between ecosystem services and human modification displays decoupling across global delta systems. Commun Earth and Environ 3(1):102. https://doi.org/10.1038/s43247-022-00431-8

Richardson CM, Davis KL, Ruiz-González C, Guimond JA, Michael HA, Paldor A, Paytan A (2024) The impacts of climate change on coastal groundwater. Nat Rev Earth and Environ 5(2):100–119. https://doi.org/10.1038/s43017-023-00500-2

Santos E, Carvalho M, Martins S (2023) Sustainable water management: Understanding the socioeconomic and cultural dimensions. Sustainability 15(17):13074. https://doi.org/10.3390/su151713074

Saravani MJ, Saadatpour M, Shahvaran AR (2024) A web GIS based integrated water resources assessment tool for Javeh Reservoir. Expert Syst Appl 252:124198. https://doi.org/10.1016/j.eswa.2024.124198

Scown MW, Dunn FE, Dekker SC, van Vuuren DP, Karabil S, Sutanudjaja EH, Middelkoop H (2023) Global change scenarios in coastal river deltas and their sustainable development implications. Glob Environ Chang 82:102736. https://doi.org/10.1016/j.gloenvcha.2023.102736

Shah AI, Dar MUD, Bhat RA, Singh JP, Singh K, Bhat SA (2020) Prospectives and challenges of wastewater treatment technologies to combat contaminants of emerging concerns. Ecol Eng 152:105882. https://doi.org/10.1016/j.ecoleng.2020.105882

Shams AK, Muhammad NS (2022) Toward sustainable water resources management: critical assessment on the implementation of integrated water resources management and water–energy–food nexus in Afghanistan. Water Policy 24(1):1–18. https://doi.org/10.2166/wp.2021.072

Sharma BM, Basu AR, Krishnan A, Sarkar SK, Bharat GK (2025) Climate change–WASH services–health nexus: the Indian case. In: Health and climate change. Academic Press, pp 457–483. https://doi.org/10.1016/B978-0-443-29240-8.00024-9

Shukla BK, Tripathi A, Nagavi JC, Prasad S, Sahoo M, Chaudary S, Verma S (2024) Integrated water resources management (IWRM) in the geospatial epoch: an ontological dive into sustainable hydrological governance. In: sustainable development and geospatial technology: Volume 2: applications and future directions. Cham: Springer Nature Switzerland, pp 1–22. https://doi.org/10.1007/978-3-031-65703-0_1

Singh L, Saravanan S (2020) Impact of climate change on hydrology components using CORDEX South Asia climate model in Wunna, Bharathpuzha, and Mahanadi. India. Environ Monit Assess 192:678. https://doi.org/10.1007/s10661-020-08637-z

Singh PK, Kumar U, Kumar I, Dwivedi A, Singh P, Mishra S, Sharma RK (2024) Critical review on toxic contaminants in surface water ecosystem: sources, monitoring, and its impact on human health. Environ Sci Pollut Res 31(45):56428–56462. https://doi.org/10.1007/s11356-024-34932-0

Singh U, Sharma PK, Ojha CSP (2021) Groundwater investigation using ground magnetic resonance and resistivity meter. ISH J Hydraul Eng 27(sup1):401–410. https://doi.org/10.1080/09715010.2019.1661802

Srinivas T (2025) Toward a Riparian Theo-Ecology: a meditation on Hinduism and climate change. J Am Acad Relig lfae077. https://doi.org/10.1093/jaarel/lfae077

Sunny MAU (2024) Sustainable water management in Urban environments. Eur J Adv Eng Technol 11(4):88–95. https://www.academia.edu/download/115833727/EJAET_11_4_88_95.pdf

Talpur Z, Choi SJ, Chung IM (2024) Enhancing the SDG 6 Index: Implementing New Directions and Trends in Priorities of Integrated Water Resources Management Principles in the Republic of Korea. Water 16(22):3259. https://doi.org/10.3390/w16223259

Tantoh HB, McKay TJM (2021) Assessing community-based water management and governance systems in North-West Cameroon using a cultural theory and systems approach. J Clean Prod 290:125804. https://doi.org/10.1016/j.jclepro.2021.125804

Teng M, Ji D, Han C, Liu P (2025) Optimizing the allocation of CO2 emission rights in the Yangtze River Delta City agglomeration Region of China based on equity, efficiency and sustainability. Int J Environ Res 19(1):1–23. https://doi.org/10.1007/s41742-024-00693-4

Tran T, Duong DV, Le TD, Loc HH, Chau LTN, Le LT, Bui XT (2025) Promoting sustainable shrimp farming: balancing environmental goals, awareness, and socio-cultural factors in the Mekong Delta aquaculture. Aquacult Int 33(1):119. https://doi.org/10.1007/s10499-024-01802-y

Tran T, Duong DV, Loc HH, Đinh VN, Trung TA, Ho VTT, Sy PC (2024) Integrated assessment of salinity intrusion and water scarcity in Ben Tre, Mekong Delta: implications for sustainable livelihoods and adaptive water management. Environ, Dev Sustain 1–30. https://doi.org/10.1007/s10668-024-05216-x

Uddin SA, He L, Hossain MJ, Nusrat N, Debi M (2024) Ganges-Brahmaputra-Meghna River Delta. In delta sustainability: a report to the mega-delta programme of the UN Ocean Decade. Singapore: Springer Nature Singapore, pp 89–116. https://doi.org/10.1007/978-981-97-7259-9_6

Yaseen ZM (2024) Flood hazards and susceptibility detection for Ganga river, Bihar state, India: Employment of remote sensing and statistical approaches. Results Eng 21:101665. https://doi.org/10.1016/j.rineng.2023.101665

Yi D, Guo J, Pueppke SG, Han Y, Ding G, Ou M, Koomen E (2025) What dominates the variation of ecosystem services across different urban expansion patterns? —Evidence from the Yangtze River Delta region. China Environ Impact Assess Rev 110:107674. https://doi.org/10.1016/j.eiar.2024.107674

Yuen KW, Hanh TT, Quynh VD, Switzer AD, Teng P, Lee JSH (2021) Interacting effects of land-use change and natural hazards on rice agriculture in the Mekong and Red River deltas in Vietnam. Nat Hazard 21(5):1473–1493. https://doi.org/10.5194/nhess-21-1473-2021

Delta Regions of India, Their Role and Significance in Water and Land Resources

2

R. V. Galkate, V. S. Yadav, Shalini Yadav, R. K. Jaiswal, and Ram Narayan Yadava

Abstract

Delta regions are highly productive fluvial landforms on Earth, covering less than 1% of the terrestrial surface yet sustaining over 500 million people globally. This chapter evaluates major Indian deltas, their ecological significance, and vulnerabilities. The Ganga–Brahmaputra Delta spans nearly 105,000 km^2. The Godavari Delta covers 7,800 km^2, the Krishna Delta 5,000 km^2, the Mahanadi Delta 9,500 km^2, and the Cauvery Delta 7,000 km^2. Collectively, they support more than 30 million people and irrigate over 6 million hectares of farmland. Irrigation efficiency ranges from 66.6 to 80%, highlighting their importance for national food security. Ecologically, deltas such as the Sundarbans, spanning 10,000 km^2 and harbouring 693 faunal species, provide critical services. They buffer cyclones, which strike the east coast five to six times annually. However, deltas face growing threats. Salinity intrusion in the Cauvery Delta exceeds 4 dS/m, while subsidence reaches 4 cm/year in the Krishna–Godavari Delta. Sea-level rise and reduced sediment inflow could cause the loss of over 2,000 km^2 of deltaic land by 2050. Water use is largely agriculture-driven, with more than 80% of available water consumed in many regions. Reservoirs like Nagarjuna Sagar (11.6 BCM) and Polavaram (8.0 BCM) support irrigation but require climate-sensitive

R. V. Galkate (✉) · V. S. Yadav · R. K. Jaiswal
National Institute of Hydrology, Central India Hydrology Regional Centre, Bhopal, Madhya Pradesh, India
e-mail: rgalkate@yahoo.co.in

S. Yadav
Rabindranath Tagore University, Raisen, Bhopal, Madhya Pradesh, India

R. N. Yadava
Mansrovar Global University, Bhopal, Madhya Pradesh, India

S. Yadav et al. (eds.), *Water Resources in Delta Region in India*, Deltas of the World,
https://doi.org/10.1007/978-3-032-06500-1_2

management. Policy interventions such as Namami Gange, the Integrated Coastal Zone Management Project, and the National Adaptation Fund for Climate Change (covering 30+ projects in 25 states) play a vital role. Delta management directly links with United Nations Sustainable Development Goals (UNSDG) as SDG 2 on zero hunger, SDG 6 on clean water and sanitation, SDG 13 on climate action, SDG 14 on life below water, and SDG 15 on life on land. The chapter highlights the need for integrated and climate-resilient strategies. Water governance, ecological restoration, and legislative protection are crucial to maintain agriculture, biodiversity, and livelihoods in Indian deltas.

2.1 Introduction

A delta is a landform that forms at the mouth of a river where it meets a standing body of water like a sea, ocean, or lake. As the river slows down, it loses the capacity to carry sediment, which then gets deposited, gradually building up a fan-shaped or triangular tract of fertile land. This depositional process gives rise to deltas, one of the most dynamic and productive natural systems on Earth (Anthony 2015). The term "delta" was first used by the Greek historian Herodotus to describe the Nile River's triangular mouth, and it remains a defining feature of riverine geography worldwide (Bhattacharya 2017).

Globally, deltas are crucial ecological, economic, and social regions. They cover only about 1% of the Earth's land area but are home to more than 500 million people, reflecting their immense habitability (Ericson et al. 2006). This is primarily due to their flat topography, rich alluvial soil and abundant water supply, which make them ideal for intensive agriculture, fisheries and human settlement. Deltas are among the world's most fertile agricultural zones; rice, wheat and a variety of vegetables and fruits are cultivated extensively in these regions. Major global deltas include the Nile Delta in Egypt, the Mekong Delta in Vietnam, the Mississippi Delta in the United States and the Ganges–Brahmaputra-Meghna Delta in South Asia, which is the world's largest and most complex delta system (Tran and Tran 2020).

The importance of deltas also lies in their biodiversity. These regions often host wetlands, estuaries, mangrove forests, and unique aquatic habitats, supporting a wide variety of flora and fauna. The wetlands act as natural filters, improving water quality, while also serving as buffers against storms and tidal surges (Barbier et al. 2011). Additionally, deltas are economic hotspots; several major cities, ports, and industrial hubs have developed in or near deltaic regions due to ease of transportation and proximity to waterways (Nicholls et al. 2007). However, deltas around the world are facing severe challenges due to climate change, sea-level rise, upstream damming, and unregulated human activities. Many deltas are sinking or eroding, and salinity intrusion is making fertile land unsuitable for farming (Syvitski et al. 2009). Yet, despite these threats, deltas remain vital for sustaining food security, economic development and environmental balance. Their strategic value, both for natural ecosystems and human societies, makes their conservation and sustainable management a global priority (Tessler et al. 2015).

India, with its vast network of rivers and a long coastline of over 7,500 km, is home to some of the most prominent and dynamic deltaic systems in the world. The delta regions of India are primarily located along the eastern coast, where major rivers like the Ganga, Brahmaputra, Mahanadi, Godavari, Krishna, and Cauvery drain into the Bay of Bengal (Rudra, 2010). These river systems, over centuries, have deposited massive volumes of sediments at their mouths, giving rise to rich alluvial deltaic plains that are vital for agriculture, biodiversity and human settlement. The most significant and largest delta in India is the Ganga–Brahmaputra Delta, also known as the Sundarbans Delta. It spans parts of West Bengal in India and a large portion in Bangladesh, covering an estimated area of around 105,000 square kilometres. This delta is not only the largest in India but also the largest in the world and it holds unparalleled ecological importance due to the presence of the Sundarbans mangrove forest, the largest contiguous halophytic mangrove forest on Earth, home to the endangered Royal Bengal Tiger (Giri et al. 2007). The region also acts as a crucial natural barrier against cyclones and coastal erosion.

Apart from the Sundarbans, several other deltas play a key role in India's physical and economic landscape. The Mahanadi Delta in Odisha, the Godavari and Krishna Deltas in Andhra Pradesh, and the Cauvery Delta in Tamil Nadu are some of the most agriculturally productive regions in the country (Mishra and Jena 2015). These deltas support intensive cultivation of rice, pulses, and horticultural crops owing to the fertile alluvial soils and well-developed irrigation systems. In fact, the Cauvery Delta is famously referred to as the "Rice Bowl of Tamil Nadu." Deltas in India are also culturally and historically significant. Many ancient civilizations and thriving trade centers have developed and flourished in these regions due to the availability of water, fertile land, and navigable routes (Chakrabarti 2001). Cities like Kolkata, Cuttack, Vijayawada, and Thanjavur owe much of their growth to their strategic deltaic locations. Moreover, these regions sustain millions of people who depend on agriculture, fishing, aquaculture, and related occupations for their livelihood.

Despite their importance, Indian deltas face growing pressures from population expansion, industrial pollution, unregulated urban development, and climate change. Rising sea levels, increasing salinity and erratic monsoon patterns pose serious threats to both livelihoods and ecosystems (Hazra et al. 2002). Therefore, the delta regions of India require well-planned, sustainable management practices to preserve their ecological integrity while continuing to serve as the backbone of the country's food, water, and economic security. In this study, a comprehensive overview of the major deltaic regions of India has been presented, focusing on their geomorphological characteristics, ecological significance, agricultural productivity, and socio-economic importance. The objective is to highlight the critical role these regions play in supporting biodiversity, livelihoods, and climate resilience while addressing the increasing challenges posed by environmental

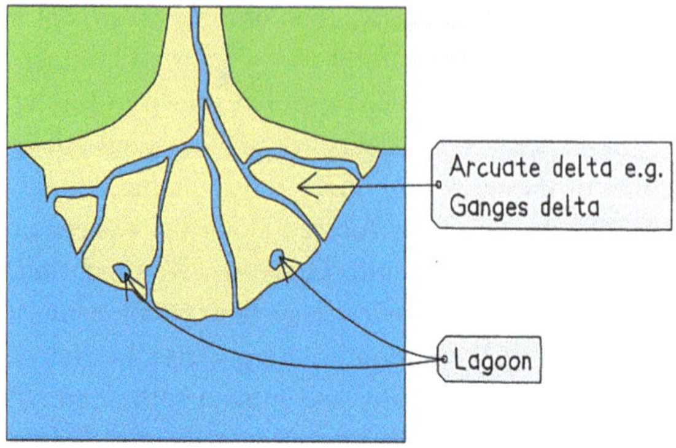

Fig. 2.1 Arcuate delta

degradation, climate change, and unsustainable development. By understanding the unique dynamics of Indian deltas, this chapter aims to promote informed strategies for their conservation and sustainable management.

2.2 Types of Deltas

Deltas are broadly classified based on the dominant geological processes shaping them such as river deposition, wave action and tidal influence. The interplay of these forces results in varying deltaic forms and structures across the world. The major types of deltas include:

2.2.1 Arcuate Delta

Arcuate deltas are fan- or arc-shaped, formed where river deposition dominates and wave action is moderate, as shown in Fig. 2.1. These deltas have a smooth convex seaward edge with numerous distributaries. They are typically very fertile and agriculturally productive (Stanley and Warne 1993). Example: The Ganga–Brahmaputra Delta (India and Bangladesh) and the Nile Delta (Egypt) are classic examples of arcuate deltas (Nicholls et al. 2007).

2.2.2 Bird's Foot Delta

This type of delta forms when river deposition is extremely dominant and there is very little wave or tidal activity, as shown in Fig. 2.2. It resembles a bird's claw, with long, narrow distributary channels extending far into the sea. These deltas tend to advance

Fig. 2.2 Bird's foot delta

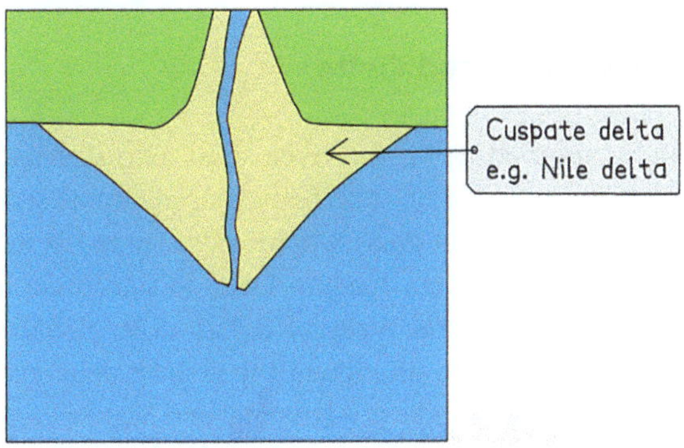

Fig. 2.3 Cuspate delta

rapidly into the water body (Wells and Coleman 1981). Example: The Mississippi River Delta in the USA is the most well-known bird's foot delta.

2.2.3 Cuspate Delta

Cuspate deltas are tooth-shaped, formed where river deposition and wave action are almost equal as shown in Fig. 2.3. The result is a pointed or triangular formation with a relatively straight coastline. These deltas are more stable and have fewer distributaries (Galloway 1975). Example: The Tiber Delta in Italy is a prominent example of a cuspate delta.

2.2.4 Estuarine Delta

This type forms in river estuaries where the river meets a sea with strong tidal currents.Sediments are deposited along the sides of the estuary, creating a funnel-shaped

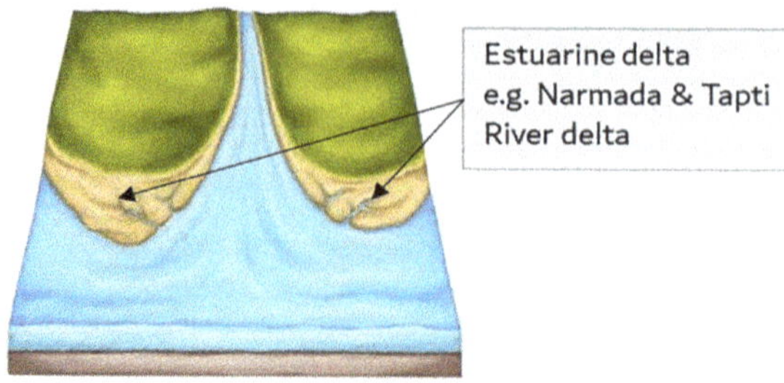

Fig. 2.4 Estuarine delta

delta, Estuarine delta has been shown in Fig. 2.4 (Dalrymple et al. 1992). Example: The Narmada and Tapti rivers in India form estuarine deltas as they drain into the Arabian Sea.

2.2.5 Inverted Delta

In an inverted delta, the river splits into distributaries before reaching the main water body, forming a delta-like feature upstream as shown in Fig. 2.5. This is a relatively rare type and occurs in areas where water spread is more inland (Kimmerer 2004). Example: The Sacramento-San Joaquin Delta in California is an example of this unique form.

Although several types of deltas exist globally, in the Indian context, primarily two types of deltas are prominent the arcuate delta and the estuarine delta. The eastern rivers like the Ganga, Godavari, Krishna, and Mahanadi form arcuate deltas due to high sediment load and relatively calmer wave action in the Bay of Bengal (Mahalik 2006). On the other hand, rivers draining into the Arabian Sea, such as the Narmada and Tapti, form estuarine deltas influenced by stronger tidal currents and narrower coastal plains (Natarajan et al. 2023). Other delta forms like bird's foot, cuspate, or inverted deltas are either rare or absent in India due to differing geological and coastal dynamics (Vörösmarty et al. 2009).

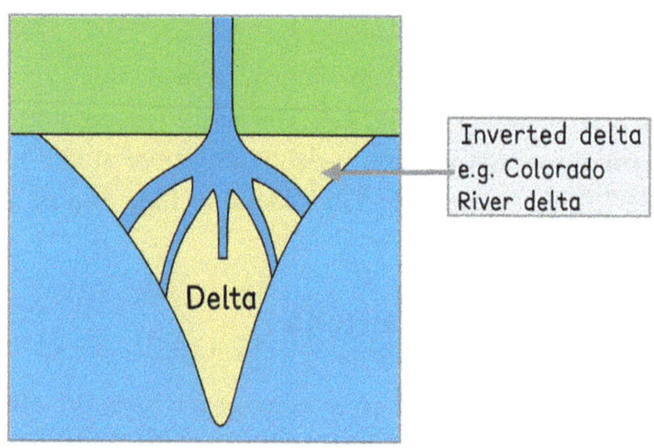

Fig. 2.5 Inverted delta

This distinction is essential to understanding the spatial distribution and behaviour of deltaic regions across the Indian subcontinent.

2.3 Major River Deltas in India

India's river deltas are among the most fertile, densely populated and ecologically vibrant regions in the country. These deltas are primarily concentrated along the eastern coastline, where major rivers discharge into the Bay of Bengal, forming large alluvial plains rich in sediments. The geomorphological and climatic conditions of peninsular and northern India favour the formation of deltas, especially due to high sediment load, lower gradient near river mouths and relatively shallow continental shelf. The western coast, in contrast, has steeper gradients and stronger tidal actions, resulting in fewer and smaller deltas (Rodrigues et al. 2024). The major river deltas of India serve as agricultural hubs, biodiversity hotspots, and centres of cultural and economic activity, playing a crucial role in the country's food security and rural economy.

2.3.1 Ganga–Brahmaputra Delta (Sundarbans Delta)

The Ganga–Brahmaputra Delta, also known as the Sundarbans Delta, is the largest delta in the world, stretching across eastern India and Bangladesh. It spans approximately 105,000 square kilometres, with a major portion lying in West Bengal and southern Bangladesh. Formed by the confluence of the Ganga, Brahmaputra and Meghna rivers, this delta is characterized by extensive mangrove forests, tidal rivers and estuarine channels (Baidya et al. 2023). The Indian portion of the delta is home to the Sundarbans mangrove forest, the world's largest tidal halophytic mangrove ecosystem and a designated UNESCO World Heritage Site. Ecologically, this delta supports unique biodiversity including the endangered Royal Bengal Tiger, saltwater crocodiles, estuarine fish species and a variety of migratory birds. Hydrologically dynamic, the delta experiences significant tidal influence, leading to regular inundation, saline water intrusion and sediment redistribution (Hazra et al. 2002). Despite being prone to cyclones, flooding and sea, level rise, the region is densely populated and agriculturally productive. Crops such as rice, jute and pulses are widely cultivated, while fishing and honey collection serve as important livelihood sources for local communities.

Beyond its ecological importance, the delta plays a vital role in disaster mitigation by acting as a natural buffer against tropical cyclones originating in the Bay of Bengal. However, this fragile region is under increasing pressure from climate change, upstream damming, deforestation and illegal encroachments, all of which disturb its hydrological equilibrium and threaten both human settlements and ecosystems.

2.3.2 Mahanadi Delta

The Mahanadi Delta is one of the most prominent and agriculturally productive deltas on India's eastern coast, located mainly in Odisha. It is formed by the Mahanadi River along with its tributaries such as the Seonath, Tel and Hasdeoriver and covers an area of approximately 9,500 square kilometres before entering the Bay of Bengal (Mahalik 2006). This fan-shaped delta begins near Cuttack and extends toward Paradeep. Over centuries, sediment deposition has enriched the alluvial soils, making the region highly fertile. Due to its intensive rice cultivation and extensive irrigation network, it is popularly known as the "Rice Granary of Odisha" and also supports crops like sugarcane, pulses and oilseeds.

Apart from agriculture, the delta sustains a strong rural economy through inland and coastal fisheries and seasonal livelihoods such as honey collection. Ecologically, it contains important wetlands, estuarine systems and river distributaries like the Kathajodi, Kuakhai and Devi. It also interacts hydrologically with Chilika Lake, which supports migratory birds, estuarine crocodiles and freshwater turtles, contributing to the region's ecological diversity (Hazra et al. 2020). Despite its productivity, the delta faces multiple threats. These include frequent floods during the monsoon, tropical cyclones, salinity intrusion from rising sea levels and contamination of groundwater. A growing concern is the reduction in downstream water flow due to dam construction and diversions in the upstream state of Chhattisgarh. This has led to tensions between Odisha and Chhattisgarh over equitable water sharing. These pressures demand integrated water governance that ensures both agricultural sustainability and ecological conservation (Joy et al. 2021).

2.3.3 Godavari Delta

The Godavari Delta lies along the eastern coastline of Andhra Pradesh, formed by the Godavari River, India's second-longest river. As the river approaches the Bay of Bengal, it bifurcates into two major distributaries, the Gautami and the Vasista, creating a fertile deltaic region that spans nearly 7,800 square kilometres (Rao et al. 2005). This area is often identified with the Konaseema region, celebrated for its lush paddy fields and scenic backwaters. Agriculturally, the Godavari Delta is one of the most productive regions in the country, often referred to as the "Rice Bowl of Andhra Pradesh." The rich alluvial soils, warm and humid climate, and well-developed irrigation infrastructure, including the historic Dowleswaram Barrage constructed by Sir Arthur Cotton, support extensive cultivation of rice, sugarcane, coconut, banana and pulses. A network of canals ensures year-round irrigation for both Kharif and Rabi crops (Ghosh and Modak 2022). In addition to farming, the region supports a thriving aquaculture and fisheries economy, particularly in brackish water zones closer to the coast. Prawn farming has become a major livelihood and export activity. The delta's landscape of distributaries, creeks and estuaries also nurtures rich biodiversity, although recent environmental stress is posing growing threats.

Salinity intrusion caused by declining freshwater flow and groundwater overextraction has started affecting soil health and agricultural productivity. The delta is frequently impacted by cyclones and storm surges during the northeast monsoon season. Furthermore, the spread of aquaculture ponds and rapid land use change have disturbed the ecological balance of the region. Concerns are also rising over upstream dam construction and increased water use, which reduce river discharge into the delta and raise questions about long-term sustainability (Mahto 2024). These issues underscore the urgent need for integrated delta management that promotes climate, resilient agriculture, ecological restoration and participatory water governance.

2.3.4 Krishna Delta

The Krishna Delta, located along the southeastern coastline of Andhra Pradesh, has evolved over centuries through sediment deposition by the Krishna River before it merges with the Bay of Bengal. Spanning roughly 5,000 square kilometres, this region stretches from Vijayawada through the districts of Krishna, Guntur and parts of Prakasam. As the river branches into distributaries, it forms a flat and fertile expanse that has supported intensive farming for generations (Madhanagopal 2023). The delta plays a central role in Andhra Pradesh's agricultural economy. At the heart of its irrigation system is the Prakasam Barrage in Vijayawada, which channels river water through an elaborate network of canals. This infrastructure allows farmers to cultivate two or more crops annually. Rice remains the dominant crop, but chillies, maize, sugarcane and horticultural produce such as bananas and flowers are also widely grown (Kumar 2017). These activities support both smallholder farmers and large cooperatives.

In recent years, aquaculture has gained momentum in the delta, particularly prawn and freshwater fish farming. Many rural households now rely on these livelihoods in addition to traditional farming. However, challenges related to water availability have become more pronounced. Reduced river flow caused by upstream dams and inter-basin water transfer projects has led to saltwater intrusion, especially in coastal parts of the delta. This salinization has affected crop productivity and drinking water sources (Murthy 2020). Climate-related risks are also increasing. The delta has faced several severe cyclones, including Hudhud and Phethai, which have caused agricultural losses and infrastructure damage. Additionally, competition over Krishna River water between Andhra Pradesh, Telangana and Karnataka intensifies during dry years, leading to policy and legal tensions. As a result, researchers and policymakers are exploring new ways to manage the delta's water more sustainably, including precision irrigation, salinity monitoring and participatory planning with local communities.

2.3.5 Cauvery Delta

The Cauvery Delta, often referred to as the "Granary of South India," is located in eastern Tamil Nadu where the Cauvery River fans out into multiple distributaries before entering the Bay of Bengal. Spanning an estimated 7,000 square kilometres, the delta covers districts such as Thanjavur, Tiruvarur, Nagapattinam and parts of Cuddalore and Ariyalur. Over centuries, the region has developed through continuous sediment deposition and river meandering, forming a fertile alluvial landscape ideal for wetland agriculture (Kumar and Karthikeyan 2022). The delta supports one of the most productive agricultural zones in India. Its traditional strength lies in paddy cultivation, sustained by a historic canal irrigation system that includes the Grand Anicut, or Kallanai, a structure dating back to the Chola period and still functional today. Apart from rice, the region grows sugarcane, banana, pulses, coconut and cotton, supported by double or even triple cropping systems under favorable conditions (Ramasamy 2006).

Culturally, the Cauvery Delta holds deep significance in Tamil history. It has nurtured temple architecture, classical arts and festivals that remain central to Tamil identity. Towns like Thanjavur and Kumbakonam reflect the region's historical richness through their monumental temples and cultural heritage. Despite its strengths, the delta has been under increasing stress due to interstate water disputes with Karnataka. Reduced river flow during dry seasons, compounded by upstream damming and erratic monsoon rainfall, has intensified uncertainty in agricultural planning. Groundwater overuse and urban, industrial expansion, particularly oil and gas extraction, have led to environmental degradation and land, use transformation (Rajavel et al. 2022). In an effort to preserve the agricultural nature of this sensitive region, the Tamil Nadu government introduced the Delta Protection Bill in 2020. This legislation declared the delta a Protected Agricultural Zone, aiming to prohibit non-agricultural industries and ensure long-term ecological and agrarian stability. It marks a significant legal approach toward sustainable development in India's deltaic ecosystems (TN Gazette 2020).

2.3.6 Other Notable Deltaic and Estuarine Systems

In addition to major deltas such as the Sundarbans, Godavari, Krishna, Mahanadi and Cauvery, several smaller deltaic and estuarine systems also hold ecological importance in and around India. The Meghna Delta, though primarily in Bangladesh, forms part of the larger Ganga, Brahmaputra, Meghna basin. Freshwater and sediments from Indian rivers in Assam and West Bengal contribute to this dynamic system, making it significant for transboundary water governance (Mirza 2004).

On the western coast, rivers like the Narmada and Tapti form estuarine deltas as they flow into the Arabian Sea. Unlike the classic fan, shaped deltas of the east, these forms narrow estuarine zones with tidal influence, supporting fisheries, ports and salt marsh

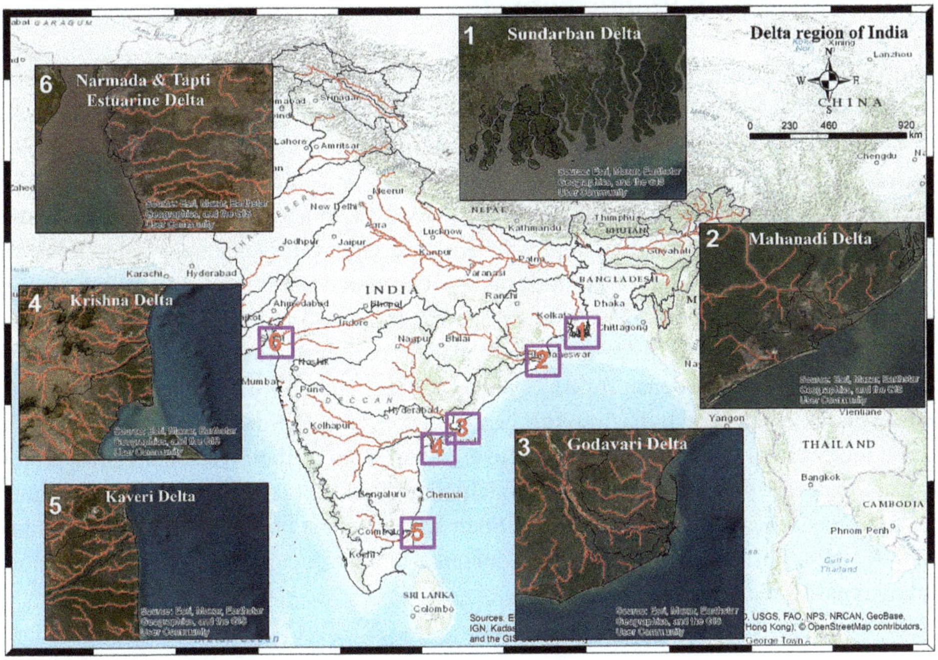

Fig. 2.6 Map of major deltaic and estuarine systems of India

habitats. Estuaries near Bharuch and Surat have become regionally vital for their economic and ecological functions (Ekka et al. 2020). Despite their smaller scale, these systems reflect India's geomorphological diversity and require region-specific conservation strategies.The spatial distribution of major and minor deltas across India is illustrated in the following Fig. 2.6, highlighting both eastern and western deltaic systems and their geographical extent.

2.4 Importance of Delta Regions

Delta regions are dynamic transitional zones between river systems and coastal environments. Their physical, ecological and cultural characteristics make them vital for agricultural production, biodiversity conservation, livelihood generation and heritage preservation in India.

2.4.1 Agricultural Productivity

Delta regions are among the most agriculturally productive landscapes in the world, primarily due to the rich alluvial soils deposited over millennia by river systems. These soils are deep, loamy and naturally replenished with nutrients through seasonal floods, making them ideal for intensive cultivation. Delta regions in India occupy less than 2.5% of the country's geographical area but contribute over 15% to national food grain production annually (MoAFW 2022). For example, the Krishna, Godavari Delta alone supports nearly 6 million hectares of agricultural land, much of it under paddy cultivation, contributing substantially to rice exports. The Cauvery Delta, often referred to as the "Rice Bowl of Tamil Nadu," accounts for over 40% of the state's total paddy output (Arun et al. 2020). These regions also host dense rural populations, more than 30 million people live in the deltaic plains of eastern India, relying heavily on irrigation, intensive farming systems, aquaculture, and allied livelihoods. Moreover, deltas such as the Sundarbans provide ecological services worth billions annually, including flood mitigation, carbon sequestration, and fish breeding grounds (UNEP 2018). Such figures reflect not only the agricultural productivity of deltas but also their socio-economic and ecological significance at regional and national levels.

2.4.2 Biodiversity

Ecologically, delta regions function as biodiversity hotspots, supporting a complex matrix of ecosystems such as mangroves, freshwater marshes, tidal estuaries and coastal wetlands. For instance, the Sundarbans Delta, covering nearly 10,000 square kilometres, harbours over 334 plant species and 693 wildlife species, including the endangered Royal Bengal Tiger, Gangetic dolphin and estuarine crocodile (Giri et al. 2015). The region supports more than 260 species of birds, many of which are migratory. The Godavari mangroves, though smaller, host around 90 bird species, including spot, billed pelicans and painted storks, while the Mahanadi estuary supports critically endangered species like the Indian skimmer and Olive Ridley turtles during nesting season (Menon et al. 2015). Vegetation in these regions not only serves as a buffer against cyclones and coastal erosion but also plays a crucial role in carbon sequestration and nutrient cycling. Moreover, delta ecosystems provide income for millions through eco-tourism, apiculture, sustainable fishing, and the collection of medicinal plants. Their conservation is thus essential for maintaining both ecological services and rural livelihoods.

2.4.3 Fisheries and Aquaculture

Deltaic regions play a critical role in India's fisheries and aquaculture sector due to their nutrient-rich, brackish water environments created by the confluence of rivers and the sea. These estuarine ecosystems support ideal conditions for fish breeding, larval development, and shrimp culture. The Krishna, Godavari and Sundarbans deltas alone contribute significantly to the country's aquaculture output. According to the ICAR, CIBA (2021), brackish water aquaculture in India covers over 1.3 million hectares, with deltas accounting for a major share. In the Sundarbans, approximately 4.5 million people are engaged in fisheries, including prawn seed collection, crabbing, and artisanal fishing (Mallik 2023). High-value species such as Penaeus monodon (tiger shrimp), Labeorohita (rohu), and Catlacatla (catla) dominate production in these areas. India's seafood exports, largely sourced from deltaic aquaculture, crossed ₹57,000 crore in 2022–23, underscoring the economic relevance of these regions (MPEDA 2023). However, unregulated expansion has led to challenges like salinity intrusion, mangrove loss, and water quality degradation, making environmentally sustainable practices imperative for the sector's future.

2.4.4 Cultural and Historical Significance

Delta regions have long served as epicenters of civilization, trade, and cultural evolution due to their fertile lands, perennial water supply, and navigable river networks. Ancient cities such as Thanjavur in the Cauvery Delta, Kalinga in the Mahanadi Delta, and Kolkata near the Sundarbans emerged as influential urban centers due to their deltaic geography. These regions hosted powerful dynasties like the Cholas, Pallavas, and Kalingas, who constructed elaborate temple complexes, irrigation systems, and maritime trade ports, many of which are now UNESCO World Heritage Sites (UNESCO 2021). Cultural festivals such as Aadi Perukku in Tamil Nadu celebrate the annual flooding of rivers, symbolizing prosperity and the sacred link between water and life. Even today, delta communities continue to preserve traditional practices, arts, and water-centric rituals that are deeply tied to seasonal river flows and agrarian rhythms (Jagannathan 2005). This enduring cultural identity highlights the profound role of deltas in shaping regional heritage and social cohesion.

2.5 Disadvantages and Vulnerabilities of Delta Regions

Despite their immense ecological and economic significance, deltaic regions remain highly vulnerable to both natural hazards and human-induced stressors. Their low elevation, proximity to coastlines, and dependence on riverine processes make them particularly susceptible to flooding, salinity intrusion, and land subsidence. These challenges threaten not only the environment but also the livelihoods of millions dependent on delta systems.

2.5.1 Flooding and Cyclones

Deltaic regions in India rank among the most disaster-prone landscapes due to their low elevation, flat morphology, and exposure to marine influences. These geophysical attributes significantly elevate their susceptibility to fluvial flooding, storm surges, and high-intensity tropical cyclones. The eastern coastline, comprising deltas such as the Sundarbans, Mahanadi, Krishna, and Godavari, encounters approximately 5–6 cyclonic systems annually, of which 1–2 escalate to severe or very severe categories (IMD 2021). Cyclone Amphan (2020), one of the most intense storms in recent decades, inflicted an estimated economic loss of ₹1.02 lakh crore (USD 13.5 billion), directly impacting over 13 million individuals and resulting in the destruction of more than 500,000 households across the Sundarbans and adjoining areas (World Bank 2021a, b, c).

Similarly, the Mahanadi Delta in Odisha faces recurrent monsoonal flooding. The 2011 flood event alone displaced over 2.5 million people, underscoring the delta's chronic exposure to hydrometeorological extremes (Odisha SRC 2012). Although structural interventions such as embankments and flood spill channels exist, they are frequently undermined by poor maintenance, encroachment, and the escalating intensity of precipitation events driven by climate variability. According to the Sixth Assessment Report of the Intergovernmental Panel on Climate Change (IPCC 2021), South Asia is projected to witness a 30–50% increase in the frequency and magnitude of extreme rainfall events by the end of the twenty-first century, further intensifying hydroclimatic risks in deltaic environments.

2.5.2 Salinity Intrusion

Salinity intrusion has emerged as a critical environmental hazard across Indian deltaic regions, primarily driven by sea, level rise, declining river discharge, and unsustainable groundwater extraction. This phenomenon results in the encroachment of saline water into freshwater aquifers and surface irrigation systems, thereby degrading water quality and impairing agricultural productivity. In the Cauvery Delta, recent monitoring indicates electrical conductivity (EC) values exceeding 4 dS/m in several zones, levels detrimental to sensitive crops like rice and pulses (CGWB, 2022).

In the Sundarbans region, mean sealevel rise has been recorded at approximately 3.14 mm/year (INCOIS 2020), facilitating frequent saline water intrusion into agricultural fields and wetlands. This process has affected nearly 1.2 million hectares of cultivable land, resulting in reported yield losses of up to 50% in areas reliant on monsoon and surface irrigation (Mandal et al. 2023). Public health is also at risk, with studies revealing that more than 20% of shallow wells in parts of the Mahanadi Delta exhibit salinity beyond acceptable drinking water standards, especially during the dry season (CGWB 2020). Such environmental degradation not only undermines food and water security but also accelerates seasonal migration and rural depopulation, further stressing urban infrastructure.

2.5.3 Land Subsidence

Land subsidence refers to the slow sinking of the ground surface over time. In India's deltaic regions, it is becoming more common due to excessive groundwater extraction, sediment compaction, and reduced silt supply from upstream dams. In the Krishna, Godavari Delta, recent interferometric synthetic aperture radar (InSAR) data from ISRO (2021) indicate annual subsidence rates ranging from 1 to 4 cm in regions with high borewell density and aquifer exploitation. Similarly, hydrogeological studies in the Cauvery Delta reveal localized subsidence of up to 1.5 cm/year, attributed to prolonged deep aquifer pumping and insufficient recharge (CGWB 2022).A critical driver of this phenomenon is sediment starvation. Large, scale hydraulic structures such as the Polavaram Project on the Godavari and the Hirakud Dam on the Mahanadi have significantly curtailed downstream sediment delivery.

This disruption reduces natural land andbuilding processes, leading to net elevation loss and heightened vulnerability to coastal inundation and saline water intrusion. The World Bank (2021a, b, c) projects that if current extraction and damming trends persist, India may experience over 2,000 sq. km of land loss in its major deltas by 2050 due to the combined effects of land subsidence and sealevel rise. The socio-economic repercussions are substantial. Subsidence contributes to the cracking of rural homes, failure of irrigation infrastructure, road damage, and a persistent threat to agricultural stability. These impacts deepen the vulnerability of agrarian communities, many of whom lack the financial resilience to adapt to long-term land degradation.

2.6 Water Resources Management in Delta Regions

Effective management of water resources in deltaic regions is crucial due to the multifaceted demands from agriculture, domestic consumption, industries, and ecosystems. These regions are increasingly stressed by upstream hydrological modifications such as dam construction, irrigation diversions, and inter-basin water transfers, which severely alter sediment delivery and flow regimes critical to delta formation and sustainability (Nagata et al. 2022). A multi-pronged, integrated approach is required to ensure both

Table 2.1 Irrigated area in major delta regions of India

Delta region	Command area (Mha)	Actual irrigated area (Mha)	Irrigation efficiency (%)
Krishna	1.5	1.2	80
Godavari	1.4	1.0	71
Cauvery	1.1	0.8	72.7
Ganga	1.0	0.75	75
Mahanadi	0.9	0.6	66.6

Source Compiled from CWC (2021), Ganugulaand Krishna (2015), and State Irrigation Department reports

human development and ecological resilience in these fragile systems. The key pillars of water resources management in the Indian delta regions are discussed below.

2.6.1 Irrigation Networks

India's deltaic zones, particularly those nourished by the Krishna, Godavari, Ganga, and Cauvery rivers, contain some of the oldest and most expansive canal irrigation systems in the country. These systems were initially developed during the British colonial period and later expanded through various state-led initiatives. They continue to play a critical role in ensuring regional food security and managing agricultural water distribution. Prakasam Barrage on the Krishna River irrigates over 1.3 million acres and sustains an extensive canal network throughout Andhra Pradesh (Ganugulaand Krishna 2015). Table 2.1 provides a summary of the total command area, actual irrigated area, and irrigation efficiency across major Indian delta regions, offering a comparative view of irrigation performance.

To complement the data presented in Table 2.1, the following Fig. 2.7 visually compares the command area and actual irrigated area across major Indian delta regions.

This graphical representation helps to quickly assess the irrigation efficiency and the gap between the designed and achieved water distribution.

2.6.2 Reservoirs and Barrages

Dams and barrages serve as essential components of delta management by regulating seasonal water flows, storing monsoonal runoff, and controlling floods. These structures also provide irrigation and drinking water to millions. However, upstream storage and over-extraction often reduce the freshwater flow reaching the delta, leading to salinity intrusion and degradation of wetland ecosystems (Van Staveren et al. 2022). Some of the most significant examples include the Nagarjuna Sagar Dam (Krishna River), Polavaram Project

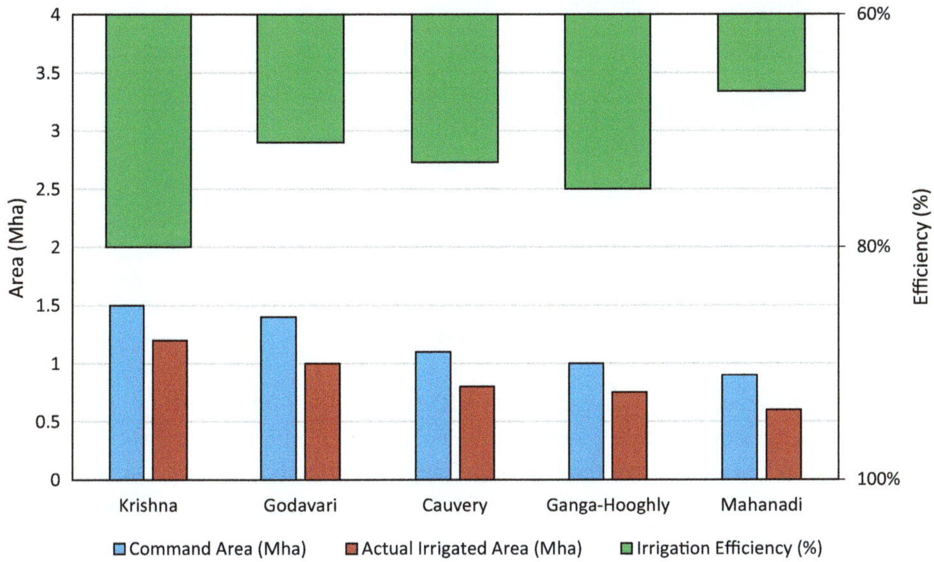

Fig. 2.7 Comparison of command vs actual irrigated area of delta region of India and their efficiencies

(Godavari River), and Farakka Barrage (Ganga River), Hirakud dam (Mahanadi), and Sardar Sarovar (Narmada),each playing a unique role in regional water distribution and delta maintenance. Table 2.2 provides a comparative overview of key reservoirs influencing Indian delta regions, highlighting their designed storage capacities and primary functional objectives.

Table 2.2 Major reservoirs in deltaic river systems

Reservoir/Barrage	River	Designed capacity (BCM)	Purpose
Nagarjuna Sagar	Krishna	11.6	Irrigationand drinking water
Polavaram (planned)	Godavari	8.0	Irrigationandhydropower
Farakka Barrage	Ganga	Regulatory	Flow regulation
Hirakud Dam	Mahanadi	8.1	Irrigationand flood control
Sardar Sarovar	Narmada	9.5	Multi-purpose

Sources NWDA (2020), Krishna Irrigation Dept. (2018), CWC (2021)

Table 2.3 Sectoral water demand in major deltas

Delta region	Agriculture (%)	Domestic (%)	Industry (%)	Environmental flow(%)
Krishna	78	12	7	3
Godavari	72	15	8	5
Ganga	68	18	9	5
Cauvery	80	10	6	4

Source MoJS (2021), NWDA (2020),Nagata et al. (2022)

2.6.3 Integrated Water Management

Integrated Water Resources Management (IWRM) provides a framework for harmonizing water use among sectors while protecting the ecological health of river basins. India has initiated basin-wide efforts like the Namami Gange Programme, Godavari–Krishna River Linking, and Delta Region Development Authorities to advance coordinated and sustainable water governance (Nagata et al. 2022). Despite this, implementation is often limited by fragmented institutional responsibilities, poor data sharing, and minimal community involvement. The sector-wise distribution of water demand in different delta regions has been shown in Table 2.3, highlighting the dominance of agriculture over other sectors.

To visually support the data presented in Table 2.3, Fig. 2.8 offers a visual breakdown of water demand patterns across major Indian deltas, reflecting the proportion of use by agriculture, domestic needs, industry, and environmental flows. Figure 2.8 provides a clearer understanding of sectoral water allocation and highlights regional variations in usage priorities.

2.7 Government Initiatives for Delta Region Management

The ecological fragility and socio-economic significance of India's deltaic systems have necessitated targeted policy interventions by both central and state governments. In response to increasing environmental vulnerabilities, especially those induced by climate change, several integrated programs have been initiated to enhance the adaptive capacity, promote sustainable livelihoods, and safeguard critical ecosystems within these regions. These initiatives encompass climate, resilient agriculture, river basin rejuvenation, coastal zone regulation, and legal safeguards aimed at long-term delta conservation.

2.7.1 National Adaptation Fund for Climate Change (NAFCC)

The National Adaptation Fund for Climate Change was established in 2015 under the Ministry of Environment, Forest and Climate Change (MoEFCC). NAFCC is designed

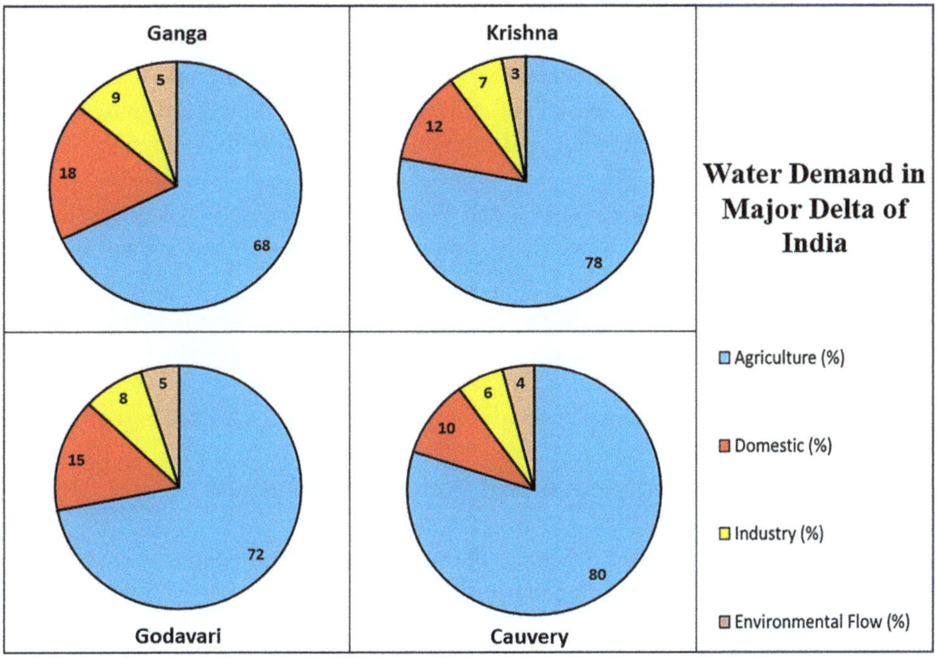

Fig. 2.8 Water demand in major delta of India

to finance climate adaptation strategies across sectors vulnerable to climate variability, particularly agriculture and water management. Delta regions such as the Godavari, Mahanadi, and Sundarbans have benefited from NAFCC, supported projects focusing on saline-tolerant crop varieties, improved irrigation techniques, and mangrove restoration. By 2023, over 30 projects across 25 Indian states had been financed, many with direct implications for deltaic ecosystem resilience (MoEFCC 2023).

2.7.2 Namami Gange Programme

This centrally sponsored flagship initiative by the Ministry of Jal Shakti seeks to rejuvenate the Ganga River basin through comprehensive pollution abatement, ecological flow restoration, and watershed-based interventions. In deltaic stretches such as the Sundarbans, the program supports the installation of sewage treatment plants, community-led wetland management, and biodiversity monitoring frameworks. Special attention is given to integrating local communities into planning and implementation processes to ensure socio-ecological sustainability (Jal Shakti Ministry 2022).

2.7.3 Integrated Coastal Zone Management Project (ICZMP)

Implemented with technical and financial assistance from the World Bank, ICZMP aims at holistic management of India's coastal and deltaic regions. The project integrates shoreline stabilization, mangrove regeneration, and promotion of eco, sensitive tourism.

- Phase I (initiated in Gujarat, Odisha, and West Bengal) focused on vulnerable deltas such as the Mahanadi and Sundarbans, where coastal erosion and saline ingress are pressing concerns.
- Phase II, launched recently, extends the project to Andhra Pradesh, Tamil Nadu, and Maharashtra, encompassing the Krishna and Godavari deltas (World Bank 2021a, b, c).

2.7.4 State, Level Legislative and Policy Interventions

In addition to centrally funded schemes, individual states have adopted legislation and region-specific policies to enhance delta protection.

- Tamil Nadu's Delta Protection Bill (2020) designates the Cauvery delta as a Protected Special Agricultural Zone, legally restricting high-risk industrial activities to safeguard fertile farmland and groundwater resources.
- Odisha's Wetland Mission targets critical wetland zones like Chilika and Bhitarkanika, emphasizing habitat restoration, real-time wetland monitoring, and inclusive governance through local stakeholder engagement (Government of Odisha 2021).

These multifaceted initiatives collectively reflect a growing national commitment to integrated delta management. However, their long-term success is contingent upon robust institutional coordination, timely policy execution, and inclusive participation of local communities and stakeholders.

2.8 Discussions

Delta regions of India are highly productive and ecologically rich. They support agriculture, fisheries, and millions of livelihoods. At the same time, they face serious challenges like floods, cyclones, salinity intrusion, and land subsidence. Climate change and human activities have increased these risks. This shows that deltas are fragile systems which need careful and integrated management. The issues of delta management are also linked to the Sustainable Development Goals (SDGs). Ensuring food production in fertile delta soils supports SDG 2 (Zero Hunger). Protecting wetlands and aquatic habitats connects to SDG 14 (Life Below Water), while conserving forests and biodiversity relates to SDG 15 (Life on Land). Building resilience against floods and cyclones supports SDG 13 (Climate Action). Access to clean water is part of SDG 6 (Clean Water and Sanitation). Since millions of poor and vulnerable people live in these regions, delta management also contributes to SDG 1 (No Poverty) and SDG 8 (Decent Work and Economic Growth).

Several government programs show progress. The National Adaptation Fund on Climate Change (NAFCC) and the Integrated Coastal Zone Management Project (ICZMP) are good examples. However, problems remain. Governance is fragmented, and local communities often have limited participation. Heavy dependence on embankments and water diversions sometimes harms ecosystems. Future strategies should focus on both technology and nature-based solutions. Community knowledge, rainwater harvesting, aquifer recharge, and sustainable aquaculture can give better results. By linking delta water management with the SDGs, India can secure food, biodiversity, and livelihoods while also improving resilience to climate change.

2.9 Summary and Conclusions

India's deltaic regions, though covering less than 2.5% of the country's landmass, contribute over 15% of national food grain production (MoAFW 2022), showing their vital agricultural role. Major deltas such as the Krishna and Godavari, with 6 Mha of farmland, and the Cauvery, producing over 40% of Tamil Nadu's paddy, highlight the fertility of alluvial soils and the benefits of canal irrigation. Irrigation efficiency varies from 66.6% in the Mahanadi to 80% in the Krishna. These fertile deltas also support dense rural populations, with more than 30 million people living in eastern India's deltaic belt.

Ecological importance is equally high. The Sundarbans, spread across 10,000 km^2, support over 693 species and deliver ecosystem services such as carbon sequestration, cyclone buffering, and fish breeding, with values running into billions annually (UNEP 2018). Brackish water aquaculture, extending over 1.3 million hectares, plays a major role in seafood exports, which exceeded ₹57,000 crore in 2022–23 (MPEDA 2023). Yet these regions face pressure from human activities. Dams like Polavaram and Hirakud, inter-basin transfers, and unregulated aquaculture, combined with climate-driven risks like sea level rise (3.14 mm/year in the Sundarbans) and land subsidence (1–4 cm/year in Krishna and Godavari deltas), are destabilizing these systems.

Water management challenges add to the problem. Agriculture consumes 68–80% of total water, while reservoirs such as Nagarjuna Sagar (11.6 BCM) and Hirakud (8.1 BCM) regulate flows but also restrict sediment, leading to salinity intrusion and wetland loss. Programs including Namami Gange, ICZMP, and the National Adaptation Fund for Climate Change, with more than 30 projects across 25 states, have improved awareness, but institutional gaps and weak community participation persist. Future strategies must align with SDG 2 (Zero Hunger), SDG 6 (Clean Water and Sanitation), SDG 13 (Climate

Action), and SDG 15 (Life on Land), ensuring ecological balance along with food and livelihood security.

If present stressors are left unchecked, over 2,000 km^2 of deltaic land could be lost by 2050 (World Bank 2021a, b, c). Addressing this requires raising irrigation efficiency above 80%, maintaining 20–30% of freshwater as environmental flow, expanding mangroves and wetlands by 10–15%, and promoting climate-resilient farming in saline-prone areas. Stronger policy measures such as Tamil Nadu's Delta Protection Act (2020), backed by SDG-focused planning, better coordination among institutions, and active community participation, are essential. These actions will help preserve the ecological integrity and socio-economic resilience of India's delta regions in a changing climate.

2.10 Recommendations

For the sustainable growth of India's delta regions, irrigation efficiency should be improved using modern techniques and better water distribution, especially in areas with low efficiency. At least 20–30% of freshwater flow should be maintained downstream to support agriculture and ecosystems. Climate-resilient farming, including salt-tolerant crops and diversified cultivation, should be promoted. Community awareness programs will help strengthen local adaptation. Expanding mangrove and wetland areas by 10–15% can protect coastlines and support biodiversity.

Water governance must be strengthened by optimizing reservoir operations, ensuring fair water distribution, and using real-time monitoring tools. Greater community involvement in planning and managing water resources will enhance resilience. Improved coordination among institutions and better data sharing can support effective planning and reduce conflicts in shared basins. These actions should align with Sustainable Development Goals, such as Zero Hunger, Clean Water, and Climate Action, to guide sustainable interventions.

Ongoing research and innovation are necessary to address the challenges facing delta ecosystems. Collaboration between government agencies, researchers, and local communities will help implement adaptive solutions. Strengthening policies and fostering cooperation can protect India's deltas, ensuring ecological balance and the livelihoods of the people who depend on them.

Acknowledgements The authors are thankful to the National Institute of Hydrology, Roorkee, and its Central India Hydrology Regional Centre, Bhopal for providing facilities to collect the relevant data and information to write the chapter. We confirm that the chapter was prepared entirely based on our own research, analysis, and interpretation of available literature. AI tools were not used in the creation of the chapter content.

References

Anthony EJ (2015) Wave influence in the construction, shaping and destruction of river deltas: a review. Marine Geology 361:53–78. https://doi.org/10.1016/j.margeo.2014.12.004

Arun D, Sunandini GP, Kumari RV, Meena A (2020) An economic analysis of different rice varieties in Cauvery Delta region of Tamil Nadu. Indian J Econ Dev 8(3):1, 5. https://doi.org/10.17485/IJED/v8.52

Baidya S, Chakraborty P, Rakkasagi S, Goyal MK, Gupta AK (2023) Pathways to build resilience toward the impact of climate change on the Indian Sunderban. In: Ecosystem restoration: towards sustainability and resilient development. Singapore: Springer Nature Singapore, pp 307, 333. https://doi.org/10.1007/978-981-99-3687-8_17

Barbier EB, Hacker SD, Kennedy C, Koch EW, Stier AC, Silliman BR (2011) The value of estuarine and coastal ecosystem services. Ecol Monogr 81(2):169, 193. https://doi.org/10.1890/10-1510.1

Bhattacharya J (2017) The sea of changes: shifting trajectories across the Bay of Bengal. Asian Polit Policy 9(2):245, 267. https://doi.org/10.1111/aspp.12321

Central Ground Water Board (CGWB) (2020) Ground Water Quality in Shallow Aquifers of India. Ministry of Jal Shakti, Government of India. https://www.cgwb.gov.in/cgwbpnm/public/uploads/documents/17068003031756689658file.pdf

Central Ground Water Board (CGWB) (2022) Ground Water Year Book–India 2021–22. Ministry of Jal Shakti. https://www.cgwb.gov.in/old_website/GW-Year-Book.html

Central Water Commission (2021) Annual report on water resources in India 2020–21. Ministry of Jal Shakti, Government of India. https://www.cwc.gov.in/sites/default/files/annual-report-2020-21.pdf

Chakrabarti DK (2001) Archaeological geography of the Ganga plain: the lower and the middle Ganga. Orient Blackswan. https://books.google.co.in/books?id=OEZe-wAIiKIC&printsec=frontcover&source=gbs_ge_summary_r&cad=0#v=onepage&q&f=false

Change IC (2014) Impacts, adaptation and vulnerability. Part A: global and sectoral aspects. Contribution of working group II to the fifth assessment report of the intergovernmental Panel on Climate Change 1132. https://www.ipcc.ch/report/ar5/wg2/

Dalrymple RW, Zaitlin BA, Boyd R (1992) Estuarine facies models; conceptual basis and stratigraphic implications. J Sediment Res 62(6):1130, 1146. https://doi.org/10.1306/D4267A69-2B26-11D7-8648000102C1865D

Ekka A, Pandit A, Sandhya KM, Sajina AM, Kumari S, Biswas DK, Das BK (2020) Economic analysis of P. monodon post larvae by, catch in Indian Sundarbans: an impasse between livelihood and conservation. https://www.preprints.org/frontend/manuscript/b3846e152e3d973f91d88f694f9ae48b/download_pub

Ericson JP, Vörösmarty CJ, Dingman SL, Ward LG, Meybeck M (2006) Effective sea, level rise and deltas: causes of change and human dimension implications. Glob Planet Chang 50(1, 2), 63, 82. https://doi.org/10.1016/j.gloplacha.2005.07.004

Galloway WE (1975) Process framework for describing the morphologic and stratigraphic evolution of deltaic depositional systems. https://archives.datapages.com/data/hgssp/data/022/022001/87_hgs0220087.htm

Ganugula TV, Krishna GS (2015) Flood frequency analysis of Prakasam Barrage reservoir, Krishna District, Andhra Pradesh using Weibull, Gringorten and L moments formula. Int J Civ Struct Eng 5(2):57–62. https://www.researchgate.net/publication/328449678

Ghosh N, Modak S (2022) MANJRA GODAVARI. Riverine systems: understanding the hydrological, hydrosocial and hydro, heritage dynamics, 267. https://doi.org/10.1007/978-3-030-87067-6_15

Giri C, Zhu Z, Tieszen LL, Singh A, Gillette S, Kelmelis JA (2008) Mangrove forest distributions and dynamics (1975–2005) of the tsunami-affected region of Asia. J Biogeogr 35(3):519, 528. https://doi.org/10.1111/j.1365-2699.2007.01806.x
Giri C, Long J, Abbas S, Murali RM, Qamer FM, Pengra B, Thau D (2015) Distribution and dynamics of mangrove forests of South Asia. J Environ Manag 148:101, 111. https://doi.org/10.1016/j.jenvman.2014.01.020
Government of Odisha (2021) Wetland conservation strategy and mission document. Odisha State Wetland Authority. https://urban.odisha.gov.in/news/wetland-restoration-and-wastewater-management-governments-top-priority?utm_source=chatgpt.com
Government of Tamil Nadu (2020) The Tamil Nadu Protected Agricultural Zone Development Act. Chennai: Department of Agriculture and Farmers Welfare. https://www.indiacode.nic.in/handle/123456789/21809?locale=en&utm
Hazra S, Ghosh T, DasGupta R, Sen G (2002) Sea level and associated changes in the Sundarbans. Sci Cult 68(9/12):309, 321. https://gyanganga.ai/admin//fileupload//Sea%20Level%20and%20associated%20changes%20in%20the%20Sundarbans.pdf?utm
Hazra S, Das S, Ghosh A, Raju PV, Patel A (2020) The Mahanadi delta: a rapidly developing delta in India. Deltas Anthropocene 53, 77. https://doi.org/10.1007/978-3-030-23517-8_3
ICAR CIBA (2021) Annual Report 2020–21. Central Institute of Brackishwater Aquaculture, Chennai. https://www.ciba.res.in/wp-content/uploads/2022/07/annualReport_english_2021.pdf?utm
India Meteorological Department (IMD) (2021) Annual Climate Summary 2020. Ministry of Earth Sciences, Government of India. https://mausam.imd.gov.in
Indian National Centre for Ocean Information Services (INCOIS) (2020) Sea Level Variability Along the Indian Coast. Ministry of Earth Sciences. https://incois.gov.in
Indian Space Research Organisation (ISRO) (2021) Monitoring Ground Deformation in Indian River Deltas Using InSAR Techniques. National Remote Sensing Centre. https://www.nrsc.gov.in
Intergovernmental Panel on Climate Change (IPCC) (2021) Climate Change 2021: The Physical Science Basis. Contribution of Working Group I to the Sixth Assessment Report of the IPCC. https://www.ipcc.ch/report/ar6/wg1/
Jagannathan M (2005) South Indian Hindu festivals and traditions. Abhinav Publications. https://openlibrary.org/books/OL13136668M/South_Indian_Hindu_Festival_and_Traditions?utm
Joy KJ, Bhadbhade N, Bhagat S, Samuel A (2021) Mahanadi Basin: an alternative approach to interstate water sharing and management. https://waterconflictforum.org/lib_docs/Report-Mahanadi-2021.pdf
Kimmerer W (2004) Open water processes of the San Francisco Estuary: from physical forcing to biological responses. San Franc Estuary Watershed Sci 2(1). https://doi.org/10.15447/sfews.2004v2iss1art1
Krishna District Irrigation Department (2018) Irrigation profile of Krishna District. Government of Andhra Pradesh. https://docslib.org/doc/12608869/irrigation-profile-of-krishna-district?utm
Kumar SA, Karthikeyan C (2022) Uzhavanapp'as a conduit to reduce the digital divide by fostering vital agricultural extension services in the state of Tamil Nadu, India. Int J Agric Innov, Technol Glob 3(1):1, 27. https://doi.org/10.1504/IJAITG.2022.126875
Kumar K (2017) Opportunities for agricultural water management interventions in the Krishna Western Delta, A case from Andhra Pradesh, India. Agribus Inf Manag 9(1):7, 17. https://www.dbpia.co.kr/Journal/articleDetail?nodeId=NODE10248781
Madhanagopal D (2023) Local adaptation to climate change in South India: challenges and the future in the Tsunami- hit coastal regions. Routledge. https://doi.org/10.4324/9781003187691
Mahalik NK (2006) A study of the morphological features and bore hole cuttings in understanding the evolution and geological processes in Mahanadi Delta, East Coast of India. Geol Soc India 67(5):595, 603. https://www.geosocindia.org/index.php/jgsi/article/view/81914?utm

Mahto MK (2024) Challenges and opportunities for water resources in India: an exploratory story of resilience. Integr Manag Water Resour India: Comput Approach: Optim Sustain Plan 59, 80. https://doi.org/10.1007/978-3-031-62079-9_4

Mallick JK (2023) A review and field verification of avian diversity and habitat use in the greater Sundarbans of India, Bangladesh and their hinterland. Probe 5(1). https://doi.org/10.18686/pas.v5i1.1778

Mandal TK, Das P, Mondal M, Khoso I (2023) Intervention of soil salinity in agriculture of Indian Sundarbans: a review. https://doi.org/10.30954/2347-9655.01.2023.12

Marine Products Export Development Authority (MPEDA) (2023) Annual seafood export performance 2022–23. Government of India. https://mpeda.gov.in/wp-content/uploads/2023/06/Marine-products-export-performance-2022-23-english.pdf?utm

Menon M, Prashanthi Devi M, Nandagopalan V, Mohanraj R (2015) Species diversity and functional assemblages of bird fauna along the riverine habitats of Tiruchirappalli, India. Environ Manag River Basin Ecosyst 729:748. https://doi.org/10.1007/978-3-319-13425-3_32

Ministry of Agriculture and Farmers Welfare (MoAFW) (2022) State-wise crop production statistics. Government of India. https://desagri.gov.in/wp-content/uploads/2023/05/Agricultural-Statistics-at-a-Glance-2022.pdf?utm

Ministry of Jal Shakti (2022) Namami Gange mission progress report 2021–22. Government of India. https://nmcg.nic.in

Ministry of Environment, Forest and Climate Change (2023) National adaptation fund for climate change: status and progress report. Government of India. https://moef.gov.in

Mirza MMQ (2004) The Ganges water diversion: environmental effects and implications—an introduction. In the ganges water diversion: environmental effects and implications. Dordrecht: Springer Netherlands, pp 1, 12. https://doi.org/10.1007/978-1-4020-2792-5_1

Mishra SP, Jena JG (2015) Morphological reconstruction of Southern Mahanadi Delta and Chilika Lagoon, India–a critical study. Int J Adv Res 3(5), 691, 702. https://www.journalijar.com/article/4592/

Murthy RVR (2020) Mapping Spatio-Temporal Cropland Changes Due To Water Stress In Krishna River Basin Using Temporal Satellite Data (Doctoral dissertation, College of Engineering (A), Andhra University, Visakhapatnam). https://oar.icrisat.org/11661/

Nagata K, Shoji I, Arima T, Otsuka T, Kato K, Matsubayashi M, Omura M (2022) Practicality of integrated water resources management (IWRM) in different contexts. Int J Water Resour Dev 38(5):897–919. https://doi.org/10.1080/07900627.2021.1921709

Natarajan K, Suresh PK, Sundaravadivelu R (2023) Shoreline Changes and Sediment. In: geospatial and soft computing techniques: proceedings of 26th international conference on hydraulics, water resources and coastal engineering (HYDRO 2021). Springer Nature, vol 339, p 141. https://doi.org/10.1007/978-981-99-1901-7_12

National Water Development Agency (2020) Godavari–Krishna River Linking Project. Ministry of Jal Shakti, Government of India. https://nwda.gov.in/content/error.php

Nicholls RJ, Wong PP, Burkett V, Codignotto JO, Hay J, Roger F, Woodroffe CD (2007) Coastal systems and low-lying areas. https://doi.org/10.1017/S0790062700000103

Odisha State Relief Commissioner (SRC) (2012) Flood Report 2011. Government of Odisha. https://prod-qt-images.s3.amazonaws.com/indiawaterportal/import/sites/default/files/iwp2/Flood_Update_II_from_WIO_2011.pdf?utm

Rajavel M, Kumar MS, Bhuvaneswari K, Vakeswaran V, Kumar SM, Sathyamoorthy N, Priyanka S (2022) Projected changes in climate over Cauvery delta zone of Tamil Nadu in Southern India. Int J Environ Clim Chang 12:1287, 1294. https://doi.org/10.9734/ijecc/2022/v12i1030927

Ramasamy SM, Saravanavel J., Selvakumar R (2006) Late Holocene geomorphic evolution of Cauvery Delta, Tamil Nadu. J Geol Soc India 67(5):653. https://www.geosocindia.org/index.php/jgsi/article/view/81941/0

Rao KN, Sadakata N, Malini BH, Takayasu K (2005) Sedimentation processes and asymmetric development of the Godavari delta, India. https://doi.org/10.2110/pec.05.83.0435

Rodrigues JN, Bhattacharya S, Sachdev BK, Cabete D (2024) Disaster management in the Sundarbans delta ecosystem through environmental governance and sustainable planning. Revista de DireitoEconômico e Socioambiental 15(2):9. https://doi.org/10.7213/revdireconsoc.v15i2.31364

Rudra K (2010) Dynamics of the Ganga in West Bengal, India (1764–2007): Implications for science–policy interaction. Quaternary International 227(2):161, 169. https://doi.org/10.1016/j.quaint.2009.10.043

Stanley DJ, Warne AG (1993) Nile Delta: recent geological evolution and human impact. Sci 260(5108):628, 634. https://doi.org/10.1126/science.260.5108.628

Syvitski JP, Kettner AJ, Overeem I, Hutton EW, Hannon MT, Brakenridge GR, Nicholls RJ (2009) Sinking deltas due to human activities. Nat Geosci 2(10):681, 686. https://doi.org/10.1038/ngeo629

Tamil Nadu Government Gazette (2020) Tamil Nadu Protected Agricultural Zone Development Act, 2020 (Act No. 10 of 2020). Chennai: Government of Tamil Nadu. https://www.tn.gov.in/acts

Tessler ZD, Vörösmarty CJ, Grossberg M, Gladkova I, Aizenman H, Syvitski JP, Foufoula Georgiou E (2015) Profiling risk and sustainability in coastal deltas of the world. Sci 349(6248):638, 643. https://doi.org/10.1126/science.aab3574

Tran TA, Pittock J, Tran DD (2020) Adaptive flood governance in the Vietnamese Mekong Delta: a policy innovation of the North Vam Nao scheme, An Giang Province. Environ Science Policy 108:45, 55. https://doi.org/10.1016/j.envsci.2020.06.004

UNESCO (2021) World Heritage Sites in India. United Nations Educational, scientific and cultural organization. https://whc.unesco.org/en/statesparties/in

United Nations Environment Programme (UNEP) (2018) The importance of deltas: ecosystem services&climate resilience. Nairobi: UNEP. https://www.unep.org/resources/report/importance-deltas-ecosystem-services-and-climate-resilience

Van Staveren M, Dewan A, Wilson E, Alam M (2022) Coastal salinity and water management practices in the Bengal Delta: A critical analysis to inform salinisation risk management strategies in Asian deltas. Earth Syst Environ 6(1):122–135. https://doi.org/10.1007/s41748-022-00335-9

Vörösmarty CJ, Syvitski J, Day J, De Sherbinin A, Giosan L, Paola C (2009) Battling to save the world's river deltas. Bull Atlc Sci 65(2):31, 43. https://doi.org/10.2968/065002005

Wells JT, Coleman JM (1981) Physical processes and fine-grained sediment dynamics, coast of Surinam, South America. J Sediment Res 51(4):1053, 1068. https://doi.org/10.1306/212F7E1E-2B24-11D7-8648000102C1865D

World Bank (2021a) Cyclone Amphan: post-disaster needs assessment for West Bengal&Odisha. https://documents.worldbank.org

World Bank (2021b) Delta vulnerability and climate resilience in South Asia: A Policy Overview. https://www.worldbank.org

World Bank (2021c) India integrated coastal zone management project: implementation completion Report. Washington. D.C.https://documents.worldbank.org/curated/en/488791507885476136/pdf/120404-WP-CountrySnapshotIndiaFinalmerged-PUBLIC.pdf?utm

Long-Term Water Allocation and Assessment Strategies for the Delta Regions

3

Manish Meena and Nitin Dutt Chaturvedi

Abstract

Water management is fundamental to the sustainability of global river deltas. Delta regions face intensifying threats from saline intrusion, flooding, and erosion, exacerbated due to climate change and human activities like unreliable water extraction, construction and management. This chapter argues that proactive, collaborative management is essential for delta resilience. This explores novel technologies and scientific frameworks for optimising water allocation and restoration, aligning resource use besides ecological limits. Examining governance reforms and promoting alliances for knowledge sharing, outlines a pathway toward integrated water management plans. This briefs the discussion on long-term water management that fosters collaboration of strategies to mitigate cascading environmental and socioeconomic risks, thereby directly support SDGs related to water security, sustainable communities, climate action, and life on land.

M. Meena (✉) · N. D. Chaturvedi (✉)
Department of Chemical and Biochemical Engineering, Indian Institute of Technology, Patna, Bihar, India
e-mail: Manish_2221cb03@iitp.ac.in

N. D. Chaturvedi
e-mail: nitind@iitp.ac.in

S. Yadav et al. (eds.), *Water Resources in Delta Region in India*, Deltas of the World,
https://doi.org/10.1007/978-3-032-06500-1_3

3.1 Introduction

River deltas are vital ecosystems that support millions of people and serve as global food baskets, yet they are among the planet's most threatened landscapes (Yan et al. 2016). Their sustainability is critical for both local livelihoods and global food security. These regions face intensifying vulnerabilities from climate change, including sea-level rise, stronger cyclones, soil and water salinization, and land subsidence (Sebesvari et al. 2016). Without urgent action, these pressures will worsen insecurity, degradation, and potentially lead to irreversible damage (Bélanger et al. 2015). The escalated risk highlights the urgent need for proactive governance focused on building productive, resilient, and inclusive deltas that must ensure sustainability and withstand environmental and climatic stressors, along with socio-ecological integrity (Zevenbergen et al. 2018).

Critical direction for sustainable delta management is the advancement of next-generation models that explicitly integrate ensemble climate projections and deep uncertainty frameworks (Bloemen et al. 2018). This requires moving beyond historical baselines to force models with downscaled climate data, improving forecasts on shifting water flow and demand under various emission scenarios (Keller et al. 2022). This enhances operational responsiveness, transition from monthly or annual allocations to high-frequency temporal resolutions (sub-daily or hourly), enabling dynamic management aligned with real-time hydrological variability (Ranjbar 2025).

Long-term water allocation in delta regions requires a balance between competing agricultural, urban, and industrial demands and the essential water needs of the environment. This balance is increasingly difficult to achieve due to climate change and upstream interventions. Sustainable water management in these complex systems depends on an integrated approach combining engineering, policy, and ecological strategies (Hommes et al. 2009). Key components for regulating upstream water flow to maintain ecological health and ensure sustainable distribution include:

- Infrastructure for Flow Regulation: Dams and reservoirs are used to store water in wet seasons and release it during dry periods, ensuring a reliable year-round supply for downstream users and ecosystems.
- Managed Water Diversion: A network of canals, weirs, and gates allows for the precise control and redirection of water to where it is most needed, while also helping to mitigate flood risks.
- Sediment Management: Strategic management through techniques like sediment bypassing or dredging is crucial. Maintaining the natural flow of sediment is vital to counter land subsidence and mitigate sea-level rise, which are existential threats to deltas.
- Floodplain Restoration: Reconnecting rivers to their natural floodplains enhances biodiversity, improves water quality through natural filtration, and provides vital capacity for absorbing and slowly releasing floodwaters.

- Integrated Water Resource Management (IWRM): This overarching framework ensures the coordinated management of water, land, and resources to maximize economic and social welfare without compromising the sustainability of vital ecosystems.

Sustainable water usage and management necessitate regulatory frameworks, monitoring mechanisms, and stakeholder engagement. The development of resilient infrastructure and the deployment of early warning systems for extreme weather events can assist in addressing the variable water flow caused by climate change. Also, the integration of distinct datasets and methodologies can facilitate the creation of a more resilient and sustainable water management in delta regions. Prior research has proposed various optimization methods for water allocation, including mathematical programming and artificial intelligence (Arto et al. 2019). Most studies, however, concentrate on water allocation issues utilising hypothetical networks and function at coarse temporal resolutions with shorter time scales. Moreover, they often depend exclusively on historical data, neglecting future climate change and providing limited attention to changes and uncertainties (Nienhuis et al. 2023).

This necessitates strategic planning with a comprehensive assessment to facilitate long-term water allocation and adjustable informaction for decision-making regarding the delta sustainability. Robust Decision Making (RDM) is a quantitative methodology designed to aid decision-making in the context of significant uncertainties (Lempert and Groves 2010). Assessing the robustness of plans under diverse climate scenarios is crucial for managing future uncertainties related to water. Historically, utilising climate model outcomes in conjunction with biophysical hydrological models is often more appropriate (Yan et al. 2015). Whereas the long-term water management plan should also encompass the interactions among dynamic variables of water resources (Karimi and Ardakanian 2010). An overview of identified essential components is presented in Fig. 3.1. The interaction among these dynamic components is essential in planning frameworks. This chapter elucidates the components and equations that provide a way to define them. The chapter examines the modelling opportunities in the management of river delta regions, highlighting the importance of preparedness, partnerships, and effective governance in water management methods.

3.2 Mathematical Strategies for Enhancing Water Management Robustness

Mathematical modelling provides a critical framework for managing water in vulnerable deltas, and offers data-driven approaches that balance human needs with environmental sustainability. Distinct data and simulation techniques enable policymakers to optimise water distribution among agricultural, industrial, and domestic users while preserving ecological health.

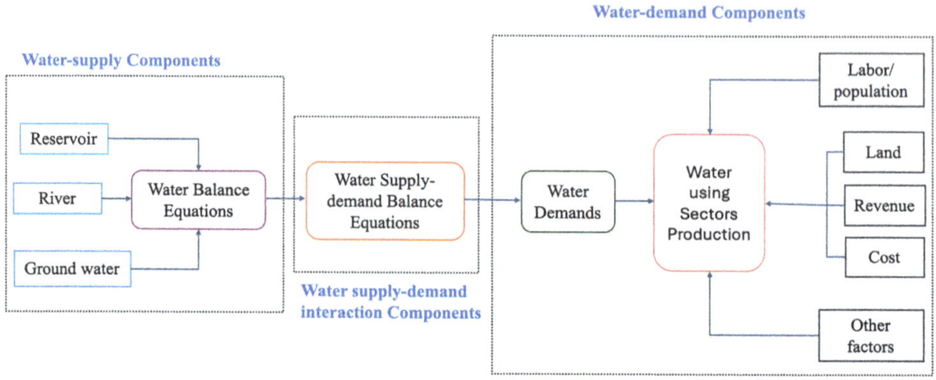

Fig. 3.1 Components of water management practice into deltas, adopted from (Karimi and Ardakanian 2010)

Developing effective long-term water management strategies is challenged by significant uncertainties, particularly those stemming from climate change. These changes are driven by a combination of environmental factors and human activities, such as deforestation, land reclamation, and siltation, which reduce natural water storage capacity and alter basin dynamics. Therefore, modelling must account for both natural processes and the consequences of human intervention to support robust decision-making under various climate scenarios.

3.2.1 Water Supply and Demand Management

Basin-wide water allocation requires strategies that are rational, scientifically-grounded, and adaptable (Kumari et al. 2024). This necessitates managing both water supply and demand while protecting ecological base flows and the minimum water volumes required to sustain ecosystem health. These needs are categorised by (Deng et al. 2024):

- **Instream Ecological Water Requirements (EWRs)**: Water needed within the river channel to maintain aquatic habitats and functions.
- **Offstream EWRs**: Water required for adjacent terrestrial ecosystems, such as riparian forests and grasslands, which are critical components of a basin's ecological framework.

Water sustainability often depends on empirical methods to implement integrated mathematical models. These models are essential for optimizing the complex distribution of water across competing domestic, agricultural, industrial, and ecological needs (Bai et al. 2025), ensuring long-term basin health and resilience (Haasnoot et al. 2012). Modelling this system is characterised by multiple objectives, demands, and constraints to tackle the traditional challenges faced in effectively directing the water distribution within basins.

$$minf_1 = \sum_{t=1}^{T}\sum_{i=1}^{I}\left|Q_{irr}^{d}(t,i) - Q_{irr}^{s}(t,i)\right|\Delta t \tag{3.1}$$

$$minf_2 = \sum_{t=1}^{T}\sum_{i=1}^{I}\left|Q_{eco}^{d}(t,i) - Q_{eco}^{s}(t,i)\right|\Delta t \tag{3.2}$$

where f_1 & f_2 represent the water shortage for agricultural and grassland, respectively. $Q_{irr}^{d}(t,i)$ and $Q_{irr}^{s}(t,i)$ represent the agricultural irrigation water demand and supply at t period and i^{th} region, respectively. $Q_{eco}^{s}(t,i)$ and $Q_{eco}^{d}(t,i)$ represent the supply flow and ecological water requirement for the agricultural forests and grasslands in the region i over a period t; Δt and T represent the period time and scheduling cycle, respectively.

Constraints

Basin water balance constraint

$$W_{surface} + W_{ground} = W_{demesticity} + W_{industry} + W_{irrigation} + W_{ecology} + W_{loss} \tag{3.3}$$

Guaranteed rate constraint

$$P(i) \geq P_s(i) \tag{3.4}$$

Reservoir water balance constraint

$$W_L(n,t) = W_G(n,t) + W_s(n,t) \mp W_{XF}(n,t) \tag{3.5}$$

Water level constraint

$$Z_{min}(n) \leq Z(n,t) \leq Z_{max}(n) \tag{3.7}$$

Reservoir storage capacity constraint

$$V_{min}(n) \leq V(n,t) \leq V_{max}(n) \tag{3.8}$$

Water supply constraint

$$S_{ecology}(t) \leq D_{ecology}(t) \tag{3.9}$$

$$S_{society}(t) \leq D_{society}(t) \tag{3.10}$$

where $W_{surface}$ and W_{ground} are the surface and ground water levels, respectively. $W_{industry}$, $W_{demesticity}$, $W_{ecology}$, and $W_{irrigation}$ are the industrial, domestic, ecological, and agricultural irrigational water demand, respectively. P and P_s are the actual and design rates of water supply, respectively; W_L, W_G, W_S and W_{XF} are the reservoir inflow, reservoir supply, storage loss, and reservoir storage variation, respectively. V and Z are reservoir storage and reservoir water level, respectively; where t is the serial number of the operation periods; $D_{ecology}$ and $S_{ecology}$ are the EWRs and supply, respectively. $D_{society}$ and $S_{society}$ are the socioeconomic water demand and supply, respectively.

3.2.2 Water Management Tool for the Administrative Water Authority

Current water management in delta regions is ineffective, lacking the adaptive capacity to respond to unforeseen events like droughts or floods (Biggs et al. 2012). Although water management techniques have the potential to improve efficiency, the development lacks adaptive allocation to modify administrative water reactions. Therefore integrated optimization frameworks are needed to reflect real-world dynamics and support decision-making across multiple governance levels. For instance, Tu et al. (2025) proposed a two-stage robust-stochastic optimization model for river basin water allocation. This hierarchical system uses administration as the primary mechanism and market-based allocation as the secondary mechanism. The model simultaneously pursues multiple objectives, including economic efficiency, pollution control, inequality reduction, and environmental sustainability, providing a scientifically robust foundation for adaptive and high-quality water governance.

$$\max_{x_k} F_0^{1\pm} = \sum_{k=1}^{K}\sum_{r=1}^{R}\sum_{(i,j)\in\Omega} P_{(i,j)} T_{kr}^{\pm} B_{kr(i,j)}^{\pm} - \sum_{k=1}^{K}\sum_{r=1}^{R}\sum_{(i,j)\in\Omega} P_{(i,j)} S_{kr(i,j)}^{\pm} C_{kr}^{\pm}$$
$$- \xi \cdot \sum_{k=1}^{K}\sum_{r=1}^{R}\sum_{(i,j)\in\Omega} P_{(i,j)} |S_{kr(i,j)}^{\pm} C_{kr}^{\pm} - P_{(i,j)} \sum_{(i,j)\in\Omega} S_{kr(i,j)}^{\pm} C_{kr}^{\pm}| \quad (3.11)$$

(Benefits in the first stage – penalty in the second stage – expected deviation between the penalty and the expected one).

Where $P_{(i,j)} \sum_{(i,j)\in\Omega} S_{kr(i,j)}^{\pm} C_{kr}^{\pm}$ represents the expected penalty for the water shortage scenario (i,j), while $S_{kr(i,j)}^{\pm} C_{kr}^{\pm}$ indicates the true penalty associated with water shortages. The expression $S_{kr(i,j)}^{\pm} C_{kr}^{\pm} - P_{(i,j)} \sum_{(i,j)\in\Omega} S_{kr(i,j)}^{\pm} C_{kr}^{\pm}$ represents the discrepancy between the actual penalty and the anticipated penalty among multiple shortage scenarios. Whereas the complete term $\sum_{(i,j)\in\Omega} P_{(i,j)} \left| S_{kr(i,j)}^{\pm} C_{kr}^{\pm} - P_{(i,j)} \sum_{(i,j)\in\Omega} S_{kr(i,j)}^{\pm} C_{kr}^{\pm} \right|$ is designed to quantify the difference between the actual penalty and the implied penalty.

The parameter ξ serves as a coefficient, indicating the fiscal and resilient characteristics of the planning scheme. When equals 0, it aligns with the traditional methodology, signifying that the decision maker prioritises economic efficiency while neglecting risk. Upon surpassing zero, it enables assessing the variability of penalty costs while striving to achieve both efficiency and the safety of the system.

Total COD Pollutant Minimization

The Chemical Oxygen Demand (COD) functions as a standard metric for evaluating emission levels. It is recommended to incorporate the COD function into an administrative water management model, represented as F_0^2:

$$\min_{x_k} F_0^2 = \sum_{k=1}^{K}\sum_{r=1}^{R}\sum_{(i,j)\in\Omega} p_{(i,j)}o_{kr}v_{kr}(T_{kr}^{\pm} - S_{kr(i,j)}^{\pm}) \tag{3.12}$$

where $T_{kr}^{\pm} - S_{kr(i,j)}^{\pm}$ represents the true water allocation for sector $r$from sub-area $k$based on the (i,j) scenario.

Gini Coefficient Minimization

The Gini coefficient, introduced byCorrado Gini (1921), serves as a significant analytical measure for assessing the extent of income distribution inequality. This method is utilised to assess the equity of resource allocation. By comparing the relative variation in water use and availability per economic unit, the Gini coefficient measures the equity of water distribution. It thus becomes useful in administrative water management mechanisms. This can be expressed as Eq. (3.13):

$$\min_{x_k} F_0^3 = \max_{(i,j)\in\Omega}\left\{\frac{1}{2K \bullet \sum_{k=1}^{K}\frac{x_{k(i,j)}}{F_{1(i,j)}^{k}}}\sum_{k=1}^{K}\sum_{q=1}^{K}\left|\frac{x_{k(i,j)}}{F_{1(i,j)}^{k}} - \frac{x_{q(i,j)}}{F_{1(i,j)}^{q}}\right|\right\} \tag{3.13}$$

where F_1^k represents the economic benefit for the region k; $X_{k(i,j)}/F_{1(i,j)}^{k}$ stands for the projected economic water use unit in the region k within the scenario (i,j).

This offers a hierarchical mixed water allocation strategy, where the primary mechanism is administrative water distribution, and the secondary mechanism is water market allocation.

3.2.3 Ecological Modelling Between Source and Reservoir Connectivity

Modelling source and reservoir connectivity in a basin involves understanding how water flows from various sources (like rainfall, snowmelt, or tributaries) through the landscape and into reservoirs (natural or man-made) (McKinney and Savitsky 2001). This is crucial for water resource management, flood control, and ecological conservation. It needed the observed ecological and geographical information over the respective delta. This information can be extracted using the developed software as:

Software: "The EPIC modelling software, documentation, and a user's manual can be downloaded from the following website: http://www.ce.utexas.edu/prof/mckinney/papers/aral/EPIC/EPICmodel.html" (Schlüter et al. 2005a, b). EPIC modelling system is capable of modelling river flow, salt transport, and energy exchange between users. EPIC offers an interface for the automated creation of networks and models, facilitates data input, reservoir and channel flow restrictions, objective weights, and result visualizations. Along with, algebraic modelling software (General Algebraic Modelling System (GAMS)) can be used to generate new results with strategies to make more sustainable connectivity.

The EPIC modelling system enables users to specify different nodes within the network. Where each node fulfils a distinct function, such as: User nodes represent locations within the system where water is extracted through consumption. Mouth nodes refer to the locations where rivers discharge into larger water bodies, including inland seas and oceans. Reservoir node represents the locations of water storage units. Simple nodes are the points that require calculations of water and salt balances, such as junctions.

This data, derived from EPIC modelling, can be enhanced through the application of a mathematical planning framework. Schlüter et al. (2005a, b) developed a model utilising the EPIC modelling system to facilitate the trade-off assessment of water allocation and the establishment of restoration objectives. The model optimises allocation within the network through a multi-objective solution across designated time intervals. The primary criterion is to minimise deficits in water delivery to each user, the planned flow to the ocean, the reservoir filling levels, and the demand for system stability. The approach streamlines and expedites calculations for the planning of a delta water system with a long-term perspective by adjusting the criteria weights. The formulated model is articulated as follows: "Part (a) of Eq. 3.1, the primary criterion, aims to optimise the ratio of water delivery to water demand for each user. Part (b) of this equation functions similarly for the river's mouth." Part (c) functions to maintain reservoirs at maximum capacity, thereby conserving stored water. Part (d) inhibits abrupt variations in water flow within the network arcs and constitutes a nonlinear term" (Schlüter et al. 2005a, b).

$$Maximize\,\rho_1\left(\frac{1}{T*N_{i_{user}}}\sum_t\sum_{i_{user}}\frac{W_{in,i,t}-W_{trans,i,t}}{W_{req,i,t}}\right)+\rho_2\left(\frac{1}{T*N_{i_{mouth}}}\sum_t\sum_{i_{mouth}}\frac{W_{in,i,t}}{W_{req,i,t}}\right)$$
$$+\rho_3\left(\frac{1}{N_{i_{res}}}\sum_t\sum_{i_{res}}\frac{Vol_{i,t}}{Vol_{i,max}}\right)+\rho_4\left(\frac{1}{T*N_{i_{user}}}\sum_t\sum_{i_{user}}\frac{W_{in,i,t}-W_{in,i,t-1}}{W_{in,i,t}+W_{in,i,t-1}}\right) \quad (3.14)$$

P_1, P_2, P_3, P_4: Dimensionless weights indicating the objective function priority.
$i_{user}, i_{mouth}, i_{res}$: $-I^{th}$ node type "user", "mouth", or "reservoir".

$N_{i_{user}}, N_{i_{mouth}}, N_{i_{res}}$: Total number of nodes "user", "mouth", or "reservoir".
t: Time step,
N: Total number of nodes.
T: Considered total number of time steps.
$W_{in,i,t}$: Water entering node i at time step t *(Million M^3).*
$W_{trans,i,t}$: Water transiting from node i t time step t *(Million M^3).*
$W_{req,i,t}$: Required water by user i in time step t *(Million M^3).*
$Vol_{i,t}$: i^{th} Reservoir volume at the end of time step t *(Million M^3).*
$Vol_{i,max}$: Capacity of reservoir t *(Million M^3).*
Followed by, the continuity between each node imposes restrictions to the solution:

Simple nodes: Water entering point that must also have an exit point

$$\sum_{i_{simple}} W_{out,i,t} = \sum_{i_{simple}} W_{in,i,t} \tag{3.15}$$

where, $W_{out,i,t}$ = water leaving node i in time step t *(Million M^3)*, and $i_{simple} = i^{th}$ node of type "simple".

User nodes: Water distributing point with a combination of transit flow, return flow, and any local source of water.

$$\sum_i W_{out.i.t} = W_{trans.i.t} + \left(\sum_i W_{in.i.t} - W_{trans.i.t}\right) ret_{i,t} + WQ_{i,t} \tag{3.16}$$

where, $ret_{i,t}$ return flow coefficient (dimensionless), 1 $0, \leq 0 \leq ret_{i,t} \leq 1$, and $WQ_{i,t}$ local source of water at point i and time t *(Million M^3)*, and

$$\sum_i W_{in.i.t} - W_{trans.i.t} \leq W_{req.i.t} \tag{3.17}$$

Reservoir node: The intake to a reservoir node, reservoir volume changes, and evaporation affect its outflow of a reservoir node.

$$\sum_i W_{in.i.t} - \sum_i W_{out.i.t} = Vol_{i,t} - Vol_{i,t-1} + e_{i,t} * \overline{A_i} \tag{3.18}$$

where $e_{i,t}$ is the evaporation rate from a reservoir i at time $t(m)$, and $\overline{A_{i,t}}$ stands for the average surface area of a reservoir i at the time step t *(Million M^3).*

McKinney and Savitsky (2001) offer a comprehensive elucidation of the EPIC modelling system for salinity and energy management, as well as its application to the basin.

This represents a nonlinear program. In the future, supplementary constraints that facilitate the improvement in the process may be implemented, contingent upon the circumstances, encompassing physical, policy, and control limitations. These scenarios for water management considers factors such as alterations in agricultural water use or a requisite flow at a certain node for ecological purposes, users may adjust the policy constraints.

3.2.4 Salinity and Water Management Practices

The salinization of delta habitats is governed with complex interplay of geomorphological, fluvial, and coastal processes. These natural dynamics are increasingly disrupted by human interventions, leading to waterlogging and the accumulation of salts in both soils and water bodies. Sustain life and livelihoods in delta regions, communities must address the growing salinization risks driven by anthropogenic activities, climate change, and socioeconomic development. Salinity management demands a robust scientific understanding of salinization processes and the implementation of multi-tiered risk management strategies at local, regional, and basin-wide levels. Crucially, stakeholder engagement is essential to ensure policy implementation and community collaboration Feist et al. (2023). Salinity distribution in deltas is shaped by a multitude of factors operating across various temporal scales, from hours to millennia. Land use changes, polder construction, and freshwater abstraction directly influence salinity levels in coastal zones (Islam et al. 2019). While human societies seek environmental stability, the inherent variability of deltaic systems often conflicts with this desire.

Agricultural productivity is particularly vulnerable. Salinity disrupts nutrient cycling and soil structure by reducing cation exchange capacity and increasing faunal mortality (Butcher et al. 2016). Salinization introduces excessive sodium (Na) into ecosystems, posing risks to biodiversity, agriculture, and human health (Scheelbeek et al. 2017). Sodium adsorption onto clay minerals causes clay expansion, reducing soil porosity and permeability, which exacerbates waterlogging (Hagage et al. 2024). High concentrations of soluble and exchangeable sodium impair plant water uptake through osmotic stress and toxicity, leading to reduced crop yields (Dasgupta et al. 2018). Salinity also threatens freshwater resources. Storm surges contaminate drinking water ponds and salinate shallow aquifers (Islam et al. 2019). Elevated sodium levels in drinking water have been linked to increased risks of hypertensive disorders, including (pre)eclampsia, infant mortality, and other cardio-metabolic conditions (Scheelbeek et al. 2017).

Innovations in Salinity Forecasting and Management

The Jampani et al. (2023) delta salinity forecasting method uses machine learning and deep learning to overcome these issues. This method predicts salinity dynamics using historical rainfall, groundwater levels and quality, river discharge and salinity, tidal height, land use, and water extraction patterns. Their AI-driven zonal threshold mapping identifies salinization-vulnerable locations. This predictive power helps farmers and water managers make quick decisions.

USGS public-domain software SEAWAT (MODFLOW/MT3DMS) simulates three-dimensional, variable-density groundwater flow and solute transport. SEAWAT is used to study saltwater intrusion and brine transport in coastal and inland aquifers (Lyu et al. 2024).

Hagage et al. (2024) created a soil salinisation, waterlogging, groundwater, and water quality monitoring scheme. Their method included chemical tests of water and soil samples, water quality index computations, statistical analysis, and satellite data interpretation.

AI/ML techniques combined with hydrochemical data from surface and groundwater interaction models may reveal environmental data trends and correlations, enabling more accurate predictions and targeted actions. These advances assist deltaic regions in managing water resources sustainably and prevent salinization's multiple effects.

3.2.5 Spatial–temporal Analysis

The multiple regions of deltas are subject to recurrent natural disasters, such as floods, droughts, earthquakes, and significant snowfall events (Cao et al. 2024). This situation is likely to worsen due to unpredictable climate change. Spatio-Temporal Analysis is a vital tool for understanding the complexities of data that changes over time and space. This is essential to understanding biodiversity conservation. This allows conservationists to track changes in habitat conditions, identify areas of ecological significance, and monitor the impacts of activities.

Understanding on spatial distribution and temporal dynamics of species and habitats, targeted conservation strategies should be developed to protect endangered species and preserve biodiversity. This provides a comprehensive understanding of the complex interactions and changes in delta regions, supporting sustainable development, disaster management, agricultural productivity, and biodiversity conservation.

For example, investigating the variations in climate patterns or human activities and their impacts on hydrological variables in river basins with historical data (Zhang et al. 2009).

3.3 The Vulnerability Assessments of Deltas

Delta vulnerability refers to the degree to which river deltas are susceptible to environmental, social, and economic threats. Deltas are incredibly fertile and often densely populated, and also among the most vulnerable ecosystems on Earth. Evaluating the vulnerability of a delta involves assessing how exposed, sensitive, and adaptive the delta is to various environmental and socioeconomic pressures (Szabo et al. 2016). This is typically done using a multi-indicator framework that combines physical, ecological, and human factors. Key indicators for delta vulnerability assessment should help identify the risks and resilience of delta regions across multidisciplinary dimensions. Driel et al. (2015) have analysed the vulnerability of various delta regions based on four indicators: relative sea level rise, population pressure, wetland ecological threat, population pressure, and delta governance. The developed results with the selection indicator is presented in Table 3.1, and the importance of these indicators in assessing delta vulnerability is mentioned as:

- Relative Sea Level Rise: Sea level rise is a critical indicator because it directly affects the physical survival of deltas.
- Wetland Ecological Threat: Wetlands are the ecological backbone of deltas, providing biodiversity, water purification, and flood control. This indicator highlights the ecological value and vulnerability of delta wetlands.
- Population Pressure: This indicator is crucial for understanding human-induced stress on delta systems and for planning resilient urban development and resource management strategies.
- Delta Governance: This includes managing transboundary water issues, enforcing environmental regulations, and coordinating disaster response. Effective governance determines how well a delta can adapt to and mitigate risks.

Table 3.1 demonstrates a wide regional distribution of indicator results. This is clear that several deltas have high and low indicator scores. This shows many deltas are vulnerable, some severely vulnerable. The Ganges–Brahmaputra-Meghna delta is most susceptible, with two 'very high' and one 'high' relative risk scores. The Niger and Volta deltas score 'very high' and 'high' for three of the four parameters. Asian deltas seem to have the greatest human vulnerability due to a high relative sea-level rise score, significant population pressure, and occasionally weak delta administration. Low vulnerability appears in the Amazon, Orinoco, and Yukon deltas. The Relative Sea Level Rise Indicator has the most 'very high' relative risk scores, followed by Delta Governance. The Wetland Ecological Threat Indicator scores highest in 'very low' and 'low'. This calibrates the indicator's information that gaps in vulnerability requires robust and adaptive measures. More detailed information on the four indicators and new vulnerability indicators would help explain delta issues and prioritise reduction. More study is needed that relate delta

Table 3.1 The risk factors and impact on delta sustainability, adapted from Driel et al. (2015)

Continents	Deltas	Risk factors that impact sustainability in delta regions			
		Ecological Risk	Sea Level Rising	Delta Governance	Population Pressure
Africa	Congo	4	2	5	2
	Limpopo	2	2	3	3
	Niger	3	5	4	4
	Nile	2	4	5	5
	Senegal	2	4	4	2
	Volta	4	4	4	3
	Zambezi	2	4	3	2
Asia	Ganges–Brahmaputra-Meghna	2	5	4	5
	Hong (Red)	1	2	3	5
	Indus	2	5	4	3
	Irrawaddy	2	5	3	4
	Mekong	2	5	3	4
	Shatt-al-Arab	2	4	5	3
Europe	Danube	5	2	4	1
	Rhine-meuse	3	2	1	4
	Rhone	4	5	2	2
	Volga	5	1	4	1
	Wisla	1	3	2	4
Americas	Amazon	2	2	3	1
	Colorado	1	4	5	2
	Grijalva	1	4	5	2
	Mississipi	1	4	2	2
	Orinoco	2	3	3	1
	Parana (la plata)	2	3	3	2
	Rio Grande	1	5	3	3
	Yukon	3	2	1	4

assessment results to river basin assessments, understanding how activities upstream in the basin affect delta systems and vice versa.

3.4 Discussions

Planning how to fairly share water in the world's river deltas is about more than just engineering; it's also securing a future for the millions who call these regions home. These low-lying areas, vital for global food production, require long-term preservation. A central theme in such planning is the inherent trade-off between consumptive uses, such as agriculture, industry, domestic supply, and environmental water requirements. In many deltas, historical over-allocation to agriculture has led to ecological degradation, saltwater intrusion, and land subsidence, undermining the very resources upon which communities depend. Long-term planning must therefore prioritise the formal recognition of water flows, both instream and off stream, as a non-negotiable component of budgets. This requires robust hydro-ecological models that can quantify the water necessary to sustain critical ecosystems, such as mangrove forests and wetland habitats.

Another critical dimension is the integration of social equity into allocation mechanisms. Water policies must not only be efficient but also fair, ensuring that vulnerable populations, such as smallholder farmers and fishing communities, are not disproportionately disadvantaged by allocation changes. Participatory modelling and inclusive governance frameworks can help incorporate local knowledge and priorities, fostering legitimacy and cooperation in the implementation of long-term plans.

The uncertainty associated with climate change necessitates a move away from static planning methods toward adaptive and flexible frameworks. Techniques such as robust decision-making and scenario-based planning that allow policymakers to test allocation strategies under a range of future climates, helping to identify solutions that perform adequately across multiple scenarios rather than optimising for a single predicted future. This is essential in precipitation patterns, monsoon intensity, and sea-level rise introduce deep uncertainties.

Moreover, delta water allocation requires coordination across jurisdictions, from upstream river basins to transnational coastal zones. Fragmented management often leads to contradictory policies, such as upstream dam operators prioritizing hydropower over downstream environmental needs. Basin-wide cooperative institutions, supported by shared data and modelling platforms, are crucial for aligning objectives and managing trade-offs across sectors and borders.

3.5 Conclusion

Long-term water allocation in deltas demands a holistic, adaptive, and participatory approach. Success required advanced sustainable modelling tools with inclusive governance, recognising environmental needs, and planning for uncertainty. Without comprehensive strategies, delta regions risk continued ecological decline, social conflict, and increased vulnerability to climate change, threatening both local livelihoods and global food security. This chapter underscores the vulnerability of delta systems to climate change and human activity, advocating for adaptive, data-driven water management. Mathematical modelling is essential for translating Sustainable Development Goals (SDGs) into actionable plans, providing a scientific basis for resilient delta futures along with critical attention to their uncertainties. Advanced models and optimisation tools help navigate trade-offs between water use and ecological preservation. Key recommendations include integrating both environmental and anthropogenic factors into high-resolution models, alongside robust uncertainty analysis and interdisciplinary collaboration. Prioritising ecological flows, salinity control, and spatiotemporal dynamics is critical for sustainability, enabling proactive and equitable water governance with the interconnections between hydrological and socioeconomic well-being.

Recommendations

The modelling approach has the ability to reveal synergies and trade-offs between different Sustainable Development Goals (SDGs). Mathematical modelling enables a critical, scientific framework for achieving the SDGs in the world's vulnerable river deltas. Creating digital replicas to complex delta systems, allow policymakers to test interventions virtually, avoiding costly real-world mistakes and navigating the inherent trade-offs between sustainability goals (Smajgl et al. 2015).

For instance, groundwater models based on Darcy's Law can predict how sea-level rise will salinise aquifers, directly informing strategies for SDG 6 (Clean Water). Similarly, crop-soil models can project the impact of this salinity on rice yields, supporting SDG 2 (Zero Hunger) by identifying resilient crop varieties or necessary irrigation changes (Minderhoud et al. 2020). Building levees protects urban areas (SDG 11) but can prevent vital sediment from nourishing sinking land, which increases long-term vulnerability. Sediment-morphodynamic models can simulate this trade-off, allowing planners to design smarter, sediment-friendly infrastructure or plan strategic areas for flooding, thereby balancing SDG 9 (Industry & Infrastructure) with SDG 13 (Climate Action) and SDG 15 (Life on Land) (Erkens et al., 2016).

The utility of models is entirely dependent on acknowledging their limitations. They are often simplifies the reality, and their accuracy is constrained by data quality and the ability to quantify natural behaviour. Rigorous uncertainty and sensitivity analysis is therefore essential to ensure that model outputs are presented with clear confidence intervals, preventing over-reliance on precise but potentially inaccurate predictions (Pianosi et al. 2016).

Acknowledgements DeepSeek AI has been used in polishing the original content.

Availability of Data and Materials The reference materials for the used data are included and cited in the text.

Declaration of Competing Interest The authors declare that they have no known competing financial or personal interests that could have appeared to influence the work.

Credit Author Statement Manish Meena: Data curation, Writing and Original draft preparation, Methodology conceptualization: Nitin Dutt Chaturvedi: -Editing, and Methodology conceptualization, Supervision.

References

Arto I et al (2019) The socioeconomic future of deltas in a changing environment. Sci Total Environ 648:1284–1296

Bai T et al (2025) Water allocation and irrigation scheme considering runoff forecasting of Ulungur River. Ecol Indic 173:113381

Bélanger V et al (2015) DELTA: an integrated indicator-based self-assessment tool for the evaluation of dairy farms sustainability in Quebec, Canada. Agroecol Sustain Food Syst 39.9:1022–1046

Biggs D et al (2012) The delta machine: water management in the Vietnamese Mekong Delta in historical and contemporary perspectives. Contested Waterscapes Mekong Reg. 203–225

Bloemen P et al (2018) Lessons learned from applying adaptation pathways in flood risk management and challenges for the further development of this approach. Mitig Adapt Strat Glob Chang 23.7: 1083–1108

Butcher K et al (2016) Soil salinity: a threat to global food security. Agronomy J 108.6: 2189–2200

Cao T et al (2024) Fast spreading of surface ozone in both temporal and spatial scale in Pearl River Delta. J Environ Sci 137: 540–552

Chan FKS et al (2024) Building resilience in Asian mega-deltas. Nat Rev Earth and Environ1–16

Deng M et al (2024) Drivers and dynamics of forest and grassland ecosystems in the Altai mountains: a framework for national park conservation. Land 14.1: 48

Dvan WFV et al (2015) Vulnerability assessment of deltas in transboundary river basins. Delta Alliance contribution to the Transboundary water assessment program, river basins assessment. Delta Alliance report 9

Erkens G et al (2015) Sinking coastal cities. Proc Int Assoc Hydrol Sci 372.372: 189–198

Feist SE, Hoque MA, Ahmed KM (2023) Coastal salinity and water management practices in the Bengal Delta: a critical analysis to inform salinisation risk management strategies in Asian Deltas. Earth Syst Environ 7.1: 171–187

Gini C (1921) Measurement of inequality of incomes. Econ J 31(121):124–126

Haasnoot M et al (2012) Exploring pathways for sustainable water management in river deltas in a changing environment. Climatic Change 115:795–819

Hagage M et al (2024) Monitoring soil salinization and waterlogging in the northeastern Nile Delta linked to shallow saline groundwater and irrigation water quality. Scientific Reports 14.1: 27838
Hommes S et al (2009) Knowledge and perceptions in participatory policy processes: lessons from the delta-region in the Netherlands. Water Resour Manag 23: 1641–1663
Islam MA et al (2019) Impact of climate change and land use on groundwater salinization in southern Bangladesh—Implications for other Asian deltas. Environ Manage 64(5):640–649
Jampani M, Sena DR, Matheswaran K (2023) Framework for advancing water resource sustainability and climate resilience through local-scale hydrological modeling in the Ganges Delta
Karimi A, Ardakanian R (2010) Development of a dynamic long-term water allocation model for agriculture and industry water demands. Water Resour Manage 24:1717–1746
Keller Arturo A et al (2022) Downscaling approaches of climate change projections for watershed modeling: review of theoretical and practical considerations. PLoS Water 1.9: e0000046
Kumari R et al. (2025) The role of artificial intelligence and its tool in water management
Lempert RJ, Groves DG (2010) Identifying and evaluating robust adaptive policy responses to climate change for water management agencies in the American west. Technol Forecast Soc Chang 77(6):960–974
Lyu P et al (2024) Integrated SEAWAT model and GALDIT method for dynamic vulnerability assessment of coastal aquifer to seawater intrusion. Sci Total Environ 925: 171740
McKinney DC, Savitsky AG (2001) EPIC Modeling System Guided Water, Salt and Energy Management Problems. http://www.ce.utexas.edu/prof/mckinney/papers/aral/EPIC/EPICmodel.html
Minderhoud PSJ et al (2019) Mekong delta much lower than previously assumed in sea-level rise impact assessments. Nat Commun 10(1):3847
Nienhuis Jaap H et al (2023) River deltas and sea-level rise. Annu Rev Earth Planet Sci 51.1: 79–104
Pianosi F et al (2016) Sensitivity analysis of environmental models: A systematic review with practical workflow. Environ Mode Softw 79:214–232
Ranjbar S (2025) Enhancing ecosystem and earth system science via remote sensing and machine learning. The University of Wisconsin-Madison
Scheelbeek PFD et al (2017) Drinking water salinity and raised blood pressure: evidence from a cohort study in coastal Bangladesh. Environ Health Perspect 125.5: 057007
Schlüter M et al (2005) Optimizing long-term water allocation in the Amudarya River delta: a water management model for ecological impact assessment. Environ Model Softw 20.5: 529–545
Schlüter M et al (2005) Optimizing long-term water allocation in the Amudarya River delta: a water management model for ecological impact assessment. Environ Model Softw20.5: 529–545
Sebesvari Z et al (2016) A review of vulnerability indicators for deltaic social–ecological systems. Sustain Sci 11: 575–590
Smajgl A et al (2015) Responding to rising sea levels in the Mekong Delta. Nat Clim Chang 5.2:167–174
Susmita D et al (2018) Climate change, salinization and high-yield rice production in coastal Bangladesh. Agric Resour Econ Rev 47.1: 66–89
Szabo S et al (2016) Population dynamics, delta vulnerability and environmental change: comparison of the Mekong, Ganges–Brahmaputra and Amazon delta regions. Sustain Sci 11: 539–554
Tu Y, Lu Y, Lev B (2025) An interval two-stage robust stochastic programming under a bi-level multi-objective framework toward river basin water resources allocation. Comput Oper Res 180:107045
Yan D et al (2016) Identifying and assessing robust water allocation plans for deltas under climate change. Water Resour Manag 30: 5421–5435
Zevenbergen et al (2018) Adaptive delta management: a comparison between the Netherlands and Bangladesh Delta Program. Internat J River Basin Manag 16(3):299–305
Zhang W et al (2009) Temporal and spatial variability of annual extreme water level in the Pearl River Delta region. China. Global and Planetary Change 69.1–2: 35–47

Assessment of Hydroclimatic and Groundwater Conditions in the Middle Ganga Basin

4

Vinay Arya and Mavidnam Someshwar Rao

Abstract

Water resource management is a critical global challenge, as increasing demand, climate variability, and pollution threaten the sustainability of surface and groundwater systems across many regions of the world. Addressing these issues requires integrated approaches that link hydrology, climate science, and socio-economic considerations to ensure water security. The Middle Ganga Basin, a vital ecological and socio-economic region in India, faces critical challenges in water resource management due to rising demand, climate variability, and pollution. This review synthesizes current knowledge on surface and groundwater availability, sectoral demand patterns, and water quality issues. It maps the spatial distribution and seasonal dynamics of water resources, identifies contamination sources, and assesses implications for resource sustainability. The study highlights that less than 33% of the basin in Uttar Pradesh is suitable for groundwater recharge, primarily in the Terai region, influenced by higher rainfall, flat terrain, and lineaments. However, high drainage density and clayey soils in the south impede infiltration. Artificial recharge potential is significant, with 92 identified sites on third-order streams for recharge structure construction. Analysis of rainfall, temperature, and groundwater level (GWL) trends reveals decreasing rainfall in zones 1 and 3 and increasing rainfall in zones 2 and 4. GWL is declining in zones 2, 3, and 4, while zone 1 shows no change. Temperature trends indicate an average rise of 0.0068 °C per year, influenced by proximity to the Himalayas and seasonal geography. Water

V. Arya (✉)
Indian Institute of Technology, Roorkee, Haridwar, Uttarakhand, India
e-mail: bigredmachine421@gmail.com

M. S. Rao
National Institute of Hydrology, Roorkee, Haridwar, Uttarakhand, India

S. Yadav et al. (eds.), *Water Resources in Delta Region in India*, Deltas of the World,
https://doi.org/10.1007/978-3-032-06500-1_4

quality analysis shows that 74% of groundwater is suitable for drinking, with areas of poor quality attributed to calcium, magnesium, and trace metals like iron, aluminum, and arsenic. Irrigation suitability indices confirm overall viability, except in districts like Jaunpur and Unnao. Health risk assessments for children and adults indicate minimal cancer risk from heavy metals. Isotopic investigations reveal varied groundwater recharge sources, including rainfall and surface water. Deeper aquifers are generally older than shallow ones, with some shallow aquifers showing surprisingly higher ages due to terrain features. Recharge mapping and isotopic data emphasize the need for localized interventions. This review positions the Middle Ganga Basin as a model for addressing water stress, advocating integrated approaches to balance conservation, efficiency, and stakeholder engagement. By contributing to global discussions, it aims to inspire sustainable water resource management in similar river basins worldwide.

4.1 Introduction

Water is a finite and unevenly distributed resource that is fundamental to sustaining life, livelihoods, food security, economic development, and environmental integrity. In India, which is home to one of the world's largest agrarian populations, water assumes a central role, particularly in agriculture. However, escalating demand across multiple sectors, combined with the adverse impacts of climate change, has placed unprecedented pressure on sustainable water supply systems. Rapid population growth and industrial expansion have further intensified freshwater demand, driving shifts in agricultural practices and leading to an overreliance on groundwater resources. At the same time, the increasing pollution of surface water bodies has added to the stress on the hydrological balance, resulting in widespread groundwater depletion. Recent studies indicate that the Ganga River Basin has been undergoing an unprecedented drying trend for more than 1,300 years, with a marked decline in groundwater recharge during the last three decades. Between 2002 and 2016, the basin reportedly lost approximately 226.57 km^3 of groundwater, severely affecting summer baseflow dynamics (Sooraj Krishnan and Indu 2023; Shaivya et al. 2026).

Satellite observations from the Gravity Recovery and Climate Experiment (GRACE) further reveal alarming depletion rates of up to 7 cm per year across regions such as Eastern Uttar Pradesh between 2002 and 2022, contributing to land subsidence in cities like Lucknow within the Ganga Basin (Aon et al. 2024; Bons 2018; Zakwan and Ahmad 2021). Because water supports health, sanitation, food production, and ecosystem stability, maintaining both its quantity and quality is essential for long-term sustainability. Integrated water resources management (IWRM), from the river basin to watershed and block levels, is therefore critical. Recent policy reviews have highlighted the importance of coherent and science-based implementation frameworks to bridge gaps in achieving Sustainable Development Goals (SDGs), particularly in the context of increasing urban water scarcity and climatic variability (Gurjar et al. 2021).

In India, water governance is closely linked to key developmental priorities, including food and energy security, rural development, urbanization, disaster risk reduction, and equitable access. India currently accounts for the largest freshwater withdrawal in the world, with an estimated annual use of about 249 km^3, representing more than one-fourth of global consumption. Groundwater alone meets over 85% of rural and nearly 50% of urban domestic water needs. Within this context, the Ganga Basin accounts for an estimated withdrawal of 28 billion cubic metres annually, raising serious concerns about irreversible aquifer depletion (Shamsudduha 2020). Persistent over-extraction threatens to push regional water systems beyond ecological and socio-economic recovery thresholds.

This chapter therefore seeks to:

1. Critically assess the hydrological status of the Ganga Basin using historical and contemporary data on water accessibility and use;
2. Identify current and emerging development needs, vulnerable subregions, and priority challenges; and
3. Review institutional frameworks, policy mechanisms, and best practices for integrated river basin management.

4.1.1 Overview of the Ganga Basin

Recognizing the increasing water stress across the Ganga Basin, the discussion focuses on evolving patterns of demand, depletion, and governance strategies. Drawing on insights from three decades of integrated water resource management across the Asia–Pacific region, this study aims to promote evidence-based policy coherence, consensus-building, and actionable pathways for sustainable water governance (Dharpure et al. 2025).

The Ganga River Basin, spanning approximately 1.1 million km^2 across India, Nepal, Bangladesh, and China, is one of the world's most densely populated and hydrologically significant river systems. Within India alone, the basin covers about 861,452 km^2, extending across 11 states, including Uttar Pradesh, Bihar, West Bengal, Uttarakhand, and Madhya Pradesh, among others (Bhat 2014; Sooraj Krishnan and Indu 2023; Shaivya et al. 2026; World Bank 2014; World Bank 2015; World Bank 2020). Protected from frozen winds from Siberia by the Himalayas in the north, the Aravalli in the west, the Vindhyas and Chotanagpur plateau in the south, and the Brahmaputra Ridge in the east, the basin supports 485 million people as per census 2011 (Bons 2018) for their water demands for agriculture, drinking water, and livelihoods.

Originating from the Gangotri Glacier in Uttarakhand as the Bhagirathi River, the Ganga flows southeast through the Indo-Gangetic plains (Anita et al. 2015) and enters Bangladesh as the Padma. Despite its vast reach and cultural reverence, the basin is rapidly under threat due to overextraction of groundwater, erratic rainfall patterns, pollution, and climate-induced glacier retreat. The basin receives an average annual rainfall of

1059.74 mm, spatially varying from 400 mm in the west to over 2000 mm in the east (Zakwan and Ahmad 2021). Yet nearly 80% of this precipitation is concentrated within the three-month monsoon period, resulting in high seasonal variability in water availability.

The total renewable water prospects of the Ganga basin are estimated at 525.02 billion cubic meters (BCM), with utilizable surface water limited to 250 BCM (Bharati et al. 2016; Gurjar et al. 2021). However, recent satellite-based assessments, including Gravity Recovery and Climate Experiment (GRACE), have raised concerns over significant groundwater depletion, making the Ganga basin one of the most rapidly depleting aquifers globally (Gurjar et al. 2021; Shamsudduha 2020; Dharpure et al. 2025; Janardhanan et al. 2023) Factors contributing to this decline include intensive irrigation, inefficient water management, unregulated groundwater abstraction, and increasing water demands due to population growth and urbanization (Dangar and Mishra 2021).

Furthermore, canal systems and dams—such as the Tehri Dam, Farakka Barrage, and the Upper, Middle, and Lower Ganga Canals—have altered natural flow regimes, often exacerbating ecological stress (Swarnkar et al. 2021; Gain and Giupponi 2014). The Ganga basin's vulnerability is further compounded by pollution from untreated sewage, industrial effluents, and agricultural runoff (Aggarwal et al. 2025; Zeshan and Saltanat 2021). Environmental degradation has also impacted key ecological zones such as floodplains and wetlands, threatening biodiversity, including species like the critically endangered Ganga River dolphin (Ruchi et al. 2025).

4.2 State of the Water Resources of the States Within the Ganga Basin

4.2.1 Uttarakhand

Uttarakhand is richly endowed with water resources, with the Ganga and Yamuna rivers originating from its glaciers. Major water sources include precipitation, glaciers, rivers, streams, lakes, springs, and groundwater. The total length of major rivers in the state is about 2,686 km (Anita et al. 2015). The state hosts 968 glaciers covering 2,857 km^2 across four river basins (Thadani et al. 2015). In 2016, total precipitation (including snowfall) was 48,351.82 MCM. In the state, the surface water availability is 22,854 MCM in the Ganga Basin and 4,090 MCM in the Sharda Basin. Additionally, 25,989 MCM is transferred annually to Uttar Pradesh through the Ganga Canal (Fig. 4.1).

In the Himalayas, groundwater appears as springs, with a total discharge of 16.63 MCM. About 10% of the state's area is covered by snow, ice, and glaciers, supporting perennial flow in the Yamuna, Bhagirathi, Alaknanda, and Kali rivers. The state has several medium and large reservoirs that cover a total area of about 310 km^2, while ponds and tanks occupy around 6.0 km^2 (IRI 2018).

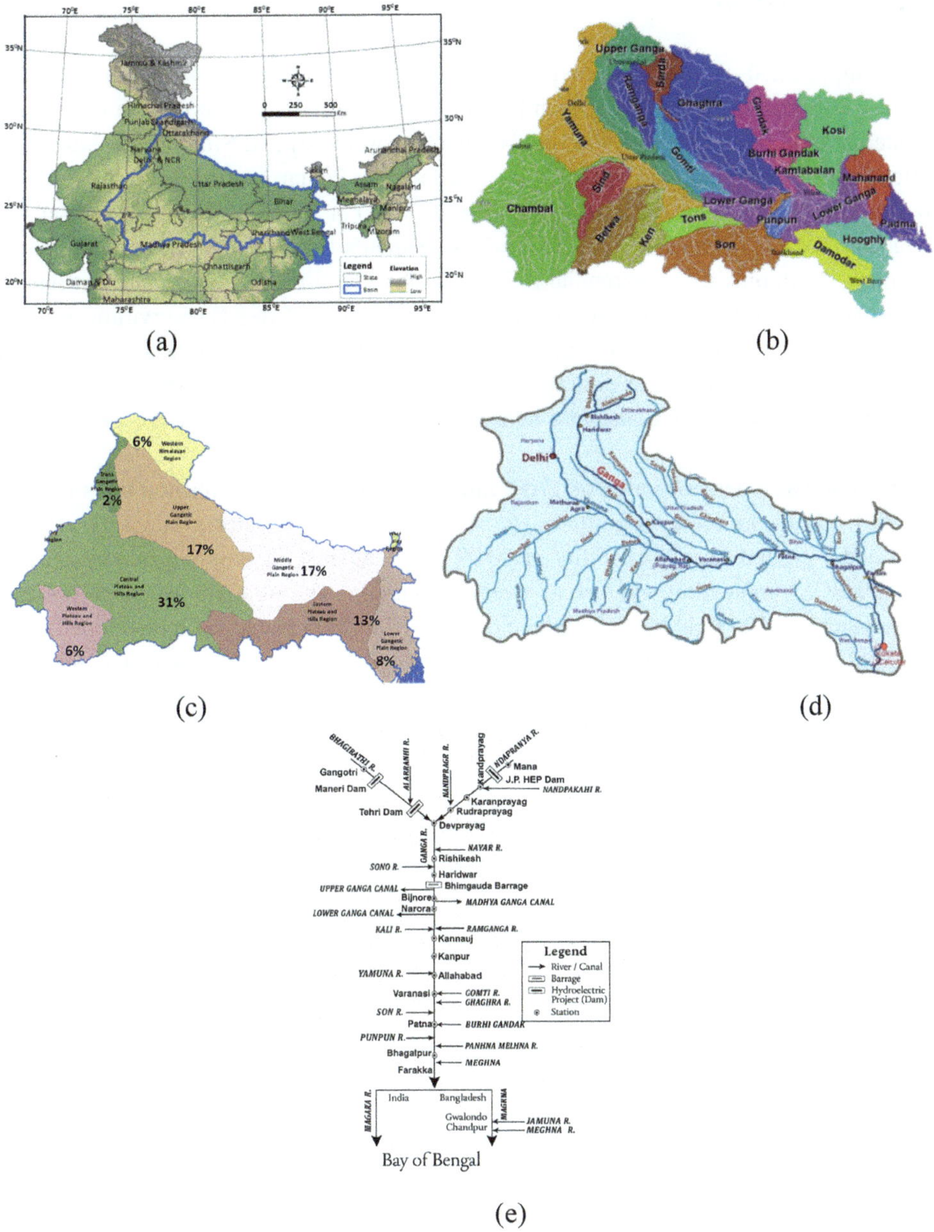

Fig. 4.1 **a** Ganga Basin in India; **b**, **c** Main Rivers and Sub-basins; **d** Schematic Diagram of the Ganga River with major tributaries and locations of water management structures (Kumar 2017); **e** Agro-climatic Zones of the Ganga Basin (Sources for a, b, c & e: CWC & NWRC 2014)

4.2.2 Rajasthan

Rajasthan, located in the western part of India, encompasses much of the arid Thar Desert—also known as the "Rajasthan Desert" or "Great Indian Desert." The state shares its international boundary with Pakistan's Punjab (northwest) and Sindh (west), along the Sutlej-Indus river valley. Domestically, it is bordered by Punjab (north), Haryana and Uttar Pradesh (northeast), Madhya Pradesh (southeast), and Gujarat (southwest). The total geographical area of the state is 342,239 km^2. Rajasthan's river systems are predominantly ephemeral, with only a few perennial rivers. Notably, the Chambal River and its tributary, the Banas River, form significant components of the Ganga River Basin within the state. The Chambal River basin covers around 31,360 km^2 in Rajasthan, while the Banas River basin—the largest river basin entirely within the state; extends over 45,833 km^2 across 11 districts. These basins contribute substantially to the hydrology of the Ganga River system.

The state has a net sown area of 17.52 million ha, of which 7.88 million ha is irrigated and 9.64 million ha remains rainfed. The state has a surface water availability of 21.71 BCM and a groundwater availability of 10.09 BCM, against an annual groundwater draft of 12.99 BCM, indicating widespread over-exploitation. The ultimate irrigation potential is estimated at 5.125 million ha, of which only 2.812 million ha has been developed so far. The total annual water demand, including irrigation, domestic, livestock, and industrial uses, stands at 91.23 BCM, far exceeding the available 31.8 BCM, reflecting an acute water deficit (WAPCOS 2018). The key implications of the state's water-resource situation are:

1. The state faces a severe water deficit, with a shortfall of nearly 59 BCM that highlights the state's chronic scarcity and dependence on inter-basin water transfers and groundwater mining.
2. Groundwater extraction consistently exceeds natural recharge, resulting in declining water tables, increasing salinity, and widespread drinking-water shortages, with many administrative blocks now classified as over-exploited.
3. More than 55% of the state's agricultural land is rainfed, making crop production highly vulnerable to drought and climatic variability; therefore, expanding micro-irrigation, soil-moisture conservation, and watershed development programmes is essential.
4. Only 55% of the ultimate irrigation potential has been developed, and completing major irrigation and canal-based projects, along with strengthened rainwater harvesting, can significantly improve irrigation coverage.
5. Since irrigation accounts for 80 to 85% of total water use, policy measures must focus on crop diversification toward less water-intensive crops such as millets, pulses, and oilseeds; reuse of treated wastewater; artificial recharge of aquifers; and revival of traditional water systems including baoris, johads, nadis, and tankas.

6. Rajasthan's arid climate and high evapotranspiration intensify water stress, making the development of climate-resilient water infrastructure, improved irrigation efficiency, and stronger community-based water governance essential.
7. In 2024, groundwater extraction reached 17.05 BCM, while natural recharge amounted to only 11.37 BCM. This imbalance has led to a sharp decline in safe groundwater blocks, from 203 in 1984 to only 37 in 2024. Groundwater across large parts of the state is affected by salinity, fluoride, and nitrate contamination. Nearly 80% of Rajasthan faces groundwater depletion, while 50% of drinking water and 60% of irrigation depend on rapidly declining aquifers.

The key interventions being implemented to address these hydrological challenges include the followings:

The state has undertaken major initiatives such as the Mukhya Mantri Jal Swavlamban Abhiyan and the Parbati-Kalisindh-Chambal (Ram Jal Setu) Link Project, which aims to transfer surplus monsoon water from the Chambal River and its tributaries to water-deficit districts in Rajasthan and Madhya Pradesh (PHED n.d.).

Achieving long-term water security will require a combination of technological, institutional, and policy measures. These include increasing water-use efficiency from the current 37% to 60% by 2047; enabling active participation of Water User Associations in irrigation management; automating dams and canal gates across medium irrigation projects; deploying AI and IoT sensor networks in irrigation and drinking-water schemes; adopting digital water metering, real-time water quality monitoring, and hydrological forecasting systems; strengthening decision support systems for water governance; expanding the use of treated wastewater for irrigation, industrial, and horticultural applications; and enhancing surface and subsurface water storage through check dams, percolation tanks, recharge shafts, and farm ponds.

4.2.3 Delhi

The National Capital Territory (NCT) of Delhi, with an areal extent of 1,484 km^2, forms the core of the National Capital Region (NCR), which includes adjoining districts of Haryana, Uttar Pradesh, and Rajasthan. Governed through the National Capital Region Planning Board (NCRPB), the region represents one of India's largest socio-economic clusters with a projected population of approximately 35 million by 2025. The Delhi Jal Board (DJB) is mandated to manage the city's water-supply and wastewater services; however, the city's hydrological dependence on external sources, combined with urban expansion and demographic pressures, has led to increasing stress on its water resources.

Delhi's water security is constrained by limited local water availability and a high dependency on inter-state surface water transfers. Approximately 86% of the water supply is derived from external sources, while internal sources such as groundwater are over-stressed. Groundwater extraction in several administrative zones exceeds 100% of recharge (CGWB 2021), leading to continuous declines in water-table levels and degradation of aquifer quality. Concurrently, projected water demand for 2025 is estimated at about 3.5 BCM based on CPHEEO norms (274 LPCD for a population of 35 million), which substantially exceeds the dependable water availability. Increasing water requirement for domestic, commercial, industrial, and institutional sectors further aggravates the deficit (Table 4.1).

Delhi receives 1,151.1 MCM of surface water annually from the Yamuna, Ganga (via Upper Ganga Canal), and Satluj (via Bhakra system). Such external dependence makes the city vulnerable to hydrological variability, interstate allocation disputes, climate-induced flow changes, and operational constraints in upstream states.

Groundwater contributes about 110.5 MCM annually, but extraction routinely exceeds recharge. Nearly 80% of NCT Delhi is categorized as "over-exploited," while peri-urban agricultural areas and unauthorized colonies remain disproportionately dependent on private borewells. Persistent over-abstraction has led to: declining water-table levels, increased salinity in deeper aquifers, nitrate contamination in high-density urban settlements, and fluoride levels above permissible limits in several peripheral zones.

Delhi receives an average annual rainfall of 658 mm, generating an estimated 0.97 BCM of potential rainwater. If even 20% of this volume is conserved via recharge trenches, revitalized lakes, floodplain reservoirs, and rooftop rainwater harvesting systems, about 194 MCM could be added to groundwater. While insufficient to meet full demand, this contribution is vital for localized recharge in critically stressed aquifers.

The NCR spans 55,083 km^2 and receives about 774.4 mm of annual rainfall, generating nearly 42.66 BCM of rainwater annually. Strategically developing managed aquifer recharge and floodplain storage structures in selected NCR districts offers an opportunity to support water availability in Delhi through regional hydrological integration.

A comprehensive and integrated approach is essential to address the structural and systemic water challenges of Delhi. Key measures suggested are (Govt. of NCTi 2017):

Table 4.1 Water resources managed by Delhi Jal Board

Water resource type	Water sources	Quantity in Million Gallon Daily; (in MCM per annum)
Surface water	Yamuna river	375 MGD (518.3 MCM)
	Ganga river	240 MGD (331.6 MCM)
	Satluj river (Bhakra Storage)	218 MGD (301.2 MCM)
Ground water	Tube well, ranney well etc.	80 MGD (110.5 MCM)
Total		913 MGD (1261.5 MCM)

Source Delhi Jal Board

1. **Strengthening Inter-Regional Water Sharing**: Development of coordinated water-allocation frameworks between DJB, NCRPB, and partnering states to ensure stable surface-water inflows. Future augmentation may require incorporating raw water supplies from major upstream multipurpose projects such as the Renuka Dam on the Giri River (Himachal Pradesh), Kishau Dam on the Tons River (Uttar Pradesh), and the Lakhwar Dam (Tons River), which together are expected to contribute approximately 720 MLD of additional raw water to Delhi's supply system.
2. **Expansion of Wastewater Treatment and Reuse**: Modernization of Sewage Treatment Plants (STPs) and Common Effluent Treatment Plants (CETPs) to produce high-quality treated wastewater suitable for industrial, horticultural, and non-potable uses. Increasing the number and efficiency of Water Treatment Plants (WTPs) specifically designed for wastewater recycling and reuse can significantly reduce freshwater demand.
3. **Managed Aquifer Recharge (MAR)**: Development of engineered recharge interventions such as floodplain storage reservoirs, infiltration basins, recharge shafts, and percolation systems along the Yamuna and Najafgarh drain basins to enhance subsurface storage. These measures are crucial for stabilizing groundwater levels in over-exploited aquifers.
4. **Reduction of Transmission and Distribution Losses**: Implementing district metered areas (DMAs), leak-detection programmes, pressure-management systems, and real-time monitoring can significantly reduce water losses in the transmission and distribution network, thereby improving effective supply without additional abstraction.
5. **Decentralized Urban Rainwater Harvesting**: Mandating rainwater harvesting infrastructure across residential, commercial, and institutional buildings; reviving lakes, urban ponds, and stormwater-fed recharge systems; and integrating decentralized recharge structures into urban planning frameworks.
6. **Digital Water Governance**: Adoption of real-time monitoring systems including digital water metering, IoT-based flow and pressure sensors, automated SCADA systems, and hydrological forecasting tools to support transparent, adaptive, and data-driven decision making.

Long-term water sustainability in Delhi will depend on regionally coordinated planning across the NCR, alignment of water resource development with land-use transitions, and the adoption of circular-water economy principles (NCRPB 2021). The following research and policy directions are likely to shape Delhi's future water security:

- Evaluating the sensitivity of upstream river flows to climate variability in the Yamuna and Ganga basins and designing adaptive water-allocation strategies.
- Expanding treated wastewater application in industrial clusters, green belts, horticulture, groundwater recharge, and construction sectors to offset freshwater dependency.
- Reducing per-capita consumption through policy incentives, behavioural interventions, and the deployment of smart metering systems coupled with consumption-based tariffs.

- Minimizing non-revenue water through systematic reduction of transmission losses, pipeline rehabilitation, and advanced leak-detection technologies.
- Achieving long-term sustainability will require synchronized short-term (operational improvements), medium-term (infrastructure expansion), and long-term (regional inter-linking and reservoir systems) investments, embedded within a coherent regulatory framework and enforceable policy architecture.

Improving aquifer mapping, groundwater modelling, and recharge-characterization studies to support scientifically grounded decision-making.

4.2.4 Water Resources Profile of Haryana

Haryana, one of the smaller states of India, spans an area of 44,212 km^2 and is hydrologically divided between two major basins: the Yamuna Sub-Basin of the Ganga Basin and the Ghaggar Sub-Basin of the Indus Basin. Approximately 40% of the state drains into the Yamuna system, while the remaining 60% drains into the Ghaggar system. The Yamuna River defines the eastern boundary of the state, while the Ghaggar River influences the northern districts. In addition, the Saraswati paleochannel traverses parts of Yamunanagar, Kurukshetra, Kaithal, Jind, Hisar, Fatehabad, and Sirsa districts, contributing to the region's paleo-hydrological significance.

Haryana faces growing pressure on both surface water and groundwater resources due to limited internal water availability, high dependence on canal irrigation, declining groundwater levels, and widespread water-quality deterioration. The state's groundwater extraction exceeds sustainable limits, leading to aquifer depletion, waterlogging in isolated pockets, and contamination by salinity, fluoride, and uranium. Rapid agricultural expansion, urbanization, and industrial development have further intensified the stress on available water resources.

Key Highlights of the Water Resources

- The state receives an average long-term annual rainfall of 527.1 mm. In 2022, rainfall was 623.6 mm, which was 18% above normal.
- Haryana has a canal irrigation network of approximately 14,085 km, supported primarily by the Yamuna Canal, Bhakra Canal System, and Gurgaon Canal.
- The first phase of lift irrigation schemes such as Jui, Sewani, Loharu, and Jawaharlal Nehru completed during 1976–1990.
- The total water resource availability of the state is estimated at 15,614.37 MCM (SIP n.d.).
  - Surface water availability: 7,475.79 MCM (47.88%).
  - Groundwater availability: 8,610 MCM (55%).

Surface Water Resources: Surface water in Haryana is primarily derived from canal systems and inter-basin transfers. The Yamuna Canal and Bhakra Canal systems form the backbone of irrigation supply. Additional sources include treated effluent water from sewage treatment plants (343.29 MCM), water bodies and rainwater harvesting structures (27.44 MCM), and untreated effluents (246.68 MCM). Despite this, spatial inequity in surface-water distribution persists, especially in southern and western districts.

Groundwater Resource: Groundwater resources are under considerable extraction pressure. The total annual groundwater recharge is assessed at 9.47 BCM, with an annual extractable groundwater resource of 8.61 BCM. Against this, the current annual groundwater extraction is 12.42 BCM, resulting in a groundwater extraction stage of 134.14%, indicating severe over-exploitation. Out of 143 groundwater assessment units in Haryana, 88 units (61.53%) are categorized as over-exploited. Waterlogging conditions (water table <1.5 m) were observed over an area of approximately 512.5 km^2 during October (HWRA 2025). These contrasting conditions reflect both excessive extraction and inadequate drainage in certain localities.

Groundwater Quality Challenges: Groundwater quality in Haryana is significantly affected by salinity, fluoride, and uranium contamination (CGWB 2023).

- **Salinity**: Marginal to high salinity levels are reported in parts of Bhiwani, Charkhi Dadri, Faridabad, Fatehabad, Hisar, Jhajjar, Jind, Kaithal, Mahendragarh, Mewat, Palwal, Rewari, Rohtak, Sirsa, and Sonipat.
- **Fluoride**: Elevated fluoride concentrations occur in groundwater in Bhiwani, Gurugram, Hisar, Jhajjar, Jind, Karnal, Mahendragarh, Palwal, Panchkula, Rohtak, Sirsa, Sonipat, Faridabad, and Fatehabad.
- **Uranium**: Uranium contamination has been detected in Faridabad, Kaithal, Fatehabad, Panipat, Sonipat, Jhajjar, Bhiwani, Mahendragarh, Mewat, Palwal, Sirsa, Hisar, and Karnal.

These quality issues limit groundwater usability for drinking and irrigation and demand targeted remediation strategies.

Addressing Haryana's growing water scarcity and quality degradation requires an integrated approach centered on modernizing canal systems, reducing conveyance losses, and promoting conjunctive use of surface and groundwater. Large-scale artificial recharge interventions, particularly along the Saraswati paleochannel and Ghaggar flood plains, combined with expanded micro-irrigation, can significantly improve water-use efficiency in agriculture. Wastewater recycling through upgraded treatment systems offers an additional avenue for augmenting non-potable supplies, especially in industrial and peri-urban areas. Managing groundwater quality challenges will require targeted defluoridation, desalination, and systematic monitoring of uranium contamination. Future water security will depend on basin-level planning across the Yamuna and Ghaggar systems, climate-resilient irrigation strategies, assessment of long-term groundwater impacts on agriculture, and scientific modeling of contamination pathways. Strengthening institutional frameworks, including water-user associations and coordinated basin governance mechanisms, along with community participation and continuous hydrological monitoring, will be essential for achieving sustainable water management in the state.

4.2.5 Water Resources Profile of Bihar

Bihar is located in the central part of the Gangetic Plain and has a geographical area of 94,163 sq km. The state experiences a sub-humid climate and receives an average annual rainfall of around 1164.4 mm, whereas in 2024, the state received 916.7 mm of rainfall which is 21% below normal (IMD 2024). Bihar comprises 14 river basins, namely:

1. Ghaghara, 2. Gandak, 3. Burhi Gandak, 4. Bagmati-Adhwara, 5. Kamla-Balan, 6. Kosi, 7. Mahananda, 8. Main Ganga river (including minor tributaries such as Kao river, Dharmawatii river, Mahi River, Baya river), 9. Karmnasa 10. Sone, 11. Punpu, 12. Kiul-Harohar, 13. Badua, and 14. Chandan River.

The total cultivable area of the state is 56,030 km^2, of which 33,510 km^2 is irrigated through various sources (Table 4.2).

During the period from 1985 to 2021, the net irrigated area by canals marginally decreased from 34.58% to 31.36% of the total irrigated area, while during the same period, the tubewell-based irrigated area increased from 33.26% to 62.41%. Bihar has utilized 90% of its groundwater irrigation potential, while only 75.8% of the potential from major and minor irrigation schemes has been harnessed; enhancing the efficiency of major and medium irrigation systems can help reduce the excessive reliance on groundwater (Tables 4.3 and 4.4).

Table 4.2 Source-wise irrigation potential created under government schemes (Sharma and Kumari 2024)

Type of irrigation potential	Ultimate potential	Created (2021)	Utilized (Utilization efficiency)
Major and Medium irrigation	5353	3722	2833 (75.8%)
Minor irrigation	6401	4527	4074 (90%)
Surface water	1544	1072	964 (90%)
Groundwater	4857	3455	3109 (90%)
Total	11,754	8247	6896 (83.6%)

Table 4.3 GW availability and utilization in Bihar (CGWB 2022)

S. No	Particulars	Quantity
1	Annual replenishable GW resources	33.15 BCM
2	Annual extractable GW resources	30.04 BCM
3	Current annual GW extraction (Total) (for use under irrigation)	13.50 BCM (10.01 BCM)
4	Stage of GW development $\{[(3)/(2)] \times 100\}$	44.94%

Table 4.4 Net irrigated area (in thousand hectares) in Bihar by source type (Sharma and Kumari 2024)

S. no	Source of irrigation	Area	Percent of the total irrigated area
1	Canal	9.67	31.36
2	Tanks	0.53	1.72
3	Tubewells	19.24	62.41
4	Other sources	1.39	4.51
5	Total	30.83	100

4.2.6 Water Resources Profile of Uttar Pradesh

Uttar Pradesh is characterized by eight major river basins and 39 sub-basins. The major river basins include the Yamuna, Ganga, Ramganga, Rapti, Ghaghra, Gandak, Gomti, and Sone. The state's agricultural landscape is dominated by sugarcane, cultivated on about 60% of the fields, while paddy and wheat cover the remaining 40%. The state produces approximately 154 million tonns of rice, and it takes about 649 L of water to produce one kilogram of rice. Notably, Uttar Pradesh is India's largest groundwater extractor for irrigation. The predominance of water-intensive crops like paddy is pushing the region toward an unsustainable water crisis. In the state, canal irrigation is provided through 75,466 km canal network.

Salient Features (Mausam n.d.).

- Geographical Area: 243,286 km^2
- Annual Rainfall (1980–2019): 844.1 mm

(Annual rainfall: West UP (765.3 mm); East UP (900.3 mm)
(Actual Rainfall 2024: West UP (798.4 mm); East UP (791 mm)

- Average Annual Precipitation: 218.11 BCM

Water Resources (IMD 2024; CGWB 2024a; Niti 2020).

- Average Annual Natural Flow through rivers: 161.64 BCM
- Utilizable Surface Water (without inter-basin transfer): 118.47 BCM

(However, the practically usable water is much lower, around **109 BCM**, due to physical constraints such as the flat terrain, which limits the feasibility of surface water storage.)

- Annual Groundwater Recharge (2024): 71.83 BCM (16.2% of India's total recharge)
- Extractable Groundwater Resource: 65.57 BCM
- Actual Annual Groundwater Extraction: 46.40 BCM

Water Utilization for Irrigation (Niti 2020).

- Surface Water: 43.8 BCM
- Groundwater: 45.8 BCM
- (An additional 27.8 BCM is associated with ongoing projects)
- Groundwater Use Breakdown:
  - Irrigation: 40.89 BCM
  - Domestic and Industrial Use: 4.95 BCM

Agricultural Statistics (UP-DES 2023)

- Gross Sown Area: 28,346 km^2 (as of 2021–22)
- Net Sown Area: 16,096 km^2 (as of 2021–22)
- Cropping Intensity: 176% (as of 2015–16)
- Gross Irrigated Area: 22,707 km^2 (as of 2015–16)
- Net Irrigated Area: 13,941 km^2 (as of 2015–16)
- Total canal length 75,466 km
- Area irrigated by canal: 15.8% (as of 2015–16)
- Area irrigated using wells/tube-wells: 84% (as of 2015–16)
- Area under Irrigation: 86.7%.

4.2.7 Madhya Pradesh

Madhya Pradesh covers a geographical area of 308,252 km^2 and receives a normal annual rainfall of approximately 1040.4 mm, with spatial variability ranging from 742.5 mm to 1471.6 mm (CGWB 2023). The rainfall distribution exhibits a distinct decreasing trend from east to west and from south to north across the state. The region is drained by several major rivers, including the Narmada, Son, Tapti, Chambal, Betwa, Ken, Pench,

Wainganga, and Mahanadi, which collectively support the surface water system of the state.

The average surface water availability at 75% dependability is estimated at 81.5 BCM, of which 56.8 BCM is allocated to Madhya Pradesh and 24.7 BCM to neighbouring states through interstate water-sharing agreements. The salient features of water resources and their utilization are summarized as follows (DoE 2022; LUS n.d.):

- annual surface water availability (75% dependability): 81.5 BCM
  - allocated to Madhya Pradesh: 56.8 BCM
  - allocated to neighbouring states: 24.7 BCM
- total annual replenishable groundwater: 36.04 BCM
- total annual extractable groundwater: 33.38 BCM
- annual groundwater extraction: 18.97 BCM
- Overall stage of groundwater development of the state: 58.75%

Agricultural land-use statistics for 2022–23 indicate that the net area sown is 158,480 km^2, of which 129,300 km^2 constitutes the net irrigated area, while 29,180 km^2 remains un-irrigated.

Out of a total of 317 groundwater assessment units spread across 269,333.27 km^2 of recharge-worthy area in the state, 8.37% are categorized as 'Over-Exploited', 1.58% as 'Critical', 19.23% as 'Semi-Critical', and 70.81% fall under the 'Safe' category. This distribution highlights the spatial variability in groundwater stress across the state and underscores the need for sustainable groundwater management practices (CGWB 2023) (Fig. 4.2).

1. CGWB 2023: Dynamic Groundwater Resources of Madhya Pradesh (as on 2023). CGWB, NCR, Bhopal, and GWS, WRD, Government of Madhya Pradesh. 2023
2. DoE & F MP (2022): Madhya Pradesh State Action Plan on Climate Change v 2.0. Department of Environment, Government of Madhya Pradesh LUS 2022–23: Land Use Statistics at a Glance 2022–23. Ministry Of Agriculture & Farmers Welfare Department Of Agriculture & Farmers Welfare Economics, Statistics And Evaluation Division New Delhi.

4.2.8 Jharkhand

Jharkhand has a geographical area of 79,716 square kilometres, and it receives 1220.7 mm of normal annual rainfall. The river basins and sub-basins of the state include the Subarnarekha, Kharkai, South Koel, Barakar, Damodar, Sankh, Punpun, Sone–Kanhar, Kao–Gangi, Mayurakshi, Falgu (also known as Niranjana), Ajay, the Ganga stem basin, Gumani, Koa–Bhena, Belsai–Chandan–Chir, Harohar, and several smaller streams that

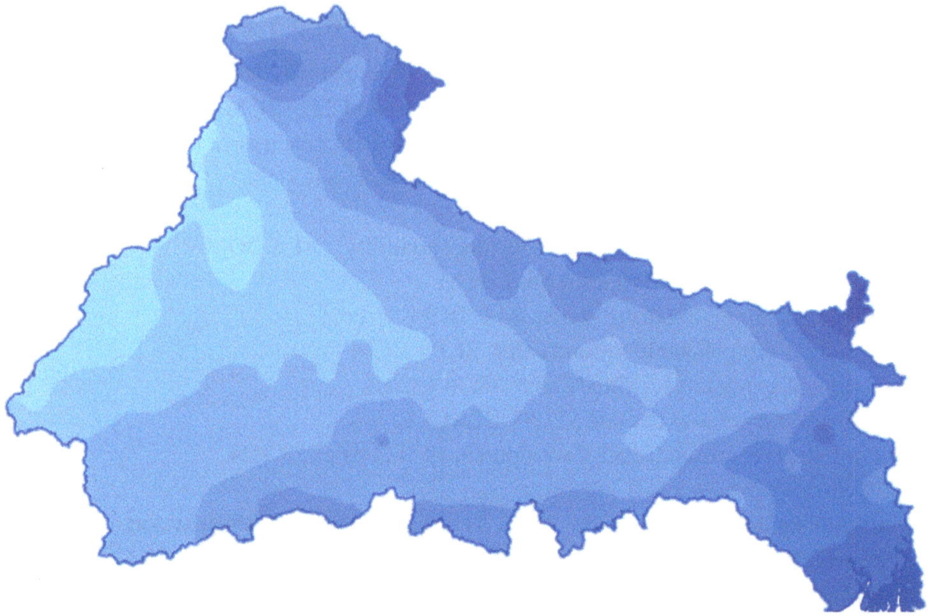

(a)

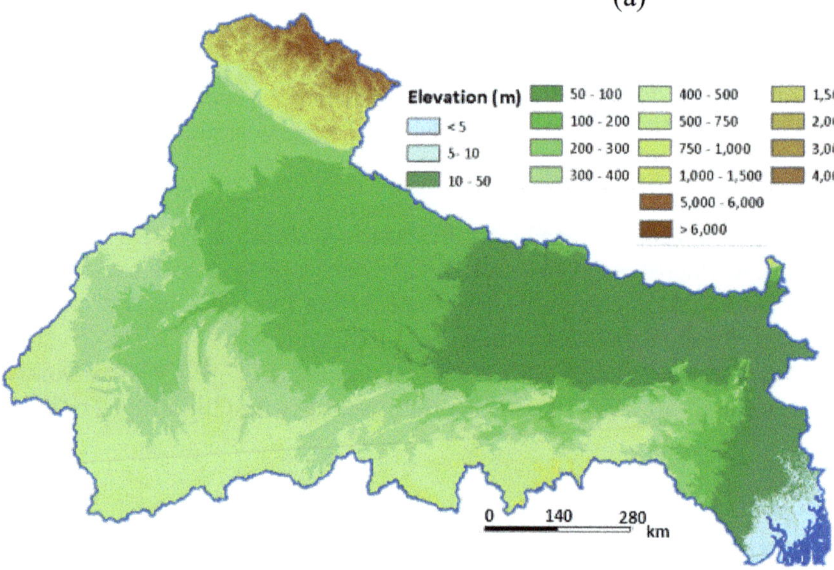

(b)

Fig. 4.2 Annual average rainfall distribution (**a**) and elevation map of the Ganga basin in (**b**) (*Source* CWC & NRSC, 2014) drain independently outside the state. The main stem of the river Ganga flows for approximately 40 kms along the northern boundary of Sahibganj district, occupying about 101 square kilometres within Jharkhand. The Damodar River, an important tributary of the Ganga, drains nearly 11.8% of the state's geographical area (Mondal et al. 2018). Rivers such as the Gumani, Punpun, and Falgu originate in Jharkhand and flow directly into the Ganga in the neighbouring state of Bihar. Similarly, the North Koel and Kanhar rivers, which also originate in Jharkhand, eventually join the Sone River, a major tributary that meets the Ganga near Patna in Bihar. Overall, approximately 41.3% of Jharkhand's geographical area lies within the Ganga River Basin (). The total allocation of surface water for the state is 20,023.25 MCM.

The ultimate irrigation potential of the state from these river basins is estimated at 5,605.4 km^2. According to the dynamic resource assessment, the current status of surface water and groundwater resources in Jharkhand is as follows: surface water availability is 25,876.98 MCM; the water requirement for irrigation (surface water and groundwater combined) is 3,813.17 MCM; and industrial water demand (surface water and groundwater) is 4,338 MCM.

As assessed by CGWB (2025a), the total annual replenishable groundwater recharge of the state is 6.28 BCM. After accounting for natural discharge of 0.52 BCM, the annual extractable groundwater resource is estimated at 5.76 BCM. The annual groundwater extraction is 1.81 BCM, comprising 0.94 BCM for irrigation, 0.22 BCM for industrial use, and 0.65 BCM for drinking water supply. The average stage of groundwater extraction for the state is 31.42%.

Out of 263 groundwater assessment units, 5 units (1.9%) have been classified as 'Over-exploited', 6 units (2.28%) as 'Critical', 12 units (4.56%) as 'Semi-critical', and the remaining 240 units (91.25%) fall under the 'Safe' category.

4.2.9 Himachal Pradesh

Himachal Pradesh covers a geographical area of 55,673 km^2, of which approximately 4,317 km^2 lie within the Ganga Basin. The state receives a normal annual rainfall of 1,245.1 mm. The Himachal Himalaya contains about 800 glaciers covering an area of 3,124.713 km^2, along with 2,679 permanent snowfields extending over 1,775.189 km^2, and nearly 20 small lakes with a combined areal extent of 134.5 ha (Bharadwaj 2014).

The Pong Dam (Maharana Pratap Sagar), constructed on the Beas River in Kangra district, covers an area of 21,712 ha. The reservoir has a total storage capacity of 8.57 BCM and an active storage capacity of 7.29 BCM. It serves multiple purposes, including hydropower generation (installed capacity of 396 MW), irrigation support to Punjab, Haryana, and Rajasthan, and moderation of downstream floodwaters; and is a Ramsar site of international importance.

Several important rivers originating in Himachal Pradesh contribute to the Upper Yamuna Basin and ultimately to the larger Ganga River system. These include the Tons River and its tributary, the Pabbar River, as well as the Giri River. The Pabbar River drains regions of Shimla and Rohru, while the Giri River flows through the districts of Shimla and Sirmaur.

The state's elevation ranges from approximately 350 m to 6,816 m above mean sea level. Rainfall patterns are strongly influenced by altitude. According to the State of Environment Report (SoE-HP 2009), areas below 800 m receive between 800 and 1,600 mm of annual rainfall; regions between 800 and 1,600 m receive around 1,800 mm; areas between 1,600 and 2,700 m receive 1,000–1,500 mm; and areas above 2,700 m receive about 200 mm annually.

Water resources in the state include snow-fed and monsoon-fed rivers, hilly springs, lakes, water-conservation structures, and groundwater from tube wells and open wells.

According to the Central Ground Water Board (CGWB 2025b), based on the 2024 groundwater resource assessment, the total annual groundwater recharge in Himachal Pradesh is 1.113 BCM. The annual extractable groundwater resource is 1.01 BCM. Annual groundwater extraction is 0.3585 BCM, comprising 0.1852 BCM for irrigation, 0.0567 BCM for industrial use, and 0.11662 BCM for domestic consumption. The stage of groundwater extraction for the state is 35.48%.

4.2.10 Chhattisgarh

Chhattisgarh covers a total geographical area of 135,194 km^2. The state is divided among four major river basins: Mahanadi (56%), Godavari (29%), Ganga (14%), and Narmada (1%). Important rivers and sub-basins in the state include the Sone, Banas, Gopad, Rihand, Kanhar, Brahmani, Mahanadi, Indrawati, Sheonath, Hasdeo, Mand, Ibb, Pairi, Sabari and Arpa. Of the 19.52 thousand cubic metres of water allocated under the Mahanadi Project in Chhattisgarh, Bhilai Steel Plant utilizes 18%, irrigation accounts for 75%, and drinking water supply accounts for 6%.

Chhattisgarh receives a normal annual rainfall of 1,266.6 mm, with spatial variation ranging between 1,112 mm and 1,481 mm. The salient features of the state's water resources are summarised below.

In terms of surface-water irrigation, the potential irrigable area is 43,000 km^2 (State of Environment Report n.d.). According to the land-use statistics of the Government of India for 2022–23 (Ministry of Agriculture Farmers Welfare 2025), the net sown area is 45,920 km^2. Of this, 8,690 km^2 is irrigated through the canal system, 5,940 km^2 is irrigated through wells and tube-wells, and the total net irrigated area from all sources is 15,350 km^2. Thus, the irrigated area constitutes 33.42% of the net sown area.

The annual surface water availability is estimated to be 48.296 BCM, of which 41.720 BCM is considered usable and 18.249 BCM is currently utilized (State of Environment Report n.d.).

According to the groundwater resource assessment by the Central Ground Water Board (CGWB 2024a) the total annual groundwater recharge is 14.18 BCM, and the natural discharge during the non-monsoon period is 1.25 BCM, yielding a net extractable groundwater resource of 12.92 BCM. The total groundwater extraction for all uses in the state is 6.11 BCM, of which irrigation accounts for more than 86%, domestic use accounts for 12%, and industrial use accounts for 2%. The stage of groundwater extraction as of March 2024 is 47.32%.

Out of 146 groundwater assessment units (blocks), 5 units (3.42%) are classified as 'Critical', 22 units (14.38%) as 'Semi-critical', and 120 units (82.19%) as 'Safe'. No units fall under the 'Over-exploited' or 'Saline' categories.

4.2.11 West Bengal

West Bengal, covering a geographical area of 88,752 km^2, extends from the Himalayan foothills in the north to the Sundarbans mangrove delta in the south. The state hosts the highest number of water bodies in India (720,000), accounting for nearly one-third of the national total (Ministry of Jal Shakti 2023).

The Ganga Basin is the hydrological backbone of West Bengal, encompassing 73,833 km^2 ($\approx$83% of the state) and 17 districts: Cooch Behar, Dakshin Dinajpur, Uttar Dinajpur, Malda, Murshidabad, Birbhum, Nadia, Purba Bardhaman, Paschim Bardhaman, Purulia, Bankura, Paschim Medinipur, Hugli, Howrah, North 24 Parganas, South 24 Parganas, and Kolkata (Das 2023). The Ganga enters the state near Rajmahal, bifurcates at Farakka Barrage into the Bhagirathi–Hooghly (serving Hugli, Howrah, and Kolkata) and the Padma (international boundary with Bangladesh).

The state exhibits a pronounced north–south rainfall gradient, with mean annual totals varying markedly by topography and proximity to the coast. In the higher-altitude northern Himalayan foothills, rainfall increases with elevation from 300 to 400 cm, reflecting orographic enhancement. In the middle Gangetic plains, it decreases southward from 160 to 110 cm, before gradually rising again to up to 170 cm in the southern coastal regions due to cyclonic influences and sea-breeze effects (India Meteorological Department 2008). Mean annual precipitation over the state yields a gross volume of 159.27 billion cubic metres (BCM) (GoWB 2020, 2021). After evapotranspiration and deep percolation losses of 52–55%, internal surface runoff amounts to 77.06 BCM and annual groundwater recharge to 31.72 BCM, resulting in a net internally generated water resource of 105.82 BCM $year^{-1}$ (GoWB 2020, 2021).

The official estimates (2024) report total annual surface water availability within West Bengal at 132.91 BCM (Opportunities, and Interconnections With Consumptive Sectors

2024) and extractable dynamic groundwater at 23.56 BCM (out of 25.89 BCM annual recharge), giving a combined total water resource availability of 156.47 BCM year^{-1} (GoWB 2024; CGWB 2024b) (Tables 4.5, 4.6 and 4.7).

At state level (2024), annual groundwater draft for irrigation is 8.99 BCM and total draft for all uses is 10.75 BCM against an extractable resource of 23.56 BCM (CGWB 2024b).

Irrigation and cropping patterns (latest estimates).

- Total cropped area: 102,590 km^2
- Net irrigated area: 33,520 km^2 (33% of cropped area) (Mandal et al. 2022)
- Irrigated by surface water: 5,530 km^2
- Irrigated by groundwater: 14,097 km^2
- Ultimate irrigation potential (surface water): 36,000 km^2
- Ultimate irrigation potential (groundwater): 33,180 km^2 (GoWB 2020; Pal et al. 2023)

Table 4.5 Stage of groundwater development in selected Ganga Basin districts of West Bengal (2024) (CGWB 2024b)

Districts	Assessment unit	Safe	Semi-critical	Critical	Saline
Murshidabad,	26	19	6	1	
Nadia,	18	3	9	6	
Hugly,	18	15	3		
Bardhaman, Purba Paschim	23 8	18 8	5		
Kolkata	1				1

Table 4.6 Annual groundwater resources (in Ha.m.) in selected Ganga Basin districts (2024) (CGWB 2024b)

Districts	Total extractable GW resource	Current GW draft
Murshidabad,	1809.104	1095.25
Nadia,	1727.785	1471.796
Hugly,	1183.741	635.1809
Bardhaman, Purba Paschim Total	1865.136 226.086 2091.222	963.917 112.3189 1076.236
Haora	209.2307	42.7039
Total	7021.082	4321.167

Table 4.7 Water availability and projected demand in selected Ganga Basin districts (2024 baseline) (GoWB 2024; Mandal et al. 2022)

Districts	Available GW resource	Availabile surface water (CGWB 2024b)	Total available water resource	Water demand	Water Gap (BCM)
Murshidabad	1.81	5.65	7.46	8.68	−1.22
Nadia	1.73	4.92	6.65	6.53	0.12
Hugly	1.18	5.24	6.42	5.95	0.47
Bardhaman,	2.09	6.31	8.4	4.94	3.46
Haora	0.21	1.86	2.07	1.98	0.09
Total	7.02	23.98	31	28.08	2.92

Water demand

- Total water demand (2015–16, all sectors): 90.54 BCM (Pal et al. 2023)
- Projected agricultural demand (2011): 84.607 BCM (GoWB 2021)
- Projected total demand (2025, all sectors): 149.872 BCM (GoWB 2021)

Although gross water availability appears substantial, the projected total demand of 166 BCM by 2025 significantly exceeds the current utilisable quantum (≈51 BCM surface + 23.56 BCM groundwater), highlighting the critical need for enhanced storage, conjunctive use, and demand-management strategies across the Ganga Basin districts of West Bengal.

4.2.12 Hydrology of Surface Water Systems

Surface water hydrology is the study of water that moves and is stored on or above the Earth's surface. It includes the study of rivers, lakes, wetlands, and soil water. It has applications in water management systems, flood forecasting, drought monitoring, and water construction. Surface water hydrology is a branch of physical hydrology. It's a core course in civil engineering.

4.2.12.1 Detailed Exploration of the Middle Ganga Basin's Hydrology

The hydrology of the middle Ganga Basin reveals critical challenges associated with groundwater and river systems, emphasizing the need for sustainable management. Groundwater serves as a crucial resource for drinking water and agriculture in the region, yet its quality and availability are on a declining trajectory (India-WRIS n.d.; Kumar et al. 2021). Between 2017 and 2022, the Upper Gangetic Basin and Yamuna Chambal Basin experienced severe groundwater drought, with groundwater levels in many parts of the basin falling at an average rate of 0.20 m per year (India-WRIS n.d.; Wisser et al. 2010). Furthermore, groundwater storage anomalies (GWSA) in these areas show a steep downward trend, highlighting the severity of depletion (Misra 2011). Groundwater is heavily relied upon for irrigation, with the Ganga Basin states accounting for nearly 50% of India's groundwater-irrigated area, placing additional strain on this vital resource (Aon et al. 2024).

The region's river systems are also shaped by seasonal and climatic factors. Approximately 80% of the annual river inflows occur during the three months of the monsoon season, underscoring the highly seasonal nature of water availability (World Bank 2020). The Ganga River is primarily fed by rainfall, subsurface flows, and snowmelt from glaciers, with major tributaries like the Yamuna, Sone, Ghagra, and Kosi contributing significantly to its flow (NMCG 2013). The basin itself spans an area of approximately 1.2 million square kilometers, making it one of the largest river basins globally (World Bank 2020). These findings underscore the pressing need for effective strategies to mitigate groundwater depletion and manage river systems sustainably, ensuring long-term water security in the middle Ganga Basin.

4.2.12.2 Description of Canal Networks, Reservoirs, and Seasonal Discharge Patterns

The Ganges Canal system is an extensive network of canals located in the middle Ganga basin, specifically designed to irrigate the Doab region, the area of land situated between the Ganges and Yamuna rivers (Wisser et al. 2010; World Bank 2015). Built between 1842 and 1854 by engineer Proby Thomas Cautley, the canal system was later extended by the British (Wisser et al. 2010; Jain and Singh 2020). The primary function of the system is to irrigate the Ganges Valley in Uttar Pradesh and Bihar, significantly boosting the production of cash crops such as sugarcane, cotton, and oilseeds (India-WRIS n.d.). The key canals in the system include the Upper Ganga Canal, which begins at Haridwar and spans a length of 5,950 miles (9,575 km), the Lower Ganga Canal that starts at Naraura with a combined length of 5,120 miles (8,240 km), and the Sarda Canal, which irrigates land near Ayodhya in Uttar Pradesh (India-WRIS n.d.). The system works by diverting water from the Ganges through barrages into the canals and includes navigation channels equipped with lock gates, facilitating both irrigation and waterway transport (Sahoo 2020; Roy and Datta 2023).

The Ganga River Basin is equipped with numerous reservoirs, dams, barrages, and canals that support water requirements for irrigation, drinking, and hydroelectric power generation (Richards et al. 2023; Maurya et al. 2019; Mukherjee et al. 2018). Some of the prominent dams in the basin include the Tehri Dam, the highest dam on the Ganga River, located in Uttarakhand; the Nanak Sagar Dam, the longest dam on the river, located in the Udham Singh Nagar district of Uttarakhand; and several irrigation dams such as the Abhaypura, Adner, Adwa, and Afzalgarh Dams (Mukherjee et al. 2018; Raju et al. 2021). Additionally, key barrages like the Adwa Barrage in Uttar Pradesh, the Ajoy and

Amanat Barrages in Jharkhand, and the Farakka Barrage in West Bengal regulate river flow (Maurya et al. 2019; Thakur et al. 2017). The canal systems, including the Upper Ganga Canal, which diverts water for irrigation at Haridwar; the Madhya Ganga Canal, which serves irrigation during the monsoon months at Bijnore; and the Lower Ganga Canal at Narora, are essential for managing water distribution (Board 2013). In response to the environmental concerns of the Ganga River Basin, the Ganga River Basin Management Plan (GRBMP) was developed in 2010 to address these issues and improve the river's health and water management practices (Raju et al. 2021; Jain et al. 2007).

The surface runoff in the middle Ganga basin constitutes 50% of the water that falls as rain, with the remaining 30% evaporating and 20% seeping into the ground (Matta et al. 2018). Several factors influence surface discharge in the Ganga basin, starting with rainfall. The Ganga basin receives an average of one million cubic meters of water per square kilometer annually from rainfall. Snow melt from glaciers also contributes to the water supply, along with subsurface flows where groundwater moves through the river's deep channel, which is bordered by high banks. The Ganga basin is subject to annual flooding, with floodplains that typically range from 0.5 to 2 km wide. The river's main tributaries—Yamuna, Sone, Ghagra, and Kosi—contribute more than half of the Ganga's annual water yield (Matta et al. 2018). Other factors influencing the Ganga include groundwater depletion, which has caused the Ganga plain aquifer to lose 30 cm of storage per year (Kumar et al. 2021), as well as a reduction in the water level of the river at different points, ranging from 0.5 to 38.1 cm per year (Kumar et al. 2021). Existing structures within the basin also impact its discharge (Matta et al. 2018).

4.2.12.3 Issues Such as Lean Flows, Flooding, and Their Water Management Implications

The lean flow in the middle Ganga basin refers to the reduced flow of the Ganga River during the dry season. Several factors contribute to this reduced flow, including the construction of dams, excessive groundwater extraction, and flow regulation for irrigation. Dams such as the Tehri dam in the upper Ganga basin decrease the water flow (Singh 2013; National Water Policy 2021), while excessive groundwater extraction in the plains causes the water table to drop, further diminishing the river's flow (National Water Policy 2021; Ministry of Jal Shakti 2022). Additionally, diverting water for irrigation purposes also reduces the flow of water downstream (Jain and Singh 2020). The impact of the reduced flow is significant, affecting aquatic species and disturbing the riverine and riparian ecology (Jain and Singh 2020). It can also make it more challenging to maintain the overall health of the river (Roy et al. 2023; National Water Policy 2021). Environmentalists recommend that no more than 30% of the river's flow should be diverted for agricultural, domestic, or industrial uses (World Bank 2014). In response, the government has set minimum flow norms for the Ganga River to help preserve its health and ecological balance (World Bank 2015).

4.2.13 Surface Water Quality

Surface water quality is the condition of bodies of water above ground, such as rivers, lakes, and streams. It's measured by the levels of physical, chemical, and biological parameters in the water. Key parameters measured include dissolved oxygen, pH, turbidity, bacteria, and toxic pollutants (Central Water Commission 2021; Pew Research Earth Genome India 2022). The quality of surface water is crucial for drinking, recreation, aquatic life, and understanding pollution impacts (Das and Bayen 2023).

4.2.13.1 Identification of Key Contaminants and Their Origins

The middle Ganga basin faces several contaminants that negatively impact water quality, including heavy metals, sewage, industrial waste, and emerging organic contaminants (EOCs). Heavy metals such as chromium, zinc, iron, and manganese pose significant health risks. Chromium is a lethal chemical pollutant, zinc exposure can lead to vomiting and nausea, excessive iron can cause gastrointestinal issues, and manganese may lead to neurological and cardiovascular toxicity (Kumar et al. 2020; Bhanja et al. 2017).

Emerging organic contaminants include pharmaceuticals like diclofenac; agrochemicals such as diuron, atrazine, and chlorpyrifos; antibiotics like sulfamethoxazole; sweeteners like saccharine, cyclamate, acesulfame K, and sucralose; and endocrine-disrupting compounds like bisphenol A and phthalates (Kumar et al. 2020; Bhanja et al. 2017; Pandey and Singh 2019). The basin is also impacted by sewage, industrial effluent, agricultural runoff, and animal waste (Bhanja et al. 2017; Das and Kaur 2019; NMCG 2020).

4.2.13.2 Seasonal Trends in Water Quality Across the Basin

Water quality in the middle Ganga basin varies seasonally due to climate and pollution. In summer, temperatures and chromium concentrations peak (Jain et al. 2007). During the monsoon, turbidity increases and dissolved oxygen decreases due to heavy runoff (Singh and Singh 2020). In winter, temperatures drop and pH levels rise (Ghosh and Das 2021). Climate influences rainfall, drought, and cold spells that alter water quality (Singh and Singh 2020). Pollution from sewage, industrial effluents, pesticides, and fertilizers further degrades water quality (Ghosh and Das 2021). Agricultural runoff during the monsoon introduces harmful chemicals (Singh 2013). Summer evaporation concentrates dissolved contaminants (Singh and Singh 2020). These factors cause low DO levels, high turbidity, and elevated nutrient loads—harmful for human use and aquatic ecosystems (Roy et al. 2023).

4.2.13.3 Implications for Agriculture, Industries, and Domestic Water Use

Seasonal and pollution-driven changes in water quality significantly affect agriculture, industry, and domestic usage. High temperatures and chromium levels in summer impair irrigation and crop health. Monsoon turbidity and low DO disrupt industrial operations and agricultural systems. Suspended solids also overload domestic filtration systems. Winter shifts in pH and temperature alter irrigation suitability for crops needing specific conditions. Continuous pollution from sewage, industries, and agriculture worsens these impacts, demanding better treatment systems and integrated water management strategies.

4.3 Groundwater Resources

4.3.1 Groundwater Distribution

4.3.1.1 Overview of the Region's Aquifers, Their Depth Ranges, and Spatial Distribution

The study area includes 27 districts of Uttar Pradesh (UP), India. The study area, lying between the Ganga River in the south and the Ramganga River and the Sharda River in the north, covers 95,900 km^2. In the north, it touches the state of Uttarakhand and Nepal, a country located in the lap of Himalayas. The boundaries of the area lie between the latitudes 29.742°N and 25.036°N, and the longitudes 78.37°E and 82.904°E. In Fig. 4.1, the location of the study area is represented. The shape of the study area is like that of a pea.

The total population of the study area is 91,060,456. Some part of the study area passes through the nearby districts of Meerut, Bulandshahr, Aligarh, Etah, Farrukhabad, Fatehpur, Mirzapur, Chandauli, Deoria, Basti and Bahraich. The other districts that are included in the study area of concern are shown in Fig. 4.1. The major rivers flowing through the study area are the river Sai, the river Gandhak, the river Ganga, the river Ghaghara, the river Gomti, the river Ramganga, the river Rapti, the river Sharda, and the river Yamuna. The river network of the study area is shown in Fig. 4.1. Many important urban centers like Lucknow (the capital city of UP), and Kanpur lie in the study area. Most of the districts of the study area fall under sensitive and very sensitive flood-prone zones (Fig. 4.3).

4.3.1.2 Long-Term Variability in Groundwater Level

As per a report by the Central Ground Water Board (CGWB), 60% of the wells in India reported a decline in the GWL (2016–2017). The middle Ganga basin is not unaffected by this decline. In a study conducted using GRACE-GPS it has been found that the groundwater is depleting at the rate of 1.46 ± 0.74 cm/year in Lucknow district.

The value of Sen's slope for pre—monsoon groundwater level and post—monsoon groundwater level is more negative than the rainfall of the study area. The values of Sen's slope are nearly the same for pre—monsoon and post—monsoon GWL. The Kendall's Tau is higher for post monsoon GWL. The temporal variation of pre monsoon GWL, and post monsoon GWL is shown in Figs. 4.4a, b and 4.5.

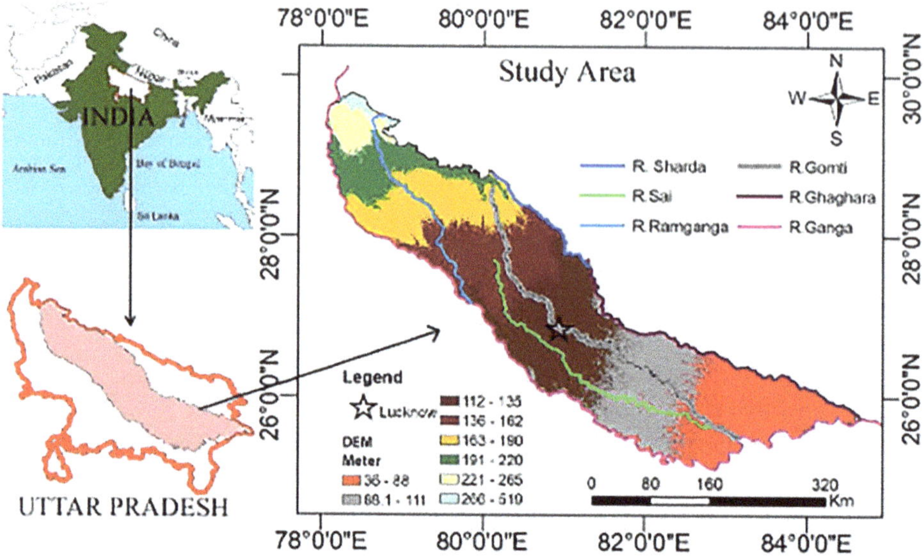

Fig. 4.3 The elevation of the study area along with rivers has been shown in the state of Uttar Pradesh, India

4.3.2 Critical Areas of Concern

4.3.2.1 Highlighting Over-Exploited Areas (Below 0 m)

Groundwater flows at right angles to the equipotential lines in the same way that water flowing down a slope would flow at right angles to the contour lines. Topography and geology are the dominant factors controlling groundwater flow. Storativity describes the property of an aquifer to store water. Hydraulic conductivity is measured by performing a pumping test, i.e. by pumping one well and observing the changes in hydraulic head in neighboring wells. Water moves underground downward and sideways, in great quantities, due to gravity and pressure. Water flows from high to low hydraulic heads. Groundwater can move upward against gravity because the hydraulic head at any point is a combination of both elevation and pressure. Hydraulic head is the level to which groundwater will rise in a well. Groundwater flows from high hydraulic head to low hydraulic head.

4.3.2.2 Analysis of Regions with Rising Groundwater Levels and Associated Challenges

Groundwater-level fluctuations in the middle Ganga basin are influenced by factors such as aquifer type, precipitation, recharge, abstraction, and regional groundwater circulation. High rainfall is a primary source of groundwater recharge, but fluctuations in water levels occur due to seasonal and yearly variations in precipitation, with a delay between infiltration and the rise in the water table. Over-extraction of groundwater, driven by

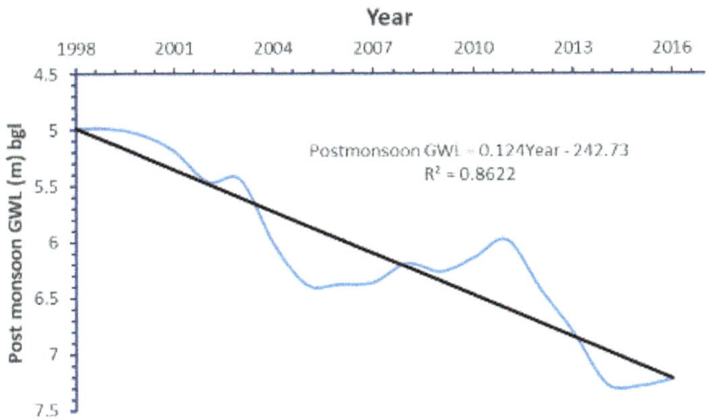

(a)

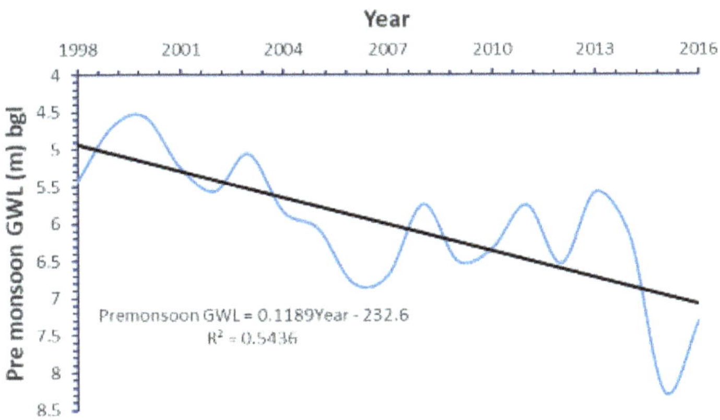

(b)

Fig. 4.4 **a** The time series of the average post monsoon GWL from 1998 to 2016 and the trend line. **b** The time series of the average pre monsoon GWL from 1998 to 2016 and the trend line

subsidies on electricity, high minimum support prices (MSP) for water-intensive crops, and increasing demand from population growth and industrialization, leads to depletion. Groundwater depletion results in various negative effects, including increased pumping costs, water quality deterioration, reduced flow in streams and lakes, and land subsidence. Spatially, groundwater levels rise in areas with higher rainfall and fall in regions with less precipitation. Terrain also plays a role, with flat areas experiencing groundwater rise, while slopes contribute to fluctuations. The study of groundwater movement is essential for understanding aquifer recharge and managing sustainable water use in the region (Fig. 4.6).

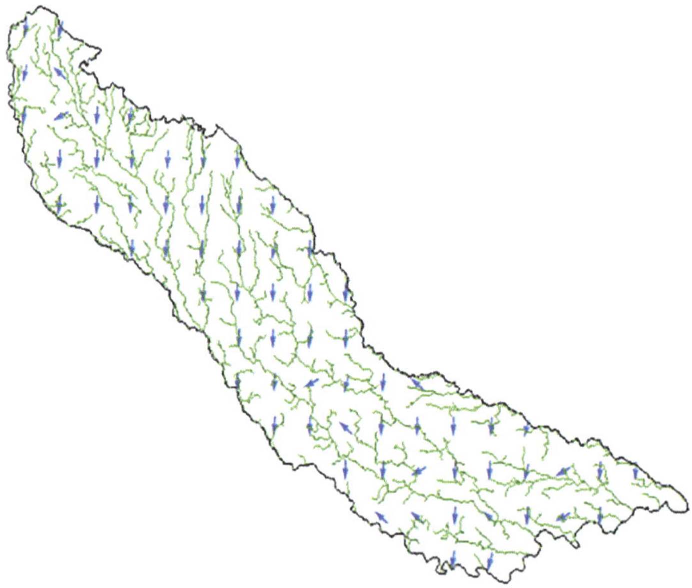

Fig. 4.5 The flow direction of the groundwater below the earth surface under the action of gravity

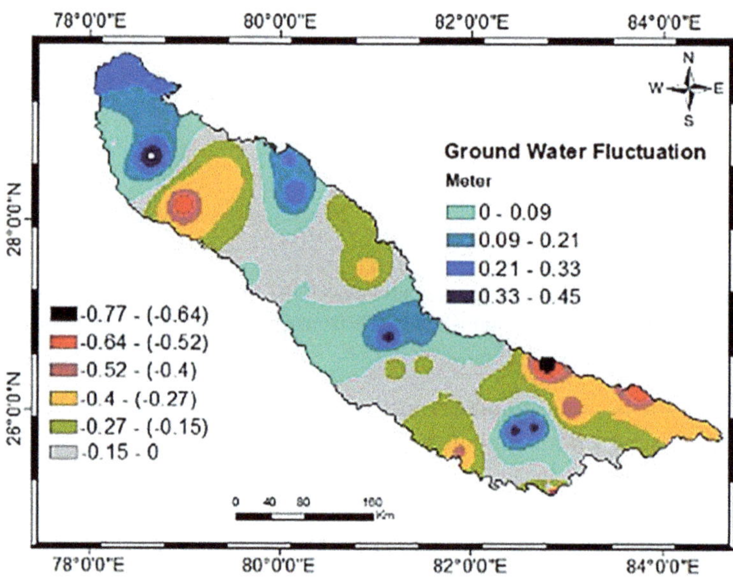

Fig. 4.6 The spatial map of groundwater fluctuation for the middle Ganga basin, UP

4.3.2.3 Discussion on Contamination Issues and Sustainability in the Face of Increasing Use

The Ganga River, a vital source of groundwater recharge, is increasingly threatened by pollution from human and industrial activities along its banks. The relentless discharge of untreated sewage into the river significantly contaminates surface and groundwater resources. This not only compromises the quality of groundwater but also poses serious health risks to communities relying on it for drinking water and agriculture. The infiltration of pollutants from the river into aquifers can lead to long-term groundwater contamination, making remediation efforts challenging and costly.

Additionally, the substantial volume of industrial and solid waste dumped into the river exacerbates the degradation of groundwater quality. Hazardous chemicals and heavy metals from industrial effluents seep into the soil and aquifers, disrupting the natural filtration process. Over time, this pollution diminishes the availability of clean groundwater and threatens the sustainability of this critical resource. Addressing these issues requires stringent regulations, effective waste management systems, and community participation to protect the Ganga and its interconnected groundwater systems.

4.4 Water Demand Across Sectors

4.4.1 Sectoral Analysis of Water Demand in the Middle Ganga Basin

4.4.1.1 Breakdown of Water Usage Across Agriculture, Industrial, and Domestic Sectors

> Water demand in the middle Ganga basin is primarily driven by the agriculture, industrial, and domestic sectors. Agriculture is the dominant consumer, accounting for a significant portion of water usage, especially for irrigation of crops like rice, sugarcane, and wheat. The industrial sector, particularly in regions such as Kanpur and Varanasi, is the second-largest consumer of water, mainly for manufacturing, cooling, and processing purposes. The domestic sector, while smaller in comparison, also places increasing pressure on water resources due to growing urban populations in cities like Patna and Allahabad.

The water used in the Ganges River basin is primarily allocated across agriculture, domestic, and industrial sectors, with significant seasonal variations due to the monsoon. Agriculture dominates water usage in the basin, with about 90% of the surface water being used for irrigation, which is crucial for growing water-intensive crops. The domestic sector also relies on water for drinking and other basic needs, while the industrial sector uses water for hydropower generation, navigation, and other industrial activities. The usage pattern varies across the seasons, particularly during the monsoon, when water availability increases, impacting the demand in these sectors (Groundwater and Climate Resilience in the Indo-Gangetic Basin 2020; Chatterjee and Ray 2018).

However, water depletion in the Ganges River basin is becoming a pressing issue. Between 1999 and 2013, the water level of the Ganga River decreased by 0.5 to 38.1 cm per year, and groundwater storage in the Ganga plain aquifer decreased by 30 cm annually. This depletion of groundwater and reduced river water levels can exacerbate food insecurity in the region, particularly in agricultural communities dependent on irrigation (Singh and Kumar 2019; Sarkar and Banerjee 2021). Factors contributing to this depletion include over-withdrawal of groundwater, which can lead to aquifer disintegration and land subsidence. Anthropogenic activities such as underground explosions, mining, and tunneling also disrupt the geological formations that support the aquatic systems in the basin. Furthermore, the diversion of water through canals can reduce the flow in the main river, further exacerbating the depletion (Unicef 2019; Kumar et al. 2017).

4.4.1.2 Trends in Seasonal Variability and Historical Demand Growth

Trends in seasonal variability show that water demand peaks during the dry summer months, as agricultural irrigation needs rise and domestic consumption increases due to higher temperatures. The monsoon season, while replenishing water sources, does not significantly reduce demand, as much of the water is used for irrigation or industrial processes. Historical demand growth has been substantial due to both population growth and the expansion of agriculture and industry in the basin. The region has seen a consistent rise in water demand, driven by agricultural expansion, industrial development, and increased urbanization.

The water demand in the middle Ganga basin fluctuates seasonally due to the monsoon season and human activities. The monsoon season, from June to September, is responsible for more than 80% of the annual rainfall, bringing a significant increase in water volume and turbidity in the river. Along with this increased flow, the monsoon season also introduces higher levels of silt and debris into the river, which leads to an increase in chemical oxygen demand and suspended solids (Water use Efficiency in Indo-Gangetic Agriculture 2018; Panda and Wahr 2016; Singh and Srivastava 2020; Gupta and Jain 2018).

In contrast, the non-monsoon season, which typically spans from October to May, brings less water into the river, resulting in water scarcity. This season also sees a decrease in dissolved oxygen levels in the river due to less light penetration into the water, further affecting the aquatic ecosystem (Water use Efficiency in Indo-Gangetic Agriculture 2018; Panda and Wahr 2016; Singh and Srivastava 2020).

4.4.2 Implications of Increasing Water Demand

4.4.2.1 Economic and Social Consequences of Rising Water Demand

The growing demand for water in the middle Ganga basin has significant economic and social consequences. Economically, the strain on water resources could limit agricultural productivity, particularly in water-intensive crops, leading to reduced income for farmers and possible food security concerns. Industrial growth could also be hindered if water resources become scarce or contaminated. Socially, increasing water demand may lead to conflicts over water allocation, particularly between agricultural and urban uses, and may exacerbate disparities in water access, especially in rural or marginalized communities.

4.4.2.2 Assessment of Long-Term Sustainability of Water Resources

The long-term sustainability of water resources in the middle Ganga basin is under considerable threat due to increasing water demand. Groundwater depletion, particularly in the eastern parts of the basin, coupled with pollution from industrial effluent and untreated sewage, poses a serious risk to water quality and availability. If current trends continue, the basin may face significant challenges in meeting the needs of all sectors, requiring more sustainable water management practices and policies to ensure that water resources can continue to support the growing population and economy of the region.

4.5 Climate Variability in the Region

4.5.1 Patterns of Climate Variability

4.5.1.1 Analysis of Precipitation Trends, Drought and Wet Periods, and Their Spatial Variability

The study area does not receive uniform rainfall. The rainfall pattern is parallel to the Himalayan Mountain ranges. Rainfall increases from south–west (350 mm to 600 mm) to north–east (1351 to 1600 mm). The northern region receives much higher rainfall (1100 mm to 1600 mm) than the southern region (350 mm to 1100 mm). The average rainfall of the study area is 867.5 mm. The variation in the rainfall is very high. The rainfall can vary from 450 mm to more than 1600 mm in the middle Ganga basin, UP. The northern region is attached with Shiwalik mountains entering the nearby state. Hence, the rainfall amount decreases from the Shiwalik to the far plains of the middle Ganga basin. The rainfall map shown in Fig. 5.1 has been divided into five classes from low rainfall to very high rainfall. The rainfall represented in this figure is the average rainfall from 1991 to 2020 (Fig. 4.7).

The lowest temperature recorded-diurnally, monthly, seasonally, or annually, or the lowest temperature of the entire record. Daily air temperature minima are recorded by the screen minimum thermometer. The process of evaporation also takes place at night but at a very slow rate. This is because even after sun set some amount of heat still exists in the environment. This small amount of heat is sufficient to carry out the process of evaporation. Atmospheric surface temperature in India has increased in the last century by about 1 °C and 1.1 °C during winter and post-monsoon months respectively (Fig. 4.8).

The Earth's surface temperature is controlled mainly by solar heating which turns on at dawn and then off at sunset. During the typical day the surface heats and during the

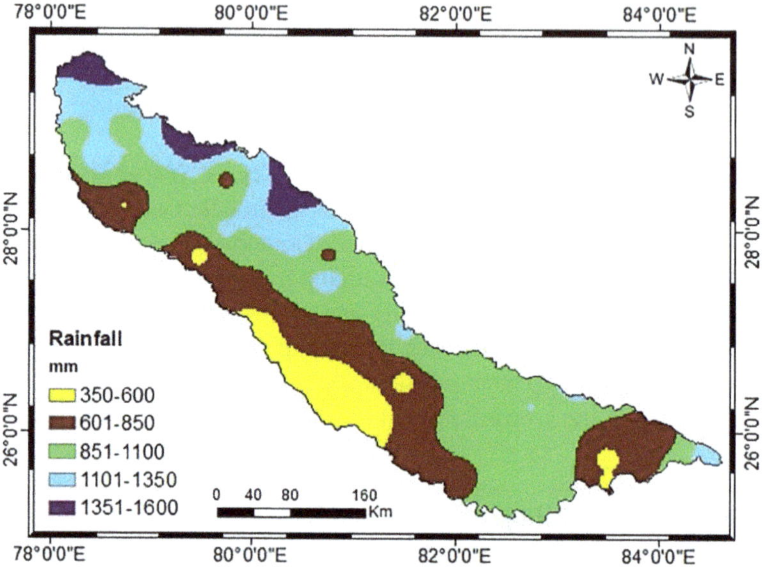

Fig. 4.7 The spatial map of rainfall showing rainfall distribution in the middle Ganga basin, UP

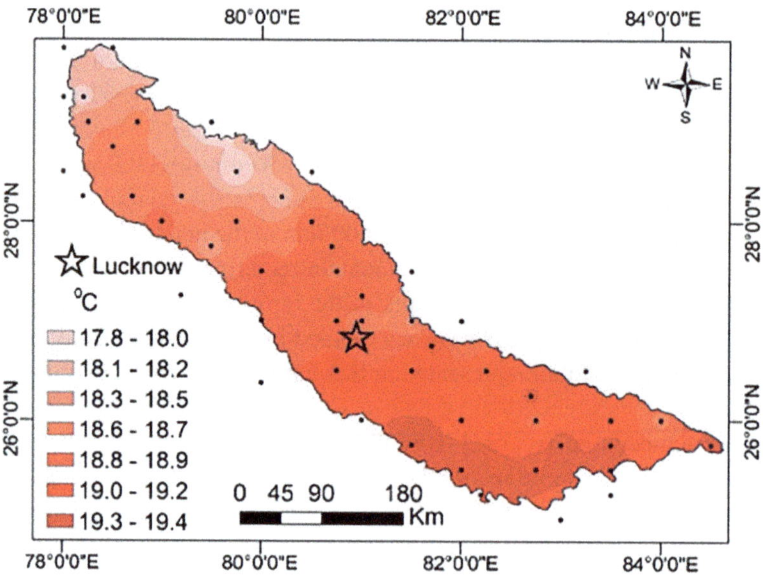

Fig. 4.8 The spatial distribution map of the minimum temperature of the middle Ganga basin, UP

night it cools. Exceptions to this happen during unusual wind events, where warm or cold air moves over the surface. As we all know, the sun will heat the Earth's surface during the day. Then at night, the sun is on the other side of the globe, so temperatures cool back down.

In the study area, the minimum temperature varies from 17.76 °C to 19.38 °C from north to south. Although water can evaporate at low temperatures, the rate of evaporation increases as the temperature increases. This makes sense because at higher temperatures, more molecules are moving faster; therefore, it is more likely for a molecule to have enough energy to break away from the liquid to become a gas.

The highest temperature of a day is termed as a 'maximum temperature'. As the land gets heated quickly at the same time it cools down quickly. So during the day due to the sun, the temperature is high. The temperatures start to increase over Uttar Pradesh in March and by April the land area becomes hot with day time maximum temperature reaching about 40 °C at many locations. During this season the range of maximum and minimum temperatures is found to be more than 15 °C at many stations. Rapid urbanization to meet the infrastructure development of increasing population has hampered city with problems of flooding cause by variability in rainfall and temperature. Thus it becomes increasingly important to study the characteristics of rainfall and temperature variables. An understanding of these variables is dominant to water resources planning and management (Figs. 4.9 and 4.10).

In the study area, the maximum temperature varies from 27.62 °C to 32.48 °C from north to south.

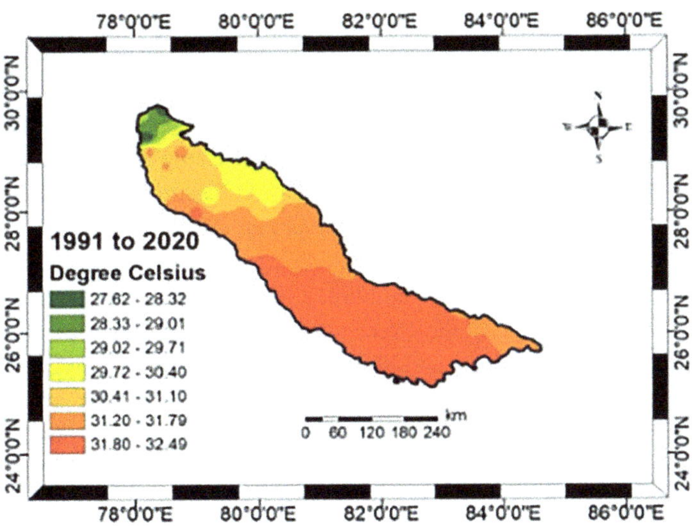

Fig. 4.9 The spatial distribution map of the maximum temperature of the middle Ganga basin, UP

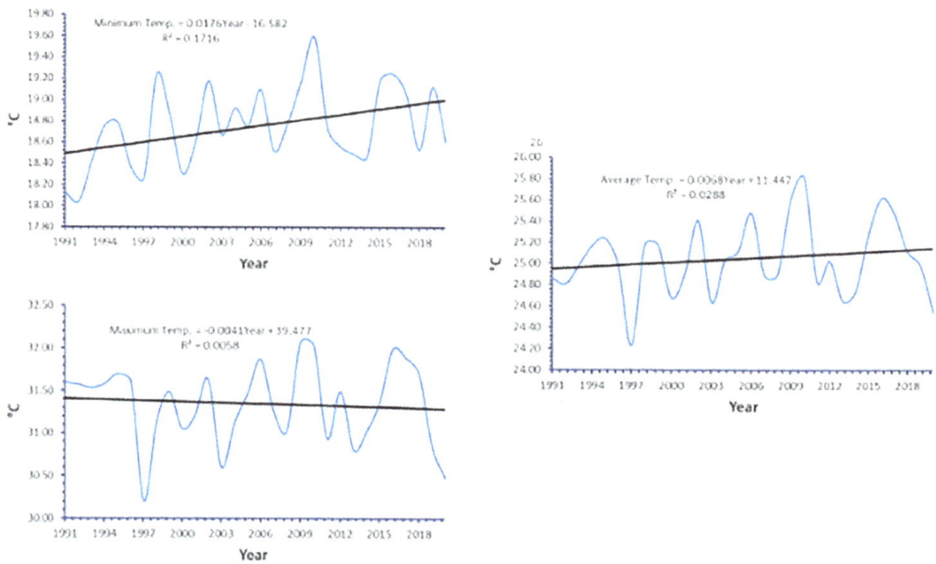

Fig. 4.10 The time series of the minimum temperature, the maximum temperature and the average temperature of the middle Ganga basin, UP and their trend equations provided separately

4.5.2 Impact of Climate Variability on Water Resources

Climate change has the potential to significantly impact water availability in the Ganga basin, including the middle Ganga basin, with potential consequences such as more frequent droughts, flooding, and higher temperatures (Jha and Shrestha 2021; Raju 2017; Jain et al. 2021; Das and Dutta 2023). Climate change can affect various aspects of the basin's water resources. Rainfall patterns may change, with some studies suggesting that reduced rainfall could be offset by increased precipitation and changes in melt runoff (Ministry of Agriculture Farmers Welfare 2022; CSE 2020). Additionally, glacier melt could be altered, potentially affecting the river's water volume. The region may also experience more droughts and flooding as a result of climate change, further impacting water availability and quality (Raju 2017; Jain et al. 2021). There may also be an increased risk of water quality deterioration due to the effects of climate change, such as rising concentrations of carbonic acid in the water and higher sediment loads (Aayog 2023; FAO 2021). The Ganga basin is particularly vulnerable to the impacts of climate change, with India being one of the most climate-vulnerable nations in the world, particularly because of its large agrarian sector (ICID 2020; Water Resources Department 2022).

4.6 Water Quality Issues

4.6.1 Identification of Major Contaminants

The identification of major contaminants is a critical step in addressing environmental pollution and ensuring water quality. In the context of groundwater and surface water systems, major contaminants include heavy metals, pesticides, nitrates, phosphates, microbial pathogens, and industrial chemicals. These pollutants originate from various sources, such as agricultural runoff, industrial discharges, untreated sewage, and leachate from landfills. Systematic identification involves conducting site-specific assessments, employing advanced analytical techniques like gas chromatography and mass spectrometry, and leveraging geospatial mapping tools to pinpoint pollution hotspots. Recognizing the type, source, and extent of contamination is essential for designing targeted remediation strategies, safeguarding ecosystems, and ensuring the sustainable use of water resources.

4.6.2 Spatial Variation of Contaminants

4.6.2.1 Insights into the Spatial Variability of Contamination Levels

A total of 159 water samples (Fig. 70) were collected for analysis of major cations and anions, including 55 deep groundwater samples (depth >120 ft) and 104 shallow groundwater samples (depth <120 ft). According to the charge balance verification of all hydro-chemical data, the ionic balance error was within the limit value of ±5%, indicating the accuracy of our data. The analysed cationic compositions follow the pattern of Na^+ Ca^{2+} Mg^{2+}>K^+. In 29% sample, the most predominant cation Na^+ exceeds the contamination level The ion exchange as well as dissolution of clay minerals and the presence of sodic plagioclase contribute to excess Na^+ ion in groundwater. Comparing all cations, a low level of K^+ (0.149–5.179 mg/L) ion signifies weak mobility in groundwater samples. High Ca^{2+} concentration than Mg^{2+} suggests carbonate mineral dissolutions.

The order of anionic compositions is HCO_3^- > SO_4^{2-} >Cl^- >NO_3^- based on sample analysis. The most prevalent anion, HCO_3^-, aids the mechanisms of the carbonate weathering process and carbonic acid dissolution in aquifers. The existence of SO_4^{2-} ions may be from sulphite ores, shales and industrial wastes. Evaporite dissolution causes the formation of Cl^- ions in groundwater SO_4^{2-} concentration beyond 400 mg/L can have a laxative impact on human organs when combined with Mg^{2+} and Na^+ (WHO 2011). The sources of NO_3^- ion are nitrogenous fertilisers, urbanisations, landfill leachate and septic tank seepage.

Water quality index (WQI) is defined as a rating reflecting the composite influence of different water quality parameters on the overall quality of water. It indicates the quality by an index number, which represents the overall quality of water for any intended use. Water quality index (WQI) was prepared using water quality parameters F, Cl, SO_4, NO_3,

Na, NH_4, K, Ca & Mg. It shows that 74% area (71,084 km^2) portion is having excellent quality of groundwater while very poor (75–100) to unsuitable (>100) quality is prevailing in small pockets in North and south west regions within area of 2795 km^2. Hardness was found to be hard in 60% area and very hard in 30% of the study area. Different water quality parameters along with minimum and maximum values are shown in Table 4.8.

Water quality parameters with the areas which are above permissible limits for drinking purposes are identified as; Flouride in Prayagraj, Ajamgarh and Jaunpur, Magnesium in border areas of Lucknow & Unnao, border areas of Hardoi & Lucknow, Pratapgarh, Nitrates in Shahjahanpur, Lucknow, Hardoi, Unnao, Pratapgarh, Prayagraj, Sulphates in Lucknow, Unnao, Jaunpur, Chloride in Lucknow, Unnao, Prayagraj, Pratapgarh, Jaunpur, Hardoi and Calcium in Lucknow, Barabanki, Sitapur, J P Nagar, Pilibhit, Shahjahanpur, Hardoi, Pratapgarh, Prayagraj, Sultanpur, Ambedkar Nagar, Mirjapur found above the acceptable limits but both were under the permissible limits (Figs. 4.11 and 4.12).

Trace elements distribution and occurrence in the groundwater depends on degree of weathering and mobility of these elements (Handa 1986). Large aluminum intake may negatively influence health. This was connected with nerve damage. Particularly people with kidney damage are susceptible to aluminum toxicity. There is a risk of allergies. Aluminum is probably mutagenic and carcinogenic. The main sources are Rock and soil leaching. Aluminum was found above permissible limit of 30 μg/l for 13% of the groundwater samples with concentration varying from 0 to 2403.73 μg/l with an average

Table 4.8 Different water quality variables with Min, Max and Avg

S. no	Variables	Min	Max	Avg	Standard deviation
1	EC	130	4700	729.364	543.1494
2	PH	6.9	10.1	7.69713	0.358292
3	HCO3	0	744.768	307.83	133.339
4	F^-	0	8.8618	0.36601	0.738188
5	Cl^-	0.3238	903.314	49.4359	108.0178
6	$SO4^{2-}$	0	1299.71	49.526	127.1596
7	NO^{3-}	0	400.972	15.2516	41.67903
8	Li^+	0	0.4701	0.01301	0.04794
9	Na^+	0	854.283	62.2863	114.5154
10	$NH4^+$	0	34.461	1.12969	4.541576
11	K^+	0.0264	106.189	6.90412	12.92646
12	Mg^{2+}	0.175	219.483	28.6728	25.77451
13	Ca^{2+}	0.9754	237.806	57.048	34.31028
14	Total hardness as CaCO3	4.05694	1202.1	260.131	174.2404

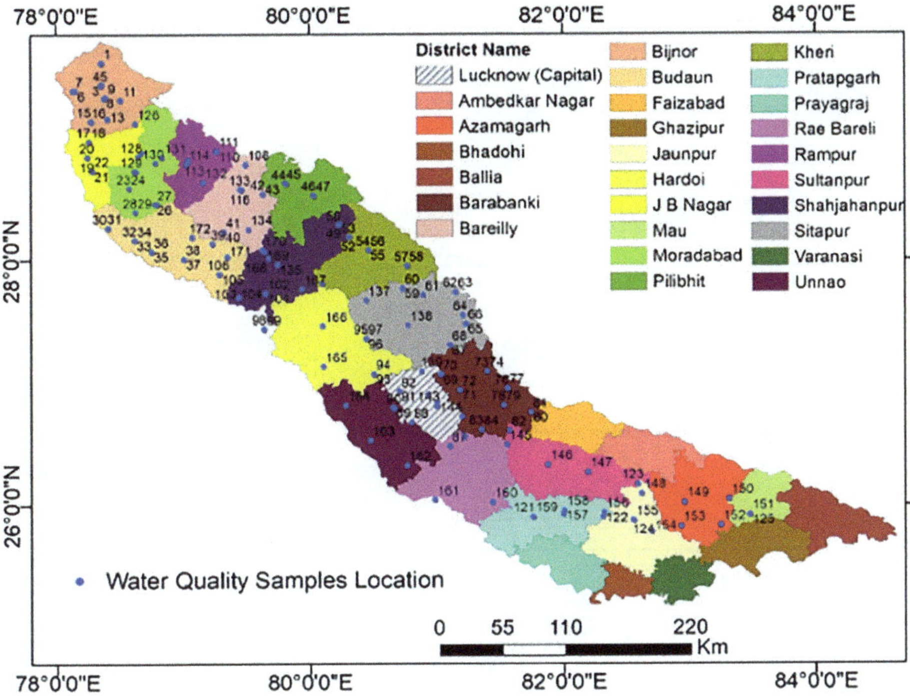

Fig. 4.11 Study area with sampling locations for water quality analysis locations along with the district

of 118.11 μg/l. The source of Arsenic in groundwater is from Leaching from natural deposits, Wood preservatives, pesticides, industrial deposits, Petroleum production, Semiconductor manufacture & Coal power plants (Aggarwal et al. 2025). Arsenic in study area varies from 0 to 147.63 μg/l with an average of 6.31 μg/l. 1% of samples were found to be above permissible limit which can cause serious skin problems, endocrine disruptor, cancer—skin, bladder, lung, kidney, liver, prostate, Harms cardiovascular & nervous systems (Figs. 4.13 and 4.14).

4.6.2.2 Sectoral Impacts of Contaminants and Interlinkages with Economic Growth

Contaminants—ranging from industrial pollutants to agricultural chemicals—impact various sectors of the economy, posing significant challenges to public health, ecosystems, and infrastructure. In agriculture, excessive use of pesticides and fertilizers leads to soil degradation, water contamination, and reduced productivity over time. This, in turn, jeopardizes food security and raises costs for farmers who must invest in soil restoration or alternative farming techniques. Similarly, industrial sectors releasing hazardous pollutants into air and water systems not only degrade environmental quality but also increase

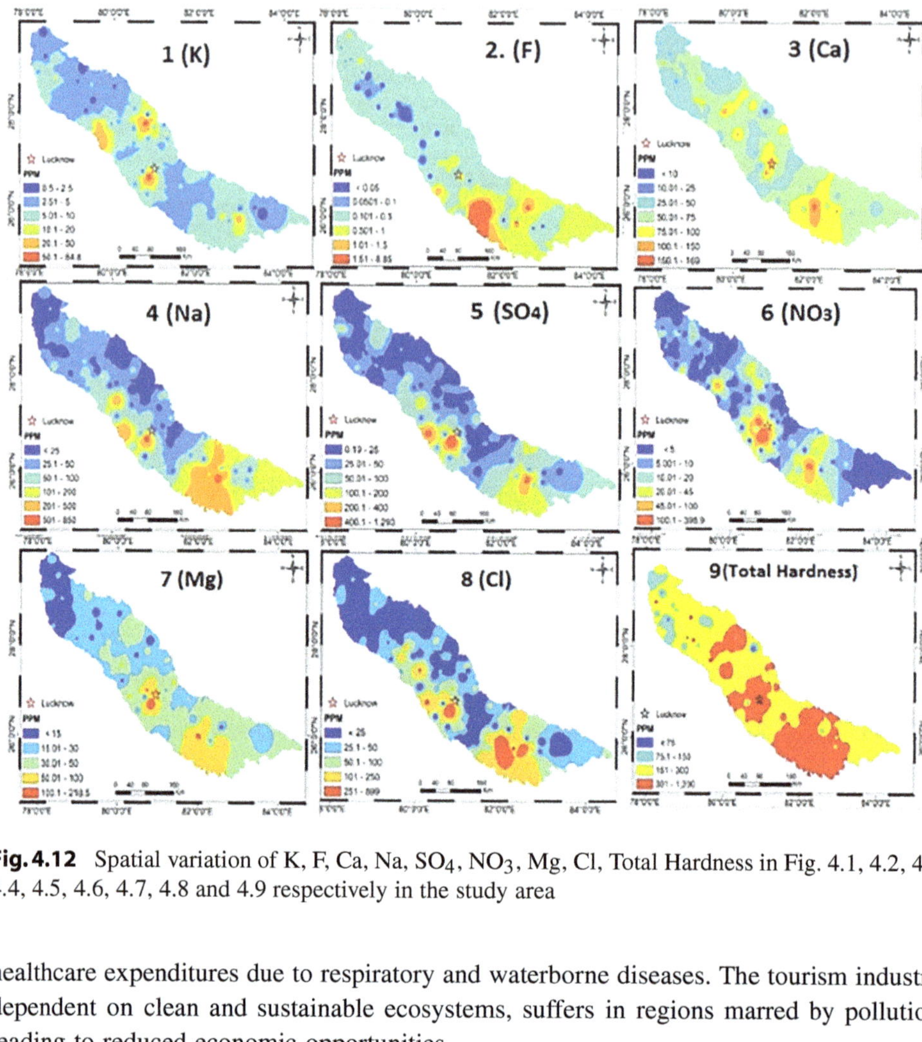

Fig. 4.12 Spatial variation of K, F, Ca, Na, SO_4, NO_3, Mg, Cl, Total Hardness in Fig. 4.1, 4.2, 4.3, 4.4, 4.5, 4.6, 4.7, 4.8 and 4.9 respectively in the study area

healthcare expenditures due to respiratory and waterborne diseases. The tourism industry, dependent on clean and sustainable ecosystems, suffers in regions marred by pollution, leading to reduced economic opportunities.

The presence of contaminants in critical sectors has a ripple effect on economic growth. Poor water quality can lead to reduced labor productivity as workers fall ill or spend time caring for affected family members. In industrial sectors, regulatory penalties and cleanup costs add to operational expenses, diverting resources from innovation and expansion. Moreover, contaminated natural resources limit the availability of raw materials, thereby increasing production costs and hampering industrial growth. While economic activities often lead to pollution, unchecked growth without sustainable practices worsens contamination, creating a feedback loop where growth itself becomes unsustainable.

Breaking the interlinkages between contaminants and negative economic outcomes requires adopting sustainable practices across sectors. Investment in green technologies,

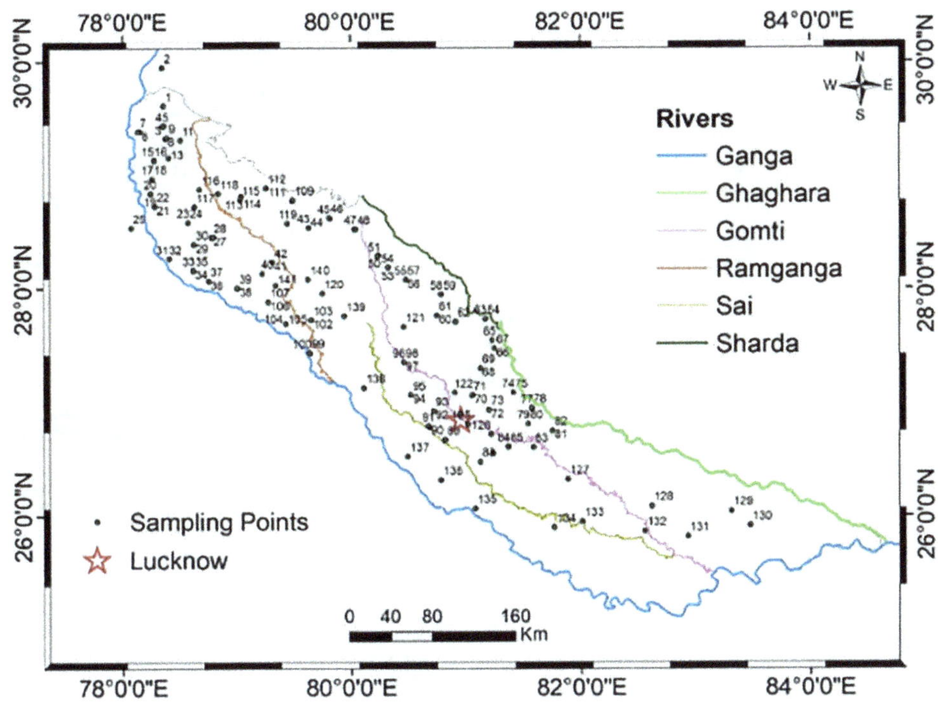

Fig. 4.13 Locations of heavy metal samples in the study area

pollution control measures, and renewable energy can mitigate the long-term impacts of contamination. For example, transitioning to organic farming can improve soil health while maintaining agricultural productivity. Likewise, enforcing stricter environmental regulations can reduce industrial emissions and promote cleaner production methods. These initiatives not only protect public health and ecosystems but also contribute to resilient economic growth by fostering innovation, reducing healthcare costs, and creating green job opportunities. In this context, addressing contamination becomes an essential strategy for achieving sustainable development.

4.7 Strategies for Water Resource Protection, Augmentation, and Management

Strategies for water resource protection, augmentation, and management include various methods that can enhance water availability, conserve resources, and ensure long-term sustainability. Rainwater harvesting, for example, helps replenish groundwater by increasing precipitation infiltration into the soil, thereby raising the water table and making more water available for wells. Water recycling and reuse involve treating wastewater for

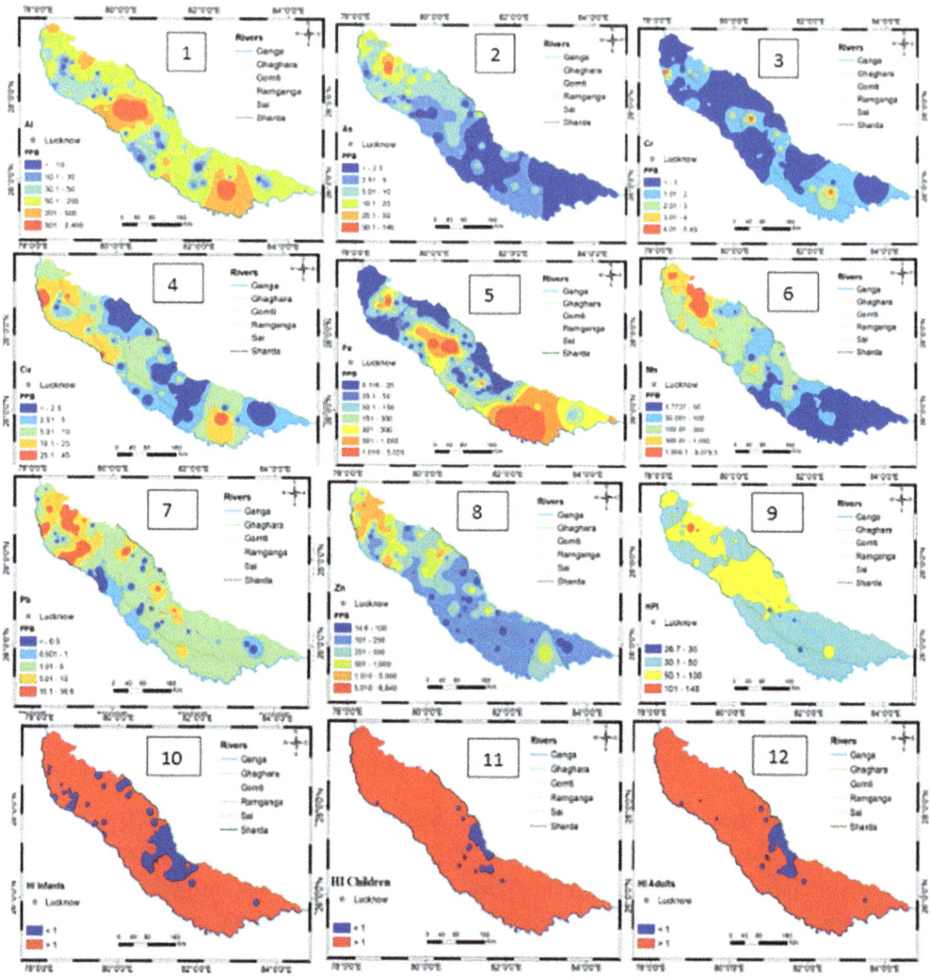

Fig. 4.14 Spatial variation of Al, As, Cr, Cu, Fe, Mn, Pb, Zn, HPI, HI(Infants), HI (Children) & HI(Adults) in Figs. 4.1, 4.2, 4.3, 4.4, 4.5, 4.6, 4.7, 4.8, 4.9, 4.10, 4.11 and 4.12 respectively

non-potable purposes such as irrigation, landscaping, and industrial processes, thus reducing the strain on fresh water resources. Efficient irrigation techniques aim to optimize water and energy usage, ensuring crop productivity while reducing water consumption, providing both environmental and economic benefits. Integrated Water Resources Management (IWRM) coordinates the development and management of water, land, and related resources in a sustainable manner. Proper water resource management involves allocating water in a sustainable, equitable, and efficient manner, balancing both demand and supply. Additionally, water quality monitoring helps assess and maintain the health of water resources. Water conservation strategies are essential for protecting water quality and managing water resources sustainably. Effective water management is crucial for ensuring access to clean water, safeguarding ecosystems, and preventing water-related conflicts.

4.7.1 Surface Water Conservation Measures

Surface water conservation measures play a crucial role in managing water resources sustainably and reducing flood risks. These measures include techniques such as rainwater harvesting, check dams, and improved storage infrastructure. Rainwater harvesting is a traditional practice that involves collecting and storing rainwater from rooftops, roads, pavements, and open fields for later use. The collected water can be stored in tanks, cisterns, or underground reservoirs, making it suitable for irrigation, drinking, and household chores. This method also helps reduce soil erosion and flood hazards (Raj and Kumar 2020; Tare et al. 2020; Annual Report on Ganga River Health Indicators 2022; Mahajan and Sharma 2021).

Check dams, built across channels or gullies, are designed to slow water flow, reduce erosion, and increase groundwater infiltration. These structures can be placed in ditches, swales, or channels to mitigate the effects of runoff (Singh and Kumar 2018; Singh and Rawat 2023). Improved storage infrastructure, such as ponds, reservoirs, or tanks, provides reliable water storage for use during droughts or dry periods, ensuring a stable water supply.

Other water conservation techniques include drip irrigation, cultivating drought-resistant crops, practicing dry farming, implementing rotational grazing, and installing water-efficient toilets and fixtures. These strategies enhance water use efficiency and support sustainable agricultural and domestic practices (Dutta and Srivastava 2019; UNEP 2020).

4.7.2 Watershed-Based Management

A watershed-level approach is crucial for balancing the needs of people, the environment, and the economy. It improves water and soil quality, increases access to drinking water, and promotes sustainable resource management (Singh and Kumar 2018; NRCG 2021; Annual report on Ganga River Health Indicators 2022; India 2020). This approach offers numerous benefits. Environmentally, it protects and restores the health of freshwater ecosystems (Singh and Kumar 2018; India 2020). Economically, it enhances local and regional economic viability, while socially, it involves local communities in decision-making, often leading to increased incomes (Mahajan and Sharma 2021; Singh and Rawat 2023). By promoting sustainable land and water management practices, it ensures resource conservation. Additionally, this approach prevents duplication of efforts, reduces conflicting actions, and simplifies predictions about permissible actions, making management more efficient and predictable (Singh and Kumar 2018; Annual report on Ganga River health indicators 2022; India 2020; Mahajan and Sharma 2021).

The watershed-level approach operates by integrating stakeholders and employing strategic conservation and management practices. It actively engages public and private sectors at local, state, tribal, and federal levels (Singh and Kumar 2018; Singh and Rawat 2023). It identifies priority areas requiring restoration and protection and ensures the conservation of critical resources like water, soil, and vegetation. By reducing water pollution, it improves water quality and enhances soil quality and water availability, which ultimately increases agricultural production (India 2020; Singh and Rawat 2023; Dutta and Srivastava 2019). This integrated strategy supports ecosystem sustainability, community development, and efficient resource management.

4.7.3 Demand-Side and Supply-Side Management

Improving water use efficiency is crucial for addressing water shortages, ensuring sustainable water management, and promoting environmental conservation. One effective strategy is rainwater harvesting, which involves collecting rainwater for later use. This method helps mitigate water shortages and improves water quality (UNEP 2020; Russo and Lall 2017). Another approach is adopting improved irrigation techniques, such as drip or sprinkler systems, which deliver water directly to plants and minimize waste (UNEP 2020).

In regions where water becomes scarce later in the season, controlled deficit irrigation can be employed, limiting water supplies during later stages of crop growth while maintaining productivity (CPCB 2021). In drought-affected areas, desalination can supplement existing water supplies by converting seawater into usable freshwater (World Bank 2022). Additionally, modifying land use by matching it with suitable soil types ensures that crops use available water more efficiently (Iwmi 2019).

Water audits are another important measure, helping to evaluate water usage in facilities. These audits guide engineers and architects in designing water-efficient structures (Escap 2021). Implementing water use efficiencies, such as volumetric pricing policies, can control water usage, encourage the adaptation of crops, and improve field practices (Raj and Kumar 2020). Lastly, community awareness programs play a vital role in educating the public about sustainable water use and conservation strategies, fostering long-term water resource management (Malik and Mishra 2021).

4.7.4 Community Participation and Public Awareness

Sustainable management requires the collective efforts of local communities, awareness campaigns, and supportive government policies. Local communities play a crucial role by using their knowledge and agency to address environmental issues, making them effective agents of change (Singh and Kumar 2018). Their participation in designing, implementing, and evaluating green living projects fosters a sense of ownership and commitment to sustainability (Annual Report on Ganga River Health Indicators 2022). By adopting a holistic approach, communities can ensure that development strategies integrate quality, values, and actionable goals for a sustainable future (India 2020).

Awareness campaigns are powerful tools for mobilizing public opinion and influencing decision-makers. Through public outreach, they can create widespread support for sustainability issues (Mahajan and Sharma 2021). Additionally, social media serves as an effective platform for disseminating messages and gathering support for environmental campaigns (Tare et al. 2020).

Government policies complement these efforts by providing financial incentives and subsidies to promote the adoption of sustainable practices and renewable energy (NRCG 2021). Governments can also foster collaboration with local communities to address their concerns and values, resulting in more effective decision-making and better outcomes (Singh and Rawat 2023).

Sustainable development is a collective endeavor that involves individuals, communities, and governments working together. The ultimate goal is to achieve social, economic, and environmental progress while preserving resources for future generations (Dutta and Srivastava 2019).

4.8 Conclusion

4.8.1 Recap of Critical Issues and Recommended Strategies from the Review

4.8.1.1 Groundwater Potential

In Uttar Pradesh, less than one-third of the middle Ganga basin is suitable for groundwater recharge, with the majority of this area found in the Terai region. Groundwater availability is influenced by factors such as higher rainfall, lineaments, and flat terrain, which are more prevalent in the eastern and northern parts of the study area, particularly in the Terai. This region experiences higher rainfall, contributing to better groundwater resources.

On the other hand, areas with higher drainage density and clayey soils have lower groundwater infiltration. The drainage density is notably higher in the southern part of the basin, where there is less vegetation and the presence of impermeable soils further reduces infiltration. To address this, 92 artificial groundwater recharge sites have been identified, particularly along third-order streams, where structures can be built to recharge the groundwater using excess runoff.

4.8.1.2 Rainfall, Temperature, and GWL Trend

The rainfall distribution in the middle Ganga basin, Uttar Pradesh, is not uniform, with a significant variation across the region. Rainfall increases from the south-west to the north-east, ranging from 350 to 600 mm in the southern region to 1351 mm to 1600 mm in the northern region. The northern area, adjacent to the Shiwalik mountains, receives significantly more rainfall than the southern region, where rainfall ranges between 350 and 1100 mm. The average rainfall across the study area is 867.5 mm, and there is a high variability, with rainfall differing from 450 mm to over 1600 mm. The rainfall distribution map classifies the area into five categories, from low to very high rainfall, based on the average rainfall data from 1991 to 2020.

Temperature patterns in the study area are influenced by solar heating, with a notable increase in surface temperatures in the last century, particularly during the winter and post-monsoon months. The minimum temperature in the region varies from 17.76 °C to 19.38 °C from north to south. Evaporation occurs at night, albeit at a slower rate, as residual heat from the day aids in the process. The Earth's surface temperature is influenced by solar radiation, warming during the day and cooling at night. However, temperature can also be affected by unusual wind events that bring warm or cold air into the region.

In terms of maximum temperatures, the area experiences significant seasonal variation. From March onwards, temperatures rise, and by April, daytime maximum temperatures often reach around 40 °C in many locations. The difference between maximum and minimum temperatures is more than 15 °C in several places during this season. Urbanization in Uttar Pradesh has exacerbated problems like flooding, especially due to rainfall and temperature variability. Studying these climatic variables is crucial for water resource planning and management, as understanding temperature and rainfall patterns is key to addressing the challenges posed by urban development and changing weather conditions.

4.8.1.3 Water Quality Analysis

The study involved the collection of 159 groundwater samples, comprising 55 deep (depth >120 ft) and 104 shallow (depth <120 ft) samples, to analyze major cations and anions. The data showed a cationic composition pattern of $Na^+ > Ca^{2+} > Mg^{2+} > K^+$, with 29% of the samples having Na^+ concentrations exceeding permissible limits due to ion exchange, clay mineral dissolution, and sodic plagioclase presence. The anionic composition followed the order $HCO_3- > SO_4{}^2- > Cl- > NO_3-$. High $SO_4{}^2-$ and NO_3-s levels were attributed to industrial waste, fertilizers, and urbanization, while evaporite dissolution contributed to Cl^- levels. Water quality index (WQI) analysis revealed that 74% of the area had excellent groundwater quality, though pockets of poor quality existed, with hardness exceeding acceptable levels in 90% of the study area.

Spatial analysis indicated regions with contaminants above permissible drinking limits, including fluoride in Prayagraj, Azamgarh, and Jaunpur, magnesium in border areas of Lucknow, Unnao, and Hardoi, and nitrates in several districts such as Shahjahanpur and Prayagraj. Sulfates, chlorides, and calcium were also identified in various locations, often linked to industrial and agricultural activities. Despite being under permissible limits, these contaminants pose long-term risks to groundwater sustainability and health.

Trace element analysis revealed aluminum concentrations exceeding permissible limits (30 μg/L) in 13% of samples, ranging from 0 to 2403.73 μg/L, with an average of 118.11 μg/L. Aluminum toxicity, primarily caused by rock and soil leaching, can lead to nerve damage, allergies, and other health issues, particularly for individuals with kidney damage. Arsenic levels, ranging from 0 to 147.63 μg/L with an average of 6.31 μg/L, exceeded safe limits in 1% of samples, posing risks such as skin cancer, cardiovascular damage, and endocrine disruption.

The spatial distribution of heavy metals such as aluminum, arsenic, chromium, and manganese highlighted specific hotspots in the study area. Human health risks, assessed through indices like HPI (Heavy Metal Pollution Index) and HI (Hazard Index), indicated varying impacts on infants, children, and adults due to these contaminants. The data underscores the critical need for targeted mitigation strategies to manage groundwater quality.

In conclusion, the study highlights significant groundwater contamination in the Ganga Basin, driven by natural and anthropogenic factors. Addressing these issues requires a multi-pronged approach, including stricter industrial regulations, promoting sustainable agricultural practices, and enhancing community awareness about groundwater conservation. Comprehensive spatial analysis and regular monitoring are essential for sustainable groundwater resource management in the region.

4.8.2 Recommendations for Policy and Practice

Sustainable management requires the collective efforts of local communities, awareness campaigns, and supportive government policies. Local communities play a crucial role by using their knowledge and agency to address environmental issues, making them effective agents of change (Water quality and sanitation in South Asia 2022). Their participation in designing, implementing, and evaluating green living projects fosters a sense of ownership and commitment to sustainability (UNEP 2020). By adopting a holistic approach, communities can ensure that development strategies integrate quality, values, and actionable goals for a sustainable future (UNEP 2020). Awareness campaigns are powerful tools for mobilizing public opinion and influencing decision-makers. Through public outreach, they can create widespread support for sustainability issues [175–178]. Additionally, social media serves as an effective platform for disseminating messages and gathering support for environmental campaigns (Kejriwal et al. 2022).

Government policies complement these efforts by providing financial incentives and subsidies to promote the adoption of sustainable practices and renewable energy (Singh and Rawat 2023). Governments can also foster collaboration with local communities to address their concerns and values, resulting in more effective decision-making and better outcomes (Global water security report 2021). Sustainable development is a collective endeavor that involves individuals, communities, and governments working together. The ultimate goal is to achieve social, economic, and environmental progress while preserving resources for future generations (Roy and Ghosh 2022).

4.8.3 Future Initiatives

The Middle Ganga Basin faces complex challenges that demand targeted research and innovative strategies for sustainable water management, ecological restoration, and socio-economic development. High-resolution hydrological and climate modeling specific to the basin is essential to predict future scenarios of water availability, droughts, floods, and glacial melt. This would enable the development of region-specific climate adaptation strategies by providing insights into the intricate interactions between climate change, river dynamics, and groundwater systems.

Efforts to manage groundwater must focus on studying the basin's aquifer systems to identify recharge zones and develop efficient techniques such as managed aquifer recharge (MAR). This can help mitigate groundwater depletion and ensure long-term sustainability. Simultaneously, implementing advanced river health monitoring systems that leverage remote sensing, IoT-based sensors, and community-driven ecological surveys can track water quality, sediment loads, and biodiversity. Such monitoring is critical for maintaining the ecological integrity of the river system.

Innovative agricultural practices must be promoted, including precision agriculture, the use of drought-resistant crop varieties, and agroforestry systems. These approaches can enhance agricultural productivity while conserving water resources and reducing soil erosion. In urban areas, research on smart water grids, wastewater recycling, and rainwater harvesting can address water scarcity and improve overall water-use efficiency in rapidly growing cities within the basin.

Community engagement and socio-economic studies are crucial for inclusive water management. Research into water usage patterns, traditional practices, and gender roles in water management can inform the design of policies that incorporate local knowledge and address specific community needs. Parallelly, technological innovations like solar desalination, advanced filtration systems, and bio-remediation techniques should be developed and tested to provide low-cost, scalable solutions for safe drinking water and reduced waterborne diseases.

To ensure effective governance, existing water governance frameworks should be analyzed to propose integrated policies that balance environmental, agricultural, and industrial water demands. Adaptive governance models can address emerging challenges more effectively. Additionally, studying sediment transport and deposition patterns is necessary to design sediment management strategies that can reduce flood risks, prevent siltation, and improve river navigability.

Finally, the restoration and conservation of the basin's wetlands and riverine biodiversity must be prioritized. This would enhance essential ecosystem services such as water purification, flood control, and carbon sequestration. By addressing these research and innovation priorities, the Middle Ganga Basin can achieve sustainable water management, improve livelihoods, and mitigate the impacts of climate change and other emerging challenges.

4.8.4 Measures to Protect and Augment Water Resources

In river basin-scale planning, surface and groundwater should be considered as a single integrated resource. This unified approach allows for better assessment of total water availability, current consumption, and reserves for future use. Such integration is essential for effective river basin water planning and management.

During the planning and operation of water resource systems, the following priority order for state water allocation is generally followed:

1. Drinking water and domestic use
2. Environmental flows
3. Irrigation
4. Hydropower and thermal power generation
5. Agro-industries and non-agricultural industries
6. Navigation and other uses

These priorities may be modified regionally based on specific socio-environmental considerations. The National Water Mission has identified several vulnerable zones that require focused attention:

- Drought-prone areas
- Flood-prone areas
- Coastal regions
- Regions with deficient rainfall
- Areas with critical, semi-critical, and over-exploited groundwater
- Water quality-affected areas
- Snow-fed river basins

Climate change is expected to further increase variability in water availability, adversely affecting both human health and livelihoods. To address this, communities should be empowered to adopt climate-resilient technologies at the local level. Variability in water availability should be managed through increased storage—soil moisture, ponds, groundwater, reservoirs (both large and small), and combinations thereof.

In light of these challenges, the following strategic measures are proposed to protect, conserve, and enhance the water resources across all domains—surface water, groundwater, quality management, agricultural practices, and governance.

1. **Surface Water Management**
   - River Interlinking: Transfer of water from surplus to deficit basins.
   - Irrigation Infrastructure: Extension, renovation, and modernization to enhance capacity and reduce losses.
   - Tank Restoration: Scientific restoration of traditional tanks and water bodies to regain lost storage capacity.
   - Micro-Irrigation Systems: Promote drip and sprinkler irrigation to increase water use efficiency.
   - Ecosystem-Based Approaches: Ensure minimum environmental flows (Aviral Dhara) and clean river water (Nirmal Dhara).
   - Wetland and Tributary Rejuvenation: Desiltation and revival through scientific methods and community participation.
   - Leakage and Evaporation Control: Shift from open channels to piped water supply to reduce losses.
   - Watershed Management: Adopt field bunding, ponds, and other hydrological interventions.
   - Piped Conveyance: Prevent contamination and evaporation losses.
   - Inter-Basin Transfers: Implement after thorough environmental, social, and economic assessment; use surplus floodwaters to recharge aquifers.
2. **Groundwater Management and Recharge**
   - Aquifer Mapping: Identify and delineate recharge zones.
   - Artificial Recharge: Implement measures in over-exploited and stressed aquifers.
   - Spring Conservation: Protect and rejuvenate springs, especially in hilly and tribal regions.
   - Aquifer-Based Management: Develop plans based on aquifer-specific characteristics and potential.
3. **Water Quality Protection and Monitoring**
   - Desalination and Treatment: Installation of WTPs, STPs, and use of appropriate desalination technologies.
   - Pollution Prevention: Prevent chemical runoff and contamination of surface and groundwater.

- Water Recycling and Reuse: Use treated wastewater for irrigation, horticulture, industry, and toilet flushing.
- Monitoring Systems: Regular tracking of water quality in lakes, rivers, and reservoirs.
- Enforcement and Governance: Introduce third-party inspections and strict penalties for polluters.

4. **Water Use Efficiency and Agricultural Practices**
   - Efficient Irrigation: Land leveling, automated systems, and water-efficient irrigation methods.
   - Appropriate Cropping: Promote cropping patterns suited to water availability in each basin or sub-basin.
   - High-Value Crops: Encourage crops that maximize return per unit of water used.
   - Climate-Resilient Agriculture: Use drought-resistant and salt-tolerant crop varieties.
   - Training and Awareness: Capacity building of farmers through ATMA and similar agencies.
   - Water Budgeting: Prepare water use plans at micro (village) to macro (basin/state) scales.
5. **Water Governance, Technology, and Policy Integration**
   - Economic Value of Water: Promote water recycling, reuse, and auditing across all user sectors with appropriate incentives.
   - Technology Adoption: Leverage tools such as AI, ML, remote sensing, GIS, isotopic tracing, real-time monitoring, DSS, and automated design and forecasting systems.
   - Data Synthesis and Forecasting: Improve data collection, integration, and modeling for decision-making and early warning systems for floods and droughts.
   - Open Data Access: All hydrological data (non-classified) should be made publicly accessible, with periodic review for declassification.
   - Institutional Collaboration: Strengthen coordination between central/state agencies and research institutions for technological advancement and field implementation.
   - Participatory Governance: Foster stakeholder involvement through programs like Jal Chetna and Atal Bhujal Yojana, including PPPs in water infrastructure development.
   - Watershed Development: Promote scientific watershed development for conservation and storage enhancement.
   - National Programs Implementation: Effectively implement schemes like Jal Jeevan Mission and Atal Bhujal Yojana.
   - Disaster Preparedness: Focus on early warning systems, flood/drought forecasting, and district/block-level mitigation strategies (Gupta and Ali 2020).

References

Aggarwal M, Anbukumar S, Vijaya Kumar T (2025) Assessment and Insights into heavy metal contamination in sediments of the Ganga River, India: sources, trends and ecological implications. Soil Sediment Contamination Int J 1–28. https://doi.org/10.1080/15320383.2025.2553105

Anita AK, Maneesh KD, Pandey NN.(2015) Scope of ornamental fisheries in snow fed rivers of Uttarakhand—overview. J Appl Nat Sci 7(2):1042–1045

Aon S, Nandi S, Sen S, Biswas S (2024) GRACE-based groundwater drought evaluation of the Ganga Basin and analysis of drought propagation using wavelet-based quantitative approach. Sci Total Environ 951:175666. https://doi.org/10.1016/j.scitotenv.2024.175666

British Geological Survey (n.d.) Southeast Asia groundwater portal. British Geological Survey. https://www2.bgs.ac.uk/groundwater/international/seasiagroundwater/home.html

Bhanja SN, Mukherjee A, Rodell M, Wada Y, Chattopadhyay S, Velicogna I, Pangaluru K, Famiglietti JS (2017) Groundwater stress and groundwater storage variability in India. Environ Res Lett 12(4):044003

Bharadwaj A (2014) Directory of water resources in Himachal Pradesh. State Centre on Climate Change, H.P. State Council for Science, Technology and Environment

Bharati L, Sharma BR, Smakhtin V (2016) Ganges river Basin. Taylor & Francis

Bhat TA (2014) An analysis of demand and supply of water in India. J Environ Earth Sci 4(11):67–72

Bons CA (Ed) (2018) Ganga River basin planning assessment report. Main volume and appendices. Deltares with AECOM and future water for the world bank and the government of India, Report 1220123-002-ZWS-0003

CGWB & UNICEF (2019) Groundwater quality hotspots of Indo-Gangetic Basin

Central Pollution Control Board (2013) Pollution assessment: River Ganga. Central Pollution Control Board. https://cpcb.nic.in/wqm/pollution-assessment-ganga-2013.pdf

Central Water Commission (2021) Water and related statistics 2021. Government of India

CGWB (2022) National Compilation on dynamic ground water resources of India, 2022. Ministry of Jal Shakti, Government of India

CGWB (2023) Dynamic groundwater resources of Haryana State (As on 31st March 2023). CGWB, Chandigarh, and GW Cell, Irrigation & Water Resources Department, Haryana

CGWB (2023) Dynamic Groundwater Resources of Madhya Pradesh (as on 2023). CGWB, NCR, Bhopal, and GWS, WRD, Government of Madhya Pradesh (2023)

CGWB (2024) groundwater resources, 2023, Uttar Pradesh. https://cgwb.gov.in/cgwbpnm/public/download/1242

CGWB (2024a) Dynamic Groundwater Resources of Chhattisgarh (2024) Ministry of Jal Shakti, Department of Water Resources, River Development & Ganga Rejuvenation, Central Ground Water Board, Raipur

CGWB (2024b) Dynamic ground water resources of West Bengal as on 31st March 2024. Central Ground Water Board, Eastern Region, Kolkata

CGWB (2025) Dynamic groundwater resource of Jharkhand. CGWB, SUO, Ranchi, Central Groundwater Board

CGWB (2025) Dynamic groundwater resources of Himachal Pradesh (March 2024). Ministry of Jal Shakti, Department of Water Resources, River Development & Ganga Rejuvenation. Central Ground Water Board, Northern Himalayan Region, Dharamsala (H.P.) and Ground Water Organisation, Jal Shakti Vibhag, Una (H.P.)

Chatterjee R, Ray R (2018) Groundwater potential assessment of the Ganga alluvial plain. Environ Earth Sci 77:623

CPCB (2021) Bio-monitoring of major Indian rivers

CSE (Centre for Science and Environment) (2020) State of India's water and policy framework
Dangar S, Mishra V (2021) Natural and anthropogenic drivers of the lost groundwater from the Ganga River basin. Environ Res Lett 16. https://doi.org/10.1088/1748-9326/ac2ceb
Das GK (2023) River systems of West Bengal—a review. Indian Sci Cruiser 37(2):11–24. https://doi.org/10.24906/isc/2023/v37/i2/223481
Das BK, Bayen S (2023) Ganga: Environmental issues, challenges and restoration initiatives. Sci Cult 89(1–2):34–42
Das S, Dutta S (2023) Spatio-temporal trends in water quality of major Indian rivers. Environ Sci Pollut Res 30:42351–42367
Das A, Kaur H (2019) Anthropogenic stress on Ganga River: A comprehensive review. Environ Monit Assess 191:706
Dharpure JK, Howat IM, Kaushik S (2025) Declining groundwater storage in the Indus basin revealed using GRACE and GRACE-FO data. Water Resour Res 61:e2024WR038279. https://doi.org/10.1029/2024WR038279
DoE & F MP (2022) Madhya Pradesh State Action Plan on Climate Change v 2.0. Department of Environment, Government of Madhya Pradesh
Dutta V, Srivastava S (2019) Urban water stress in Indo-Gangetic cities. Sustain Cities Soc 48:101532
Encyclopaedia Britannica (n.d.) Ganges River: economy. Encyclopaedia Britannica. https://www.britannica.com/place/Ganges-River/Economy
FAO (2021) Climate-resilient agriculture in South Asia
FAO & IWMI (2019) Groundwater management in South Asia
Gain A, Giupponi C (2014) Impact of the Farakka Dam on thresholds of the hydrologic flow regime in the lower Ganges River Basin (Bangladesh). Water 6:2501–2518. https://doi.org/10.3390/w6082501
Ghirardelli A, Tarolli P, Rajasekaran MK, Mudbhatkal A, Macklin MG, Masin R (2021) Organic contaminants in the Ganga Basin: from the green revolution to the emerging concerns of modern India. iScience 24(3):102122. https://doi.org/10.1016/j.isci.2021.102122
Ghosh A, Das S (2021) Hydrogeochemical evaluation of shallow groundwater in the Ganga alluvial plain. Groundw Sustain Dev 14:100623
Gogu R, Carabin G, Hallet V, Peters V, Dassargues A (2001) GIS-based hydrogeological databases and groundwater modelling. Hydrogeol J 9:555–569. https://doi.org/10.1007/s10040-001-0177-0
Government of India, Ministry of Jal Shakti (2021) National Water Policy (Revised Draft). Government of India
Government of India, NMCG (2020) Ganga water quality improvement report 2020. Ministry of Jal Shakti
Govt. of NCT, Delhi (2017) Water policy for Delhi. Prepared by Indian National Trust for Culture and Heritage
GoWB (2020) West Bengal state action plan on climate change (Version 2.0). Department of Environment, Government of West Bengal, Kolkata
GoWB (2021) State of environment report West Bengal 2021, Vol. I. West Bengal Pollution Control Board, Kolkata
GoWB (2024) Water resource availability of West Bengal (composite estimate). State Water Investigation Directorate & Irrigation & Waterways Directorate
Gupta S, Ali S (2020) Water infrastructure degradation in India. Environ Res Lett 15(6):064041
Gupta V, Jain SK (2018) Flood hazards in the Ganga Basin: a geomorphic assessment. Nat Hazards 93:425–440
Gurjar GN, Sanjay-Swami, Singh S, Sheshma MK, Sharma RL (2021) Natural resources of India: water at a glance. In: Sanjay-Swami, Hasan W, Chanyal PC, Singh H, Naz H (eds), Climate

change and its impact on agriculture. Biotech Books, New Delhi, India. pp 105–112. ISBN: 978-81-7622-473-4
HWRA 2025: Haryana Water Resources Atlas (2025). https://hwra.org.in/Atlas2025/HWRAAtlas2025.pdf
ICID (2020) Water productivity enhancement in Indian irrigation systems
International Water Management Institute (2018) Water use efficiency in Indo-Gangetic agriculture
IMD (2024) Rainfall statistics of India-2024, IMD, REPORT NO. ESSO/MoES/IMD/HS/Rainfall Report/02(2025)/66
India Meteorological Department (2008) Climate of West Bengal. National Climate Centre, IMD, Pune (Fig. 6)
India-WRIS (National Water Resources Information System) (n.d.) Barrages and weirs in Ganga Basin. National Water Informatics Centre. https://indiawris.gov.in/wiki/doku.php?id=barrages_weirs_in_ganga_basin
Institution of Civil Engineers (ICE) (n.d.) The Ganges Canal. ICE. https://www.ice.org.uk/what-is-civil-engineering/infrastructure-projects/the-ganges-canal
IRI (2018) A report on state specific action plan (SSAP) for water sector Uttarakhand, Report prepared by Irrigation Research Institute, Roorkee under National Water Mission, Govt of India
Jain CK, Singh S (2020) Impact of climate change on the hydrological dynamics of River Ganga, India. J Water Climate Change 11(1):274–289. https://doi.org/10.2166/wcc.2019.183
Jain SK, Agarwal PK, Singh VP (2007) Hydrology and water resources of India. Springer
Jain SK, Agarwal PK, Kumar P (2021) Hydrology of Himalayan river basins. Springer
Janardhanan S, Nair AS, Indu J et al (2023) Estimation of groundwater storage loss for the Indian Ganga Basin using multiple lines of evidence. Sci Rep 13:1797. https://doi.org/10.1038/s41598-023-28615-y
Jeuland M, Harshadeep N, Escurra J, Blackmore D, Sadoff C (2013) Implications of climate change for water resources development in the Ganges basin. Water Policy 15(S1):26–50. https://doi.org/10.2166/wp.2013.009
Jha SK, Shrestha RR (2021) River health assessment of the Ganga Basin. Int J River Basin Manag 19(4):491–503
Kejriwal R, Sharma P, Singh S (2022) Groundwater sustainability challenges in the Ganges alluvium. Groundw Sustain Dev 17:100724
Krishnan S, Indu J (2023) Assessing the potential of temperature/vegetation index space to infer soil moisture over Ganga Basin. J Hydrol 621:129611. https://doi.org/10.1016/j.jhydrol.2023.129611
Kumar A, Yadav BK, Gupta A (2017) Agricultural water footprint of Ganga Basin cropping systems. Ecol Ind 77:323–331
Kumar M, Das A, Shankar N (2020) Assessment of groundwater recharge methods in Indo-Gangetic alluvium. J Environ Manage 268:110423
Kumar C, Ghosh A, Yash D, Manojit, Bhadury P (2021) Seasonal dynamicity of environmental variables and water quality index in the lower stretch of the River Ganga. Environ Res Commun 3:075008. https://doi.org/10.1088/2515-7620/ac10fd
LUS 2022-–23: Land Use Statistics at a Glance 2022–23. Ministry of agriculture & farmers welfare department of agriculture & farmers welfare economics, statistics and evaluation division New Delhi
Mahajan R, Sharma R (2021) Soil erosion and sedimentation in the Ganga Basin. CATENA 203:105336
Maheswaran R, Khosa R, Gosain AK, Lahari S, Sinha SK, Chahar BR, Dhanya CT (2016) Regional scale groundwater modelling study for the Ganga River basin. J Hydrol 541:727–741. https://doi.org/10.1016/j.jhydrol.2016.07.060
Malik GC, Mishra A (2021) Water accounting of Ganga Basin states. J Environ Manage 285:112170

Mandal KG, Thakur A, Mohanty R, Mishra A, Sinha S, Biswas B (2022) Policy perspectives on agricultural water management. Current Sci 122:386–395. https://doi.org/10.18520/cs/v122/i4/386-395

Matta G, Naik P, Kumar A et al (2018) Comparative study on seasonal variation in hydrochemical parameters of Ganga River water using comprehensive pollution index at Rishikesh (Uttarakhand), India. Desalin Water Treat 118:87–95. https://doi.org/10.5004/dwt.2018.22541

Maurya PK, Malik DS, Yadav KK, Kumar A, Kumar S, Kamyab H (2019) Bioaccumulation and potential sources of heavy metal contamination in fish species in River Ganga basin: possible human health risks evaluation. Toxicol Rep 6:472–481. https://doi.org/10.1016/j.toxrep.2019.03.011

Mausam. https://mausam.imd.gov.in/responsive/rainfallinformation_state.php

Ministry of Agriculture & Farmers Welfare (2022) Irrigation efficiency and agriculture water budget for India

Ministry of Water Resources, River Development & Ganga Rejuvenation (2014) Ganga Basin: Version 2.0. Government of India.

Ministry of Agriculture & Farmers Welfare. Land Use Statistics at a Glance 2023–24. Department of Agriculture & Farmers Welfare, Economics, Statistics and Evaluation Division, New Delhi July (2025)

Ministry of Jal Shakti (2022) Compendium of water recharge structures in India. Government of India

Ministry of Jal Shakti (2023) 1st Census of water bodies—national report. Government of India, New Delhi

Mishra V, Thirumalai K, Jain S (2021) Droughts in the Ganga Basin under future climate. Sci Total Environ 784:147126

Misra AK (2011) Impact of urbanization on the hydrology of Ganga Basin (India). Water Resour Manage 25:705–719. https://doi.org/10.1007/s11269-010-9750-0

Mondal GC, Singh AK, Singh TB (2018) Damodar River Basin: Storehouse of Indian Coal. In: Singh DS (ed), The Indian River. Springer Hydrogeology. https://doi.org/10.1007/978-981-10-2984-4_21

Mukherjee A, Bhanja SN, Ojha C, Kumar B, Scanlon BR, Panda DK (2018) Groundwater depletion in India: implications, management and policies. J Hydro Regional Studies 20:33–52

National Mission for Ganga (2022) Annual report on Ganga River health indicators

Nandimandalam JR (2012) Evaluation of hydrogeochemical processes in the Pleistocene aquifers of middle Ganga Plain, Uttar Pradesh, India. Environ Earth Sci 65:1291–1308. https://doi.org/10.1007/s12665-011-1251-3

National Mission for Clean Ganga (NMCG) (2010) Ganga River Basin Management Plan (GRBMP): Interim Report. NMCG. https://nmcg.nic.in/writereaddata/fileupload/25_GRBMPInterim_Rep.pdf

National Mission for Clean Ganga (NMCG) (2013) Ganga River Basin Management Plan (GRBMP): Interim Report (Vol. 5). NMCG & IIT Consortium. https://nmcg.nic.in/writereaddata/fileupload/25_GRBMPInterim_Rep.pdf

NITI Aayog (2023) District-level water vulnerability assessment

NCRPB (2021) Functional plan for groundwater recharge in national capital region. https://ncrpb.nic.in/pdf_files/FunctionalPlanongroundforNCR.pdf

Niti 2020: Uttar Pradesh State Water Policy (2020). https://www.nitiforstates.gov.in/public-assets/Policy/policy_files/PSSNAE000086.pdf

NRCG (National River Conservation Directorate) (2021) National river conservation plan: Progress report

Pal S, Mallick P, Jha A, Chakraborty SK, Rahaman M, Nandi AK (2023). A Study of Agriculture and Groundwater Utilization in West Bengal. Scientist (Philadelphia, Pa.). 3:550–562. https://doi.org/10.5281/zenodo.7783784

Panda DK, Wahr J (2016) Satellite-based monitoring of groundwater depletion in India. Remote Sens Environ 187:476–485

Pandey PK, Singh VP (2019) Water accounting framework for Ganga's agricultural water use. Agric Water Manag 213:1116–1126

Pandey DN, Singh R (2020) Nature-based solutions for Ganga rejuvenation. Curr Sci 118(8):1235–1243

Patel A, Rai SP, Akpataku KV, Puthiyottil N, Singh AK, Pant N, Singh R, Rai P, Noble J (2023) Hydrogeochemical characterization of groundwater in the shallow aquifer system of Middle Ganga Basin. India. Groundwater Sustain Dev 21:100934. https://doi.org/10.1016/j.gsd.2023.100934

Pew Research & Earth Genome India (2022) Water risk index for Indian districts

PHED, Rajasthan. Making Water Secure Rajasthan Through Improved Water Governance. https://phedwater.rajasthan.gov.in/

Raj A, Kumar P (2020) Emerging pollutants in Indian rivers. Environ Chem Lett 18:1467–1490

Raju NJ (2017) River pollution and restoration in India: a case of the Ganga. Environ Earth Sci 76:509

Raju NJ, Reddy TVK, Sreekanth P (2021) Groundwater quality in the Indo-Gangetic Plains: assessment and implications. Environ Geochem Health 43:1985–2001

Richards LA, Guo S, Lapworth DJ, White D, Civil W, Wilson GJL, Lu C, Kumar A, Ghosh A, Khamis K, Krause S, Polya DA, Gooddy DC (2023) Emerging organic contaminants in the River Ganga and key tributaries in the middle Gangetic Plain, India: characterization, distribution & controls. Environ Pollut 327:121626. https://doi.org/10.1016/j.envpol.2023.121626

Roy P, Datta SP (2023) Water contamination in Indo-Gangetic basin. Environ Pollut 322:121234. https://doi.org/10.1016/j.envpol.2023.121234

Roy A, Ghosh D (2022) Water governance challenges in India. J Water Policy 24(3):345–362

Roy T, Nandi S, Sen S, Pal S (2023) Water quality analysis of selected sites in the Ganges River. Environ Sci Pollut Res 30:10178–10190

Ruchi B, Goura CD, Surya PS, Aftab AU, Shivani B, Srishti B, Syed AH (2025) Mapping hotspots and unveiling drivers of mortality in the endangered Gangetic Dolphin (Platanista gangetica) to mitigate human-mediated conservation conflicts in the Ganga River Basin. India, Ecol Inf 90:103239. https://doi.org/10.1016/j.ecoinf.2025.103239

Russo T, Lall U (2017) Water stress in India: a multi-sectoral review. Hydrol Earth Syst Sci 21:367–378

Saba N, Umar R, Ahmed S (2016) Assessment of groundwater quality of a major industrial city of central Ganga plain, western Uttar Pradesh, India through mass transport modeling using chloride as contaminant. Groundw Sustain Dev 2:154–168. https://doi.org/10.1016/j.gsd.2016.08.002

Sahoo P (2020) Water crisis and its impact on Indian economy. Int J Creative Res Thoughts 8(3):9–15. https://ijcrt.org/papers/IJCRT2003002.pdf

Sarkar S, Banerjee S (2021) Microplastic pollution in the Ganga River: Review and impacts. Sci Total Environ 782:146808

Shaivya A, Hari BS, Brajesh KD (2026) Plastic pollution in the Ganga River: Modeling the town-level impact of population and waste management systems using region-specific parameters. Mar Pollut Bullet 222, Part 1:118577. https://doi.org/10.1016/j.marpolbul.2025.118577

Shamsudduha M (2020) Groundwater storage dynamics in the Himalayan river basins and impacts of global change in the Anthropocene. In: Global groundwater source, scarcity, sustainability, security, and solutions, 47–63.https://doi.org/10.1016/B978-0-12-818172-0.00004-9

Sharma Y, Kumari M (2024) A study on groundwater irrigation expansion in Bihar, India. Asian J Agric Extension Econ Sociol 42(53–60):2024
Shukla AK, Ojha CSP, Shukla S, Garg RD (2021) Water quality challenges in the Ganga River Basin. In: KS (Ed), The Ganga River basin: a Hydrometeorological approach, pp 1–9. Springer
Singh O (2013) Climate change and its impact on groundwater resources in the Ganga Basin. Environ Earth Sci 70:1221–1231
Singh S, Kumar A (2018) Arsenic in the Ganga Basin: Sources, distribution, and health risk. Environ Geochem Health 40(5):2005–2023
Singh RB, Kumar P (2019) Water contamination hotspots in the Ganga. Geograph Rev India 81(3):235–250
Singh M, Rawat S (2023) Assessment of district water balance in northern India. Environ Monit Assess 195:543
Singh R, Singh SK (2020) Spatio-temporal analysis of water quality in the Ganga Basin. Environ Monit Assess 192(4):265
Singh SK, Srivastava P (2020) Water security challenges of the Ganga River Basin. Sustain Water Resour Manag 6:12
SIP 2016-22: State Irrigation Plan, Haryana 2016–22 for implementation of pradhan Mantri Krishi Sinchayee Yojna (PMKSY). NABARD Consult. Pvt. Ltd
SoE-HP (2009) State of the environment report, Himachal Pradesh. Department of Environment, Science and Technology, Government of Himachal Pradesh
Swarnkar S, Mujumdar P, Sinha R (2021) Modified hydrologic regime of upper ganga basin induced by natural and anthropogenic stressors. Sci Rep 11. https://doi.org/10.1038/s41598-021-98827-7
Tare V, Bose P, Mishra A (2020) Status and restoration of Ganga river water quality. J Environ Eng 146(9):04020082
Thadani R, Singh V, Chauhan D, Dwivedi V, Pandey A (2015) Climate change in Uttarakhand: current State of Knowledge and Way Forward. ISBN: 978-81-211-0928-4
Thakur JK, Singh SK, Thomas T (2017) Water security in the Ganga Basin: a geospatial assessment. Water Resour Manage 31(5):1563–1580
UNEP (2020) Water pollution in Asia: assessment and outlook
UP-DES (2023) Uttar Pradesh at a Glance, Economic & Statistics Division, State Planning Institute Planning Department, Uttar Pradesh (updes.up.nic.in)
United Nations University—INWEH (2022) Water quality and sanitation in South Asia
UN ESCAP (2021) Policy pathways for resilient water systems in Asia
United Nations University (2020) Groundwater and climate resilience in the Indo-Gangetic Basin
Water Resources, State of Environment Report, Chhattisgarh. Chhattisgarh Environment Conservation Board. https://www.enviscecb.org/Envis%20data%20sheet/WaterRes_SOE_Chhatisgarh.pdf
Water Management in West Bengal: Challenges, Opportunities, and Interconnections With Consumptive Sectors. Switch On Foundation (2024)
WAPCOS (2018) A journey to water security for all and for ever. Department of agriculture state irrigation plan (SIP) for Rajasthan under PMKSY (2016–17 to 2022–23), Government of Rajasthan
Water Resources Department, Government of Bihar (2022) Bihar water resources report
World Economic Forum (2021) Global water security report
WWF India (2020) State of Himalayan rivers report
Wildlife Institute of India—Ganga Aqualife Conservation Monitoring Centre (WII-GACMC) (2017) Aquatic fauna of the Ganga River: Status and conservation. Wildlife Institute of India, Dehradun
WII (Wildlife Institute of India) (2020) Assessment of aquatic biodiversity in the Ganga River

Wisser D, Frolking S, Douglas EM, Fekete BM, Vörösmarty CJ, Schumann AH (2010) Water use in the Ganges River basin: past trends and projections. Environ Sci Policy 13(8):665–678. https://doi.org/10.1016/j.envsci.2010.09.003
World Bank (2014) Water resources management in the Ganges basin: a regional perspective. World Bank. https://documents.worldbank.org/
World Bank (2015) India: The National Ganga River Basin Project. World Bank. https://www.worldbank.org/
World Bank (2020) Hydrologic and water resources modelling for the ganges basin: a compilation and bibliography. World Bank. https://documents.worldbank.org/
World Bank (2022) Climate resilience in South Asian river basins
WRD-Jh: Water resource Department, Govt of Jharkhand (https://wrd.jharkhand.gov.in/river-basins-of-jharkhand; https://wrd.jharkhand.gov.in/sites/default/files/2022-01/53_15.01.15.pdf)
Zakwan M, Ahmad Z (2021) Trend analysis of hydrological parameters of Ganga River. Arab J Geosci 14(3):163. https://doi.org/10.1007/s12517-021-06453-4
Zeshan US, Saltanat P (2021) Pesticides pollution and risk assessment of river Ganga: a review. Heliyon 7:e07726. https://doi.org/10.1016/j.heliyon.2021.e07726

Stream Flow Impacts of Inter-Basin Water Transfer: A Case Study of the Nethravathi River

5

V. V. Sachin Ramesh, C. Deepa, and H. Ramesh

Abstract

The river basin is the ideal unit of analysis for water audits and hydrological research. One of the most effective and popular methods for hydrological research is basin-scale distributed modeling. Water distribution for different sectors would be seriously unbalanced due to the temporal and geographical volatility of rainfall events throughout the region. Every year, the area receives varying precipitation depths in different parts of the region. Therefore, "inter-basin water transfer" may be necessary to balance regions with excess rainfall against those with deficit rainfall through inter-basin water transfer (IBWT). It can be applied to the upstream of the river and the deltas downstream to provide a sustainable water supply for domestic, agricultural, and environmental needs. The present study is linked to the United Nations sustainable development goals SDG 6 (Clean Water and Sanitation), which addresses water scarcity receiving basin. However, they also have a significant impact on other SDGs, such as SDG 1 (No Poverty) and SDG 2 (Zero Hunger) through increased agricultural productivity, and SDG 12 (Responsible Consumption and Production) through their influence on water resource

V. V. Sachin Ramesh · H. Ramesh (✉)
Department of Water Resources and Ocean Engineering, National Institute of Technology Karnataka, Surathkal, Mangalore, India
e-mail: hramesh@nitk.edu.in

V. V. Sachin Ramesh
e-mail: sachin.jagannath@gmail.com

C. Deepa
Department of Information Technology, National Institute of Technology Karnataka, Surathkal, Mangalore, India
e-mail: deeparamesh88@yahoo.com

S. Yadav et al. (eds.), *Water Resources in Delta Region in India*, Deltas of the World,
https://doi.org/10.1007/978-3-032-06500-1_5

management of the donor basin. However, this could change the donor basin's hydrological and environmental regimes. Inter-basin water transfer can significantly impact the streamflow regime of rivers, as illustrated by studies on the Nethravathi River in Southwest India in the state of Karnataka, which originates in the Western Ghats. The river basin receives an average annual rainfall between 3000 and 5000 mm. However, the eastern parts of Karnataka state, including Kolar, Chikkaballapur, Ramanagar, and Tumkur, receive around 600–800 mm of rainfall. 24 TMC is diverted from this basin in an easterly direction, whereas the flow in Nethravathi is westward. The implications of rerouting 24 TMC of water from the Nethravathi River into the Cauvery basin's beneficiary regions were simulated using the SWAT model. The objective of the study was to examine the diversion of 24 TMC of water from the upper watersheds of the Nethravathi basin on downstream flows of the donor basin. Merely 20 km upstream from the estuary, at the Bantwal gauging station, the ArcSWAT model was verified and calibrated. The coefficient of determination (R^2) and the Nash–Sutcliffe (N-S) coefficient were used to evaluate the model's performance between observed and simulated flows. The N-S values for calibration and validation were found to be 0.88 and 0.891, respectively, with coefficient determination values of 0.91 and 0.9. Using the Flow Duration Curve (FDC), the impact of diverting 24 TMC of water on downstream flows was carefully examined. FDC deviation was analyzed following the predicted diversion. Based on the model findings, the project impacts the higher altitude sub-basins (up to second-order streams) of the Nethravathi basin. The Nethravathi river has not been affected much overall since both the simulated and actual flows were almost identical to those at the Bantwal gauging station. This reason was due to the higher contribution of baseflow into the downstream sub-basins. The results, therefore, indicate that the extreme outflow of the basin is not considerably affected by the inter-basin water transfer of 24 TMC from the Nethravathi river. However, a detailed study integrating ecological and hydrological climate change parameters needs to be carried out to assess the long-term impact.

5.1 Introduction

Rivers and deltas are the primary source of fresh water on the earth's surface. An increase in population and changes in lifestyle result in a drastic increase in water requirements. This made society to explore all possibilities of water availability, leading to an imbalance in river flow. Water is a fundamental need of life, and sufficient water must be provided for drinking, agriculture, industries, and environmental sustainability (Berhanu and Bisrat 2020). Therefore, proper planning and knowledge about water resource availability in the region is a primary objective of any government. Hydrologists recommend river basins as the area of land draining into the river (Davis et al. 1992; Gippel and Stewardson 1198; Chow et al. 1964) and the best scale to study water resources. All the available water is easily accounted for within the river basin for water auditing and budgeting. The hydrologic model uses water balance principles on the basin scale and simulates the runoff at the given outlet. People living along the riverbanks depend on the river, called riparians (Arnold and Allen, 1996). When a dam is built or takes away water from a basin, one should ensure the water requirement of the riparian and should not disturb the river ecology. Thus, the environmental flow should be maintained in the river before any dams are constructed or inter-basin water transfer. The idea of large-scale water transfer is the result of the interest that several researchers have demonstrated in managing water resources. The Ministry of Water Resources and the Central Water Commission developed a National Perspective Plan (NPP) in 1980 to identify several large-scale inter-basin water transfer links to move excess water to basins with water deficits and reduce India's flood and drought issues (Kumar and Verma 2019). Environmental flow in India has historically been defined as the minimal flow that must be released downstream from a dam in order to compensate for riparian rights, without taking into account the ecosystem's influence on the river (Babar and Ramesh 2015; Sharad and Pradeep 2014). With significant climate change issues, people have started considering ecological dependability and sustainability (Acreman et al. 2005). Thus, by entering data on precipitation, temperature, land use, soil, and other factors, hydrological models can simulate runoff at a specific river outlet and consider potential effects on streamflows from dam building and/or inter-basin water transfer (Zing et al. 2021; Rollason et al. 2022). The current study has endeavoured to investigate the impact of water transfer across basins on the downstream flows of the Nethravathi river basin. The Indian state of Karnataka is transferring 24 TMC of inter-basin water from the Nethravathi River to the eastern sub-basin of the Cauvery River basin.

The eastern Cauvery sub-basins receive less rainfall of about 600 to 1000 mm, and the groundwater table has significantly depleted to a depth of more than 1000 feet. In contrast, the Nethravathi basin receives an annual average rainfall of about 3750 mm, and the majority of this water flows joins the ocean after traveling for about 100 km. Therefore, the research has focused more on studying the impacts of the anticipated inter-basin water transfer on the donor Nethravathi river basin. This would help to understand the interrelationship between various hydrological regimes. The present study was more about the water resources than the ecological perspective; hence, ecological issues such as flora, fauna, etc., are not considered.

5.2 Study Area

5.2.1 The Nethravathi River Basin

The Nethravathi river basin in southern India was chosen for study. The River Nethravathi rises in the Kudremukh range of the Western Ghats in Karnataka, India, at a height of 1200 m above sea level. The height of the basin ranges from 0 to 1200 msl. The Nethravathi river flows 103 kms from its source to the sea, draining 3,657 km^2. The most prominent of the Nethravathi River's several significant tributaries is Kumaradhara. Belthangadi Nallas, Shishla Uppar, and Mundaja Neriya are the others. In Fig. 5.1, the Nethravathi river basin lies between longitudes E74°49′08″ and E75°47′53″ and latitudes N12°29′11″ and N13°11′11″.

The coastal tract, midlands, and highlands are the three main regions that make up the Nethravathi river basin. The western coastal tract is a very fragmented terrain. The midland comprises hillocks and mounds and is an undulating landscape. Reserve woods and plantations mostly cover the mountainous areas of the basin's eastern highland zone. Main crops including rice, areca nut, banana, cocoa, cashew, rubber, etc., are heavily grown in the basin's middle reaches (Reghunath et al. 2005). The lowest portions of the basin are also used for the cultivation of paddy and areca nuts. With a population of over 0.8 million people (Census of India 2011) and significant petrochemical and other businesses, the basin's coastline tract is completely urbanized.

About 80% of the 3500 mm of rainfall that the region receives on average comes from the southwest monsoon, which runs from June to October. The Central Water Commission (CWC) measures the Nethravathi river flows at the Bantwal, which is just 20 km upstream from the river mouth. Figure 5.2 displays the Nethravathi River's yearly discharge. This river provides water for about 1 million people, including Mangaluru city, Bantwal, and nearby villages, towns, industries, and agricultural activities. No major dams have been constructed across this river. Still, a few vented dams (dams with openings and operated seasonally) were built across the river to store water to the maximum flood level. The vents will be kept open during the monsoon season so that excess floodwater passes through them. Vents will be closed during the non-monsoon period to store the base flow.

5.2.2 The Proposed Diversion Project

During the monsoon season, the Karnataka government has decided to divert 24 TMC of water from the upper Nethravathi River reaches. This water is being transferred to water-stressed regions of the eastern Cauvery sub-basin, including Kolar, Chikaballapur, Tumkur, Bangalore Rural, Ramanagara, and Bangalore. As a result, this project is called "the Yettinahole diversion project." It aims to divert eight streams that contribute

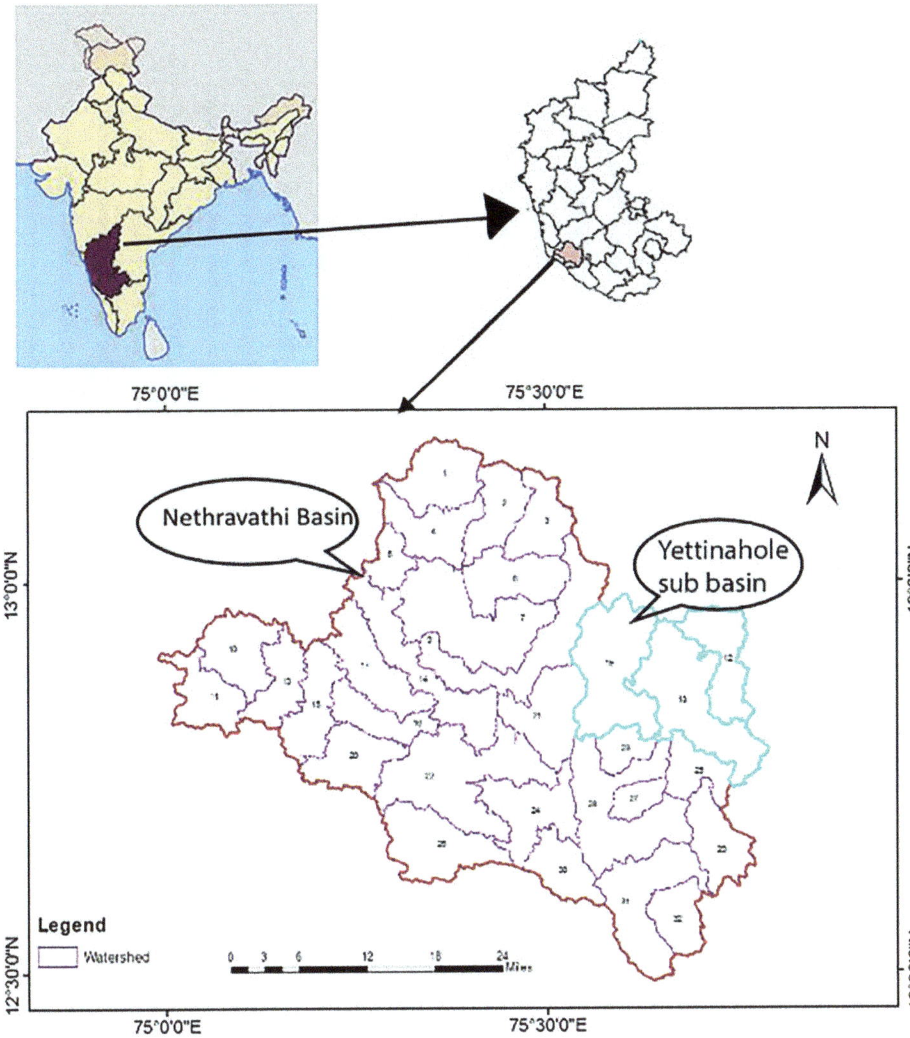

Fig. 5.1 Location of the Nethravathi basin along with sub-basins

to the river Kumaradhara, a sub-basin of the Nethravathi river: Yettinahole, Yettinahole Tributary-1, Yettinahole Tributary-2, Kadumanehole-1, Kadumanehole-2, Yettinahole lower reach, Kerihole, and Hongadahalla hole, as depicted in Fig. 5.3. About 607 km^2, or almost 16% of the whole Nethravathi basin, are covered by the Yettinahole sub-basin where the diversion is being carried out.

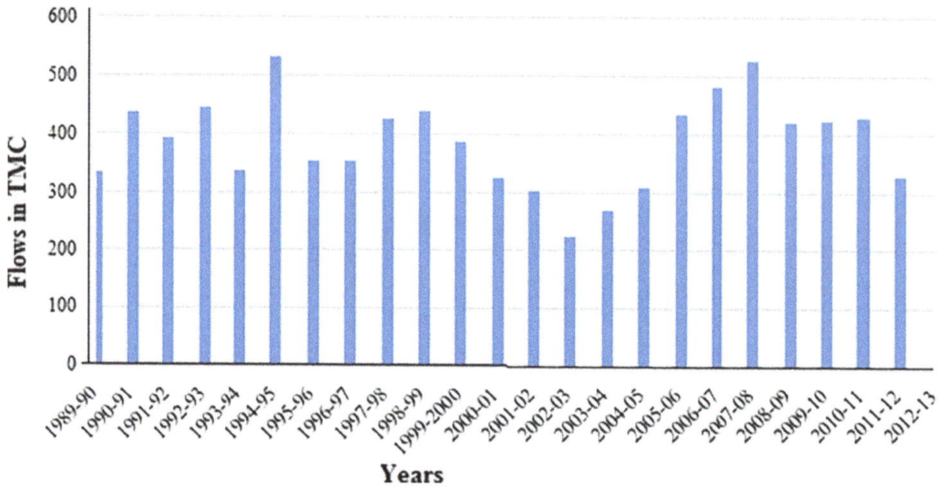

Fig. 5.2 Annual discharge of the Nethravathi river (CWC)

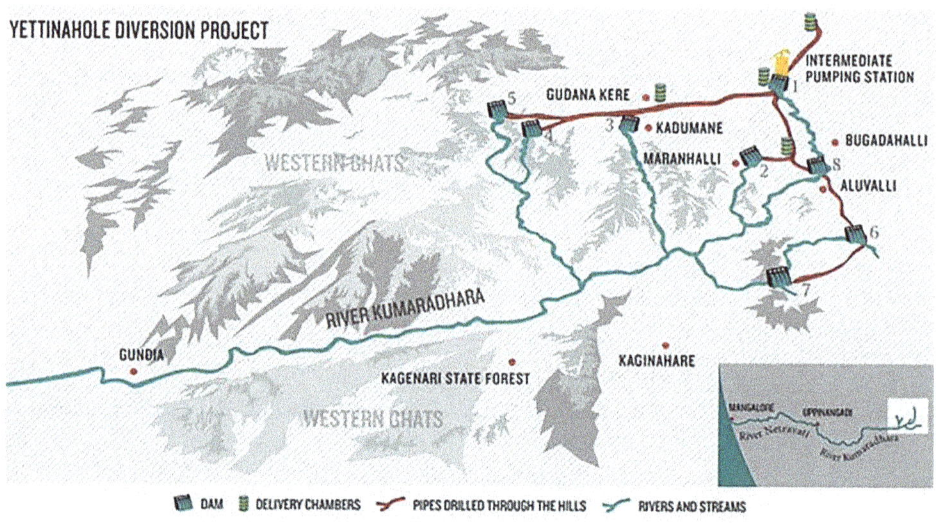

Fig. 5.3 Location of diversion dams in Nethravathi basin (Bhat Shivachandra 2017)

5.2.3 Data Used

Several climate stations in and around the study region provide temperature and precipitation data. 15 m resolution ASTER Digital Elevation Model; USGS Landsat ETM + picture from February 2013; Rainfall data from nine field stations in the basin; soil map from NBSS and LUP and field station data on maximum and lowest temperatures were utilized in this investigation. DEM is processed to generate a streamflow of the region. Ridgelines are identified and the watershed is delineated. The land use map from the Landsat image was classified using "ERDAS Imagine" software (see Fig. 5.4). According to the site condition, it is classified into six classes: evergreen forest, deciduous forest, agricultural land, urban settlement, water and barren land. Seventeen soil classifications are identified in the Nethravathi basin according to the NBSS and LUP. The model uses data from one temperature station and nine rainfall gauge stations spanning 14 years. Calibration and validation are done using streamflow data that is tracked by the Central Water Commission (CWC) at the Bantwal station. A distributed modeling approach is used in the study. The basin area is first delineated from the DEM. It is further subdivided into sub-basins. Sub-basins give the flow pattern and the flow breakup of the desired streams in the river basin. Sub-basins are further classified into hydrological response units (HRUs) based on soil, land use, and slope classes. SWAT was used to develop the hydrologic model of the Nethravathi river basin.

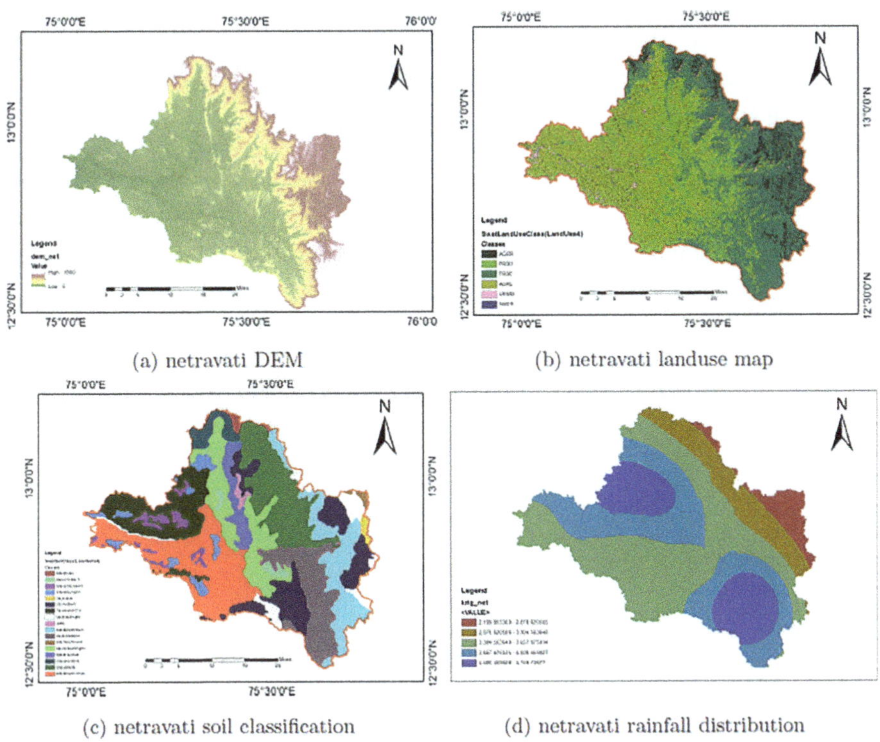

(a) netravati DEM

(b) netravati landuse map

(c) netravati soil classification

(d) netravati rainfall distribution

Fig. 5.4 Hydrological input parameter: DEM, Land use land cover, soil and rainfall

5.3 Methodology

5.3.1 The SWAT Model

SWAT (Soil and Water Assessment Tool) is a continuous, semi-distributed, time-dependent watershed simulator that operates on a daily time step (Arnold et al. 1998). SWAT is frequently used to assess the effects of management and climate change on sediment, water resources, and agricultural chemical yields in watersheds and larger river basins (Abbaspour et al. 2007, 2015). The semi-physically based model allows for the modelling of a high level of spatial complexity by dividing the watershed into many sub-watersheds (Jayakrishnan et al. 2005). The SWAT model divides the river basin into sub-basins, which are further separated into HRUs, based on soil, land, and slope classifications (Jing et al. 2008; Hossein 2010; Hughes and Louw 2010). SWAT provides user choices of equations to find the evapotranspiration, such as Penman' s-Monteith, Priestly Taylor, Hargreaves, or Green-Ampts equation. In this study, the Hargreaves equation was used to find evapotranspiration. It requires less data than other methods, which require solar radiation and relative humidity on daily time steps. SWAT employs the SCS-Curve number to find runoff of the basin. This method is developed by Soil Conservation Services (SCS) of the USA. It is one of the most popular methods used to find the runoff of a basin using Eq. 5.1. The curve number (CN) value, which is determined by parameter S in Eq. 5.2, is dependent on the basin's soil, vegetation and land use complex as well as the antecedent soil moisture state.

$$Q = \frac{(P - \lambda S)^2}{(P + (1 - \lambda S))} \tag{5.1}$$

where P = total precipitation; S = potential maximum retention; λ = factor

$$S = \frac{25400}{CN} - 254 \tag{5.2}$$

Precipitation and the highest and lowest temperatures are the input data needed for SWAT. The soil maps and the land use/cover map are the additional inputs needed. Figure 5.5 indicates the flow chart for the SWAT model used in this investigation.

5.3.2 Model Calibration and Validation

Once the model with a necessary database is developed, the model is ready for simulation. The model will be calibrated to assess its performance. So, the calibration is the process of determining the rate of change of model output for model input parameters (Setegn et al. 2008; Mike et al. 2014). During the calibration process, the model's sensitive parameters

Fig. 5.5 Hydrological SWAT model architecture

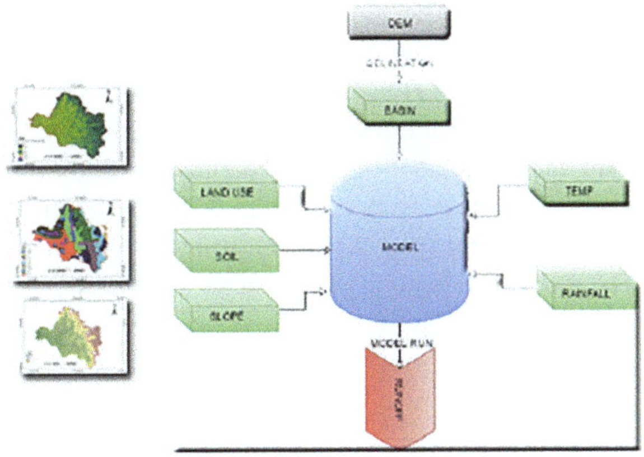

are identified. Model calibration is carried out by carefully choosing model input parameter values and comparing actual data with model predictions (output) for a specific set of output circumstances. Model output uncertainty is measured using the 95% prediction uncertainty (95 PPU), which is computed at the 2.5% and 97.5% levels of the cumulative distribution of output variables produced from Latin Hypercube Sampling. Sequential Uncertainty Fitting 2 (SUFI 2) is used to calibrate the model and represent input parameter uncertainty. Steps involved in SWAT-CUP are the list of parameters that are selected from the SWAT output (Xu et al. 2009). The model was run by fixing the number of simulations required with an input observed streamflow and specifying the calibration period. To evaluate the model's performance, performance indices like R^2, Nash–Sutcliffe efficiency (N-S), etc., will be chosen (Steinfield et al. 2015). Although the NSE is often regarded as the most common performance metric in hydrology (e.g., Klotz et al. 2024; Nelson et al. 2024). It is a popular technique for evaluating the model's functionality. A score of 1 indicates that the model fits the data exactly, whereas a value of 0 indicates that the model predictors are as accurate as the mean of the observed data. The observed mean is a more accurate predictor than the model if N-S is smaller than 0. Nash–Sutcliffe model efficiency coefficient denoted by E, which is calculated by Eq. (5.3).

$$E = 1 - \frac{\sum_{i=1}^{n} [O_i - S_i]^2}{\sum_{i=1}^{n} [O_i - O]^2} \tag{5.3}$$

where O_i = Observed values; S_i = Simulated values; O = Mean of the observed discharge.

Figure 5.6 shows the 95 PPU (Percentage Prediction Uncertainty) graph for the calibration. Here, the blue line shows the estimated value, the red line shows the best prediction of the model, and the green band shows the 95 PPU plot. Estimation is completed monthly in this model. The model is calibrated using monthly average rainfall measurements. Streamflow over the Nethravathi river observed at the Bantwal gauging station between 2001 and 2009 was used to calibrate the SWAT model.

In Table 5.1, the performance coefficients, including R^2 and N-S, are tabulated. Both have shown numbers that make sense when taking into account their true range. The model is verified using data from 2010 to 2012 following the calibration procedure. The process of comparing the model's output with the measured data is known as validation. The model's validation performance is assessed by computing statistical metrics like R^2 and N-S.

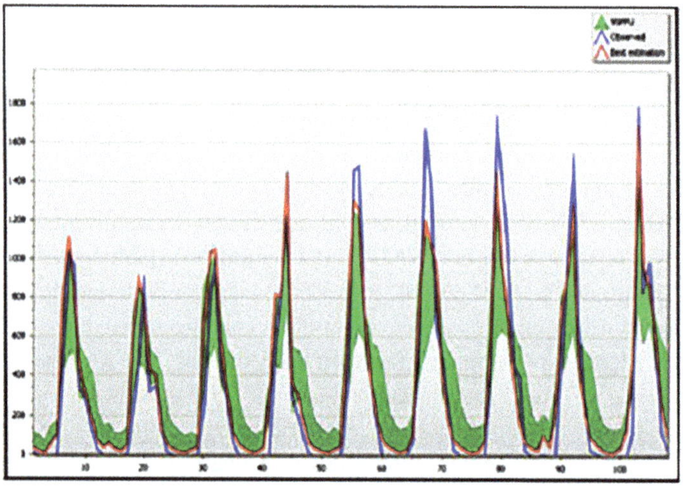

Fig. 5.6 95-PPU graph for the calibration

Table 5.1 SWAT performance coefficients

Performance coefficient	Calibration	Validation	Criteria
R^2	0.90	0.913	0 and 1; A value of 1 indicates a perfect agreement and zero indicates no agreement
N-S	0.89	0.889	$-\infty$ and 1; A value of 1 indicates a perfect agreement and a value of zero indicates that the model does not explain any part of the initial variance

5.4 Results and Discussions

The SWAT model for predicting runoff in the Nethravathi river basin was successfully created, calibrated, and verified. There were thirty-two sub-basins within the river basin. The model can analyze water balance at the sub-basin scale by using the flow at each outlet. The proposed diversion project diverts water from Yettinahole, Yettinahole Tributary 1, Yettinahole Tributary 2, Yettinahole Lower reach, Kadumanehole 1, Kadumanehole 2, Kerihole and Hongadahole areas of the Nethravathi Basin. These eight streams are tributaries of the river Kumaradhara, the major tributary of the river Nethravathi. Sub-basin outlet 19 of the basin provides the combined flow of all these streams located at the Bantwal, a gauging station. The bar chart shown in Fig. 5.7 indicates the flow at outlet 19 from June to November 2001–2013. The proposed project aims to divert 24 TMC of water per year from the basin during monsoon months, covering June to October. The present study has three strategies to divert this 24 TMC in different periods. Considering 5 months from June to October, the quantity of water to divert daily is estimated to be 51 m^3/s. Because it shows the proportion of time equalled or surpassed for a specific flow, the Flow Duration Curve (FDC) is used to evaluate the amount of water available in the sub-basin. Additionally, Table 5.2 contains the sensitivity parameters for the basin.

However, the drawback of FDC is that it does not provide any idea about the temporal variation of the flow, i.e., at what point of the year a particular flow occurs cannot be identified from the FDC. FDC was derived and plotted using monthly average flow for each sub-basin. It is found that the peak flow always occurs during the monsoon season (June–September). As seen in Fig. 5.8, the basin's peak flow only happens in July and

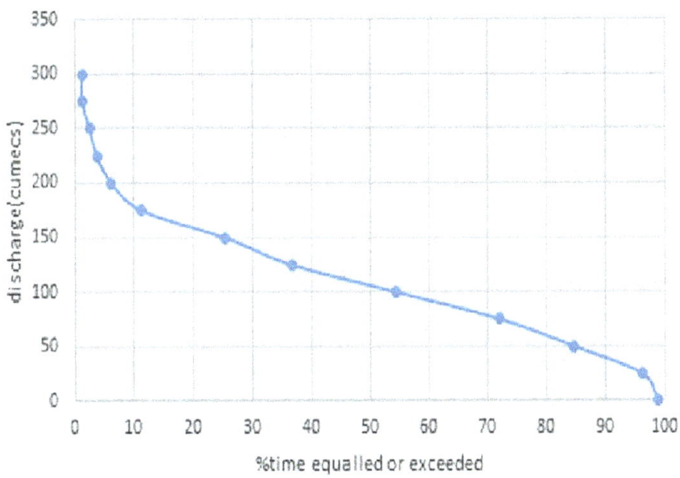

Fig. 5.7 Flow in sub-basin 19 for June to November from 2001–2013

Table 5.2 Sensitive parameters by SWAT

Parameter	Description
CN2 SCS	Curve number for moisture condition 2
GW	Delay Ground water delay time (days)
GW QMN	Threshold depth of water in shallow aquifers required for return flow to occur
SOL AWC	Available water capacity of soil layer
CH N2	Manning's n value for main channel
CH K2	Effective hydraulic conductivity in main channel
ESCO	Soil evaporation compensation factor
EPCO	Plant uptake compensation factor

August. FDC curves were developed for the 24 TMC of water being diverted at a sub-basin level and the whole basin separately. Both FDC's, i.e., before and after the diversion of 24 TMC, were compared. This plot clearly defines the change in the flow of the stream once the project is implemented. Suppose the diversion of water is planned for all six months. In that case, the estimated divertible quantity of water per second per day is about 42.94 m^3/s, which is the minimum, and it is presumed that the rainfall in all six months is expected to occur at least monthly average rainfall. Otherwise, this strategy is very difficult to execute as the monsoon will be completed by October. This strategy can be executed if there is a chance of a post-monsoon occurrence.

Figure 5.9 shows the FDC before and after the diversion at the sub-basin outlet-19 (Yettinahole sub-basin) from June to November. The gap between the two curves shows

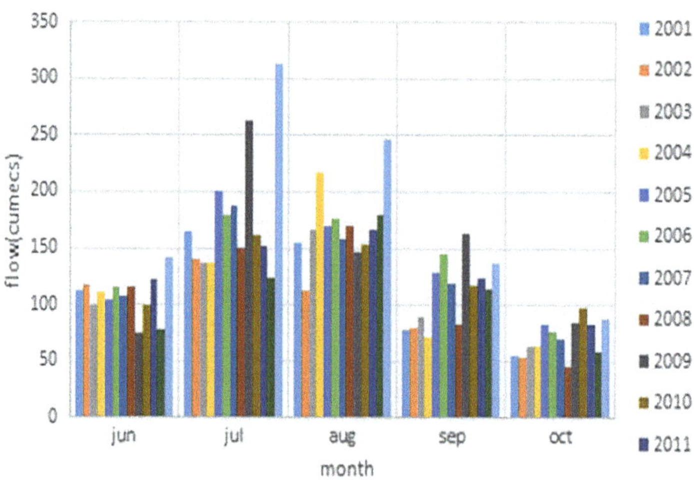

Fig. 5.8 Flows in monsoon from June–October

the change in flow after the diversion. The downstream area of the basin, which lies adjacent to the Yettinahole sub-basin (i.e., sub-basin 19), is the one most affected by the diversion, as indicated by FDC. FDC results have shown a flow greater than or equal to 200 cumecs available for 20% of the time and for 100 cumecs for 40% of the time. The Nethravathi basin is also flood-prone if rainfall exceeds 150 mm over the basin. When the quantity of water being diverted is planned for only 3 months (July, August, September), then the estimated divertible quantity of water is about 85.45 m^3/s, which is the maximum. Further, the FDC of the sub-basins downstream is generated and compared with FDC without diversion and after diversion till the last sub-basin.

The sub-basin immediately downstream of sub-basin 19 is sub-basin 21; for this area, there are no branch streams other than from sub-basin 19. Thus, sub-basin 21 is most affected by the water diversion as it receives water from sub-basin 19. Further downstream, it is seen that the gap between FDC decreases for sub-basin 22 and sub-basin 16. This indicates that the anticipated diversion of 24 TMC does not have much impact downstream. This is mainly because flow to the downstream sub-basin is also from the main Nethravathi river course, which is higher than the river Kumaradhara river. The impact is considerably lower in the last sub-basin downstream, as proved by the gap between the

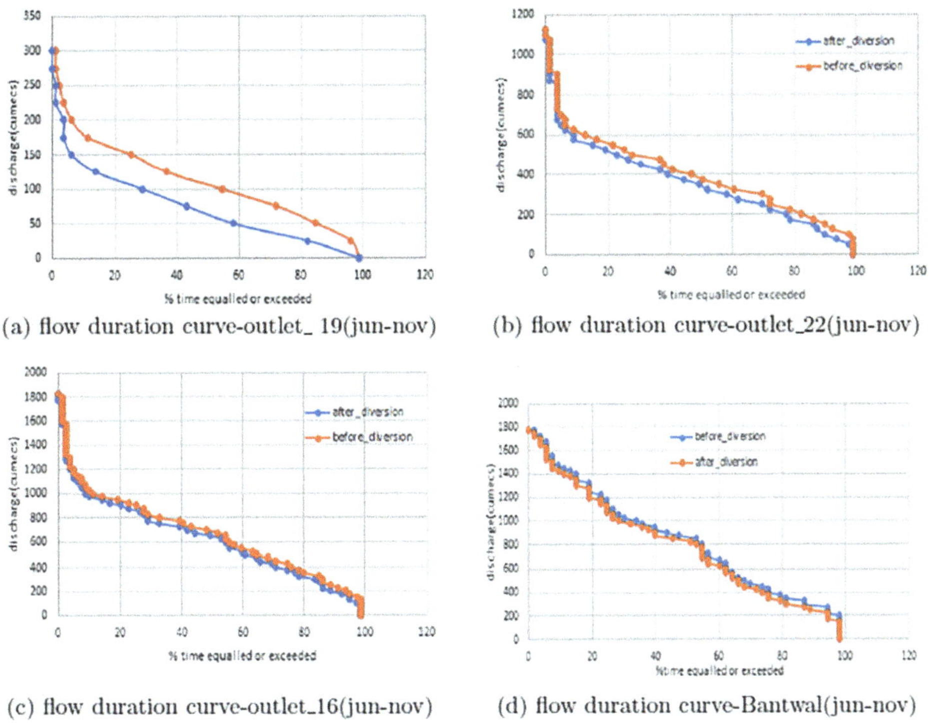

(a) flow duration curve-outlet_ 19(jun-nov)

(b) flow duration curve-outlet_22(jun-nov)

(c) flow duration curve-outlet_16(jun-nov)

(d) flow duration curve-Bantwal(jun-nov)

Fig. 5.9 FDC for diversion at the sub-basin -19

Table 5.3 The volume of water to be diverted

No. of months diversion planned	Pumping/diversion required (m^3/s)
6	42.94
5	51.38
4	63.91
3	85.45

FDC of sub-basin 14 and the whole Nethravathi basin. The flow quantiles are also the same at sub-basin 14 before and after the anticipated diversion. Therefore, the anticipated diversion of 24 TMC has practically no impact on the extreme outlet at Bantwal. Table 4 shows the flow quantiles of the Yettinahole sub-basin (outlet-19). Yettinahole outlet has a flow of 26.28 m^3/s, and the Bantwal outlet has 193.34 m^3/s for the flows from June to November. From the FDC plotted before and after the diversion, it is evident that the project is not affecting the flow at the Bantwal outlet. However, most of the time, the requisite amount of water is unavailable at the Yettinahole outlet. The flow at the Yettinahole sub-basin is analyzed so far. The volume of water available in sub-basin 19 is at 50% time availability between July and September, and the estimated volume of water is about 0.7685 TMC. However, the water required for the project is about 0.6792 TMC. The difference in these volumes of water is 0.0893 TMC, which will remain available in the basin after diversion. The volume of water to be diverted from the Yettinahole watershed is computed and tabulated in Table 5.3.

Flow calculation shows that if diversion is planned for 50% of the time in the monsoon, then the 88.38% volume of the water available in the sub-basin needs to be diverted to gain 0.6792 24 TMC of water.

5.5 Discussions

Any river its flow has to satisfy the needs of not only the people living around but also the flora, fauna, environmental requirements, and the total development of the area. The present study on interbasin water transfers (IBWT) is primarily related to the UN Sustainable Development Goals SDG 6 (Clean Water and Sanitation), which addresses water scarcity receiving basin. However, they also have a significant impact on other SDGs, such as SDG 1 (No Poverty) and SDG 2 (Zero Hunger) through increased agricultural productivity, and SDG 12 (Responsible Consumption and Production) through their influence on water resource management of donor basin.

The quantity of water required for this ecosystem, and the safely divertable quantity, is open to discussion and a further thorough integrated study. This is the limitation of the present study as environmental and biodiversity issues are not considered. However, the proposed project area lies in the evergreen forest area of the Western Ghats, which would cause some damage and needs more studies, as the Western Ghats are protected and declared as hotspots of biodiversity in India. However, on a priority basis, drinking water must be made available through this scheme if there are no other water sources. Since this anticipated diversion project intends to supply drinking water to the drought-prone region in the rain shadow area of the Western Ghats, no alternate water source is available. In order to make up for the harm that the projected diversion would do, the current analysis suggests that the development of the Nethravathi river basin in terms of water conservation be prioritized.

5.6 Conclusions

The goal of the current study is to examine the potential effects of transferring 24 TMC of water from the Nethravathi basin to the Cauvery basin's drought-prone regions in order to supply drinking water. The average rainfall in India is 1083 mm per year. Nethravathi basin receives an average rainfall greater than 3000 mm per year. This rainfall is almost three times higher than the national average. So, compared to the national context, the Nethravathi basin is a rainfall surplus basin. The recipient area of the Yettinahole project is the rain shadow region, where rainfall is less than the national average and is about 700–800 mm. Moreover, the groundwater levels are drastically depleted to a range of 250–400 m, which leads to high concentrations of TDS, fluoride, and other water quality problems. Therefore, the study used the SWAT tool to develop a hydrological model. The limitation of the study is that only hydrological streamflow was studied before diversion and anticipated diversion, and no environmental and ecological parameters were considered. Statistical measures such as the Nash–Sutcliffe coefficient and the coefficient of determination were used to assess the SWAT model's efficacy. N-S values for validation and calibration were 0.889 and 0.89, respectively.

The simulated and real streamflows are the same since the calculated N-S value is almost 1. Similarly, the simulated and real streamflow had calibration and validation R^2 values of 0.91 and 0.90, respectively. As a result, the simulated and real streamflows are nearly identical. Consequently, it can be said that the model's performance is satisfactory in terms of both N-S and R^2. Then, the model was used to predict streamflow for an anticipated divertible quantity of 24 TMC. Three strategies were set in the present study to divert a total quantity of 24 TMC. The first strategy is to divert 24 TMC in three monsoon months: July, August, and September. The divertible flow was estimated to be 85.45 m^3/s, which is the maximum. The second strategy considers five monsoon months from June to October; the quantity of water to divert daily is estimated to be 51 m^3/s. The third strategy is to consider six months from June- November, then the estimated divertible quantity of water per second per day is about 42.94 m^3/s, which is the minimum. The studies show that the second strategy is more suitable than the first or third one. A study has been extended to assess the impact of this 24 TMC at the extreme downstream sub-basin. The results showed little impact, as indicated by FDC at the Bantwal gauging station. However, as proved by FDC, this diversion from sub-basin 19 (Yettinahole) impacts adjoining downstream sub-basins 21 and 22. The study results conclude that the overall impact of the diversion of 24 TMC from sub-basin-19 on the Nethravathi river flow at the Bantwal is insignificant.

5.7 Recommendations

The Soil and Water Assessment Tool (SWAT), coupled with GIS and land use and land cover derived from satellite images, could be effectively used to assess the stream flow due to the diversion of 24 TMC on the downstream.

Based on the results obtained from the SWAT model indicate that the excess outflow of the basin at Bantwal is not considerably affected by the inter-basin water transfer of 24 TMC from the Nethravathi river. However, a detailed study integrating ecological, hydrological and climate change parameters needs to be carried out to assess the long-term impact.

Acknowledgements The authors would like to acknowledge the concerned Government agencies for sharing data for conducting this study.

State Whether You Used AI or not in Preparing AI was not used for preparing the contents of this chapter. The entire chapter is written by the authors based on the results obtained from the study.

References

Abbaspour KC, Yang J, Maximov I, Siber R, Bogner K, Mieleitner J, Zobrist J, Srinivasan R (2007) Using SWAT, modeling hydrology and water quality in the pre-Alpine/Alpine Thur watershed. J Hydrol 333:413–430. https://doi.org/10.1016/j.jhydrol.2006.09.014

Abbaspour KC, Rouholahnejad EVS, Srinivasan R, Yang H, Kløve B (2015) A continental-scale hydrology and water quality model for Europe: Calibration and uncertainty of a high-resolution large-scale SWAT model. J Hydrol 524:733–752. https://doi.org/10.1016/j.jhydrol.2015.03.027

Acreman M (2005) Linking science and decision-making: features and experience from environmental river flow setting, Environ Modell Softw 20:99–109. https://doi.org/10.1016/j.envsoft.2003.08.019

Arnold JG, Allen PM (1996) Estimating hydrologic budgets for three Illinois watersheds. Hydrol 176: 57–77. https://doi.org/10.1016/0022-1694(95)02782-3

Arnold JG, Srinivasan P, Muttiah RS, Williams JR (1998) Large area hydrologic modeling and assessment. Part I. Model development. J American Water Resour Assoc 34:73–89. https://doi.org/10.1111/j.1752-1688.1998.tb05961.x

Babar S, Ramesh H (2015) Streamflow response to land use land cover change over the Nethravathi River Basin, India. J Hydrol Engg ASCE, 05015002–1–11. https://doi.org/10.1061/(ASCE)HE.1943-5584.0001177

Berhanu B, Bisrat E (2020) Alleviating water scarcity in the central rift valley lakes through an Inter-Basin Water Transfer, Ethiopia. Nat Resour 11:554–568. https://doi.org/10.4236/nr.2020.1112033

Bhat Shivachandra MS (2017) What are the environmental impacts of the Yettinahole river diversion projects?, https://www.quora.com viewed on 17th November 2017

Chow T Van (1964) Water-resources technology: handbook of applied hydrology. A compendium of water-resources technology. Ed. McGraw-Hill, New York, 1964. 1418 pp

Davies BR, Thoms M, Meador M (1992) An assessment of the ecological impacts of inter-basin water transfers, and their threats to river basin integrity and conservation aquatic conservation: marine freshwater ecosystems 2:325–49. https://doi.org/10.1002/aqc.3270020404

Gippel C, Stewardson M (1998) Use of wetted perimeters in defining environmental flows. Regul River 14:53–67

Hughes AD, Hannart P (2002) A desktop model used to provide an initial estimate of the ecological in streamflow requirements of rivers in South Africa. J Hydrol 270(3):167–181. https://doi.org/10.1016/S0022-1694(02)00290-1

Hossein T (2010) Evaluation of reference crop evapotranspiration equations in various climates. Water Resour Manage 24(10):2311–2337. https://doi.org/10.1007/s11269-009-9553-8

Hughes DA, Louw D (2010) Integrating hydrology, hydraulics and ecological response into a flexible approach to the determination of environmental water requirements for rivers. J. Environ Model Software 25(8):910–918. https://doi.org/10.1016/j.envsoft.2010.03.004

Jayakrishnan RSRS, Srinivasan R, Santhi C, Arnold JG (2005) Advances in the application of the SWAT model for water resources management. Hydrol Process 19:749–762. https://doi.org/10.1002/hyp.5624

Jing Y, Peter R, Abbaspour KC, Jun X, Hong Y (2008) Comparing uncertainty analysis techniques for a SWAT application to the Chaohe Basin in China. J Hydrol 358:1–23. https://doi.org/10.1016/j.jhydrol.2008.05.012

Klotz D, Gauch M, Kratzert F, Nearing G, Zscheischler J (2024) Technical note: The divide and measure nonconformity. Hydrol Earth Syst Sci, 1–25. https://doi.org/10.5194/hess-28-2895-2024

Kumar N, Verma A (2019) Inter-basin water transfer and policies of water resource management. In: Environmental concerns and sustainable development: Vol 1: Air, Water and Energy Resources (pp 257–274). Singapore: Springer Singapore. https://doi.org/10.1007/978-981-13-5889-0_13

Mike A, Angela AH, Mathew CJ, Carol C, Neville CD, Fiona D, Ian O, Carmel PA, Michael SJ, William Y (2014) Environmental flows for natural, hybrid, and novel riverine ecosystems in a changing world. Front Ecol Environ 12(8):466–473. https://doi.org/10.1890/130134

Nelson AM, Maskey ML, Northup BK, Moriasi DN (2024) Calibrating agro-hydrological model under grazing activities and its challenges and implications. Hydrol 11(4):42 https://doi.org/10.3390/hydrology11040042

Reghunath R, Murthy TRS, Raghavan BR (2005) Time series analysis to monitor and assess water resources: a moving average approach. Environ Monit Assess 109(1–3):65–72. https://doi.org/10.1007/s10661-005-5838-4

Rollason E, Sinha P, Bracken LJ (2022) Interbasin water transfer in a changing world: a new conceptual model. Progress Phys Geogr Earth Environ 46(3):371–397. https://doi.org/10.1177/03091333211065004

Setegn SG, Srinivasan R, Dargahi B (2008) Hydrological modeling in the Lake Tana Basin, Ethiopia using SWAT model. Open Hydrol J 2:49–62. https://doi.org/10.2174/1874378100802010049

Sharad KJ, Kumar P (2014) Environmental flows in India: towards sustainable water management. Hydrol Sci J 59(3–4):751–769. https://doi.org/10.1080/02626667.2014.896996

Steinfield CMM, Kingsford RT, Webster CE, Sharma A (2015) A simulation tool for managing environmental flows in regulated rivers. J. Enviro Model Software 73(C):117–132. https://doi.org/10.1016/j.envsoft.2015.08.006

Xu ZX, Pang JP, Liu CM, Li JY (2009) Assessment of runoff and sediment yield in the Miyun Reservoir catchment by using SWAT model. Hydrol Process 23:3619–3630. https://doi.org/10.1002/hyp.7475

Zhang M, Wang X, Zhou W (2021) Effects of water-saving irrigation on hydrological cycle in an irrigation district of Northern China. Sustainability 13(15):8488. https://doi.org/10.3390/su13158488

Surface Displacement Analysis of Krishna Godavari (KG) Basin Delta Plain Using the Synthetic Aperture RADAR (SAR) and GRACE Sensor Interoperability

6

Akshar Tripathi, Sangeeta Singh, Kapil Malik, and Shailendra Kumar Singh

Abstract

The increased human population has triggered a rise for fuel and energy needs. As a result, new sources of petroleum are being located and identified both on and off-shore worldwide. The KG basin delta region in India is one of the newest sources of petroleum and natural gas in the country and has been a source of huge debate amongst the environmental activists of the country in terms of the exploration and exploitation of fuel oil and gas. The region is bound by the mighty Godavari, and Krishna rivers of peninsular India, and KG basin, is highly fertile and thousands of agricultural families depend upon the crop yield of the region for food and livelihood. The exploitation of petroleum and natural gas in the region has led to the issues of dislocation of many families and is soon to cause unprecedented land subsidence and

A. Tripathi (✉) · S. K. Singh
Department of Civil and Environmental Engineering, Spatial Analytics and Remote Sensing Hub (SPARSH), Indian Institute of Technology (IIT) Patna, Patna, Bihar, India
e-mail: akshar@iitp.ac.in

S. K. Singh
e-mail: shail@iitp.ac.in

S. Singh
CMP Degree College, University of Allahabad, Allahabad, Uttar Pradesh, India
e-mail: sangeeta.phy@cmpcollege.ac.in

K. Malik
Department of Mining Engineering, Indian Institute of Technology (IIT), Indian School of Mines (ISM) Dhanbad, Dhanbad, India

S. Yadav et al. (eds.), *Water Resources in Delta Region in India*, Deltas of the World,
https://doi.org/10.1007/978-3-032-06500-1_6

seawater ingress. The use of active SAR remote sensing has eased the process of surface displacement and subsidence mapping and monitoring. For this study, the C-band Sentinel-1 Differential SAR Interferometry (DInSAR) technique has been used, to estimate the surface subsidence of 11,000 km^2 area of the KG Basin region from 2015 to 2024. The study also relates the findings of the DInSAR-based surface subsidence with Gravity Recovery and Climate Experiment (GRACE) in order to establish good sensor interoperability from 2015 to 2017 and proves a high sensor interoperability with a correlation coefficient value of 0.68. Further, the study goes on to use the gravimetric mean-monthly average values from GRACE and GRACE-FO missions along with the Sentinel-1 DInSAR-based surface subsidence mean monthly values to a Deep Learning Neural Network (DLNN) model to estimate the gravimetric anomaly for the entire study region of the KG Basin in a way carrying out an indirect downscaling. The DLNN model gave satisfactorily with R^2-statistics, of 0.833 and 0.801, in the training and testing stages respectively. In this study, Sentinel-1 SAR and GRACE remote sensing data integration provides an important framework for environmental monitoring under a changing climate and thus aligning well with Sustainable Development Goal (SDG) 13: Climate Action.

6.1 Introduction

The eastern coast of India has always been a rich reserve of minerals and diverse life forms. Most of the Indian rivers originating both from Himalayas and from peninsular India meet their final abode, the sea in the Bay of Bengal. Apart from the numerous deltas located on the eastern coast of India, the KG basin delta is one of the most diverse and most vulnerable from the ecological perspective (Reshma et al. 2025). The KG delta region is undergoing significant surface subsidence due to the overexploitation of petroleum and natural gas (Tripathi et al. 2023a, b). The vulnerability of the KG basin region owing to the present scenario of surfacer subsidence and climate change and global warming-induced sea-water level rise is posing another threat to the vast mlangrove cover that stretches across the region's coastline (Das and Chatterjee 2018). In order to effectively study and monitor the changes happening in the region both due to natural and anthropogenic causes in terms of surface subsidence and seawater ingress, sufficient sensor interoperability is important using multitude of remote sensing data (Tripathi et al. 2022a, b). The active SAR remote sensing based DInSAR LOS surface subsidence has been found to be highly accurate over the years (Das and Chatterjee 2018). However, since the GRACE data is available at a highly coarse spatial resolution, it is important to downscale the gravimetric anomaly values at a higher spatial resolution through indirect downscaling methods using multi-sensor remote sensing data integration (Kumar et al. 2014).

The main driving force behind the surface subsidence being faced by the KG Basin region is the petroleum and natural gas exploitation that began only a decade ago (Richa et al. 2022). Apart from the reserves of petroleum, the region is an agricultural region and due to the underdeveloped canal system, most of the farmers utilize the groundwater from borewells for agriculture and domestic usage, this supplementing the surface subsidence scenario (Reshma et al. 2025), which in turn makes the region even more ecologically sensitive and fragile (Singha et al. 2014). The region is also home to a rich and dense mangrove forest patch along the coastline and in the event of land subsidence and sea level rise over the last few decades, there is also a risk to the precious aquamarine life in these mangroves of the region (Riedel et al. 2010). The numerous risk factors which if unchecked are sure and soon to cause a huge ecological and human disaster in the region (Ramu et al. 2021). The conventional terrestrial surface subsidence measurement techniques are often labor and time-intensive and often cause errors in observations (Mani et al. 2016). On the contrary, the highly calibrated active SAR measurements based on DInSAR have proved to be a highly accurate and reliable tool for measurement and mapping of surface subsidence worldwide (Das and Chatterjee 2017).

The C-band SAR-based DInSAR measurements have a proven track record of making surface subsidence measurements in millimeter levels of accuracy (Tripathi and Tiwari 2021b). The DInSAR- based LOS surface subsidence measurements are made on the basis of the phase differences in SAR signals (Tripathi and Kumar 2018; Tripathi et al. 2018b). Since the KG Basin region on the eastern coast of India receives rainfall from the ongoing and retreating monsoons twice a year, there are very few cloud free time periods in the region, therefore the weather independent SAR datasets become an obvious choice for the mapping monitoring of any changes in the Land Use/Land Cover (LULC) for the region (Kumar et al. 2024a, b). SAR provides the details mapping and measurement of LOS surface subsidence due to its high resolution along with regularity in monitoring owing to the high temporal resolution (Tripathi and Tiwari 2019c). However, the inherent causes and impacts of the surface subsidence are not provided or explained by SAR data alone. Therefore, the SAR data must be integrated with GRACE datasets to study and analyse the changes in gravimetry along with the LOS surface subsidence (Tripathi et al. 2018a, b, c). GRACE data is highly useful to understand issues such as groundwater overexploitation, changes in groundwater, and mass changes in the Earth (Tripathi et al. 2023a, b). The integration of SAR and GRACE data is crucial for the overall analysis of the causes and impacts of the LOS surface subsidence and changes in the gravimetry of different regions which might help in resource conservation and environmental protection (Reshi et al. 2022).

The interoperability of the different remote sensing sensors increases the accuracy, versatility and reliability of the measurements taken by one sensor without much need of ground truth validation (Tripathi et al. 2018a). The synergistic sensor interoperability approach solves the problem of a particular sensor's limitations while giving additional information which gives the researchers an insight into the deeper causes and impacts of environmental issues (Tripathi and Tiwari 2019b). This study paves the way for the interoperability between the SAR and GRACE satellite sensors with a view for indirect data downscaling based gravimetric anomaly estimation at a higher spatial resolution which was otherwise impossible with the highly coarse spatial resolution of the GRACE data alone (Reshi et al. 2023). The study could be useful to take adequate steps to combat the estimated environmental issues that might result from the long term exploitation of the petroleum and natural gas in the KG Basin region (Moniruzzaman et al. 2022). The chapter further brings to the forefront of academic research that development is important, but it must be sustainable and inclusive with the appropriate usage of technology. The chapter also lays the foundation for deeper research into the other impacts and causes of the subsidence of KG Basin region in the coming future with the synergistic integration of more remote sensing datasets.

6.2 Materials and Methods

6.2.1 Study Area

Approximately 11,000 km^2 was taken as the study area in the KG Basin region for this study. The study area is rich in agriculture, rocky terrain, mangrove forests along the coastline with some patches of barren or uncultivated land (varying from season to season). The study area experiences a monsoonal type of climate with rainfall from June to September and again fromDecember to January. The study area map is shown in Fig. 6.1.

6.2.2 Datasets

The details of the datasets used is mentioned in Table 6.1.

6.2.3 Methodology

The time series Sentinel-1 SLC SAR datasets were deburst since the acquisition is done in burst or beam mode. Thereafter the datasets were split to delineate the study area in the respective Interferometric Wide (IW) sub-swath. This reduces the speckle noise in the data and makes the interpretation and feature identification easier. The split datasets were then calibrated to relate the Digital Number (DN) values, to SAR backscatter in -dB (degree decibels). The datasets at 12-day repeat pass were then stacked together to co-register them on a temporal scale with the first image of 2015. Thereafter interferogram formation was carried out by forming the interferometric phase as shown by Eq. 6.1 (Kumar et al. 2024a, b; Rahaman et al. 2025)

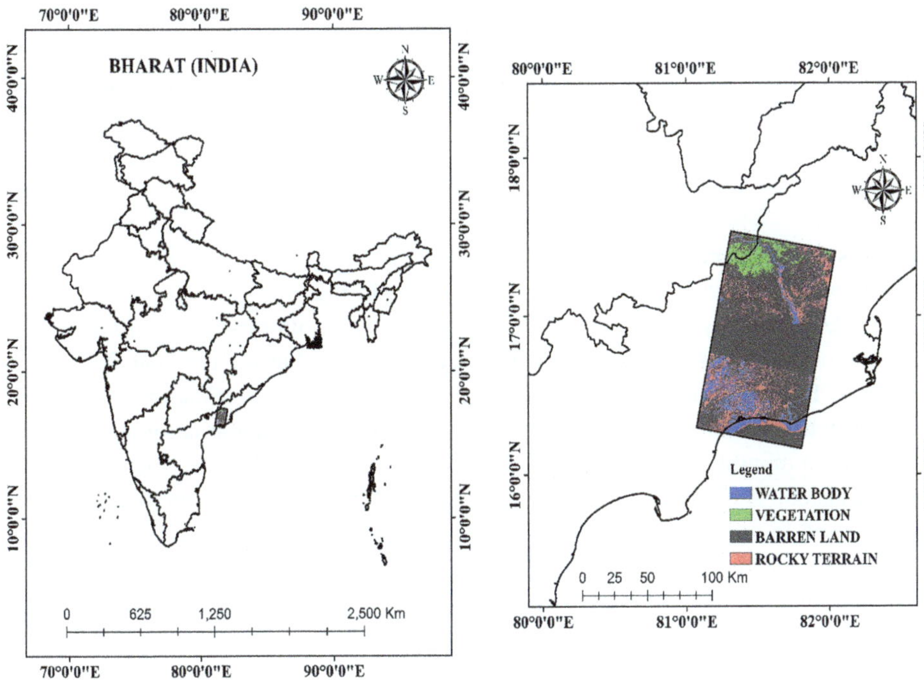

Fig. 6.1 Study area land use/land cover (LULC) map

Table 6.1 Datasets used

Sl. no	Datasets used	Period	Purpose
1	Sentinel-1 Single Look Complex (SLC) data	2015 to 2024	DInSAR based Surface subsidence mapping
2	GRACE data	2015 to 2017	Correlation with surface subsidence
3	GRACE-FO	2018 to 2024	For estimation of the gravimetric anomaly using the DLNN model

$$\varphi_{int} = \frac{4\pi}{\lambda} \cdot B_{\perp} \cdot \frac{h}{r} - \frac{4\pi}{\lambda} \cdot \mathrm{d} \tag{6.1}$$

where φ_{int} is the interferometric phase between two SAR images, λ is the wavelength, $B\perp$ is the perpendicular baseline of the pair of SAR images, h is the height difference, r is the slant range distance, d is the ground displacement in the radar line-of-sight (LOS). The displacement equation along the SAR line of Sight (LOS) is given in Eq. (6.2) (Tripathi 2025)

$$d_{LOS} = \frac{\lambda}{4\pi} \cdot \varphi_{defo} \tag{6.2}$$

The vertical displacement equation is as follows, Eq. (6.3) (Hussain et al. 2025; Kumar et al. 2025).

$$d_{vert} = \frac{d_{LOS}}{cos\theta} \tag{6.3}$$

where θ is the SAR incidence angle.

The overall methodology flow diagram is shown, in Fig. 6.2.

The GRACE data is available till 2017 and post 2017, with the launch of GRACE-FO (GRACE follower) mission, the GRACE FO data is available for the year 2017–18 till date. GRACE and Sentinel-1 have different spatial resolutions, and to estimate the changes in the gravimetric values owing to the sub-surface petroleum exploitation and sea level rise, the mean monthly values of GRACE and Differential SAR Interferometry (DInSAR) were correlated. The correlation check is important for good sensor interoperability that could be used in case of gap in remote sensing data availability by the time a successor mission is launched.

The per-pixel mean monthly values of LOS vertical surface displacement from Sentinel-1 SLC SAR and GRACE-FO from 2018 to 2024 were fed into the DLNN model

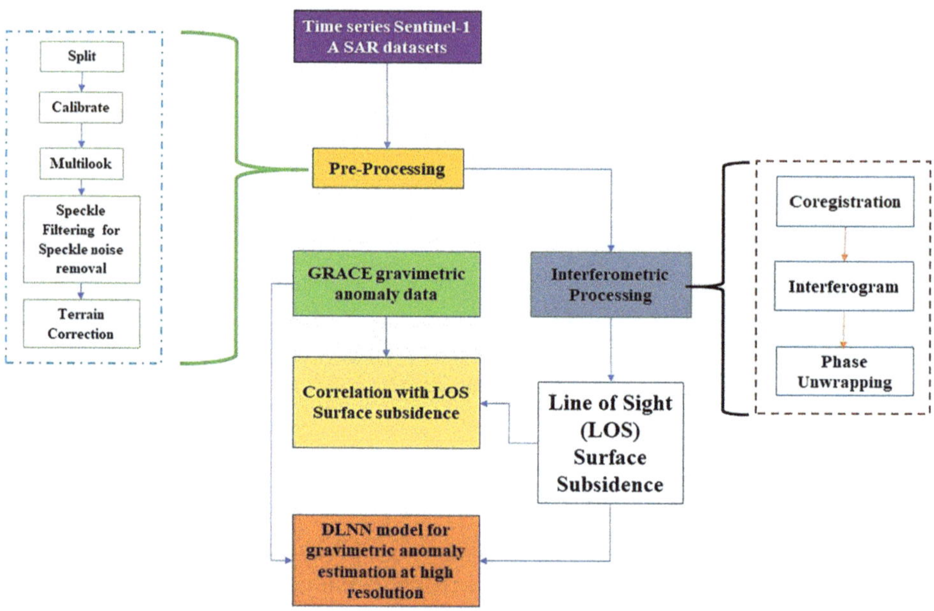

Fig. 6.2 Methodology flow diagram

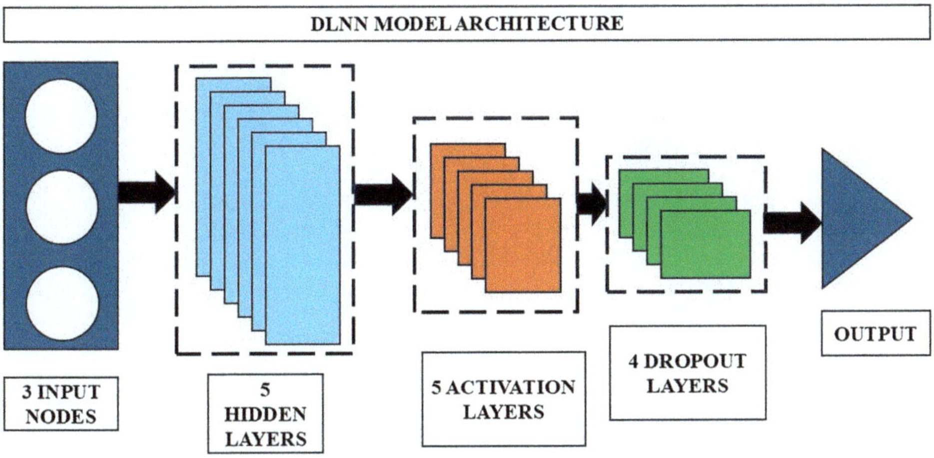

Fig. 6.3 DLNN model for gravimetric anomaly estimation

for the estimation of the gravimetric anomaly in the entire region at 14 m by 14 m spatial resolution (same as that of Sentinel-1). The DLNN model had three input nodes, six hidden layers, five activation layers and four drop-out layers with Rectified Linear Unit (ReLu) as the activation function using a 70:30 split where 70% of the data was used for training the model and 30% for testing and validation. The schematic diagram of the DLNN model is shown in Fig. 6.3.

6.3 Results and Discussion

From the analysis, it was observed that the DLNN model gave sufficient values of R^2-statistics values of 0.833 and 0.801 respectively and Root Mean Squared Error (RMSE), and Mean Absolute Error (MAE), of 0.18 and 0.07 respectively. Also, the correlation results between the observed temporal LOS subsidence, and the gravimetric anomaly values from 2015 to 2017 with mean monthly average values gave a high correlation, thus indicating the changes in the gravimetric values owing to the petroleum and natural gas extraction from the KG basin.

6.3.1 Results

The temporal plots for LOS surface subscience in cm and gravimetric changes (pure ratio hence unit not mentioned) are shown, in Figs. 6.4 and 6.5.

To relate the Sentinel-1 SAR data-based LOS displacements with the gravimetric data, mean monthly values of both data were used. The correlation plot between the mean

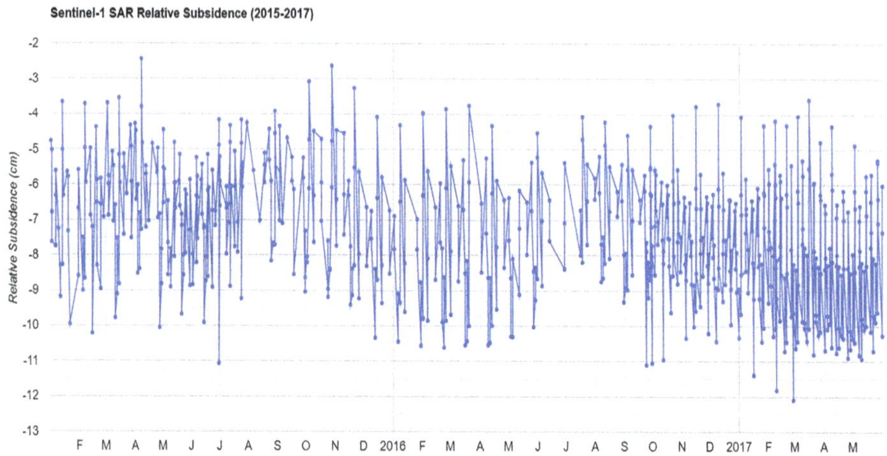

Fig. 6.4 Temporal LOS subsidence (in cm), from 2015 to 2017

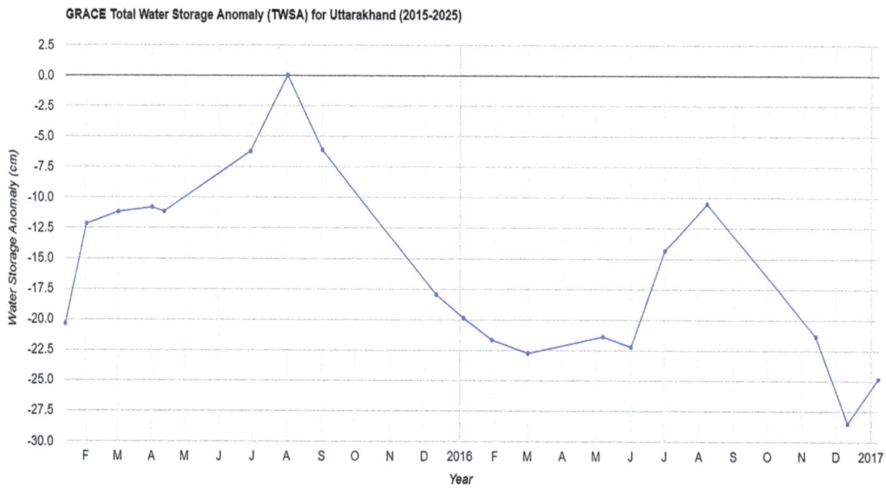

Fig. 6.5 Temporal gravimetric anomaly from 2015 to 2017

monthly gravimetric values from 2015 to 2017 and LOS surface subsidence is shown in Fig. 6.6.

The DLNN model-based gravimetric anomaly map, is shown in Fig. 6.7.

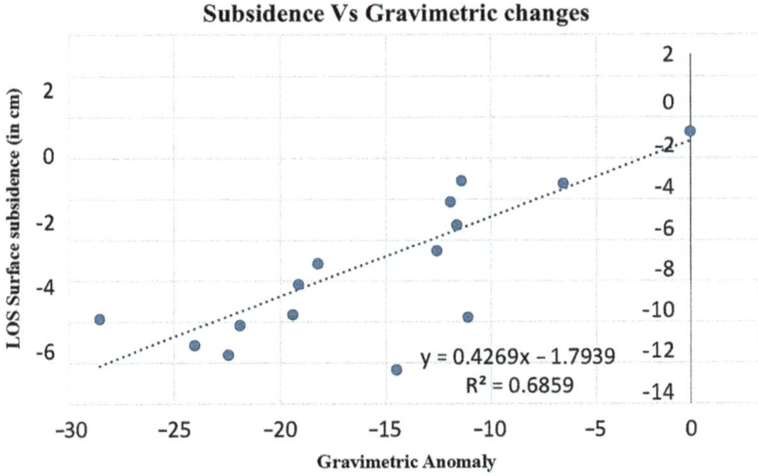

Fig. 6.6 Correlation plot of temporal LOS surface subsidence (in cm) Vs gravimetric anomaly (changes)

6.3.2 Discussion

From the results, it is observed that the KG basin in India is one of the ecologically most sensitive zones that a home to rich flora and fauna. Apart from the biodiversity, the land is fertile along with a rich presence of mangrove forests on the coast. With the oil and natural gas exploration being carried out at an immense pace in the region along with the global sea-level rise, the KG basin region is at a huge risk and is already suffering surface subsidence (Tripathi et al. 2021; Tripathi and Tiwari 2021a), as explained and shown in this study and many other studies (Reshma et al. 2025). Remote sensing is a non-invasive technology which4 has been in use since a long time with varied applications in a number of applicable domains of scientific research (Aziz et al. 2022; Tripathi and Tiwari 2022b). However, since the conventional multi-spectral remote sensing has a disadvantage of being opaque to cloud cover, therefore active SAR remote sensing is utilized for Land Use/ Land Cover (LULC) classification and mapping, as active SAR remote sensing is weather independent and can penetrate through clouds (Tripathi et al. 2022a, b).

Apart from LULC classification, the active SAR remote sensing has been utilized for a number of LOS subsidence studies worldwide (Tripathi and Tiwari 2022a). This study utilized the C-band SAR remote sensing data from Sentinel-1 and estimated the LOS subsidence in the KG Basin, with high accuracy. The study also established an interoperability with GRACE data thus paving the way for high -resolution gravimetric anomaly estimation through indirect data downscaling as shown in Fig. 6.6.

The findings in this study align well and have potential implications in the achievement of United Nations Sustainable Development Goals (SDGs). With the utilization of freely

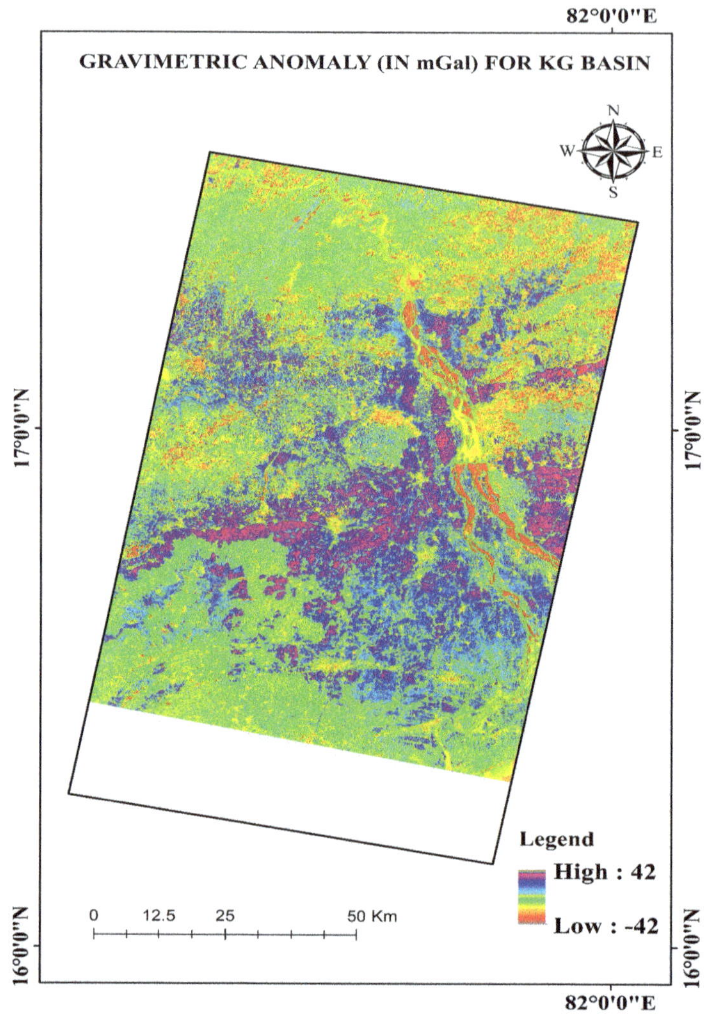

Fig. 6.7 DLNN estimated gravimetric anomaly map for the KG basin

available remote sensing data from Sentinel-1 and GRACE, the study is cost-effective with and accurate (with high correlation between gravimetric changes and surface subsidence), and thus directly aligns with SDG 13: Climate Action, by giving a research framework for climate change induced rising sea-levels and surface subsidence impacts on coastal regions. Further, the study has potential for research directed towards SDG 6: Clean Water and Sanitation where the impacts of overexploitation of groundwater on gravimetric balance and surface subsidence can be analyzed for urban set-ups all over the world.

The focus of this study on the ecologically sensitive mangrove forests and agricultural lands in the KG basin under rising sea-levels and subsiding land surface, also makes this research align with SDG 15: Life on Land.

6.4 Conclusions

The interoperability between the DInSAR-based LOS surface subsidence measurements from SAR and GRACE satellite sensors, as established by this study, provides an important insight into the inherent causes of surface subsidence as well as gravimetric changes. The primary causes for the surface subsidence though appear to be only petroleum and natural gas exploitation from the region. However there might be many other causes, such as the terrestrial water storage changes which need deeper investigation. The study is a landmark in terms of sensor interoperability and synergistic multisensory remote sensing data integration using sophisticated machine learning techniques such as the DLNN model.

This study shows substantial surface subsidence in the region which varies from season and time to time. This subsidence is a combined effect of both petroleum exploitation and groundwater tapping from underground aquifers, which is attributed to a combination of climate change-induced sea-level rise, petroleum exploitation, and groundwater extraction (Aziz et al. 2022; Tripathi and Tiwari 2022b). The DInSAR-based surface subsidence measurements are attributed to highly localized phenomena, but the inherent causes are often spread over long time spans and large areas (Bal et al. 2024; Reshi et al. 2024; Tripathi and Tiwari 2019a). The established interoperability of the SAR data with GRACE further concretes the findings in terms of the seasonal groundwater storage changes and petroleum exploitation in the region (Tripathi et al. 2018a, b, c). The sensors' interoperability establishes the versatility and reliability of the results, which show that the groundwater storage changes and petroleum exploitation is one of the many causes of the gravimetric anomalies and associated surface subsidence variations (Tripathi et al. 2018a).

The novel features of the study can be summarized as:

- The study establishes the first of its kind interoperability between the SAR and GRACE satellite sensors, and paves for research beyond the obvious causes of surface subsidence.
- The study indirectly downscales the gravimetric changes to a higher resolution as compared to the coarse resolution data of the GRACE satellite sensor.
- It is a cost-effective study, as it utilizes the freely available satellite data from Sentinel-1 and GRACE, and reduces the ground truth validation.

Ultimately, the study establishes a technological foundation for evidence-based policy framing, that is essential for the advancement of key SDGs related to climate action, water security and life on land.

6.5 Recommendations

Based on the findings of this study, the following recommendations are proposed for policymakers, researchers and environmental enthusiasts-

- Continuous remote sensing-based monitoring of the KG basin using the proposed methodology.
- Integrate the subsidence and gravimetric change data with climate data parameters for better analysis of the climate change impacts in the KG basin region.
- Integration of multi-sensor remote sensing and ground-based datasets for an early warning disaster management and ecological monitoring system for KG basin and a novel Coastal Vulnerability Index (CVI).
- Enforcement of stricter regulations in terms of groundwater exploitation and petroleum and natural gas extraction from the KG basin region.
- Create open-source data repository for further research and detailed Land Use/Land Cover (LULC) change analysis of the region at regular intervals.

Acknowledgements The work was supported by the Spatial Analytics and Remote Sensing Hub (SPARSH), Indian Institute of Technology (IIT) Patna.

State Whether You Used AI or not in Preparing There was **NO AI tool** used in the chapter for drawing or creating any figures or in generation of results. In the writing of the text and grammar check, Microsoft Office Copilot and Grammarly were utilized at few places where sentences needed clarity or had grammatical errors.

Author Contribution Akshar Tripathi: Conceptualization, Methodology, Software, Writing-Reviewing and Editing.

Sangeeta Singh: Data curation, Writing- Original draft preparation.

Kapil Malik: Figures Preparation, Software.

Shailendra Kumar Singh: Validation, Editing.

Data Availability The data would be made available by the corresponding author upon reasonable request.

Declarations

Ethics Approval All authors have read, understood, and complied as applicable with the statement on "Ethical responsibilities of Authors" as found in the Instructions for Authors and are aware that, with minor exceptions, no changes can be made to authorship once the chapter is submitted.

Competing Interests The authors declare no conflict of interest.

References

Aziz MA, Moniruzzaman M, Tripathi A, Hossain MI, Ahmed S, Rahaman KR, Rahman F, Ahmed R (2022) Delineating flood zones upon employing synthetic aperture data for the 2020 flood in Bangladesh. Earth Syst Environ. https://doi.org/10.1007/s41748-022-00295-0

Bal A, Misra S, Sen D (2024) Nanopore heterogeneity and accessibility in oil and gas bearing cretaceous KG (Raghampuram) Shale, KG Basin, India: an advanced multi-analytical study. Nat Resour Res 33(3):1131–1154. https://doi.org/10.1007/s11053-024-10319-3

Das B, Chatterjee R (2017) Wellbore stability analysis and prediction of minimum mud weight for few wells in Krishna-Godavari Basin, India. Int J Rock Mech Mining Sci 93:30–37. https://doi.org/10.1016/j.ijrmms.2016.12.018

Das B, Chatterjee R (2018) Well log data analysis for lithology and fluid identification in Krishna-Godavari Basin. India. Arabian J Geosci 11(10):231. https://doi.org/10.1007/s12517-018-3587-2

Hussain SA, Tripathi A, Tiwari SP, Reshi OR, Rahman SM, Kumar P (2025) Climate change induced vulnerability analysis of the Saudi coast: a multi-parametric remote sensing approach. Environ Earth Sci 84(12):337. https://doi.org/10.1007/s12665-025-12297-y

Kumar P, Attri L, Ranjan S, Tripathi A, Verma AK, Kumar P, Singh TN (2024) A novel modified goldstein phase filter for psinsar based line of sight (los) displacement mapping of Jakarta, Indonesia. In: 2024 IEEE India geoscience and remote sensing symposium (InGARSS), pp 1–4. IEEE. https://doi.org/10.1109/InGARSS61818.2024.10984216

Kumar P, Chandrakant SC, Ranjan S, Tripathi A (2025) Agricultural drought assessment using deep learning and multi-sensor remote sensing data integration. Phys Chem Earth Parts a/b/c 104006. https://doi.org/10.1016/j.pce.2025.104006

Kumar P, Collett TS, Boswell R, Cochran JR, Lall M, Mazumdar A, Ramana MV, Ramprasad T, Riedel M, Sain K, Sathe AV, Vishwanath K, Yadav US (2014) Geologic implications of gas hydrates in the offshore of India: Krishna–Godavari Basin, Mahanadi Basin, Andaman Sea, Kerala–Konkan Basin. Marine Petrol Geol 58:29–98. https://doi.org/10.1016/j.marpetgeo.2014.07.031

Kumar P, Tripathi A, Moniruzzaman M, Chandrakant SC (2024b) Multi-Sensor remote sensing data integration for agricultural droughts: a PRISMA-based review in the Indian context. Earth Syst Environ. https://doi.org/10.1007/s41748-024-00540-8

Mani D, Ratnam B, Kalpana MS, Patil DJ, Dayal AM (2016) Elemental and organic geochemistry of Gondwana sediments from the Krishna–Godavari Basin, India. Geochemistry 76(1):117–131. https://doi.org/10.1016/j.chemer.2016.01.002

Moniruzzaman M, Uddin MS, Akhter MA, Tripathi A, Rahaman KR (2022) Application of geospatial techniques in evaluating spatial variability of commercially harvested mangoes in Bangladesh. Sustainability 14(20). https://doi.org/10.3390/su142013495

Rahaman KR, Moniruzzaman M, Islam GT, Hasan MM, Tripathi A (2025) Post-DORIAN forest damage assessment in the prince Edward island national park, Canada employing multi-sensor satellite data. Remote Sens Appl Soc Environ 101729. https://doi.org/10.1016/j.rsase.2025.101729

Ramu C, Sunkara SL, Ramu R, Sain K (2021) An ANN-based identification of geological features using multi-attributes: a case study from Krishna-Godavari basin. India. Arabian J Geosci 14(4):299. https://doi.org/10.1007/s12517-021-06652-z

Reshi AR, Moniruzzaman M, Tripathi A, Tiwari RK, Rahaman KR (2022) A remote sensing based study of tropospheric ozone concentration amid COVID-19 lockdown over India using Sentinel-5P satellite data. Geocarto Int 1–20. https://doi.org/10.1080/10106049.2022.2123957

Reshi AR, Pichuka S, Tripathi A (2024) Applications of Sentinel-5P TROPOMI satellite sensor: a review. IEEE Sens J 1. https://doi.org/10.1109/JSEN.2024.3355714

Reshi AR, Sandhu HA, Cherubini C, Tripathi A (2023) Estimating land subsidence and gravimetric anomaly induced by aquifer overexploitation in the Chandigarh Tri-City Region, India by coupling remote sensing with a deep learning neural network model. In Water (Vol 15, Issue 6). https://doi.org/10.3390/w15061206

Reshma KN, Murali RM, Kumar SS, Agrawal R, Ramakrishnan R, Rajawat AS (2025) Spatial variability of ground deformation of coastal regions of the Krishna Delta, East Coast of India Using SAR Interferometry. J Indian Soc Remote Sens 53(1):149–167. https://doi.org/10.1007/s12524-024-02026-6

Richa, Maurya SP, Singh KH, Singh R, Kumar R, Kushwaha PK (2022) Application of maximum likelihood and model-based seismic inversion techniques: a case study from K-G basin, India. J Petrol Exploration Prod Technol 12(5):1403–1421. https://doi.org/10.1007/s13202-021-01401-0

Riedel M, Collett TS, Kumar P, Sathe AV, Cook A (2010) Seismic imaging of a fractured gas hydrate system in the Krishna–Godavari Basin offshore India. Marine Petrol Geol 27(7):1476–1493. https://doi.org/10.1016/j.marpetgeo.2010.06.002

Singha DK, Chatterjee R, Sen MK, Sain K (2014) Pore pressure prediction in gas-hydrate bearing sediments of Krishna–Godavari basin, India. Marine Geol 357:1–11. https://doi.org/10.1016/j.margeo.2014.07.003

Tripathi A, Kumar S (2018) Effect of phase filtering on interferometry based displacement analysis of cultural heritage sites. 2018 5th IEEE Uttar Pradesh section international conference on electrical, electronics and computer engineering (UPCON), 1, 1–5. https://doi.org/10.1109/UPCON.2018.8597027

Tripathi A (2025) Remote sensing-based analysis of methane (CH_4) and ozone (O_3) concentration variations with agricultural growth: a multivariate support vector approach. Environ Monit Assess 197(7):1–21. https://doi.org/10.1007/s10661-025-14232-x

Tripathi A, Attri L, Tiwari RK (2021) Spaceborne C-band SAR remote sensing–based flood mapping and runoff estimation for 2019 flood scenario in Rupnagar, Punjab, India. Environ Monitoring Assess 193(3). https://doi.org/10.1007/s10661-021-08902-9

Tripathi A, Kumar S, Maithani S (2018a) Spaceborne bistatic polarimetrie SAR for scattering analysis and classification of man-made and natural features. 2018 3rd international conference on microwave and photonics (ICMAP), pp 1–2. https://doi.org/10.1109/ICMAP.2018.8354494

Tripathi A, Maithani S, Kumar S (2018b) Minimization of the ambiguity of merging of urban builtup and fallow land features by generating 'C2' covariance matrix using Spaceborne Bistatic dual pol SAR data. IEEE XPLORE. https://doi.org/10.1109/RAIT.2018.8389058

Tripathi A, Maithani S, Kumar S (2018b) X-band persistent SAR interferometry for surface subsidence detection in Rudrapur City, India. Proc SPIE, 10793. https://doi.org/10.1117/12.2326267

Tripathi A, Malik K, Reshi AR, Moniruzzaman M, Tiwari RK (2023a) Multi-temporal SAR Interferometry (MTInSAR)-based study of surface subsidence and its impact on Krishna Godavari (KG) basin in India: a support vector approach. Environ Monit Assess 195(11):1298. https://doi.org/10.1007/s10661-023-11896-1

Tripathi A, Moniruzzaman M, Reshi AR, Malik K, Tiwari RK, Bhatt CM, Rahaman KR (2023b) Chamoli flash floods of 7th February, 2021 and recent deformation: a PSInSAR and deep learning neural network (DLNN) based perspective. Nat Hazards Res. https://doi.org/10.1016/j.nhres.2023.03.003

Tripathi A, Reshi AR, Moniruzzaman M, Rahaman KR, Tiwari RK, Malik K (2022a) Interoperability of C-band Sentinel-1 SAR and GRACE satellite sensors on PSInSAR based urban surface subsidence mapping of Varanasi. India. IEEE Sensors J 1. https://doi.org/10.1109/JSEN.2022.3208117

Tripathi A, Tiwari RK (2019a) C-band SAR Interferometry based flood inundation mapping for Gorakhpur and adjoining areas. 2019 International conference on computer, electrical & communication engineering (ICCECE), 1–6. https://doi.org/10.1109/ICCECE44727.2019.9001870

Tripathi A, Tiwari RK (2019b) Mapping of deflection caused due to hydrostatic pressure using Differential SAR Interferometry (DInSAR) on Bhakhra dam. IEEE XPLORE. https://doi.org/10.1109/UPCON47278.2019.8980117

Tripathi A, Tiwari RK (2019c) Utilization of spaceborne C-band SAR data for analysis of flood impact on agriculture and its management. International archives of the photogrammetry, remote sensing and spatial information sciences—ISPRS Archives, 42(3/W6). https://doi.org/10.5194/isprs-archives-XLII-3-W6-521-2019

Tripathi A, Tiwari RK (2021a) A simplified subsurface soil salinity estimation using synergy of SENTINEL-1 SAR and SENTINEL-2 multispectral satellite data, for early stages of wheat crop growth in Rupnagar, Punjab, India. Land Degradation Dev 32(14):3905–3919. https://doi.org/10.1002/ldr.4009

Tripathi A, Tiwari RK (2021b) Role of space-borne remote sensing technology for monitoring of urban and environmental hazards. In: Rai PK, Singh P, Mishra VN (Eds), Recent technologies for disaster management and risk reduction: sustainable community resilience & responses, pp 295–317. Springer International Publishing. https://doi.org/10.1007/978-3-030-76116-5_18

Tripathi A, Tiwari RK (2022a) Synergetic utilization of sentinel-1 SAR and sentinel-2 optical remote sensing data for surface soil moisture estimation for Rupnagar, Punjab. India. Geocarto Int 37(8):2215–2236. https://doi.org/10.1080/10106049.2020.1815865

Tripathi A, Tiwari RK (2022b) Utilisation of spaceborne C-band dual pol Sentinel-1 SAR data for simplified regression-based soil organic carbon estimation in Rupnagar, Punjab, India. Adv Space Res 69(4):1786–1798. https://doi.org/10.1016/j.asr.2021.08.007

Tripathi A, Tiwari RK, Tiwari SP (2022) A deep learning multi-layer perceptron and remote sensing approach for soil health based crop yield estimation. Int J Appl Earth Observ Geoinf 113:102959. https://doi.org/10.1016/j.jag.2022.102959

7 Water, Inequality, and Development in India's Delta Regions: A Socio-Ecological Analysis

Sahil Wadhwa, Amrit Kumar, and Abul Amir Khan

Abstract

Water scarcity and unequal access are among the most pressing challenges of the twenty-first century, with profound implications for global food security, public health, and social stability. Delta regions, which support millions of people and serve as critical ecological and agricultural hubs worldwide, are especially vulnerable to climate change, environmental degradation, and governance failures. This chapter explores the complex relationships among water resource management, social inequality, and economic development across four major Indian deltas: the Kaveri, Krishna-Godavari, Mahanadi, and Sundarbans. Analysis of secondary data from government reports, water resource studies, and scholarly research reveals how these ecologically vital and agriculturally productive landscapes face mounting pressures from climate change, environmental degradation, and socio-political marginalization. The research demonstrates that water-related vulnerabilities in these regions stem not merely from physical scarcity, but from structural inequities embedded in governance frameworks that disproportionately impact lower-caste communities, women, and gender minorities. The chapter systematically examines how sectoral water use across agriculture, aquaculture, industry, and domestic domains shapes economic outcomes and creates differentiated

S. Wadhwa · A. Kumar · A. A. Khan (✉)
Amity Centre for Air Pollution Control (ACAPC)/Amity Centre for Ocean-Atmospheric Science and Technology (ACOAST), Amity University Haryana, Gurugram 122413, India
e-mail: aakhan@ggn.amity.edu

S. Wadhwa
e-mail: swadhwa@ggn.amity.edu

A. Kumar
e-mail: akumar20@ggn.amity.edu

S. Yadav et al. (eds.), *Water Resources in Delta Region in India*, Deltas of the World,
https://doi.org/10.1007/978-3-032-06500-1_7

vulnerabilities. Each delta faces distinct challenges: the Kaveri suffers with declining freshwater availability and inter-state conflicts; the Krishna-Godavari from fractured governance and rising coastal salinity; the Mahanadi struggles from industrial competition for water and frequent flooding; while the Sundarbans endures accelerated sea-level rise and livelihood disruptions. Further, the study emphasizes how intersections of caste and gender aggravate water access disparities, leaving underprivileged groups to bear the greatest environmental hazards and resource constraints. Finally, the chapter advocates for measures to support fair and sustainable water management, advancing environmental justice, integrating multiple knowledge systems, and promoting participatory decision-making. Although such approaches face considerable challenges, they offer pathways to reduce entrenched inequalities and enhance the ecological and economic resilience of India's delta ecosystems.

7.1 Introduction

Water, the lifeblood of human civilization and a fundamental natural resource, is increasingly becoming a source of inequity and vulnerability in India, particularly in its ecologically fragile and densely populated delta regions. (Kumar and Saleth 2018; Bera et al. 2022). In a country where nearly 60% of the population depends on agriculture and where water-intensive sectors such as industry and aquaculture are rapidly expanding, freshwater systems face mounting pressures (Jayasankar 2018; Adhikari et al. 2018). These pressures are especially acute in India's deltaic regions, areas that are agriculturally productive, ecologically sensitive, and densely populated, yet highly vulnerable to water stress, pollution, salinization, and administrative neglect (Rahman et al. 2019; Nath et al. 2021; Gupta et al. 2024). Climate change and environmental damage aggravate these weaknesses even more, so generating a vicious cycle of risk, particularly for underprivileged populations.

This chapter focuses on four major delta regions in India, the Kaveri, Krishna-Godavari, Mahanadi, and Sundarbans deltas to examine how water resource management and distribution intersect with economic development and social inequality. While delta regions are hydrologically dynamic and agriculturally rich, they remain ecologically fragile and home to millions who depend on water for farming, fishing, domestic use, and industry (Lauria et al. 2018; Rahman et al. 2020; Singha and Pal 2023). But these areas experience seasonal water scarcity, saline intrusion, and flooding—vulnerabilities exacerbated by poor government and climate change (Nath et al. 2021; Rahman et al. 2019; Gupta et al. 2024).

Recent research highlights structural mismanagement, ineffective allocation, and sociopolitical exclusion as well as physical scarcity driving India's water crisis (Mehta 2005; Joy et al. 2020). These issues disproportionately affect marginalized communities, particularly women, lower-caste groups, and transgender individuals, with caste hierarchies often determining access to wells and public water infrastructure, and gender norms assigning water collection burdens to women, exposing them to health risks, unpaid labor, and exclusion from decision-making (Agarwal 2010; Narain 2014). Lack of gender-inclusive facilities and social stigma cause transgender people to be excluded more especially, which increases their particular health risks and further marginalization.

Inter-state water-sharing conflicts in the Kaveri delta have caused chronic uncertainty in irrigation access, so upsetting rice production and farm incomes (Iyer Ramasamy 2003; Nair et al. 2024). Unregulated groundwater extraction and industrial pollution that deplete aquifers and contaminate surface waters damage the Krishna-Godavari basin's smallholder farmers and tribal communities (Jain et al. 2012; CPCB 2021). The Sundarbans suffer extensive land salinization from shrimp aquaculture, displacing traditional farming and pushing vulnerable groups into informal labor markets (Nayak and Berkes 2014). Likewise, industrial development and dam construction in the Mahanadi delta have changed water flows, so depriving downstream users of basic resources without enough compensation or consultation (Parhi et al. 2012; Ghosh et al. 2019). These cases of administrative neglect, characterized by inadequate policy attention, financing, and enforcement exacerbate regional vulnerabilities.

Rising sea levels, erratic rainfall, and more frequent extreme weather events brought on by climate change exacerbate some water-related vulnerabilities. Marginalized groups suffer most from these effects and have a limited ability to recover. In the low-lying Sundarbans and Mahanadi deltas, where marginalized communities form a significant population share, flooding often leads to displacement, crop loss, and disease outbreaks, burdens disproportionately shouldered by women and gender minorities due to existing social and economic disadvantages (Ray-Bennett 2009; IPCC 2022). Despite these challenges, India's water governance mechanisms tend to prioritize engineering and infrastructure solutions, such as large dams, river interlinking, and canal systems, while neglecting the socio-political dynamics of access and equity (Sangameswaran 2009). Official datasets frequently lack caste- and gender-disaggregated data, hindering inclusive policy design and implementation (Ahmed and Araral 2019).

In this context, sustainable water management must be reinvented not only as a tool for furthering social justice but also as an environmental need. This calls for realizing and correcting the unequal power relations influencing water access by inclusive, participatory, and context-sensitive policies (Boelens et al. 2010). In deltaic areas, where water shortage directly affects food security, employment, and health, an integrated approach balancing ecological sustainability and social equity is crucial. By examining secondary data from government sources, water assessments, and academic studies to investigate how water use across agriculture, aquaculture, industry, and household sectors shapes economic outcomes and social vulnerabilities in India's delta regions. Using a multidisciplinary approach, this chapter pursues four main objectives: (1) to investigate how water shapes economic activities across deltaic landscapes; (2) to analyze how social identities, particularly caste and gender, influence access to water and decision-making power; (3)

to compare the four major deltas in terms of their distinctive vulnerabilities and adaptive capacities; and (4) to propose evidence-based strategies for inclusive and sustainable water governance that center on the needs of marginalized communities. Five linked pieces make up the study. The analysis is structured into five interconnected sections. The first provides a comprehensive overview of the physical characteristics, economic significance, and historical development of India's delta regions, highlighting geomorphological and hydrological processes underpinning their productivity. The second explores the complex relationships between water availability, quality, and key economic sectors, examining how changing water use patterns affect agricultural sustainability, aquaculture expansion, and industrial growth. The third analyzes how social hierarchies, especially caste and gender norms, mediate access to water infrastructure, governance participation, and vulnerability to water-related hazards. The fourth offers a systematic comparison of the Kaveri, Krishna-Godavari, Mahanadi, and Sundarbans deltas, identifying shared challenges and region-specific dynamics requiring tailored policy responses. The final section evaluates governance frameworks and proposes alternative approaches grounded in environmental justice, knowledge integration, and participatory decision-making. By integrating ecological, economic, and social perspectives on water governance, this chapter seeks to advance both scholarly understanding and practical strategies for addressing the complex challenges facing India's delta regions amid climate change and growing resource competition.

7.2 Water and Development in India's Delta Regions

7.2.1 Significance and Features of the Deltas of India

Among the most ecologically important, densely populated, and most fertile landscapes in India are its delta areas. Formed by major river systems including the Ganga, Brahmaputra, Krishna-Godavari, Mahanadi, and Kaveri, whose continuous deposition of alluvial sediments sustains intensive agriculture, aquaculture, and freshwater provisioning, these dynamic zones have historically supported dense human settlements. These deltas act as vital linkages between land and sea, contribute to national food security, stimulate regional economies, and embody a distinctive cultural and ecological legacy. India's deltaic areas support millions of rural livelihoods (Dar et al. 2013; Venkatesan et al. 2023) and collectively are essential for national food security especially through rice farming in flood-prone areas. These plains have developed from intricate geomorphological and hydrological processes working over millennia. Accumulating at river mouths, sediment carried from Himalayan and peninsular watersheds produces nutrient-rich soils with great water retention capacity, perfect for growing rice, pulses, and oilseeds (Kuehl et al. 2005; Galy et al. 2008). Every delta has unique characteristics that define its agricultural output, hazard exposure, and course of development. Comprising almost 105,000 km^2 across

India and Bangladesh, the Ganga-Brahmaputra Delta including the Sundarbans gets some of the highest rainfall in the world: 1500 to 2000 mm annually (Mandal et al. 2020). By contrast, the Kaveri Delta depends more on irrigation since it spans roughly 8800 km^2 in Tamil Nadu and gets much less rainfall (800–1200 mm annually) (Selvakumar et al. 2014).

7.2.2 Pressures and Environmental Value

While preserving great biodiversity, the vast networks of rivers, distributaries, and wetlands improve agricultural potential. Functioning as a major carbon sink and natural storm buffer, the Sundarbans alone host over 330 plant species, 250 fish species, and endangered wildlife including the Royal Bengal tiger (*Panthera tigris*) (Ghosh et al. 2015). Analogous brackish water mangrove ecosystems supported by the Krishna-Godavari Delta stabilize shorelines and act as fish nurseries (Kubo et al. 2018). Beyond simple resource provision, these deltaic ecosystems provide a wide spectrum of ecosystem services. Shielding communities from environmental shocks depends much on regulating services including carbon sequestration, water purification, and flood control. For example, the Sundarbans mangroves' storm-protection function was estimated to prevent an average of USD 396 in damage cost per household during Cyclone Sidr for communities protected by an intact, one-kilometer-wide forest (Hossain et al. 2020). Deltas remain hubs of religious and cultural life. Thus, cultural services are also quite important. Nevertheless, many times in development planning, these non-market values are overlooked, leading to choices compromising ecological integrity. Including ecosystem service valuation into water allocation systems and environmental assessments will help to greatly enhance governance results. But rising pressures from upstream hydrological changes challenge these ecosystems. Major dams and barrages, such as the Hirakud on the Mahanadi, Mettur on the Kaveri, and Nagarjunasagar on the Krishna, have altered seasonal flows and sediment transport. While providing irrigation benefits, they also reduce freshwater discharge and sediment supply downstream, contributing to coastal erosion, saltwater intrusion, and land subsidence (Lal 2005; Ekka et al. 2022). For instance, over the past century, sediment loads to the Krishna-Godavari Delta have dropped by almost 75%, so compromising natural delta-building mechanisms (Rao et al. 2010).

7.2.3 Major Deltas' Financial Profiles

Every one of the four deltaic areas studied in this chapter: the Kaveri, Krishna-Godavari, Mahanadi, and Sundarbans showcase a unique economic profile shaped by hydrological regimes, colonial legacies, and post-independence development priorities (D'Souza 2006;

Chowdhury et al. 2008). Though sometimes underappreciated in policy debate, their contributions to regional and national economies are significant. The ancient Grand Anicut (Kallanai) has shaped the Kaveri Delta, for example, and its high-intensity farming makes it sometimes known as "the Rice Bowl of Tamil Nadu". It supports almost 4.5 million agricultural livelihoods and explains roughly 30–35% of Tamil Nadu's rice output (Arun et al. 2020; Ravikumar et al. 2024). But aging infrastructure, declining groundwater, and inter-state water disputes with Karnataka all add to the stress this legacy system now bears. The average annual water deficit in the basin can reach 80 billion cubic meters during drought years, which forces farmers to choose less water-intensive but less profitable crops or to leave agriculture completely (Karthick et al. 2020; Khandekar and Srinivasan 2021). On the other hand, the Krishna-Godavari Delta has grown to be a major center for brackish water aquaculture, especially in relation to fish and shrimp farming, so supporting rural employment as well as exports. Comprising more than 40% of India's aquaculture output, the combined delta generates almost USD 1.5 billion in export earnings annually. However, this growth comes at a cost: during the past three decades, roughly 25–30% of coastal Andhra Pradesh's agricultural land has been turned to aquaculture ponds, which has resulted in soil salinization and shrinking cropland (Bhatta 2013; Nagaraju et al. 2022). Furthermore, by means of antibiotic runoff, disease outbreaks, and declining coastal biodiversity, intensive aquaculture has led to environmental degradation. These dynamics call for a more nuanced assessment of trade-offs between development and ecological conservation.

Globally recognized as the largest contiguous mangrove forest and a UNESCO World Heritage Site, the Sundarbans Delta is part of the transboundary Ganga-Brahmaputra-Meghna system. By means of fishing, small-scale agriculture, and forest product harvesting, the Indian Sundarbans support approximately 4.5 million people annually, generating over INR 350 crore (USD 47 million) (Kabir and Baten 2012) Its mangroves offer important ecosystem functions. Worth USD 600–900 million in potential carbon credits, they sequester almost 4.15 million tonnes of carbon annually, so reducing cyclone damage and greatly lowering disaster recovery costs (Das and Vincent 2009). Rising sea levels, more frequent storms, and unsustainable human activity including deforestation and overuse of resources, all threaten these vital services, though.

Apart from farming and aquaculture, other livelihood industries play major role in the delta economy. Small-scale businesses and crafts, especially in textiles, offer job possibilities in the Mahanadi Delta, especially for women. Studies indicate that women in the Mahanadi Delta are predominantly engaged in services, manufacturing, and fishing sectors, highlighting the importance of these industries in local livelihoods. (Cazcarro et al. 2018). Though usually under dangerous conditions, honey collecting and gathering of non-timber forest products provide additional income in the Sundarbans (Biswas et al. 2024). For years, traditional honey collectors, known as moulis, have gathered honey from mangrove forests despite facing dangers such as tiger attacks during their journeys (Samanta et al. 2013). Particularly for landless or underprivileged groups, these sectors, though less visible in official economic assessments, are crucial for livelihood diversification and resilience. Strengthening such sectors can enhance climate adaptation and reduce pressure on overstressed agricultural systems.

Covering almost 9500 km^2 in Odisha, the Mahanadi Delta reflects the conflict between industrial growth and traditional way of life. Thermal power plants, steel buildings, and aluminum smelters have fast industrialized the area since the 1990s. Although these developments have strengthened the state's economy, manufacturing today accounts for roughly 28% of Odisha's GDP (Hazra et al. 2016; Hazra et al. 2020). Industrial water demand in the basin has surged nearly 300% since 1990, while agriculture's share has declined, despite supporting a far larger population (Asokan and Dutta 2008). This shift raises important concerns about equity, sustainability, and the future of water governance in delta regions. These complex dynamics illustrate the urgent need to examine how sectoral water use patterns intersect with socio-ecological vulnerabilities. The next section analyzes these sectoral patterns in greater detail.

7.3 Sectoral Water Use Patterns: Consequences

7.3.1 Agriculture: A Lifeline Under Pressure

Traditionally sustained by nutrient-rich alluvial soils and extensive riverine networks, agriculture has long formed the backbone of delta economies. Often referred to as the "Rice Bowl of Tamil Nadu," the Kaveri Delta has historically profited from ideal agro-climatic conditions and lots of irrigation water (Arun et al. 2020; Ravikumar et al. 2024). But erratic monsoons, upstream diversions, and ongoing use of ineffective irrigation systems (Gooptu et al. 2017; Ghosh et al. 2023) are progressively endangering this agrarian dominance. When compared to international standards, irrigation efficiency is clearly poor in the major agricultural center of the Krishna-Godavari Delta (Pramanik-Nayak 2010; Reshma and Murali 2018). Particularly during the dry season, these inefficiencies lead to significant conveyance and application losses, thereby directly affecting crop yields and rural livelihoods. Changing water availability and income uncertainty aggravate the vulnerability of smallholder farmers. Addressing this requires a shift toward micro-irrigation technologies like drip and sprinkler systems, alongside capacity-building programs to promote efficient water use and adaptive farming techniques (Iyer n.d.).

7.3.2 Aquaculture: Environmental Trade-offs and Economic Returns

In recent decades, the rapid expansion of brackish water aquaculture, especially shrimp farming, has transformed parts of the Krishna-Godavari Delta. Now generating almost USD 1.5 billion yearly, this industry offers jobs in coastal districts (MPEDA 2020). Over the past thirty years, 25–30% of agricultural land in Andhra Pradesh is thought to have been turned to aquaculture ponds (Sarangi et al. 2004; Kannan et al. 2022). Although profitable, this change creates serious environmental problems. Often permanent, salinization of soil lowers the future viability of land for food crops. Moreover, high stocking densities help to frequent disease outbreaks while the strong use of antibiotics and chemicals threatens nearby ecosystems by contaminated effluent (Ruban and Gunaseelan 2011; Kaur et al. 2017). These consequences can be felt in nearby towns that depend on pure water for household consumption as well as for agriculture. Achieving sustainability in aquaculture calls for land zoning rules, effluent treatment standards, and support of eco-certified farming methods.

7.3.3 Industrial Development: Rival Needs for Limited Water

Thanks in part to strategic location, resource availability, and infrastructure investments, the deltas have also grown focal points for industrial development. Hosting thermal power plants, aluminum refineries, and steel industries, the Mahanadi Delta in Odisha epitomizes this trend. Although the industry has raised regional GDP, the tripling of industrial water demand since 1990 (Panda 2019) has sharpened competition in the domestic and agricultural sectors. Most of the population still depends on agriculture, yet rapid industrialization has led to water reallocations away from the sector, intensifying competition and insecurity. Such trade-offs beg challenging issues about long-term resource sustainability, livelihood security, and water fairness. Further taxing supplies are many industrial plants lacking sophisticated water recycling systems. Policy tools needed to balance these conflicting needs are integrated water auditing, sectoral water budgets, and required recycling targets (Hazra et al. 2019).

7.3.4 Use at Home: The Increasing Load of Urbanization

In delta cities like Kakinada (Andhra Pradesh) and Cuttack (Odisha), urbanization is speeding forward, and domestic water demand is rising dramatically. Many of these cities, however, have antiquated or overburdened water infrastructure that results in sporadic supply, unfair access, and reliance on groundwater and private tankers. For instance, delayed upgrades to the aging pipeline network (Gundala and Tanikonda 2023; Rout and Tete 2023) have contributed to public complaints Cuttack Municipal Corporation has received over water shortage during dry months. As urban populations keep growing, such pressures are probably going to get worse without timely infrastructure spending. Promising remedies at the ward and municipal levels are groundwater recharge projects, greywater reuse, and rainwater collection.

7.3.5 Towards Combined Water Management

India's deltaic landscapes profoundly entwine the water demands of industry, aquaculture, agriculture, and household use. Decisions made in one field sometimes have unexpected effects on another. This emphasizes how urgently Integrated Water Resources Management (IWRM) systems that span sectoral silos and support cross-sectoral coordination, stakeholder involvement, and ecological protections (Samuel et al. 2017; Arunima et al. 2023) are needed. Protecting the future of these essential ecosystems and the millions of people who depend on them depends on establishing river basin authorities with regulatory authority, guaranteeing real-time water monitoring, and funding climate-resilient infrastructure.

7.4 Social Hierarchies and Water Availability in Delta Regions of India

India's delta regions, such as the Kaveri, Krishna-Godavari, Mahanadi, and Sundarbans, are ecologically vital and agriculturally rich landscapes. Yet, they also remain deeply embedded in socio-cultural hierarchies that determine who has access to water and who does not. Caste and gender continue to play decisive roles in shaping not just access to water, but also participation in governance and exposure to water-related vulnerabilities (Biswas 2020; Keiko 2008; Lindberg et al. 2016). This section examines how these social structures intersect to produce and reinforce inequalities in water access and management across India's deltaic regions.

7.4.1 Social Reversals in Water Availability

Despite constitutional protections, caste remains a powerful determinant of access to natural resources in India. Legal protections notwithstanding, Dalits and other underprivileged groups often find themselves excluded from dependable water sources. Land ownership patterns and power disparities that let upper-caste groups control important water infrastructure, including wells, boreholes, and irrigation channels, help to explain this (Tiwary and Phansalkar 2007; Dutta et al. 2018; Ramakrishnan 2021). Discrimination can be both overt and covert, as seen in practices such as separate water points, limited access times, or requiring some to wait until others have finished, in many villages. Dalits are compelled to rely on dominant castes for water, a situation sometimes described as "water feudalism." Lower-caste voices are often absent or disregarded even in local government institutions, including water user associations and gram panchayats. Another very important factor is gender. Water collection falls mostly on women, particularly those from poorer households and lower castes. Long, taxing trips that take time away from education, employment, or caregiving are common requirements of this work (Kulkarni et al. 2017; Ghosh and Sarkar 2023). Women may also be at risk from harassment or violence on these travels. Though they consume most of the water, women are hardly included in water management. When they are, their involvement is sometimes symbolic and has minimal actual impact. For Dalit women, who must deal with the combined difficulties of caste and gender, the situation is even more trying. Because of their "intersectional disadvantage," they are more likely to be harassed or denied access to water when attempting to use common sources (Sabharwal and Sonalkar 2015; Dutta et al. 2018). The time and effort required to guarantee water limits their chances for civic involvement, education, and employment. Their health suffers because of inadequate access to clean water as well; diseases, maternal health problems, and chronic fatigue result. Nevertheless, most water policies ignore how caste and gender interact and treat them as separate concerns.

7.4.2 Towards Inclusive and Fair Water Management

Solving these disparities calls for more than just mechanical repairs. It demands structural reform and inclusive governance frameworks rooted in equity and justice. Water policies must explicitly address the overlapping impacts of caste and gender; equity audits and social impact assessments should be mandatory for water-related interventions in delta regions. Deliberate efforts are needed to support the participation of marginalized groups, especially Dalit women, in water governance. This includes leadership training, legal awareness, and quotas in water user committees (Joshi 2011; Johns 2012; Dutta et al. 2018). Furthermore, anti-discrimination provisions related to water access must be enforced rigorously. Finally, water management ought to be community-driven and participative. The results are more likely to be fair and sustainable when all stakeholders, from small farmers to women household heads, are involved in planning and monitoring water systems.

7.5 Comparative Vulnerabilities and Adaptive Capability of the Principal Delta Regions of India

For millions of people, India's four main delta regions, namely the Kaveri, Krishna-Godavari, Mahanadi, and Sundarbans, serve as lifelines, by providing fertile agricultural land, freshwater, and vital ecological services. However, these deltas face distinct yet overlapping threats from hydrological alteration, environmental degradation, and socio-economic pressures (Deb and Ferreira 2017; Reshma and Murali 2018; Panda et al. 2022; Gupta et al. 2024). This section discusses these dynamics in detail, highlighting key quantitative impacts where data are available.

7.5.1 Kaveri Delta: Environmental Stress and Agricultural Decline

Reduced freshwater availability and environmental damage have made the Kaveri Delta increasingly vulnerable (Arun et al. 2020; Ravikumar et al. 2024). Mettur and other thermal power plants consume about 184 million litres of water every day, so drastically lowering downstream flows (Bijli et al. 2016; Solomon et al. 2021). Particularly close to the Bhavani River, industrial pollution from tanneries and dyeing facilities has degraded water quality (Ramakrishnan et al. 2021). In coastal areas, sediment starvation from upstream dams has accelerated coastal erosion and salinity intrusion greatly lowering rice yields in impacted areas (Gnanachandrasamy et al. 2016; Vasanthakumar 2017). Adaptive capacities in the area comprise efforts at crop diversification led by emerging farmers and functional but strained traditional irrigation systems (anicuts). However, weak institutional coordination between Tamil Nadu and Karnataka exacerbates water allocation disputes, thereby undermining the effectiveness of adaptive measures (Ferdin et al. 2010; Ghosh et al. 2018).

7.5.2 Krishna-Godavari Delta: Conflicts Between States and Difficulties with Water Distribution

Supporting many agricultural and aquacultural activities, the Krishna-Godavari (KG) Delta is an essential agro-ecological area in India. But it also has major governance problems, including institutional fragmentation and growing interstate water conflicts, especially between Andhra Pradesh and Telangana. With roughly 70% of the Krishna basin's catchment area located in Telangana, the state's heightened demand for a larger share of water resources has fueled recurrent conflicts over project approvals and distribution (Venkateshwarlu 2023; Khanna 2025). Projects involving upstream irrigation and dam building have changed natural flow patterns, so drastically lowering seasonal water availability in the delta's downstream regions and so affecting agricultural output. In response, the area has seen the slow acceptance of micro-irrigation technologies to increase water-use efficiency and the development of adaptive strategies, most notably the extension of aquaculture, especially shrimp farming, as an alternative livelihood (Muralidhar et al. 2012). Notwithstanding these initiatives, water-sharing rules are still not strictly enforced, and salinity intrusion resulting from lower freshwater flow still jeopardizes soil and crop health. These difficulties draw attention to how closely unresolved governance problems and political frictions impede integrated water resource management in the basin and hence reflect limited adaptive capacity.

7.5.3 Mahanadi Delta: Flood Risk and Industry

Rapid industrial growth in Odisha's Mahanadi Delta including steel, power, and aluminum sectors has sharply raised water demand, so aggravating competition with agriculture. Siltation and the aging of the Hirakud Dam have reduced both flood control capacity and irrigation efficiency (Mishra et al. 2022; Gupta et al. 2024). Regular floods endanger agriculture and communities; they are made worse by wetland loss and deforestation. Adaptive capacities are stronger in disaster preparedness, with advanced early warning systems and community-based flood shelters. Nonetheless, institutional water management remains largely reactive, focusing on post-disaster response rather than proactive risk reduction. Industrial policy and poor watershed governance significantly constrain adaptive capacity.

7.5.4 Sundarbans Delta: Climate Change and Disturbances of Livelihood

Accelerated sea-level rise, saline intrusion, and stronger cyclones are just a few of the climate vulnerabilities the Sundarbans suffers. Over half of the landmass of islands like Ghoramara has been lost in three decades, displacing thousands and seriously affecting livelihoods and agriculture (Mandal and Hosaka 2020; Giri et al. 2022; Kanan et al. 2023). Promising adaptive strategies are provided by traditional knowledge of saline-tolerant rice varieties and community mangrove restoration (Vyas and Sengupta 2012; Padhy et al. 2022); But remoteness and extreme poverty restrict access to infrastructure and government support, so undermining the effectiveness of adaptation initiatives. This delta represents the intricate interaction among environmental and socioeconomic vulnerabilities.

7.6 Reevaluating Water Governance in India's Delta Areas: Equity, Complexity, and Future Directions

India's delta areas are social landscapes marked by great inequality as well as ecological hotspots. The demand for inclusive and sustainable water governance becomes critical as development needs rise and climate stresses get more intense. Building on earlier debates of social vulnerability and ecological stress, we seek to critically review current governance structures and offer forward-looking ideas. It advocates a shift toward participatory, justice-oriented approaches grounded in both scientific and traditional knowledge systems. Still, this shift comes with challenges: institutional inertia, trade-offs, and power conflicts all must be addressed.

7.6.1 Where Current Governance Fails

Delta area water management today is disjointed and top-heavy. Many times, without coordination, several agencies run with overlapping agendas (Pasi 2012; Mahto 2024). Policy agendas still center large-scale infrastructure projects including dams, irrigation canals, and embankments, so serving strong industrial and agricultural interests (Shah and Gibson 2013). Smallholder farmers, fishermen, and women, meanwhile, still face unequal access to water and are mostly left out of decision-making (Adams et al. 2018). Structures of centralized government are not suited to handle the progressively erratic water cycles brought on by climate change (Birchall et al. 2023). These systemic gaps have resulted in rising conflicts, environmental degradation, and widening social divisions. Meaningful change therefore requires not only institutional reform but also a rethinking of whose voices are heard and what forms of knowledge are valued.

7.6.2 Pathways towards Inclusive and Sustainable Government

7.6.2.1 Center Justice: Rightful Action

Governance must start with environmental justice if we are to correct past injustices (Ryder 2017). This entails giving long-marginalized communities: Dalits, women, and indigenous fisherfolk rights and needs top priority. Strengthened local councils and water user associations should serve as mechanisms to ensure these groups have genuine decision-making power (Gandhi et al. 2020). Legal literacy initiatives, focused subsidies, and leadership development courses can all help to increase the capacity required for significant involvement.

7.6.2.2 Linking Knowledge Systems

From tidal fishing to rainwater collecting, Delta people have rich customs of water management. These techniques provide insightful information that would help scientific methods, including hydrological modeling (Kumar et al. 2021), complement each other. Using tools like participatory mapping, citizen science, and community monitoring (Danielsen et al. 2017), governance models should inspire the co-creation of knowledge. Such integration enhances local ownership and responsibility as well as ecological results.

7.6.2.3 Giving Participation Real Meaning

Participation goes beyond mere consultation. Good government calls for venues where many stakeholders including farmers, fishermen, businesses, civil society, and government can deliberate, negotiate, and co-create answers (Antwi et al. 2025). Especially as water gets more limited and competition gets fiercer, these areas should also feature dispute resolution systems (Mutahara et al. 2020). When implemented effectively, participatory governance improves openness, fosters trust, and produces more long-lasting results.

7.6.2.4 Negotiating the Trade-offs

Using these techniques is not simple. Participatory processes call for time, education, and money (Sebunya and Gichuki 2024). Governments must be ready to allocate funds and develop institutional capacity. Power relations present difficulties as well; dominant actors could object to changes endangering their interests (Touch et al. 2024). Ecological protections can also contradict immediate financial objectives. For example, stricter industrial pollution controls in the Mahanadi Delta may reduce employment opportunities in the short term but are essential for maintaining long-term water quality (Cao et al. 2021). In the same vein, maintaining mangroves might restrict the spread of aquaculture but improve resistance to sea-level rise and storms. Scaling up successful local models, such as community-led mangrove restoration in the Sundarbans, requires careful adaptation to diverse ecological and social context (Mfitumukiza et al. 2020).

7.7 Discussion

India's delta regions are dynamic socio-ecological systems that sustain dense populations and diverse livelihoods through agriculture, aquaculture, and industry. They play a critical role in national food security, economic growth, and ecological resilience (Kuenzer and Renaud 2012; Gupta et al. 2024). Yet, their ability to perform these functions is under increasing strain. Hydrological disturbances from large-scale infrastructure projects, combined with climate stress and entrenched social inequalities, are progressively undermining both ecosystems and livelihoods. Deltas are especially vulnerable to climate change due to their low-lying location and exposure to storm surges, flooding, sea-level rise, and subsidence, with consequences that affect human well-being and economic activities in the deltas and linked regions (Milliman et al. 1989; Raju et al. 2007; Syvitski et al. 2009; Nicholls et al. 2016; Arto et al. 2019; Begum et al. 2021; Singh and Naik 2024).

Interlinkages between Ecology, Economy, and Society

Ecological processes such as sediment transport, freshwater flows, and mangrove regeneration are directly linked to agricultural productivity, aquaculture success, and industrial expansion (Sarangi et al. 2004). However, these ecological foundations are destabilized by upstream dams, irrigation systems, and industrial discharge, which reduce freshwater availability and accelerate saline intrusion (Penmetsa et al. 2013; Nagaraju et al. 2022). In the Krishna-Godavari Delta, for example, the expansion of aquaculture has boosted export earnings but depleted soils and freshwater reserves, creating fragile livelihood systems (Muralidhar et al. 2013; Penmetsa et al. 2013; Nagaraju et al. 2022). Similarly, the Sundarbans' mangroves function simultaneously as biodiversity reservoirs, livelihood bases, and storm buffers, but overexploitation diminishes both ecological resilience and human security (Das and Crépin 2013; Nagaraju et al. 2022).

Social Inequalities and Intersectionality

Water access and governance in delta regions are profoundly shaped by caste, class, and gender hierarchies. Dalit households often lack reliable access to water points, while women, particularly from marginalized groups, are forced to walk long distances to fetch water under unsafe conditions (Sengupta and Hazra n.d.; Padala and Madhu 2025). These burdens compound risks to health, education, and economic opportunity (Chowdhury and Brahma 2019; Prati 2024). Water scarcity in areas like the Krishna River basin has also reduced crop production, elevated unemployment, and driven migration, further impacting women agricultural wage workers through higher workloads and lower earnings (Padala and Madhu 2025). Transgender communities, absent from most policy frameworks, are excluded altogether from water infrastructure (Sahoo et al. 2025; Benjamin and Hueso 2017; Mishra 2017). Such inequalities are not incidental but embedded in governance systems that fail to collect disaggregated data or prioritize equity (Prati 2024; Sahoo et al. 2025).

Climate Change as a Risk Multiplier

Climate change intensifies existing vulnerabilities in delta areas. Rising sea levels, erratic monsoons, and stronger cyclones increase exposure to flooding, erosion, and salinity (Nicholls et al. 1999; Myers 2002; Ericson et al. 2006; Syvitski 2008; Eissa and Zaki 2011; Foufoula-Georgiou et al. 2011; Field et al. 2012; De Souza et al. 2015). In the Mahanadi, industrial zones are regularly disrupted by floods, while in the Sundarbans, entire islands are being lost to erosion and sea-level rise (Ghosh et al. 2019). These risks disproportionately affect marginalized groups lacking financial buffers and political representation (Day et al. 2016; Szabo et al. 2016; Tessler et al. 2015). Women, for example, bear additional responsibilities during climate crises, including displacement management, caregiving, and food provisioning (Hazra et al. 2021; Prati et al. 2022). Without climate-sensitive governance, adaptation efforts risk remaining piecemeal and reactive (Agrawal et al. 2024).

Policy and Governance Shortcomings

The persistence of water insecurity is also rooted in governance choices that prioritize engineering and economic output over ecological and social sustainability. Large dams, embankments, and interlinking projects are framed as solutions but frequently displace communities, redistribute benefits unequally, and disrupt ecological flows (Raju et al. 2007; Pradhan and Srinivasan 2022). Governance remains fragmented across multiple institutions, with interstate disputes such as those between Tamil Nadu and Karnataka over the Kaveri, further eroding cooperative management (Agrawal et al. 2024). The neglect of smallholder farmers, fisherfolk, women's collectives, and indigenous groups reinforces inequities by treating them as passive recipients of policy rather than active stakeholders (Begum et al. 2021; Singh and Naik 2024).

Comparative Insights across the Four Deltas

While the persistence of water insecurity is rooted in systemic governance shortcomings, the specific combination of physical and socio-political pressures varies significantly by region. Understanding these regional differences is critical for moving beyond generalized critiques toward locally informed policy. Each delta embodies a distinctive set of pressures:

- **Kaveri**: Agricultural decline tied to reduced freshwater flows and interstate conflicts (Katyaini and Barua 2017; Ghosh et al. 2018; Agrawal et al. 2024).
- **Krishna-Godavari**: Aquaculture-driven growth with ecological trade-offs and governance disputes (Sarangi et al. 2004; Sekovski et al. 2012; Kantamaneni et al. 2019).
- **Mahanadi**: Industrial expansion intensifying water conflicts and flood risks (Behera et al. 2018; Das et al. 2019; Ghosh et al. 2019).
- **Sundarbans**: Climate-driven displacement and fragile adaptation capacity (Das and Crépin 2013; Anurag Danda et al. 2019).

Despite these differences, three consistent patterns emerge: ecological degradation undermines livelihoods, caste and gender hierarchies magnify vulnerability, and governance responses remain fragmented and technocratic (Prati et al. 2022). Together, these dynamics point to the need for systemic reform rather than isolated interventions (Haasnoot et al. 2012; Kwakkel et al. 2015; Raju et al. 2007; Begum et al. 2021; Prati et al. 2022; Singh and Naik 2024; Chapman et al. 2016).

Pathways towards Inclusive and Sustainable Governance

Encouragingly, there are cases where inclusive approaches have produced tangible benefits. Participatory Irrigation Management in Andhra Pradesh and Telangana has enabled farmer associations to achieve fairer water distribution and improved productivity (Raju et al. 2007; Singh et al. 2009). Community-led mangrove conservation in the Sundarbans has blended formal policy with traditional knowledge, supporting both biodiversity and livelihoods (Begum et al. 2021; Sen et al. 2024). National programs such as the Jal Jeevan Mission have demonstrated how large-scale interventions can incorporate community participation in water delivery (Haasnoot et al. 2012; Kwakkel et al. 2015; Singh and Naik 2024; Chapman et al. 2016). These examples illustrate that inclusive water governance is not merely aspirational but already functioning in diverse contexts across India. They also show that sustainability requires recognizing water as a basic right, valuing local knowledge alongside scientific expertise, and ensuring meaningful participation of marginalized voices (Agrawal et al. 2024). Equity, flexibility, and collaboration must therefore form the foundation of future water management in delta regions.

7.8 Conclusions

Water management in India's delta regions is not solely a question of physical scarcity but a reflection of deeper governance, social, and ecological inequities. The comparative analysis of the Kaveri, Krishna-Godavari, Mahanadi, and Sundarbans deltas demonstrates how diverse hydro-ecological settings interact with entrenched hierarchies of caste, class, and gender to shape access to water and livelihood security. Climate change has further intensified these structural challenges, manifesting through erratic rainfall, saline intrusion, and the increasing frequency of floods and droughts.

Although each delta presents distinct vulnerabilities, a common pattern emerges: administrative neglect, fragmented institutions, and sectoral competition for water continue to undermine resilience and inclusivity. Agricultural distress in the Kaveri Delta, aquaculture-induced degradation in the Krishna-Godavari basin, industrial expansion in the Mahanadi Delta, and climate-induced livelihood disruptions in the Sundarbans collectively highlight the multidimensional nature of water insecurity. These outcomes reaffirm that technical or infrastructure-driven approaches alone cannot resolve the crisis, as the roots of water stress are embedded in social and institutional frameworks.

Sustainable water governance must therefore be grounded in environmental justice, transparency, and participatory engagement. Socially inclusive management, recognizing the knowledge, needs, and agency of marginalized communities offers a more enduring pathway to resilience. Empowering local communities to participate in decision-making, ensuring gender and caste equity, and improving coordination between administrative levels can foster adaptive and context-specific solutions.

In essence, the future of India's delta regions will depend not merely on managing rivers and reservoirs but on transforming governance systems that determine who benefit from water and who bears the cost of its depletion. Sustainable and inclusive water management must integrate ecological restoration with social equity and institutional reform. Strengthening knowledge-sharing, promoting community stewardship, and aligning development priorities with environmental limits can collectively secure both human and ecological well-being. The lessons from these deltas thus extend beyond regional boundaries, offering valuable insights for managing complex socio-hydrological systems in a changing climate.

7.9 Recommendations

Drawing from the comparative analysis of India's major delta regions, it is evident that achieving water security requires systemic reform which involves balancing ecological restoration, social equity, and institutional accountability. The following recommendations outline key strategies to strengthen resilience, inclusivity, and sustainability in delta water governance.

Strengthen Participatory Water Governance: Empower decentralized and community-based water user associations that ensure equitable representation of women, marginalized castes, and minority groups. Institutional frameworks must guarantee their active involvement in decision-making, monitoring, and evaluation to enhance accountability and inclusivity.

Integrate Social Equity into Policy and Planning: All delta-level water management plans should incorporate social and gender impact assessments. Government agencies should collect disaggregated data on caste, class, and gender to identify disparities in water access and inform equitable policy interventions.

Promote Integrated Water Resources Management (IWRM): Foster coordination among agriculture, aquaculture, industrial, and domestic sectors through basin-level authorities capable of regulating water extraction, ensuring ecological flow maintenance, and minimizing inter-sectoral conflicts. Policy coherence across ministries and states is essential to prevent fragmented governance.

Enhance Climate Adaptation and Resilience: Develop localized adaptation strategies focused on flood control, salinity management, and sustainable agriculture. Indigenous knowledge such as traditional irrigation, crop diversification, and rainwater harvesting should be integrated with scientific innovations to strengthen adaptive capacity.

Ensure Accountability and Transparency: Establish mechanisms for independent water audits, regular environmental impact assessments, and public disclosure of industrial and agricultural water usage. Transparent monitoring systems can build community trust and ensure compliance with environmental standards.

Invest in Ecosystem Restoration: Prioritize the conservation and restoration of wetlands, mangroves, and riverine ecosystems to enhance ecological resilience and livelihood security. Nature-based solutions, including sediment management and green infrastructure, can mitigate climate risks while supporting biodiversity and local economies.

Foster Cross-Regional and Interdisciplinary Collaboration: Encourage cooperation among deltaic states, research institutions, and international river-basin organizations to exchange knowledge, share best practices, and co-develop sustainable management models. Such collaboration can help integrate science, policy, and community action for long-term resilience.

Acknowledgement The authors sincerely thank the Honourable Founder President, Honourable Chancellor, Vice Chancellor, and Pro-Vice Chancellor of Amity University Haryana (AUH), Manesar-Gurugram, India, for their ongoing support of this research. We also wish to acknowledge the valuable collaboration and support of Prof. Dr. P. C. S Devara, Director and Head, Amity Centre of Excellence in Ocean-Atmospheric Science and Technology (ACOAST) & Environmental Science and Health (ACESH), Amity University Haryana, which was instrumental to this chapter.

State whether you used AI or not in preparing the chapter.

Author's Note In preparing this chapter, AI-based language tools were employed to enhance sentence structure, correct grammatical errors, and improve overall clarity and readability.

References

Adams EA, Juran L, Ajibade I (2018) 'Spaces of exclusion' in community water governance: a feminist political ecology of gender and participation in malawi's urban water user associations. Geoforum 95:133–142. https://doi.org/10.1016/j.geoforum.2018.06.016

Adhikari S, Chaudhury AK, Gangadhar B, Ramesh R, Mandal RN, Sarosh I, Saha GS et al (2018) Adaptation and mitigation strategies of climate change impact in freshwater aquaculture in some states of India, pp 16–21. https://doi.org/10.21767/1307-234x.1000142

Agrawal T, Pompoes R, Verzijl A, Srinivasan V, Nair J, Huijbens E, Murugan V (2024) Creating Kaveri Delta beneath our feet: an experiment in grounding socio-hydrology in Tamil Nadu, India. J Hydrol 644:131896. https://doi.org/10.1016/j.jhydrol.2024.131896

Agarwal B (2010) Gender and green governance: the political economy of women's presence within and beyond community forestry. Oxford University Press

Ahmed M, Araral E (2019) Water governance in India: evidence on water law, policy, and administration from eight Indian states. Water 11(10):2071. https://doi.org/10.3390/w11102071

Antwi SH, Stephens CG, Rolston A, Getty D, Linnane S (2025) Public participation in environmental decision-making: a water sector perspective. Environ Sustain Indic 26:100656. https://doi.org/10.1016/j.indic.2025.100656

Anurag Danda A, Ghosh N, Bandyopadhyay J, Hazra S (2019) Managed retreat: adaptation to climate change in the Sundarbans ecoregion in the Bengal Delta. J Indian Ocean Reg 15(3):317–335. https://doi.org/10.1080/19480881.2019.1652974

Arto I, García-Muros X, Cazcarro I, González-Eguino M, Markandya A, Hazra S (2019) The socioeconomic future of deltas in a changing environment. Sci Total Environ 648:1284–1296. https://doi.org/10.1016/j.scitotenv.2018.08.139

Arun D, Sunandini GP, Kumari RV, Meena A (2020) An economic analysis of different Rice varieties in Cauvery Delta region of Tamil Nadu. Indian J Econ Dev 8(3):1–5. https://doi.org/10.17485/IJED/v8.52

Arunima KT, Firoz MC, Kasthurba AK, Gupta P (2023) Strategic spatial planning based on ecosystem services (ES): a case of Cauvery Basin. Urban India 43:162–178

Asokan SM, Dutta D (2008) Analysis of water resources in the Mahanadi River basin, India under projected climate conditions. Hydrol Process Int J 22(18):3589–3603. https://doi.org/10.1002/hyp.6962

Begum F, de Bruyn LL, Kristiansen P, Islam MA (2021) Institutionalising co-management activities for conservation of forest resources: evidence from the Sundarban mangrove forest management of Bangladesh. J Environ Manag 298:113504. https://doi.org/10.1016/j.jenvman.2021.113504

Behera MD, Tripathi P, Das P, Srivastava SK, Roy PS, Joshi C et al (2018) Remote sensing based deforestation analysis in Mahanadi and Brahmaputra River basin in India since 1985. J Environ Manag 206:1192–1203. https://doi.org/10.1016/j.jenvman.2017.10.015

Benjamin C, Hueso A (2017) LGBTI and sanitation: what we know and what the gaps are. In: 40th WEDC international conference, Loughborough, UK

Bera A, Meraj G, Kanga S, Farooq M, Singh SK, Sahu N, Kumar P (2022) Vulnerability and risk assessment to climate change in Sagar Island, India. Water 14(5):823. https://doi.org/10.3390/w14050823

Bijli MMS, Paneerselvam T, Alahiyanambi K, Jafar Basha MA (2016) Optimization of boiler efficiency at Mettur thermal Power Station. In: Proceedings of the international conference on soft computing systems: ICSCS 2015, vol 1. Springer India, pp 581–592

Birchall SJ, Bonnett N, Kehler S (2023) The influence of governance structure on local resilience: enabling and constraining factors for climate change adaptation in practice. Urban Clim 47:101348. https://doi.org/10.1016/j.uclim.2022.101348

Bhatta R (2013) Economics of sustainable shrimp aquaculture in India. In: Mukherjee S, Chakraborty D (eds) Environmental scenario in India, pp 45–64

Biswas C, Channarayapatna S, Pandey CN (2024) Multifaceted livelihood and beyond relations of local communities with Sundarban mangroves. Hum Ecol 53(2):241–257. https://doi.org/10.1007/s10745-025-00596-8

Biswas S (2020) Structure and functions of social-ecological systems: a case study from Indian sundarbans. Institute for Social and Economic Change

Boelens R, Getches D, Guevara-Gil A (2010) Water struggles and the politics of identity. In: Boelens RA, Getches D, Guevara-Gil A (eds) Out of the mainstream: water rights, politics and identity, pp 3–25

Cao S, Liu Z, Li W, Xian J (2021) Balancing ecological conservation with socioeconomic development. Ambio 50(5):1117–1122. https://doi.org/10.1007/s13280-020-01448-z

Cazcarro I, Arto I, Hazra S, Bhattacharya RN, Osei-Wusu Adjei P, Ofori-Danson PK, Hossen Z (2018) Biophysical and socioeconomic state and links of deltaic areas vulnerable to climate change: Volta (Ghana), Mahanadi (India), and Ganges-Brahmaputra-Meghna (India and Bangladesh). Sustainability 10(3):893. https://doi.org/10.3390/su10030893

Chapman A, Tompkins EL, Vincent K, Day S (2016) A framework for the design and evaluation of adaptation pathways in large river deltas

Chowdhury AN, Brahma A (2019) Environment and wellbeing: eco-psychiatry in Sundarban delta, India. Int J Manag Soc Sci 14(2):37–53. https://doi.org/10.21013/jmss.v14.n2sp.p5

Chowdhury AN, Mondal R, Brahma A, Biswas MK (2008) Eco-psychiatry and environmental conservation: study from Sundarban Delta, India. Environ Health Insights 2:EHI-S935. https://doi.org/10.4137/EHI.S935

Central Pollution Control Board (CPCB) (2021) Status of water quality in India. https://cpcb.nic.in/status-of-water-quality-in-india/

Danielsen F, Enghoff M, Magnussen E, Mustonen T, Degteva A, Hansen KK, Plieninger T (2017) Citizen science tools for engaging local stakeholders and promoting local and traditional knowledge in landscape stewardship. In: The science and practice of landscape stewardship, pp 80–98

Dar MH, De Janvry A, Emerick K, Raitzer D, Sadoulet E (2013) Flood-tolerant rice reduces yield variability and raises expected yield, differentially benefitting socially disadvantaged groups. Sci Rep 3(1):3315

Das P, Behera MD, Pal S, Chowdary VM, Behera PR, Singh TP (2019) Studying land use dynamics using decadal satellite images and dyna-CLUE model in the Mahanadi River basin, India. Environ Monit Assess 191(3):804. https://doi.org/10.1007/s10661-019-7698-3

Das S, Vincent JR (2009) Mangroves protected villages and reduced death toll during Indian super cyclone. Proc Natl Acad Sci 106(18):7357–7360. https://doi.org/10.1073/pnas.0810440106

Day JW, Agboola J, Chen Z, D'Elia C, Forbes DL, Giosan L, Kemp P, Kuenzer C, Lane RR, Ramachandran R, Syvitski J, Yañez-Arancibia A (2016) Approaches to defining deltaic sustainability in the 21st century. Estuar Coast Shelf Sci 183:275–291. https://doi.org/10.1016/j.ecss.2016.06.018

Deb M, Ferreira CM (2017) Potential impacts of the Sunderban mangrove degradation on future coastal flooding in Bangladesh. J Hydro Environ Res 17:30–46. https://doi.org/10.1016/j.jher.2016.11.005

D'Souza R (2006) Water in British India: the making of a 'colonial hydrology'. Hist Compass 4(4):621–628. https://doi.org/10.1111/j.1478-0542.2006.00336.x

De Souza K, Kituyi E, Harvey B, Leone M, Murali KS, Ford JD (2015) Vulnerability to climate change in three hot spots in Africa and Asia: key issues for policy-relevant adaptation and resilience-building research. Reg Environ Chang 15(5):747–753. https://doi.org/10.1007/s10113-015-0755-8

Dutta S, Sinha I, Parashar A (2018) Dalit women and water: availability, access and discrimination in rural India. J Soc Incl Stud 4(1):62–79. https://doi.org/10.1177/2394481118774487

Eissa AE, Zaki MM (2011) The impact of global climatic changes on the aquatic environment. Procedia Environ Sci 4:251–259. https://doi.org/10.1016/j.proenv.2011.03.030

Ekka A, Keshav S, Pande S, van der Zaag P, Jiang Y (2022) Dam-induced hydrological alterations in the upper Cauvery river basin, India. J Hydrol RegNal Stud 44:101231. https://doi.org/10.1016/j.ejrh.2022.101231

Ericson J, Vörösmarty C, Dingman S, Ward L, Meybeck M (2006) Effective Sea-level rise and deltas: causes of change and human dimension implications. Glob Planet Chang 50(1–2):63–82. https://doi.org/10.1016/j.gloplacha.2005.07.004

Ferdin M, Görlitz S, Schwörer S (2010) Water stress in the Cauvery basin, South India. How current water management approaches and allocation conflict constrain reform. ASIEN Ger J Polit Econ Cult 117:27–44

Field CB, Barros V, Stocker TF, Dahe Q (2012) Managing the risks of extreme events and disasters to advance climate change adaptation: special report of the intergovernmental panel on climate change. Cambridge University Press. https://doi.org/10.1017/cbo9781139177245

Foufoula-Georgiou E, Syvitski J, Paola C, Hoanh CT, Tuong P, Vörösmarty C, Kremer H, Brondizio E, Saito Y, Twilley R (2011) International year of deltas 2013: a proposal. EOS Trans Am Geophys Union 92(40):340–341. https://doi.org/10.1029/2011EO400006

Galy V, France-Lanord C, Lartiges B (2008) Loading and fate of particulate organic carbon from the Himalaya to the ganga–Brahmaputra delta. Geochim Cosmochim Acta 72(7):1767–1787

Gandhi VP, Johnson N, Neog K, Jain D (2020) Institutional structure, participation, and devolution in water institutions of eastern India. Water 12(2):476. https://doi.org/10.3390/w12020476

Ghosh P, Sarkar S (2023) Female water fetchers: analyzing the role of women in collecting drinking water in India. Glob Soc Welf 10(4):359–369. https://doi.org/10.1007/s40609-023-00283-w

Ghosh S, Barik DK, Renganayaki P, Kang B, Gumber S, Venkatesh S, Akunuri S (2023) The impact of short-duration precipitation events over the historic Cauvery basin: a study on altered water resource patterns and associated threats. Sci Rep 13(1):14095. https://doi.org/10.1038/s41598-023-41417-6

Ghosh A, Das S, Ghosh T, Hazra S (2019) Risk of extreme events in delta environment: a case study of the Mahanadi delta. Sci Total Environ 664:713–723. https://doi.org/10.1016/j.scitotenv.2019.01.390

Ghosh N, Bandyopadhyay J, Thakur J (2018) Conflict over Cauvery waters: imperatives for innovative policy options. Observer Research Foundation

Giri S, Daw TM, Hazra S, Troell M, Samanta S, Basu O, Chanda A (2022) Economic incentives drive the conversion of agriculture to aquaculture in the Indian Sundarbans: livelihood and environmental implications of different aquaculture types. Ambio 51(9):1963–1977. https://doi.org/10.1007/s13280-022-01720-4

Gnanachandrasamy G, Ramkumar T, Venkatramanan S, Chung SY, Vasudevan S (2016) Identification of saline water intrusion in part of Cauvery deltaic region, Tamil Nadu, southern India: using

GIS and VES methods. Mar Geophys Res 37(2):113–126. https://doi.org/10.1007/s11001-016-9271-6

Gooptu N, Parry J, Lindberg S, Athreya VB, Djurfeldt G, Rajagopal A, Vidyasagar R (2017) Progress over the long haul: dynamics of agrarian change in the Kaveri Delta. In: Gooptu N, Parry J (eds) Persistence of poverty in India. Routledge, pp 344–369. https://doi.org/10.4324/9781315146973

Gundala V, Tanikonda VR (2023) Delineation of seawater intrusion into freshwater aquifers by using VES & GIS in the Kakinada region, east Godavari District, Andhra Pradesh, India. In: Etikala B, Yadav GV, Prasad M (eds) Emerging Technologies for Water Supply, conservation and management. Springer, pp 333–341. https://doi.org/10.1007/978-3-031-35279-9_16

Gupta A, Barik SK, Chauhan PS (2024) Towards sustainable delta ecosystems: pollution mitigation for achieving SDGs in Indian delta region. In: Solid waste Management in Delta Region for SDGs fulfillment. Springer Nature Switzerland, pp 27–46. https://doi.org/10.1007/978-3-031-58253-0_2

Haasnoot M, Middelkoop H, Offermans A, van Beek E, van Deursen WPA (2012) Exploring pathways for sustainable water management in river deltas in a changing environment. Clim Chang 115(3):795–819. https://doi.org/10.1007/s10584-012-0444-2

Hazra S, Patel A, Das S, Hans A, Ghosh A, Giri J (2021) Women-headed households, migration and adaptation to climate change in the Mahanadi Delta, India. In: Hans A, Rao N, Prakash A, Patel A (eds) Engendering climate change: learnings from South Asia. Routledge, pp 172–196. https://doi.org/10.4324/9781003142409-11

Hazra S, Das S, Ghosh A, Raju PV, Patel A (2020) The Mahanadi delta: a rapidly developing delta in India. In: Deltas in the Anthropocene. Springer, pp 53–77. https://doi.org/10.1007/978-3-030-23517-8_3

Hazra S, Cazcarro I, Arto I, Bhattacharya R (2016) Biophysical and socioeconomic state of the Mahanadi delta region of India from the perspectives of gender and spatial relations. DECCMA Working Paper

Hossain MA, Halim A, Alam MS (2020) Value of the storm-protection function of Sundarban mangroves in Bangladesh. J Sustain Dev 13(3):128–138

Intergovernmental Panel on Climate Change (2022) Climate change 2022: impacts, adaptation and vulnerability

Iyer Ramasamy R (2003) Water: perspectives, issues, concerns. Sage Publications, New Delhi

Iyer JC (n.d.) Genesis of micro irrigation: need for micro irrigation in command areas

Jain CK, Yerramilli SS, Yerramilli RC (2012) A case study on blowout and its control in Krishna-Godavari (KG) basin, east coast of India: safety and environmental perspective. J Environ Earth Sci 2(1):49–60

Jayasankar P (2018) Present status of freshwater aquaculture in India–a review. Indian J Fish 65(4):157–165. https://doi.org/10.21077/ijf.2018.65.4.81300-20

Johns H (2012) Stigmatization of Dalits in access to water and sanitation in India. International Dalit Solidarity Network. https://idsn.org/wp-content/uploads/user_folder/pdf/New_files/UN/HRC/Stigmatization_of_dalits_in_access_to_water_sanitation.pdf

Joshi D (2011) Caste, gender and the rhetoric of reform in India's drinking water sector. Econ Polit Wkly 46(18):56–63

Joy KJ, Paranjape S, Gujja B, Goud V, Vispute S (eds) (2020) Water conflicts in India: a million revolts in the making. Routledge

Kabir MH, Baten MA (2012) The economics of the biodiversity and ecosystem services of the Sundarbans. Unpublished Research Report. Unnayan Onneshan

Kanan AH, Pirotti F, Masiero M, Rahman MM (2023) Mapping inundation from sea level rise and its interaction with land cover in the Sundarbans mangrove forest. Clim Chang 176(8):104. https://doi.org/10.1007/s10584-023-03574-5

Kannan R, Naidu KA, Kanungo A, Murty MR, Avishek K, Ramana KV (2022) Hydrochemical characteristics of groundwater—assessment of saltwater intrusion along Krishna and Godavari delta region, Andhra Pradesh, India. In: Geospatial technology for landscape and environmental management: sustainable assessment and planning. Springer, pp 99–120. https://doi.org/10.1007/978-981-16-7373-3_5

Kantamaneni K, Sudha Rani NNV, Rice L, Sur K, Thayaparan M, Kulatunga U, Rege R, Yenneti K, Campos LC (2019) A systematic review of coastal vulnerability assessment studies along Andhra Pradesh, India: a critical evaluation of data gathering, risk levels and mitigation strategies. Water 11(2):393. https://doi.org/10.3390/w11020393

Karthick S, Saminathan R, Gopinath R (2020) A study on the problems faced by farmers in paddy marketing of Cauvery Delta region, Tamil Nadu. Int J Manag 11(10):2155–2164

Katyaini S, Barua A (2017) Assessment of interstate virtual water flows embedded in agriculture to mitigate water scarcity in India (1996–2014). Water Resour Res 53(8):7382–7400. https://doi.org/10.1002/2016WR020247

Kaur G, Katyal P, Singh D (2017) Physico-chemical and microbiological profile of effluents from common effluent treatment plants (CETPs). Int J Curr Microbiol App Sci 6(9):1403–1413

Keiko Y (2008) Politics and representation of caste identity in regional historiography: a case study of Kammas in Andhra. Indian Econ Soc Hist Rev 45(3):353–380. https://doi.org/10.1177/001946460804500302

Khandekar N, Srinivasan V (2021) Dispute resolution in the Cauvery Basin, India. In: Ferrier R, Jenkins A (eds) Handbook of catchment management, 2nd edn. Wiley, pp 549–577. https://doi.org/10.1002/9781119531241.ch22

Khanna A (2025) Inter–state river disputes and the Indian federal order: constitutional challenges and political realities. SSRN. https://doi.org/10.2139/ssrn.5250695

Kubo S, Kumar KCVN, Demudu G, Malini BH, Rao KN, Agrawal R, Rajawat AS (2018) Monitoring of eco-restoration of mangrove wetlands through time series satellite images: a study on Krishna-Godavari Delta region, east coast of India. Geogr Rev Jpn Ser B 90(2):66–75

Kuehl SA, Allison MA, Goodbred SL, Kudrass HERMANN (2005) The ganges-brahmaputra delta

Kuenzer C, Renaud FG (2012) Climate and environmental change in river deltas globally: expected impacts, resilience, and adaptation. In: The Mekong Delta System. Springer, pp 7–46. https://doi.org/10.1007/978-94-007-3962-8_2

Kulkarni S, O'Reilly K, Bhat S (2017) No relief: lived experiences of inadequate sanitation access of poor urban women in India. Gend Dev 25(2):167–183. https://doi.org/10.1080/13552074.2017.1331531

Kumar MD, Saleth RM (2018) Inequality in the Indian water sector: challenges and policy options. Indian J Hum Dev 12(2):265–281. https://doi.org/10.1177/0973703018793727

Kumar P, Dasgupta R, Dhyani S, Kadaverugu R, Johnson BA, Hashimoto S, Mishra BK (2021) Scenario-based hydrological modeling for designing climate-resilient coastal water resource management measures: lessons from Brahmani River, Odisha, eastern India. Sustainability 13(11):6339. https://doi.org/10.3390/su13116339

Kwakkel JH, Haasnoot M, Walker WE (2015) Developing dynamic adaptive policy pathways: a computer-assisted approach for developing adaptive strategies for a deeply uncertain world. Clim Chang 132:373–386. https://doi.org/10.1007/s10584-014-1210-4

Lal M (2005) Climatic change—implications for India's water resources. In: Climatic change and water resources in South Asia. AA Balkema (Taylor & Francis), pp 155–193

Lauria V, Das I, Hazra S, Cazcarro I, Arto I, Kay S, Fernandes JA (2018) Importance of fisheries for food security across three climate change vulnerable deltas. Sci Total Environ 640:1566–1577. https://doi.org/10.1016/j.scitotenv.2018.06.011

Lindberg S, Athreya VB, Vidyasagar R, Rajagopal A, Djurfeldt G (2016) Changing social and political relations in the Kaveri delta

Mahto MK (2024) Challenges and opportunities for water resources in India: an exploratory story of resilience. In: Integrated Management of Water Resources in India: a computational approach: optimizing for sustainability and planning, pp 59–80

Mandal MSH, Hosaka T (2020) Assessing cyclone disturbances (1988–2016) in the Sundarbans mangrove forests using Landsat and Google earth engine. Nat Hazards 102:133–150

Mehta L (2005) The politics and poetics of water: the naturalisation of scarcity in Western India. Orient Blackswan

Mfitumukiza D, Roy AS, Simane B, Hammill A, Rahman MF, Huq S (2020) Scaling local and community-based adaptation. Global commission on adaptation background paper, 2020

Milliman JD, Broadus JM, Gable F (1989) Environmental and economic implications of rising sea level and subsiding deltas: the Nile and Bengal examples. Ambio 18:340–345

Mishra SP, Mishra A, Kumar C, Mishra S (2022) Geospatial sedimentation of Hirakud reservoir, Odisha

Mishra VK (2017) Key challenges of marginalised communities on sanitation and hygiene and recommendations to clean India

MPEDA (Marine Products Export Development Authority) (2020) Indian seafood statistics. Ministry of Commerce and Industry, government of India

Muralidhar M, Kumaran M, Jayanthi M, Muniyandi B, Ponniah AG, Nagothu US, Eknath A (2012) Case study on the impacts of climate change on shrimp farming and developing adaptation measures for small-scale shrimp farmers in Krishna District, Andhra Pradesh, India (network of aquaculture centers in Asia-Pacific, 126 p)

Muralidhar M, Kumaran M, Kumar JA, Dayal JS, Jayanthi M, Saraswathy R, Lalitha N, Panigrahi A, Sreenivasa Rao A, Muniyandi B, Murugan P (2013) Climate change and coastal aquaculture in west Godavari District, Andhra Pradesh: impacts, vulnerability, adaptations and mitigations for resilience. J Agrometeorol 15(Special Issue II):116–122

Mutahara M, Warner JF, Khan MSA (2020) Multi-stakeholder participation for sustainable delta management: a challenge of the socio-technical transformation in the management practices in Bangladesh. Int J Sust Dev World 27(7):611–624

Myers N (2002) Environmental refugees: a growing phenomenon of the 21st century. Philos Trans R Soc B Biol Sci 357(1420):609–613. https://doi.org/10.1098/rstb.2001.0953

Nagaraju TV, Malegole SB, Chaudhary B, Ravindran G (2022) Assessment of environmental impact of aquaculture ponds in the western delta region of Andhra Pradesh. Sustainability 14(20):13035. https://doi.org/10.3390/su142013035

Nair J, Thomas BK, Bahinipati CS (2024) Cropping decisions under water stress: evidence from Cauvery Delta region, India. World Water Policy 10(3):711–729

Narain V (2014) Whose land? Whose water? Water rights, equity and justice in a peri-urban context. Local Environ 19(9):974–989. https://doi.org/10.1080/13549839.2014.907248

Nath A, Samanta S, Banerjee S, Danda AA, Hazra S (2021) Threat of arsenic contamination, salinity and water pollution in agricultural practices of Sundarban Delta, India, and mitigation strategies. SN Appl Sci 3:140. https://doi.org/10.1007/s42452-021-04544-1

Nayak PK, Berkes F (2014) Linking global drivers with local and regional change: a social-ecological system approach in Chilika lagoon, bay of Bengal. Reg Environ Chang 14(5):2067–2079. https://doi.org/10.1007/s10113-012-0369-3

Nicholls RJ, Hutton CW, Lazar AN, Allan A, Adger WN, Adams H, Wolf J, Rahman M, Salehin M (2016) Integrated assessment of social and environmental sustainability dynamics in the Ganges-Brahmaputra-Meghna delta, Bangladesh. Estuar Coast Shelf Sci 183:370–381. https://doi.org/10.1016/j.ecss.2016.08.017

Nicholls RJ, Hoozemans FM, Marchand M (1999) Increasing flood risk and wetland losses due to global sea-level rise: regional and global analyses. Glob Environ Chang 9(Supplement):S69–S87. https://doi.org/10.1016/S0959-3780(99)00019-9

Padala KK, Madhu BR (2025) Inequitable water distribution: socio-legal implications of Krishna River allocation on agricultural labour in Andhra Pradesh. In: Rivers unbound. Routledge, pp 183–189

Padhy SR, Dash PK, Bhattacharyya P (2022) Challenges, opportunities, and climate change adaptation strategies of mangrove-agriculture ecosystem in the Sundarbans, India: a review. Wetl Ecol Manag 30(1):191–206. https://doi.org/10.1007/s11273-021-09844-2

Panda GK, Chatterjee U, Mishra M, Mohanty A (2022) Vulnerability assessment of environmental degradation due to coastal hazards using geospatial technology—a study of Puri coast along the Mahanadi Delta, India. Arab J Geosci 15(15):1322. https://doi.org/10.1007/s12517-022-10578-5

Panda RK (2019) Mahanadi: looking beyond coal. Heinrich boll Stiftung India, 8

Parhi PK, Mishra SK, Singh R, Tripathi VK (2012) Floods in Mahanadi River basin, Orissa (India): a critical review. In: India water week, water, energy and food security: call for solutions, pp 10–14

Pasi N (2012) Rethinking water resources management, governance, and decision-making: a case study of interlinking of Rivers project, India [doctoral dissertation, State University of New York College of environmental science and forestry]

Penmetsa ARKR, Muppidi SRR, Popuri R, Golla SB, Tenneti R (2013) Aquaculture and its impact on ground water in east Godavari District Andhra Pradesh, India–a case study. Int Res J Environ Sci 2(10):101–106

Pradhan A, Srinivasan V (2022) Do dams improve water security in India? A review of post facto assessments. Water Secur 15:100112. https://doi.org/10.1016/j.wasec.2022.100112

Pramanik-Nayak S (2010) Agriculture and irrigation in the Krishna-Godavari tract: an epigraphical analysis from the 8th to the 10th century CE. In: Proceedings of the Indian history congress, vol vol 71. Indian History Congress, pp 80–87

Prati G (2024) Migration and adaptation: the impact of migration on women's adaptive capacity in the Mahanadi Delta, India [doctoral dissertation, University of Southampton]

Prati G, Cazcarro I, Hazra S (2022) Gender dimensions of the migration, sustainability and care nexus: the case study of the Mahanadi delta, India. Curr Res Environ Sustain 4:100104. https://doi.org/10.1016/j.crsust.2021.100104

Rahman MM, Ghosh T, Salehin M, Ghosh A, Haque A, Hossain MA, Kleinn J, Lazar AN, Lázár AN, Salehin M, Shaw R, Szczuciński W, Landis WG, Hutton CW (2020) Ganges-Brahmaputra-Meghna delta, Bangladesh and India: A transnational mega-delta. In: Deltas in the Anthropocene. Palgrave Macmillan, pp 23–51

Rahman MM, Penny G, Mondal MS, Zaman MH, Kryston A, Salehin M, Nahar Q, Islam MS, Bolster D, Tank JL, Müller MF (2019) Salinization in large river deltas: drivers, impacts and socio-hydrological feedbacks. Water Secur 6:100024. https://doi.org/10.1016/j.wasec.2019.100024

Raju KV, Shashidhara HL, Reddy NN, Babu N (2007) Participatory Irrigation Management in Andhra Pradesh

Ramakrishnan A (2021) Caste realities and the struggles of India's Adivasi and Dalit population in accessing groundwater. In: Groundwater law and management in India: from an elitist to an egalitarian paradigm. Routledge, pp 31–48

Ramakrishnan S, Anusha G, Kirupasankar K, Venkatesh C, Pradeep S (2021) A study on the water quality assessment of Bhavani River in Tamil Nadu. In: IOP conference series: materials science

and engineering vol 1145, no 1. IOP Publishing, p 012068. https://doi.org/10.1088/1757-899X/1145/1/012068
Rao KN, Subraelu P, Kumar KCVN, Demudu G, Malini BH, Rajawat AS, Ajai (2010) Impacts of sediment retention by dams on delta shoreline recession: evidences from the Krishna and Godavari deltas, India. Earth Surf Process LandfS J Br Geomorphol Res Group 35(7):817–827
Ravikumar P, Chandrasekar V, Bhaskaran G (2024) Real time rice acreage estimation in delta region of Thanjavur District, Tamil Nadu, India: based on RISAT-1 (C band) SAR data
Ray-Bennett NS (2009) Multiple disasters and policy responses in pre-and post-independence Orissa, India. Disasters 33(2):274–290. https://doi.org/10.1111/j.1467-7717.2008.01074.x
Reshma KN, Murali RM (2018) Current status and decadal growth analysis of Krishna-Godavari delta regions using remote sensing. J Coast Res 85(sp1):1416–1420. https://doi.org/10.2112/SI85-284.1
Rout S, Tete N (2023) Household perception on water consumption in urban area: a case study of Cuttack city, Odisha, India. Sch Res J Interdiscip Stud 12(78):470–486. https://doi.org/10.5281/zenodo.8300927
Ruban P, Gunaseelan C (2011) Antibiotic resistance of bacteria from Krishna Godavari Basin, bay of Bengal, India. Environ Exp Biol 9(1):33–136
Ryder SS (2017) A bridge to challenging environmental inequality: intersectionality, environmental justice, and disaster vulnerability. In: Social thought and research: a continuation of the mid-American review of sociology, pp 85–115
Sabharwal NS, Sonalkar W (2015) Dalit women in India: at the crossroads of gender, class, and caste. Glob Justice Theory Pract Rhetor 8(1). https://doi.org/10.21248/gjn.8.1.54
Sahoo KC, Suman SS, Mishra M, Sinha A, Das D, Pati S (2025) Water, sanitation, and hygiene among transgender population living in urban informal settlements: a qualitative study in Odisha, India. Int J Transgender Health:1–12. https://doi.org/10.1080/26895269.2025.2452186
Samanta A, Chakraborti K, Bandopadhyay M, Sengupta R (2013) Moule, honey collectors of Sundarbans and their ITKs. Am J Med Sci 1(2):1–6
Samuel A, Joy KJ, Bhagat S (2017) Integrated water management of the Mahanadi Basin
Sangameswaran P (2009) Water: perspectives of governance
Sarangi N, Kumaraiah P, Rangarcharyulu PV, Giri BS (2004) Status of freshwater aquaculture in Krishna-Godavari delta: a profile. Regional Research Centre, Central Institute of Freshwater Aquaculture
Sebunya J, Gichuki A (2024) The impact of participatory planning on sustainable development: a literature review. J Strat Manag 4(4):1–9
Sekovski I, Newton A, Dennison WC (2012) Megacities in the coastal zone: using a driver-pressure-state-impact-response framework to address complex environmental problems. Estuar Coast Shelf Sci 96:48–59. https://doi.org/10.1016/j.ecss.2011.07.011
Selvakumar S, Chandrasekar N, Srinivas Y, Simon-peter T, Magesh NS (2014) Evaluation of the groundwater quality along coastal stretch between Vembar and Taruvaikulam, Tamil Nadu, India; a statistical approach. J Coast Sci 1:22–26
Sen A, Mukherjee J, Ghosh A, Fickert T (2024) For, of, and by the community: critical place-based reflections on mangroves regeneration in the Indian Sundarbans. Community Dev J bsae066. https://doi.org/10.1093/cdj/bsae066
Sengupta A, Hazra A (n.d.) Man, nature and adversity of livelihood: a study of the deltaic Sundarbans of West Bengal. Vols XIII–XIV, p 76
Shah SH, Gibson RB (2013) Large dam development in India: sustainability criteria for the assessment of critical river basin infrastructure. Int J River Basin Manag 11(1):33–53. https://doi.org/10.1080/15715124.2012.754445

Singh A, Naik G (2024) Rural drinking water supply program and societal development: evidence from the early implementation phase of India's Jal Jeevan Mission. PLoS One 19(11):e0312144. https://doi.org/10.1371/journal.pone.0312144

Singh A, Singai CB, Srivastava S, Sivam S (2009) Inclusive water governance: a global necessity. Lessons from India. Transit Stud Rev 16(2):598–608. https://doi.org/10.1007/s11300-009-0091-0

Singha P, Pal S (2023) Wetland transformation and its impact on the livelihood of the fishing community in a flood plain river basin of India. Sci Total Environ 858:159547. https://doi.org/10.1016/j.scitotenv.2022.159547

Solomon GR, Balaji RB, Ilayaperumal K (2021) Performance analysis and efficiency enhancement of cooling tower in 210 MW thermal unit. J Phys Conf Ser 2054(1):012062. https://doi.org/10.1088/1742-6596/2054/1/012062

Syvitski JP (2008) Deltas at risk. Sustain Sci 3(1):23–32. https://doi.org/10.1007/s11625-008-0043-3

Syvitski JPM, Kettner AJ, Overeem I, Hutton EWH, Hannon MT, Brakenridge GR, Day J, Vörösmarty C, Saito Y, Giosan L, Nicholls RJ (2009) Sinking deltas due to human activities. Nat Geosci 2(10):681–686. https://doi.org/10.1038/ngeo629

Szabo S, Nicholls RJ, Neumann B, Renaud FG, Matthews Z, Sebesvari Z, AghaKouchak A, Bales R, Ruktanonchai CW, Kloos J, Foufoula-Georgiou E, Wester P, New M, Rhyner J, Hutton C (2016) Making SDGs work for climate change hotspots. Environ Sci Policy Sustain Dev 58(6):24–33. https://doi.org/10.1080/00139157.2016.1209016

Tessler ZD, Vörösmarty CJ, Grossberg M, Gladkova I, Aizenman H, Syvitski JPM, Foufoula-Georgiou E (2015) Profiling risk and sustainability in coastal deltas of the world. Science 349(6248):638–643. https://doi.org/10.1126/science.aab3574

Touch V, Tan DK, Cook BR, Liu D, Cross R, Tran TA, Utomo A, Yous S, Grunbuhel C, Cowie A (2024) Smallholder farmers' challenges and opportunities: implications for agricultural production, environment and food security. J Environ Manag 370:122536. https://doi.org/10.1016/j.jenvman.2024.122536

Tiwary R, Phansalkar SJ (2007) Dalits' access to water: patterns of deprivation and discrimination. Int J Rural Manag 3(1):43–67. https://doi.org/10.1177/097300520700300103

Vasanthakumar J (2017) Yield gap and constraints limiting rice productivity in Cauvery delta zone of Tamil Nadu. Indian J Nat Sci 8(43):1–5

Venkateshwarlu C (2023) Policy perspectives and state responses on water and politics in two states (Telangana state and Andhra Pradesh). In: Promoting sustainability through water management and climate change adaptation. Springer Nature Singapore, pp 111–117. https://doi.org/10.1007/978-981-99-5479-7_8

Vyas P, Sengupta K (2012) Mangrove conservation and restoration in the Indian Sundarbans. In: Sharing lessons on mangrove restoration, p 93

8 Rapid Coastal Flood Extent Mapping with Sentinel-1 SAR Polarization Combinations: Thresholding and Change Detection Methods

Ashrumochan Mohanty, Bhabagrahi Sahoo, and Ravindra Vitthal Kale

Abstract

Real-time flood extent maps are essential for emergency response and disaster management in coastal regions which are frequently impacted by inland rainfall and tropical cyclones. Reliable flood extents enable efficient relief efforts and help reduce the impacts of such disasters. Despite advancements, challenges remain in automating the processing of Sentinel-1 Synthetic Aperture Radar (SAR) imagery to produce consistent, precise inundation maps. SAR's resilience in all weather conditions, including cloud cover, makes it highly suitable for rapid flood assessment in coastal areas. This study addresses the gap in utilizing various SAR polarization combinations for coastal flood mapping. Specifically, ten combinations derived from VH and VV polarizations are evaluated for their effectiveness in delineating flood extents using Thresholding and Change Detection methods. Among the tested combinations, multiplication, squared multiplication, addition, and squared addition show the highest potential for accurate flood mapping in the Brahmani-Baitarani river basin in eastern India. A zero-depth

A. Mohanty · B. Sahoo (✉)
School of Water Resources, Indian Institute of Technology Kharagpur, Kharagpur, India
e-mail: bsahoo2003@yahoo.com; bsahoo@swr.iitkgp.ac.in

A. Mohanty
e-mail: ashrumochan05@gmail.com

R. V. Kale
Surface Water Hydrology Division, National Institute of Hydrology (Department of Water Resources, River Development and Ganga Rejuvenation, Ministry of Jal Shakti, Govt. of India), Roorkee, Uttarakhand, India
e-mail: rvkale.nihr@gov.in

S. Yadav et al. (eds.), *Water Resources in Delta Region in India*, Deltas of the World,
https://doi.org/10.1007/978-3-032-06500-1_8

threshold is applied to further refine flood mapping, reducing overestimated flooded areas and enhancing accuracy by about 7%. Both methods were implemented using the Google Earth Engine platform to successfully identify coastal flood extents with minimal user involvement. Notably, the squared addition polarization performs consistently well, enhancing flood mapping precision. The findings of this study support local authorities in rapid flood assessment, aiding decision-making and resource allocation during flood disasters. This study supports SDG 11 (Sustainable Cities and Communities) and SDG 13 (Climate Action) by enabling rapid and accurate flood delineation for early warning and response.

8.1 Introduction

Widespread flooding results in severe destruction, significant economic losses, and tragic human casualties. With climate change intensifying, the frequency and magnitude of flood are projected to worsen in the coming decades (Hirabayashi et al. 2013; Pati and Sahoo 2022). Flood impacts have surged in recent decades, driven by population growth and economic expansion in vulnerable areas, with global damages averaging USD 651 billion annually during 2000–2019 (Pralle 2019; Tellman et al. 2021). A study by the World Resources Institute (2020) envisages that the financial burden and the number of people affected by floods are expected to escalate by the year 2030; and climate change could increase the annual number of those impacted to 147 million, doubling from the current 72 million. The alarming rise in flood risks demands a better understanding of flood timing and location to reduce losses (Mosavi et al. 2018).

Satellite remote sensing plays a vital role, providing crucial data on flood extent and severity across diverse regions. Advanced flood mapping technologies, such as synthetic aperture radar (SAR), have become essential for monitoring the risks and enhancing preparedness in vulnerable regions (Rawat et al. 2025). Despite advancements in satellite-based flood mapping, the existing methods often face limitations, particularly in regions with dense cloud cover areas (Huang et al. 2025; Zhuang et al. 2025). Addressing these challenges requires innovative approaches that harness the unique capabilities of advanced remote sensing technologies like SAR (Mohanty et al. 2025; Tejaswi et al. 2025). Flood mapping using remote sensing began with the launch of ERTS-1 (Earth Resources Technology Satellite, later Landsat-1) in 1972 (Ruggles 1975). Passive satellite imagery, like the United States Geological Survey (USGS) Landsat and European Space Agency (ESA) Sentinel-2, has proven effective in mapping floods at the catchment scale (Goffi et al. 2020; Kordelas et al. 2019). Further, the Landsat archives have enabled the analysis of past floods and the validation of hydraulic models (Ogilvie et al. 2018). Cloud cover during flood events often hampers the effectiveness of satellite-based data collection, particularly for passive remote sensing systems that rely on visible and infrared light. These systems are further limited in detecting flooding beneath forest canopies, where water pixels are obscured by dense tree cover. Even longer wavelengths, such as those in the L-band, struggle to fully penetrate vegetation, making accurate flood extent mapping in such environments challenging (Tazmul Islam and Meng 2022). To overcome these limitations in flood extent mapping, Sentinel-1 Synthetic Aperture Radar (SAR) polarization combinations are advocated. The vertical-horizontal (VH) and vertical-vertical (VV) polarizations in Sentinel-1 imageries could help in distinguishing between water and non-water surfaces. By leveraging the cloud-based capabilities of Google Earth Engine (GEE), these polarizations ensure scalability and applicability across diverse regions (Matgen et al. 2016; Westerhoff et al. 2013). Unlike traditional multispectral systems, SAR remains fully operational during critical flood events, providing consistent and reliable data. Moreover, SAR is uniquely suited for flood mapping due to its ability to penetrate cloud cover and operate both day and night in all weather conditions. This resilience to adverse conditions makes SAR an indispensable tool for accurate and timely flood monitoring, particularly in regions prone to extreme weather events. The deployment of ESA's Copernicus Programme Sentinel-1 satellites has further popularized SAR for flood studies (Cao et al. 2019; Martinis et al. 2015; Uddin et al. 2019). However, challenges such as distinguishing water from water-like surfaces and underutilization of SAR polarization combinations persist, creating opportunities for innovative solutions. Traditional SAR-based flood mapping methods include histogram thresholding, active contour modelling, region growing, change detection, classification, and segmentation (Landuyt et al. 2019). However, distinguishing water from water-like surfaces remains a challenge (Shen et al. 2019). Despite this, Sentinel-1A (launched in 2014) and Sentinel-1B (launched in 2016) based public domain high-resolution (10 m) data has become a preferred solution for large-scale flood mapping (Zhang et al. 2020). With a 6-day revisit cycle in Europe and mid-latitudes and 12 days globally, these satellites enable consistent monitoring of surface changes across vast regions (DeVries et al. 2020). SAR sensors detect floods by leveraging the unique scattering behaviour of open water surfaces. Microwave energy transmitted at off-nadir angles produces specular reflection, directing most energy away from the sensor and resulting in low backscatter signals. Flood detection algorithms utilize this principle by applying backscatter thresholds to classify water pixels, often combined with object-based methods for improved accuracy (Chini et al. 2017). Common techniques for flood analysis using SAR imagery include: (a) histogram thresholding, which uses a single backscatter threshold to differentiate flooded from non-flooded areas; (b) RGB composition, highlighting features with colour coding for visual flood detection; (c) change detection, comparing pre- and post-flood backscatter values to identify affected areas; (d) data fusion with SAR and optical images; and (e) classification methods, both supervised and unsupervised, to delineate flooded regions (Tazmul Islam and Meng 2022). Polarization combinations, such as VV and VH, provide complementary information that enhances flood detection by improving the distinction between water and non-water surfaces in complex environments. This capability significantly increases the precision and reliability of flood extent mapping, addressing critical gaps in current methods.

By addressing gaps in current flood mapping methods, the study advocated herein provides a foundation for developing operational tools that can be applied globally, particularly in regions prone to recurrent flooding. This can also aid in real-time flood monitoring systems to mitigate the impacts of climate change on vulnerable populations. For flood extent mapping, this approach uses various polarization combinations of Sentinel-1 SAR imageries. Two methods, thresholding and change detection, are applied to map flooded areas, and their outputs with different polarization combinations are compared. By leveraging the cloud computing capabilities of GEE, this approach is tested for scalability and ease of application across diverse regions.

8.2 Materials and Methods

8.2.1 Study Area

The Brahmani-Baitarani River delta (see Fig. 8.1), adjacent to Bengal, is one of the most flood-prone regions in India. Formed by the confluence of both the rivers, i.e., Brahmani and Baitarani, this delta region is classified as a high-vulnerable flood zone (Sahoo et al. 2020), resulting in widespread disruption, economic losses, and damage to infrastructure. Due to its low-lying landscape, climate-induced altered weather patterns, and extensive paddy-dominated agricultural practices, this delta is exposed to both riverine and tropical cyclonic floodings. The delta region receives streamflow from the Brahmani River at the Jenapur outlet and from the Baitarani River at the Anandpur outlet (Mohanty et al. 2024). The Bay of Bengal witnesses around four cyclones yearly (Mondal et al. 2022), contributing 5–6% of the world's tropical storms. In the Brahmani-Baitarani River delta, flood risks are intensified by tidal backwater flow from the Bay of Bengal, which, during high tides, elevates water levels and worsens flooding. These tidal effects, combined with seasonal monsoon floods, create complex dynamics where the water level fluctuates due to both river discharge and tidal phases.

This unique interplay of factors makes the delta an ideal basin to demonstrate advanced flood mapping techniques. The Land Use Land Cover (LULC) classification indicates that agricultural land (34.2%) and barren land (31.2%) dominate the region, together covering over 65% of the area. The Brahmani-Baitarani River delta consists of agricultural land (34.2%) and barren land (31.2%) as the dominant land cover types, together making up over 65% of the area. It also includes dense forests (13%) and open forests (8.5%), contributing to the region's green cover. Additionally, the region is covered by built-up areas (9.6%) and water bodies (3.5%). The LULC distribution is represented in Fig. 8.2. This study utilizes the Sentinel-1 SAR dataset, specifically designed for Earth observation and disaster monitoring. The dataset used in this study includes dual-polarization bands, VH and VV, acquired in Interferometric Wide (IW) swath mode with a spatial resolution of approximately 10 m. These polarizations are particularly effective for flood mapping,

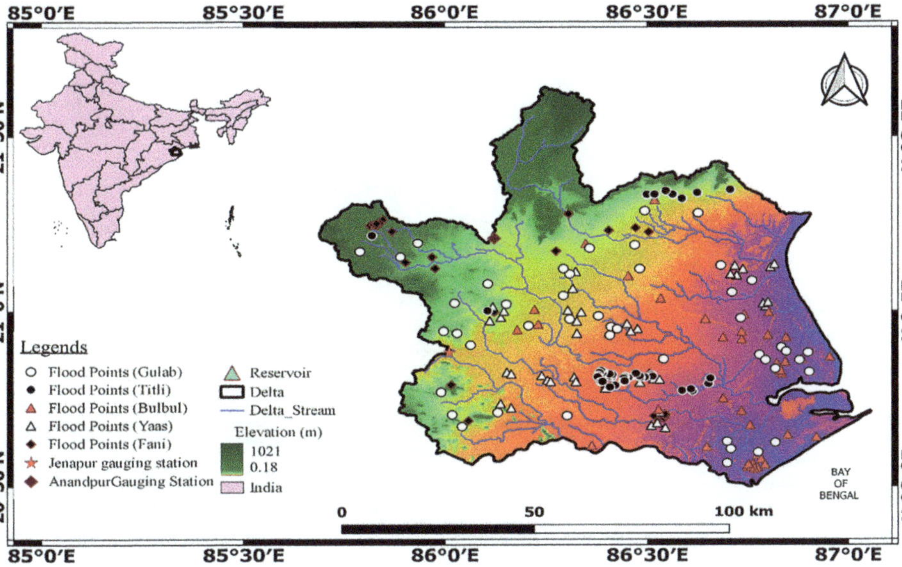

Fig. 8.1 Brahmani-Baitarani river basin map showing elevation, reservoirs, gauging stations, and flood validation points derived from social media for differnt cyclone events

as they use differences in backscatter characteristics to accurately distinguish between water and non-water surfaces.

8.2.2 Method

This study uses Sentinel-1 Synthetic SAR imagery to derive flood extents, using GEE. Dual polarizations (VH and VV) and their derived combinations are used to enhance flood detection by exploring the unique backscatter characteristics of water and non-water surfaces. The methodology consists of three main stages *viz.*, pre-processing, flood extent mapping, and refinement through flood depth estimation. Pre-processing includes filtering Sentinel-1 imageries based on the study area, polarization, and temporal criteria, followed by speckle filtering to reduce noise and ensure data quality. The flood extent mapping is carried out by the Threshold Method, which classifies non-flooded and flooded areas by applying a backscatter threshold, and the Change Detection Method, which identifies changes in backscatter values between pre-flood and flood images to detect inundated regions. Both these methods are refined using a zero-depth flood estimation approach by employing the Digital Elevation Model (DEM) data to eliminate overestimated areas, such as roads and pavements with similar backscatter characteristics to water.

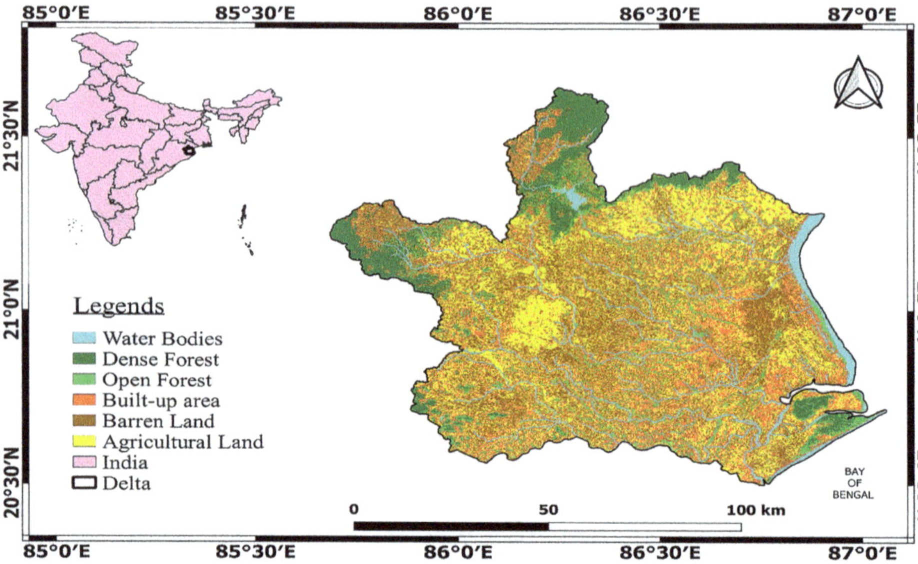

Fig. 8.2 Land use and land cover map of the Brahmani-Baitarani river delta for the year 2023, derived from Landsat-8 imagery

8.2.2.1 Image Pre-processing

Though pre-processed Sentinel-1 imageries are available in the GEE data catalog, several additional steps are necessary to prepare the dataset for flood extent mapping. These steps include defining the study area, filtering the Sentinel-1 image collection, and applying a speckle filter to reduce noise and enhance image quality. First, the study area boundary is identified either directly within GEE using geometry tools or uploaded as a shapefile to represent the area of interest. Once the boundary is established, Sentinel-1 images are filtered to include only those relevant to the selected study area and time period. The filtering process starts with loading the complete Sentinel-1 image collection into GEE. First, the dataset is filtered by instrument mode, with the Interferometric Wide Swath mode being selected for its suitability in flood mapping applications. Then, the dataset is further refined by selecting images based on orbiting pass direction, either ascending or descending. For this study, ascending orbit images were chosen due to better coverage of flood dates in the study area. The filtering process also accounts for spatial resolution, ensuring that only 10 m × 10 m resolution satellite imageries are used for analysis. Subsequently, the next filter is used to include images which fully cover the study area, as defined by the uploaded boundary shapefile. Finally, the dataset is restricted to include only the VV and VH polarizations, which are critical for flood detection. Sentinel-1 SAR imagery inherently exhibits low backscatter values, which are pre-processed by changing the backscatter intensity from linear to a logarithmic scale in decibels (dB) using the standard equation provided in the GEE data catalog as:

$$\sigma_0(\text{dB}) = 10 \log 10(\sigma_0) \quad (1)$$

where, σ_0 (dB) represents the backscatter image in decibels.

The amount of backscatter signal received by the sensor depends on the structural properties of the observed surface. Variations in backscatter values are influenced by the polarization mode, with VH providing higher returns in volume scattering regions, such as vegetated areas, due to interactions with multiple branches. On the other hand, VV polarization typically shows stronger returns in specular scattering environments, such as smooth surfaces. The urban buildings exhibit higher backscatter values in VV polarization due to the double-bounce effect, making them appear brighter in the imagery. Conversely, smooth, flat surfaces like water bodies or airport runways produce low magnitude of backscatter values because of specular reflection, directing most of the transmitted energy away from the radar. In addition to analyzing individual VH and VV polarizations, this study explores ten derived combinations of these channels to evaluate their potential for urban flood mapping. These combinations include basic arithmetic operations (e.g., addition, subtraction) as well as squared and division operations, as summarized in Table 8.1. These combinations are designed to enhance the spectral information available for flood detection, allowing for better differentiation of flooded areas. By leveraging these combinations, it is possible to highlight specific geographical features and improve the accuracy of flood area identification.

The addition and multiplication polarization combinations highlight areas with higher backscatter values, such as urban structures, while subtraction combinations are better suited for identifying low-backscatter areas like water bodies. This comprehensive approach allows for enhanced analysis of flooded areas by leveraging the complementary strengths of individual and combined polarizations. After filtering, a speckle filter is applied to mitigate the inherent graininess or speckle effect, a characteristic of radar images. This noise, if left unaddressed, could result in misclassification during the analysis. The smoothing filter with a circular radius of 50 m is used to reduce this noise while preserving the essential backscatter characteristics of the imagery. This pre-processing workflow ensures that the Sentinel-1 SAR images are appropriately prepared for accurate and reliable flood extent mapping.

Table 8.1 Different polarization combinations of Sentinal-1 imageries

Sl. No	Operation	Combination	Sl. No	Operation	Combination
1	Single	VH	6	Division	VV/VH
2	Single	VV	7	Squared addition	$VH^2 + VV^2$
3	Addition	VH + VV	8	Squared subtraction	$VH^2 – VV^2$
4	Subtraction	VH–VV	9	Squared multiplication	$VH^2 \times VV^2$
5	Multiplication	VH × VV	10	Squared division	VV^2/VH^2

8.2.2.2 Computing Flood Extent

The VH and VV base polarizations from Sentinel-1 SAR imagery, along with eight derived polarization combinations calculated using GEE, are utilized in this study using two distinct methods: thresholding and change detection. These methods are selected for their computational efficiency, scalability and ease of implementation, making them well-suited for larger-scale flood mapping applications. Despite their frequent use in flood mapping studies, a comparative analysis of these techniques in a single study area remains limited.

Threshold Method

The pre-processing and polarization combinations are complete, and the next step is to differentiate flooded areas from non-flooded regions. For this purpose, a histogram thresholding technique, as described by Otsu et al. (1979), is applied. Histogram thresholding is a widely used method in digital image processing to separate two features with distinct spectral characteristics. The objective is to identify an optimal threshold value that can categorize the satellite imageries into non-flooded and flooded areas. In VV and VH polarizations, flooded regions exhibit low backscatter values due to specular reflection on the water surface, while non-flooded areas display higher backscatter values. A histogram with a bimodal distribution is ideal for identifying the threshold value, as it presents two distinct peaks separated by a valley. One peak corresponds to flooded pixels with low backscatter values, and the other represents non-flooded features with higher backscatter values. A representative subset of the study area, encompassing both flooded and non-flooded regions, is selected for threshold determination. Each polarisation combination's threshold value is then calculated based on the pixel value distribution within the subset. This process is automated in GEE using Otsu's thresholding method, which ensures consistent and precise classification.

Change Detection Method

Change detection is a widely used remote sensing technique for analysing variations in the Earth's surface due to natural events or human activities. In this study, the method focuses on detecting changes in backscatter values between pre-flood and flood periods to identify flooded areas. The analysis relies on co-registered Sentinel-1 SAR images, where the backscatter coefficient is calculated for both pre-flood and flood imagery across all polarization combinations. The pixel-based change detection approach is employed by subtracting the backscatter values of the flood-period images from the pre-flood images, highlighting areas where significant changes have occurred. Flooded areas are characterized by a substantial reduction in backscatter values compared to their pre-flood state. This reduction is attributed to specular reflection caused by the presence of water, which is particularly noticeable in open areas and smooth surfaces such as roads. These features exhibit lower backscatter coefficients during flood periods, enabling clear differentiation between flooded and non-flooded regions. Following the classification, a visually identified subset of the study area, containing representative samples of both flooded and non-flooded pixels, is used. Otsu's method is applied to this subset to determine the threshold, ensuring accurate classification tailored to the specific flood event. This approach avoids using arbitrary cut-off values and provides a robust method for identifying flood-induced changes.

Following the classification of flooded and non-flooded areas, three refinement steps are implemented to remove potential misclassifications and improve the accuracy of the flood extent maps. First, permanent and semi-permanent water bodies, defined as areas with water coverage exceeding 10 months annually, were excluded to eliminate false-positive flood detections. The Joint Research Centre Global Surface Water dataset, which provides temporal water distribution data from 1984 to 2018, is utilized for this refinement. Second, regions with land slopes greater than 5% are removed, as these areas are unlikely to retain water. For this step, the Shuttle Radar Topography Mission Digital Elevation Model available in the GEE catalog is used (UN-SPIDER, 2021). Lastly, areas containing fewer than eight connected pixels, classified as flooded areas, are eliminated as these small regions are considered noise and not indicative of actual flooding. After applying these refinement steps, the final flood layer for each polarization combination is generated and designated as the initial flood map. This approach ensured the production of accurate and reliable flood extent maps, minimizing false positives and enhancing the overall quality of the results.

8.2.2.3 Validation Using Social Media

In the absence of real-time aerial photographs, social media data offers a viable alternative for validating flood extent maps generated by Sentinel-1 SAR imagery (Guo et al. 2023). Geotagged posts, images, and videos shared on platforms like X, Facebook, and Instagram provide near-real-time insights into flood events, particularly in urban areas where social media activity is higher. Studies have demonstrated the efficacy of social media in flood mapping by extracting spatial and temporal information from public posts. For instance, location-specific posts during the 2015 Chennai (India) floods are utilized to validate water height interpolations, achieving high accuracy when combined with field survey data (Karmegam et al. 2021). Similarly, social media-derived flood spots are integrated with hydrodynamic models to validate flood extent maps for the 2020 Chengdu flood, revealing a strong correlation with precipitation patterns (Guo et al. 2023). Despite challenges like noise and irrelevant content, filtering and keyword-based extraction methodologies have enabled precise identification of flood-related posts. By combining this data with SAR-derived maps, this study aims to validate flood extents in the Brahmani-Baitarani River basin, supporting robust decision-making and efficient disaster management. In this study, social media posts, including geotagged images and videos (see Fig. 8.3), are collected from platforms like X, Facebook, and Instagram to validate flood extent maps for the Brahmani-Baitarani River delta.

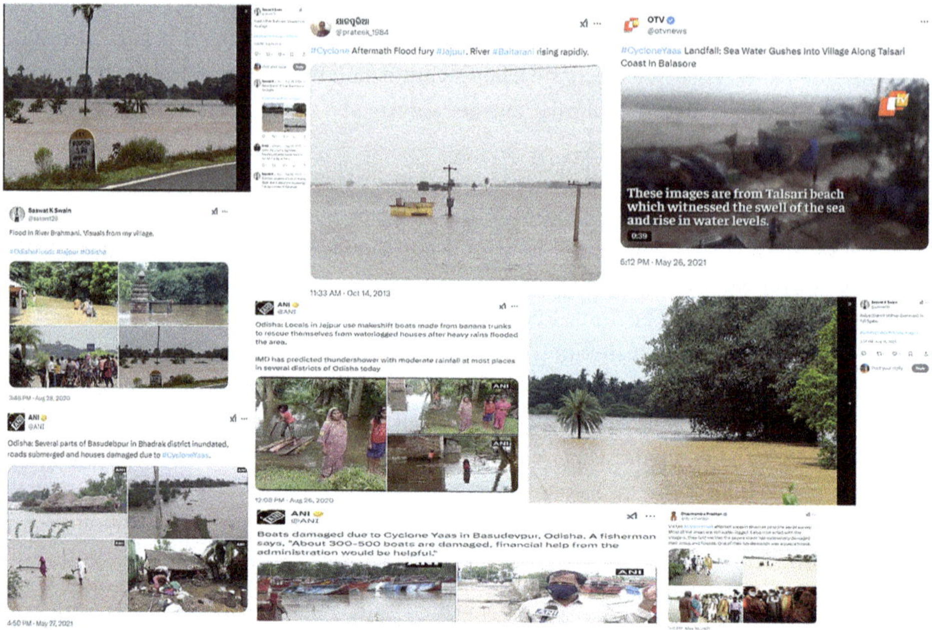

Fig. 8.3 Social media posts with geotagged images and videos from X, Facebook, and Instagram used to validate flood inundation maps during Cyclone events

To assess the accuracy of the SAR-derived flood maps, the Overall Accuracy (OA) metric is applied. OA is calculated as the ratio of correctly classified flood and non-flood areas to the total number of validation points, defined by:

$$OA = \frac{TP + TN}{TP + TN + FP + FN} \tag{2}$$

where True Positives (TP) represent social media-reported flood spots that are correctly identified as flooded in the SAR-derived map, ensuring alignment between the two datasets. True Negatives (TN) denote areas accurately classified as non-flooded in both the SAR maps and social media data, demonstrating agreement in identifying unaffected regions. Conversely, False Positives (FP) refer to areas mistakenly classified as flooded in the SAR-derived map, even though social media data indicates they are not flooded. False Negatives (FN) occur when the flood spots reported on social media are not recognized as flooded in the SAR-derived map, highlighting missed detections. These metrics collectively provide a robust framework for assessing the reliability of SAR-derived flood maps when validated against social media data.

8.3 Results

The analysis focuses on flood events associated with five significant tropical cyclones that impacted the eastern coast of India in recent years. Cyclone Fani (April 25–May 4, 2019) and Cyclone Yaas (May 23–28, 2021) caused widespread flooding in the Brahmani-Baitarani river basin due to heavy rainfall and storm surges. Similarly, Cyclones Titli (October 7–14, 2018), Bulbul (November 5–12, 2019), and Gulab (September 27–28, 2021) also contributed to substantial inundation in the region, exacerbating existing flood risks in low-lying and coastal areas. These events are selected for analysis due to their varying meteorological characteristics, spatial extents and impacts, providing a comprehensive dataset for evaluating flood mapping. Hereafter, the Cyclones Fani, Yaas, Titli, Bulbul and Gulab induced floods are named as Flood Event [1], [2], [3], [4] and [5], respectively. The flood check-in points for social media validation, derived from geotagged posts and videos, are shown in Fig. 8.3.

Figure 8.4 demonstrates the backscatter coefficient values for VH and VV polarizations, where dark regions indicate low backscatter values corresponding to water bodies and roads; while bright areas represent high backscatter from buildings and flooded vegetation. The reference image Fig. 8.4a, captured on a date closely matching the Sentinel-1 images on 26-Sep. 2021, has been used to validate the flood mapping results of Flood event [5]. The visual assessment of various polarization combinations derived from Sentinel-1 SAR imagery indicates that most combinations offer limited additional insights into flooded areas compared to the primary VH and VV polarizations. Notably, the VH + VV combination enhances the visibility of high-backscatter regions, particularly urban structures like buildings, which stand out as bright areas. Both VH and VV polarizations effectively identify water bodies as dark regions due to low backscatter values. However, VV polarization provides better delineation of urban structures, enhancing its utility in distinguishing high-backscatter features like buildings and flooded vegetation. While additional combinations, such as VH − VV and VH/VV are explored, their visual contribution is found to be minimal. These combinations, along with others like Multiplication, Squared Addition, Squared Multiplication, Squared Division and Squared Subtraction, do not yield substantial visual differences but are nonetheless utilized as inputs for further analysis using thresholding and change detection methods. Other combinations (not displayed in figure) are excluded from discussion as they failed to exhibit any distinctive characteristics or provide meaningful identification of flooded or non-flooded features.

8.3.1 Accuracy Assessment of Threshold Method

The histogram distributions of backscatter coefficient values for VH, VV, and their derived combinations (VH + VV), (VH − VV), (VH × VV), (VV/VH), ($VH^2 + VV^2$), (VH^2 −

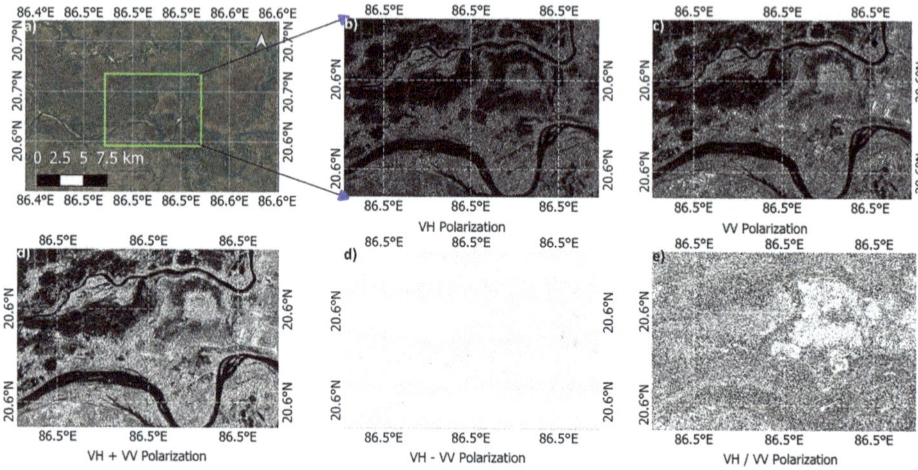

Fig. 8.4 Comparison of Sentinel-1 SAR polarizations: **a** Reflectance image, **b** VH polarization, **c** VV polarization, **d** VH + VV polarization, **d** VH–VV polarization, and **e** VH/VV polarization highlighting flooded and non-flooded features

VV^2), ($VH^2 \times VV^2$), and (VV^2/VH^2) are analyzed to distinguish the flooded and non-flooded areas. In Fig. 8.5, the histograms are shown with the x-axis denoting backscatter coefficient values, and the y-axis denoting their frequencies. The identified threshold values for each polarization combination, derived using bimodal distributions, serve as the basis for classification. The thresholds for VH and VV polarizations are estimated to be 18.257 dB and 21.738 dB, respectively; while the derived combinations VH + VV and VH − VV yield thresholds of 39.344 dB and −3.894 dB, respectively. The pixels below these threshold values are classified as flooded, while those above are marked as non-flooded. The thresholds for other derived combinations are presented in Table 8.2. To validate the results, this study utilizes a social media validation technique. Additionally, potential misclassifications caused by the features, such as permanent water bodies, high-slope regions and isolated pixels with fewer than eight connected neighbours are addressed. These refinements could effectively reduce false positives, especially in urban environments where smooth surfaces like roads and pavements often mimic the backscatter characteristics of flooded areas.

Among the tested combinations, VV × VH exhibits the highest performance in identifying flooded regions, with an average accuracy of 82% across all flood events. This is followed by VV + VH) and $VH^2 \times VV^2$), which could achieve the average accuracies of 77% and 76.6%, respectively. The (VH^2 VV^2) combination also demonstrates competitive performance with an average accuracy of 74.6%, which is comparable with the individual polarization of VH (74.2%). Conversely, the (VV/VH) and (VV^2/VH^2) combinations exhibit relatively lower average accuracies of 70.6% and 65.2%, respectively. Similarly,

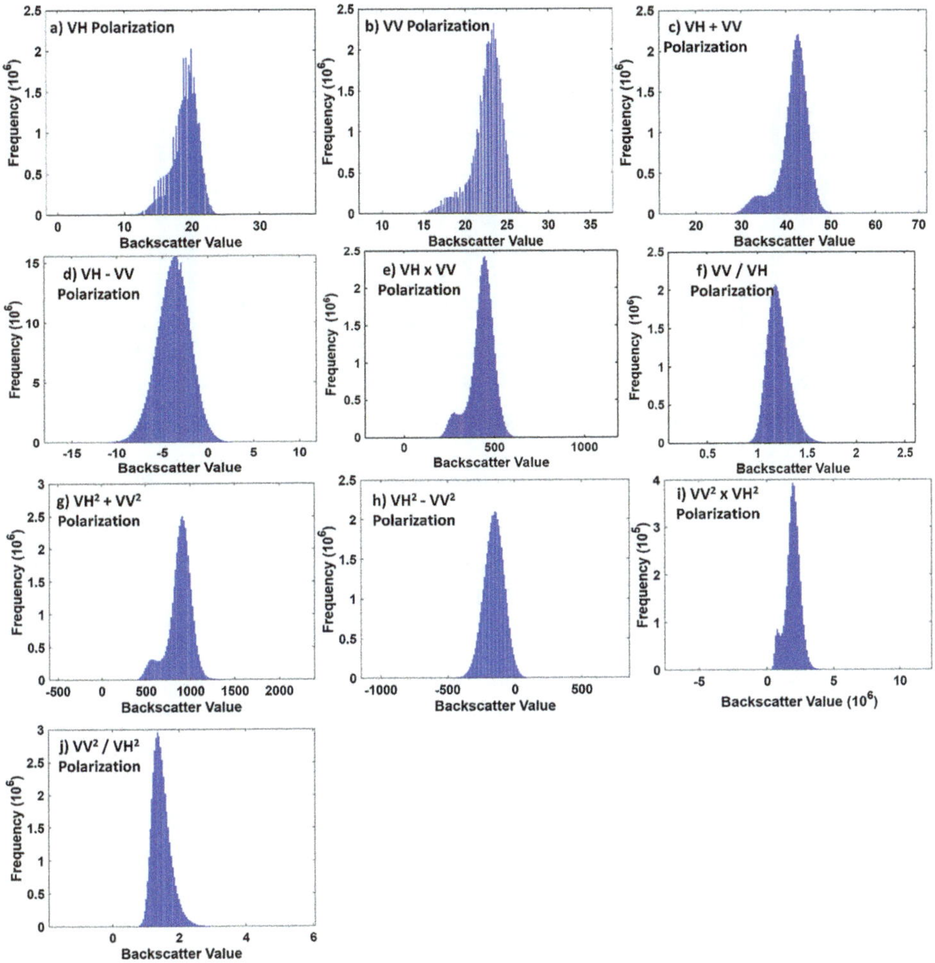

Fig. 8.5 Histogram distributions of backscatter coefficient values for VH, VV, and their derived polarization combinations used for flood mapping using threshold method

Table 8.2 Threshold values for VH, VV, and derived polarization combinations in flood mapping using threshold method

Polarization	Threshold (dB)	Polarization	Threshold (dB)
VH	18.257	VV/VH	1.226
VV	21.738	$VH^2 - VV^2$	−164.385
VH + VV	39.344	VV^2/VH^2	1.531
VH − VV	−3.894	$VH^2 + VV^2$	805.275
VH × VV	398.030	$VH^2 \times VV^2$	179825.000

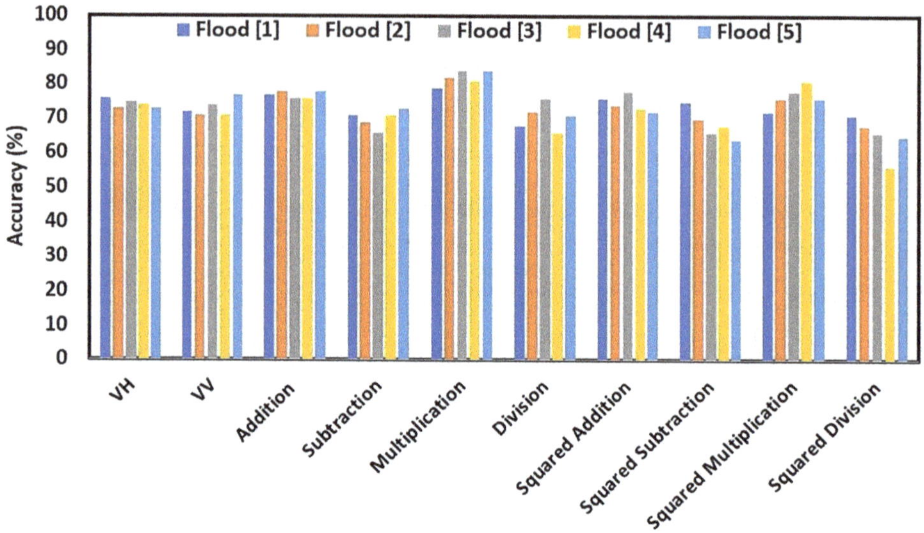

Fig. 8.6 Accuracy of polarization combinations for five cyclonic flood events with different polarization combinations using the threshold method

the (VH − VV) and ($VH^2 - VV^2$) combinations show moderate performance, with the average accuracies of 70% and 68.6%, respectively. Among the individual polarizations, VV could achieve an average accuracy of 73%, slightly below that of VH. The detailed OA for different flood events and polarization combinations is shown in Fig. 8.6. These findings highlight the superior capability of Multiplication, Addition, and their squared variants in distinguishing flooded areas with higher precision. These results underscore the potential of advanced polarization combinations in operational disaster management and flood response applications.

8.3.2 Accuracy Assessment of the Change Detection Method

The change detection approach relies on the comparison of pre-flood and post-flood Sentinel-1 SAR images to identify flooded areas. By analyzing the differences in backscatter coefficients between these two time periods, the regions impacted by flooding can be effectively delineated. For this analysis, the pre-flood imagery is chosen as the baseline, representing the dry state of the landscape, while the post-flood imagery captured the inundated conditions. The differences in backscatter values for each polarization combination (VH, VV, and their derived metrics) are computed with the flooded areas typically exhibiting a significant decrease in backscatter due to specular reflection from water surfaces. To determine the threshold values required for classification, a representative subset of the study area is selected. This subset encompasses a mix of flooded and non-flooded

Table 8.3 Threshold values for VH, VV, and derived polarization combinations in flood mapping using change detection method

Polarization	Threshold (dB)	Polarization	Threshold (dB)
VH	17.639	VV/VH	1.226
VV	20.927	$VH^2 - VV^2$	−148.815
VH + VV	38.447	VV^2/VH^2	1.543
VH − VV	−3.603	$VH^2 + VV^2$	763.620
VH × VV	375.725	$VH^2 \times VV^2$	158830.000

regions, ensuring that the threshold values are robust and reflective of the broader study area. Using Otsu's method, the thresholds are calculated automatically for each combination, leveraging the bimodal distribution of pixel backscatter values within the subset. The flooded pixels characterized by lower backscatter values are classified as below the threshold, while non-flooded pixels are classified as above (Table 8.3).

The histogram distributions of backscatter coefficient values for VH, VV, and their derived combinations (VH + VV), (VH − VV), (VH × VV), (VV/VH), ($VH^2 + VV^2$), ($VH^2 - VV^2$), ($VH^2 \times VV^2$), and (VV^2/VH^2) are analyzed to classify flooded and non-flooded regions effectively. The bimodal distributions in the histograms illustrated in Fig. 8.7 enable the identification of distinct threshold values for classification. For instance, the thresholds for VH and VV polarizations are calculated as 17.639 dB and 20.927 dB, respectively; while the derived combinations of (VH + VV) and (VH-VV) have the thresholds of 38.447 dB and −3.603 dB, respectively. The pixels with backscatter values below the thresholds are classified as flooded, while those above are labelled as non-flooded. Refinements are implemented to reduce misclassifications caused by permanent water bodies, high-slope regions, and isolated pixels with fewer than eight connected neighbours. These refinements are particularly beneficial in urban environments, where smooth surfaces like roads and pavements often exhibit backscatter characteristics similar to flooded areas. The accuracy of different polarization combinations is derived using social media validation, leveraging geotagged flood points from platforms such as X, Instagram, and Facebook. Geotagged images, videos, and audio recordings are analysed to extract spatial information, enabling a comparison with SAR-derived flood maps based on Eq. (2). Among the tested combinations, Multiplication (VH × VV) exhibits the highest performance, with accuracies averaging 85% across all the selected flood events. This is closely followed by Addition (VH + VV) and Squared Addition ($VH^2 + VV^2$), with average accuracies of 82% and 78%, respectively. Base polarizations, such as VH achieved an average accuracy of 74.8%, while VV achieved 73.8%. Further, Subtraction (VH − VV) and Squared Subtraction ($VH^2 - VV^2$) demonstrate moderate performance, with average accuracies of 65% and 67.6%, respectively. Similarly, Division (VV/VH) and Squared Division (VV^2/VH^2) exhibited relatively lower accuracies of 68.4% and 64.2%, respectively. The detailed OA for different flood events and polarization combinations with the Threshold Method is shown in Fig. 8.8.

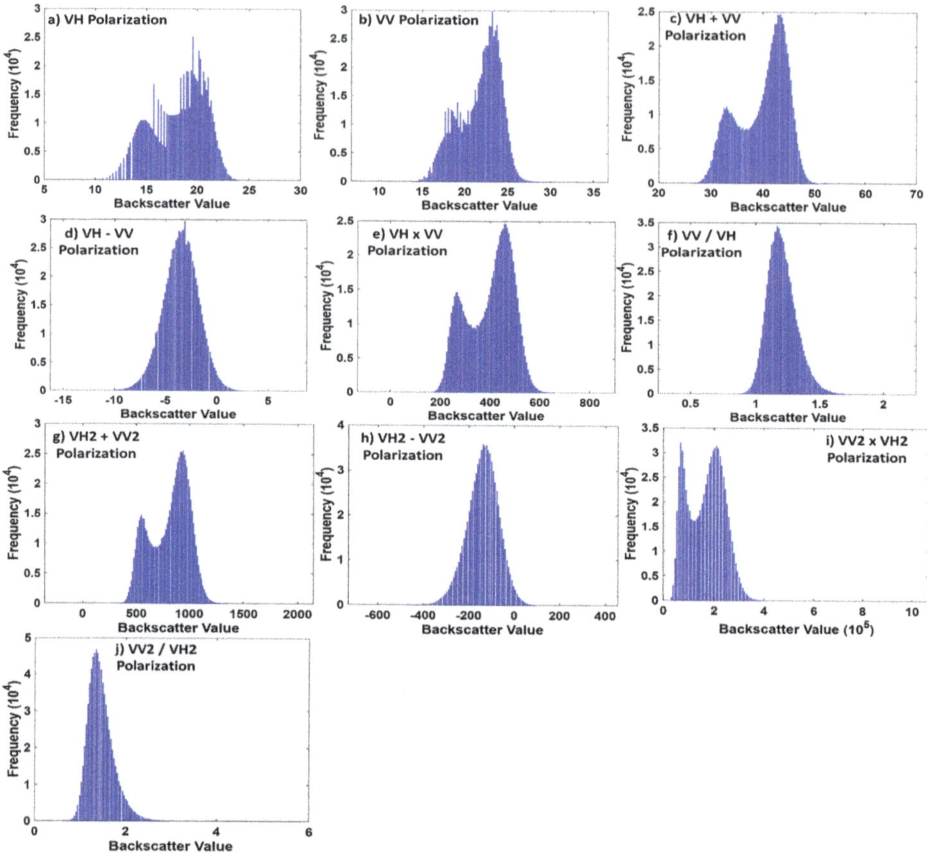

Fig. 8.7 Histogram distributions of backscatter coefficient values for VH, VV, and their derived polarization combinations used for flood mapping using the change detection method

These findings emphasize the superior performance of Multiplication, Addition, and Squared Addition polarizations in delineating flooded regions, further underscoring the utility of advanced polarization combinations in flood mapping. The maps presented in Fig. 8.9 showcase the deltaic flood inundation maps as derived with the Multiplication (VV $\times$ VH) polarization combination using the change detection method, identified as the most effective approach herein. This method consistently provided accurate flood delineation across all five cyclonic events, reinforcing its applicability for operational flood mapping.

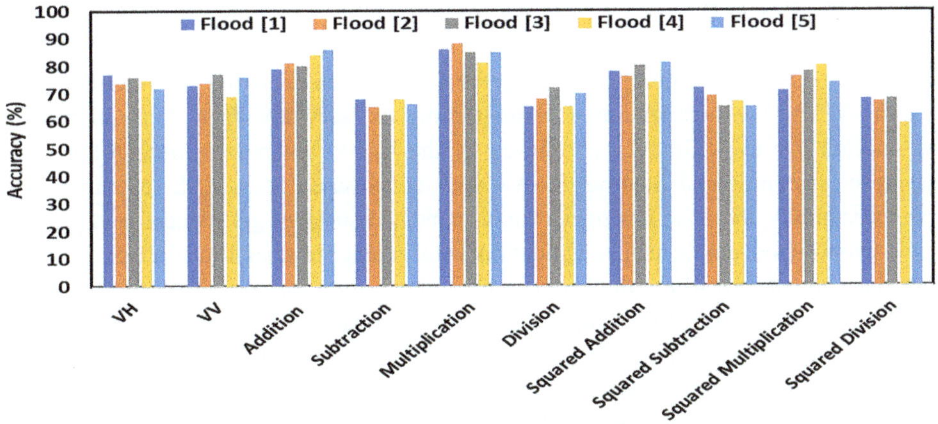

Fig. 8.8 Accuracy of polarization combinations for five different typical cyclonic flood events with different polarization combinations using the threshold method

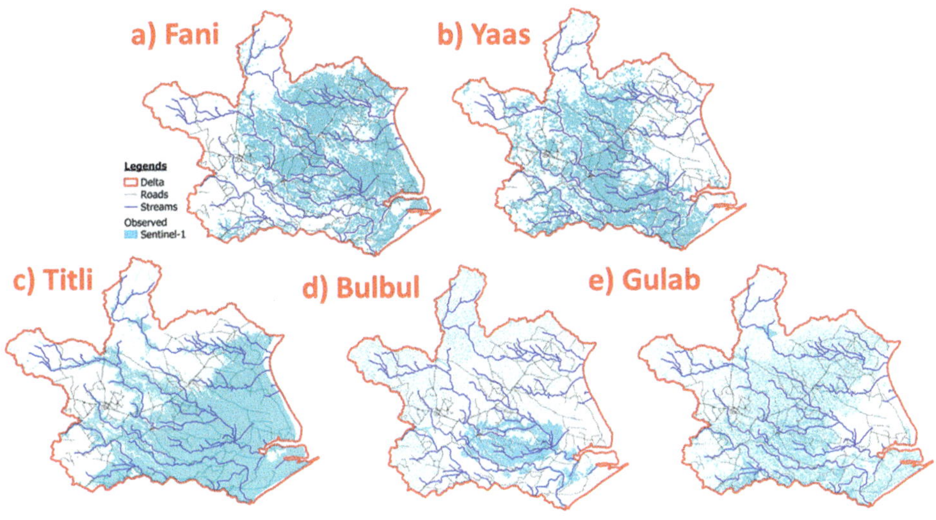

Fig. 8.9 Flood inundation maps derived with the multiplication (VV × VH) polarization combination using the change detection method for the cyclonic events of **a**) Fani, **b**) Yaas, **c**) Titli, **d**) Bulbul, and **e**) Gulab

8.4 Discussion

This study demonstrates that, while using the Change Detection and Thresholding methods, the VH × VV polarization provides the highest accuracy flood delineation across the selected five cyclonic events followed by the VH × VV polarization. Additionally, due to the integration of Sentinel-1 SAR with DEM-based zero-depth screening, slope and water masks, the false positives of non-inundated surfaces get reduced and produces stable event-based usable maps. Validation with geotagged social-media posts corroborates mapped patterns where the aerial or field observations are unavailable. Collectively, these outcomes advance Sustainable Development Goals (SDG) 11 (Sustainable Cities and Communities) by enabling faster, higher-fidelity situational awareness for warnings, evacuation planning, and protection of critical urban assets, and strengthen SDG 13 (Climate Action) by providing a practical, scalable approach to monitor climate-driven flood hazards. Because the workflow runs on GEE and remains lightweight, it can be replicated across regions and events, supporting institutional disaster-risk practices and integration with existing modelling and decision systems; thereby reinforcing city resilience and adaptive capacity.

8.5 Conclusion

The study highlights the potential of polarization combinations such as squared addition, squared multiplication, simple addition, and multiplication of VH and VV polarizations in improving the accuracy of flood mapping using thresholding and change detection methods. The study further improves classification accuracy by eliminating misclassified pixels in areas like roads and pavements, which often share similar backscatter properties with water. This methodology reduces overestimation in flood maps, ensuring that only areas with significant water depth are classified as flooded. The integration of Google Earth Engine (GEE) provides a robust computational platform for processing large-scale Sentinel-1 SAR datasets efficiently, offering scalability and ease of implementation for emergency response and flood management. Additionally, the integration of social media data and real-time validation techniques could enhance the reliability of flood maps, supporting timely and effective decision-making in disaster management and mitigation strategies. The VV × VH polarization combination demonstrates the highest effectiveness, achieving an average accuracy of 85% with the change detection method in deriving reliable flood inundation maps using Sentinel-1 SAR data. This is followed by the VH + VV combination with an average accuracy of 82%, highlighting their potential for reliable flood mapping in disaster management. These findings underline the significance of SAR data for operational flood mapping and emergency response, particularly in regions with limited access to optical data due to cloud cover or other constraints. This study reinforces the value of Sentinel-1 SAR data in addressing global flood challenges and advancing urban resilience. Overall, the workflow advances SDG 11 (Sustainable Cities and Communities) and SDG 13 (Climate Action) by providing scalable, near-real-time flood mapping that helps cities reduce disaster impacts and build climate resilience.

8.6 Recommendation

The Change Detection with VH × VV polarization results in the most accurate flood delineation in the selected tests (with the average Overall Accuracy (OA) ≈ 85%), while the Threshold approach with VH × VV polarization also performs strongly (≈ 82%). The VH + VV and $VH^2 + VV^2$ polarizations further enhance separability between the flooded and non-flooded pixels, consistently outperforming the single-polarization baselines (VH ≈ 74.8%, VV ≈ 73.8%). Further, the maps can be refined by applying DEM-based zero-depth screening, excluding the steep slopes (>5%) and masking the permanent water and removing small isolated components, which reduces false positives over the smooth, non-inundated surfaces such as roads. Social-media geotags effectively validate mapped extents in the Brahmani-Baitarani delta, and the GEE-based workflow scales reliably across multiple cyclonic flood events, indicating portability to similar tide–river systems.

Acknowledgement The first author acknowledges the student fellowship received from the Ministry of Education (MoE), Government of India.

Use of AI in Preparing the Chapter During the preparation of this work, the authors used Perplexity.ai and Grammarly (provided by Indian Institute of Technology Kharagpur, India) in order to improve readability of the sentences and correct grammatical error. After using this tool/service, the authors reviewed and edited the content as needed and take full responsibility for the content of the published article.

References

Cao H, Zhang H, Wang C, Zhang B (2019) Operational flood detection using Sentinel-1 SAR data over large areas. Water 11. https://doi.org/10.3390/w11040786

Chini M, Hostache R, Giustarini L, Matgen P (2017) A hierarchical split-based approach for parametric thresholding of SAR images: flood inundation as a test case. IEEE Trans Geosci Remote Sens 55:6975–6988. https://doi.org/10.1109/TGRS.2017.2737664

DeVries B, Huang C, Armston J, Huang W, Jones JW, Lang MW (2020) Rapid and robust monitoring of flood events using Sentinel-1 and landsat data on the google earth engine. Remote Sens Environ 240:111664. https://doi.org/10.1016/j.rse.2020.111664

Goffi A, Stroppiana D, Brivio PA, Bordogna G, Boschetti M (2020) Towards an automated approach to map flooded areas from Sentinel-2 MSI data and soft integration of water spectral features. Int J Appl Earth Obs Geoinf 84:101951. https://doi.org/10.1016/j.jag.2019.101951

Guo K, Guan M, Yan H (2023) Utilising social media data to evaluate urban flood impact in data scarce cities. Int J Disaster Risk Reduct 93:103780. https://doi.org/10.1016/j.ijdrr.2023.103780
Hirabayashi Y, Mahendran R, Koirala S, Konoshima L, Yamazaki D, Watanabe S, Kim H, Kanae S (2013) Global flood risk under climate change. Nat Clim Chang 3:816–821. https://doi.org/10.1038/nclimate1911
Huang H, Yu Y, Chen T, Yokoya N, Li J, Plaza A (2025) SAR and social media-based change detection with dual threshold fusion for flood inundation mapping. IEEE J Sel Top Appl Earth Obs Remote Sens
Karmegam D, Ramamoorthy S, Mappillairaju B (2021) Near real time flood inundation mapping using social media data as an information source: a case study of 2015 Chennai flood. Geoenviron Disasters 8. https://doi.org/10.1186/s40677-021-00195-x
Kordelas GA, Manakos I, Lefebvre G, Poulin B (2019) Automatic inundation mapping using sentinel-2 data applicable to both camargue and do'ana biosphere reserves. Remote Sens 11. https://doi.org/10.3390/rs11192251
Landuyt L, Van Wesemael A, Schumann GJP, Hostache R, Verhoest NEC, Van Coillie FMB (2019) Flood mapping based on synthetic aperture radar: an assessment of established approaches. IEEE Trans Geosci Remote Sens 57:722–739. https://doi.org/10.1109/TGRS.2018.2860054
Martinis S, Kersten J, Twele A (2015) A fully automated TerraSAR-X based flood service. ISPRS J Photogramm Remote Sens 104:203–212. https://doi.org/10.1016/j.isprsjprs.2014.07.014
Matgen P, Giustarini L, Chini M, Hostache R, Wood M, Schlaffer S (2016) Creating a water depth map from SAR flood extent and topography data. In: IEEE international geoscience and remote sensing symposium, pp 7635–7638. https://doi.org/10.1109/IGARSS.2016.7730991
Mohanty A, Sahoo B, Kale RV (2024) A hybrid model enhancing streamflow forecasts in paddy land use-dominated catchments with numerical weather prediction model-based meteorological forcings. J Hydrol 635:131225
Mohanty A, Sahoo B, Vitthal R (2025) Real-time forecasting of coastal flood inundations under regulated reservoir and storm-tide influences. Adv Water Resour 199:104920. https://doi.org/10.1016/j.advwatres.2025.104920
Mondal M, Biswas A, Haldar S, Mandal S, Bhattacharya S, Paul S (2022) Spatio-temporal behaviours of tropical cyclones over the bay of Bengal Basin in last five decades. Trop Cyclone Res Rev 11:1–15. https://doi.org/10.1016/j.tcrr.2021.11.004
Mosavi A, Ozturk P, Chau KW (2018) Flood prediction using machine learning models: literature review. Water 10:1–40. https://doi.org/10.3390/w10111536
Ogilvie A, Belaud G, Massuel S, Mulligan M, Le Goulven P, Malaterre PO, Calvez R (2018) Combining Landsat observations with hydrological modelling for improved surface water monitoring of small lakes. J Hydrol 566:109–121. https://doi.org/10.1016/j.jhydrol.2018.08.076
Otsu N, Smith PL, Reid DB, Environment C, Palo L, Alto P, Smith PL (1979) Otsu 1979 Otsu Method. IEEE Trans Syst Man Cybern Part C 62–66
Pati A, Sahoo B (2022) Effect of low-impact development scenarios on pluvial flood susceptibility in a scantily gauged urban–peri-urban catchment. J Hydrol Eng 27(1):05021034
Pralle S (2019) Drawing lines: FEMA and the politics of mapping flood zones. Clim Chang 152:227–237. https://doi.org/10.1007/s10584-018-2287-y
Rawat K, Sahoo B, Sahoo DP (2025) Entropy-based streamflow estimation in data-scarce Rivers using MODIS–Landsat fusion imagery. J Hydrol Eng 30(4):04025022
Ruggles F (1975) Optical data processing and projected applications of the Erts- 1 imagery covering the 1973 Mississippi river valley floods' during the spring of 1973, the Mississippi River valley experienced some of the most disastrous flooding in recorded history. T 10:1023–1039

Sahoo DP, Sahoo B, Tiwari MK (2020) Copula-based probabilistic spectral algorithms for high-frequent streamflow estimation. Remote Sens Environ 251:112092. https://doi.org/10.1016/j.rse.2020.112092

Shen X, Anagnostou EN, Allen GH, Robert Brakenridge G, Kettner AJ (2019) Near-real-time non-obstructed flood inundation mapping using synthetic aperture radar. Remote Sens Environ 221:302–315. https://doi.org/10.1016/j.rse.2018.11.008

Tazmul Islam M, Meng Q (2022) An exploratory study of Sentinel-1 SAR for rapid urban flood mapping on google earth engine. Int J Appl Earth Obs Geoinf 113:103002. https://doi.org/10.1016/j.jag.2022.103002

Tejaswi A, Sahoo B, Kale RV, Murty PLN, Mohanty A (2025) A method for simulating coastal inundation and inland flooding by integrating inland hydrology with ocean hydrodynamics. Nat Hazards. https://doi.org/10.1007/s11069-025-07149-8

Tellman B, Sullivan JA, Kuhn C, Kettner AJ, Doyle CS, Brakenridge GR, Erickson TA, Slayback DA (2021) Satellite imaging reveals increased proportion of population exposed to floods. Nature 596:80–86. https://doi.org/10.1038/s41586-021-03695-w

Uddin K, Matin MA, Meyer FJ (2019) Operational flood mapping using multi-temporal Sentinel-1 SAR images: a case study from Bangladesh. Remote Sens 11. https://doi.org/10.3390/rs11131581

UN-SPIDER Knowledge Portal (2021) Recommended practice: flood mapping and damage assessment using sentinel-1 SAR data in google earth engine. Retrieved from: https://www.un-spider.org/advisory-support/recommendedpractices/recommended-practice-google-earth-engine-flood-mapping. Accessed 23 Dec 2021

World Resources Institute (2020) RELEASE, new data shows millions of people, trillions in property at risk from flooding–but infrastructure investments now can significantly lower flood risk

Westerhoff RS, Kleuskens MPH, Winsemius HC, Huizinga HJ, Brakenridge GR, Bishop C (2013) Automated global water mapping based on wide-swath orbital synthetic-aperture radar. Hydrol Earth Syst Sci 17:651–663. https://doi.org/10.5194/hess-17-651-2013

Zhang M, Chen F, Liang D, Tian B, Yang A (2020) Use of sentinel-1 grd Sar images to delineate flood extent in Pakistan. Sustainability 12:1–19. https://doi.org/10.3390/su12145784

Zhuang H, Wang P, Hao M, Fan H, Tan Z (2025) Flood inundation mapping in SAR images based on nonlocal polarization combination features. J Hydrol 646:132326

9 Recent Studies on the Northeast Indian Rivers: A Crucial Water Resource of the Region

Adrija Roy and Subhadeep Biswas

Abstract

Rivers are one of the most important sources of water in Northeastern India. This chapter reviews recent research on major river systems in Northeast India, including the Brahmaputra, Barak, Lohit, and Dibang rivers. Water obtained from these rivers is utilized for drinking as well as for other important activities for ages. However, with the due course of time, rapid industrialization as well as climate change have severely impacted water quality and ecosystem health. The focus of this chapter is to highlight the recent research works that have been carried out for determining the suitability of the water obtained from these rivers against specific purpose. Water quality index as well as different physical, chemical, biological parameters of the river water from this region has been compared. Hydrogeological and hydrochemical characteristics of several river basins in the recent times in this area are focused. This chapter also highlights the use of hydrological and remote sensing models for discharge prediction and climate impact assessment. Lastly, the review offers recommendations for sustainable water management and policymaking.

A. Roy
Leibniz-Centre for Agricultural Landscape Research (ZALF) e.V, Muncheberg, Germany
e-mail: adrija.roy@zalf.de

S. Biswas (✉)
Civil Engineering Department, National Institute of Technology, Silchar, Assam, India
e-mail: sdeep0819@gmail.com; subhadeep@civil.nits.ac.in

S. Yadav et al. (eds.), *Water Resources in Delta Region in India*, Deltas of the World,
https://doi.org/10.1007/978-3-032-06500-1_9

9.1 Introduction

The north-eastern Indian region is rich in biodiversity. It consists of eight states, namely Assam, Arunachal Pradesh, Manipur, Meghalaya, Nagaland, Mizoram, Tripura and Sikkim. The first seven states are collectively called the seven sisters while Sikkim is called their brother. The region is well known for its distinguished geographical characteristics containing eastern Himalayas and various plateaus, hills and valleys such as the Meghalaya plateau, Lushai hills and Barak and Brahmaputra valley. About 67% area of this place is hilly in nature, while the rest is covered by plains and valleys. The Northeast India consists of a wide variety of flora and fauna and is rich in mineral resources, too.

Different articles described the beauty and distinct landscape features of this region (Bargohain and Mokashi-Punekar 2022). The whole Northeast basically falls under the eastern Himalayas, which is one of the biodiversity hotspots among the two hotspots in the Indian territory. This region is also a part of the Indo-Burma biodiversity hotspot and contains about 17.2 million ha area forest land (Pandey et al. 2022). It ultimately covers nearly one-fourth of the total forest area in India. Besides, this part of the country is the homeland of many endemic species such as one-horned rhino, hornbills, Namdapha flying squirrel etc. Moreover, many ethnic communities reside in this place and through ages a symbiotic relationship has developed between them and the natural surroundings. Das and Das (2015) also highlighted the high plant diversity existing in the districts of Barak valley, South Assam. For diversity analysis, the authors carried out an extensive study on the 181 home gardens from 38 villages. A home garden can be defined as a specific land use pattern for the management of various trees, shrubs, and perennial crops, within the compound of an individual house, and the whole labor is provided by the family members. The results revealed that there existed 161 species from 47 families.

However, to maintain high biodiversity and ecological balance, water is undoubtedly considered an essential item. It is extremely important for the improvement and sustainable development of any community residing at a particular location. This region is fortunate to have abundant natural water resources. Khangenbam and Kshetrimayum (2019) in one of their recent articles, discussed about the natural water resources of the Barak valley, South Assam. The authors described that this region consists of a huge number of surface water bodies including ponds, lakes, rivers, paleochannels, which ultimately constitutes the low-lying topography of this place (22–35 m above mean sea level). The Meghalaya plateau, one of the most distinguished spots in this area is the birthplace of many rivers. On its north, lies the Assam valley and the southern part is covered by the plains of Bangladesh. Due to its unique geographical location, it receives the maximum amount of rainfall in the world. However, due to the unavailability of any suitable water resource management plan, this huge volume of runoff water often causes devastating floods in this area of the subcontinent (Nath et al. 2023). Some rivers such as Umiam, Umkhen, arising from this plateau, flow north eastward to join the Barhamputra river in

Assam, while others, including Umngi, Umiew, Wahrew etc., flow south to merge with the Meghna river in Bangladesh.

The present chapter describes the significance of the Northeastern Indian rivers in constituting the overall natural water resources of the region. After a brief introduction, some of the major rivers flowing through this zone are highlighted along with their important tributaries. For ages, these rivers have served as an important drinking water source as well as offering water for various other important tasks. However, due to rapid industrialization and other anthropogenic activities, the water quality of these rivers is degrading with time. Some of the relevant reported studies are discussed. Deterioration of the river health is also associated with methane and other greenhouse gases generation which ultimately may hamper the overall climate pattern of the place. Hence, conservation of these river water resources and water pollution remediation is one of the most crucial works to the scientists and engineers of our society.

9.2 Important Northeast Indian Rivers

Numerous significant rivers flow through the northeastern region of India, serving as major water resources for the area. Starting from the extreme east, the states of Arunachal Pradesh, Nagaland, Manipur and Mizoram demarcate the border of the Indian territory sharing an international border with Myanmar. The state of Nagaland contains some important rivers like Doyang, Dikhu, Dhansiri and Tizu. The Doyang river flowing through Wokha district is one of the most significant rivers for the local people (Lkr et al. 2020). It plays an important role in economic development as well as offering congenial grounds for fishing and hunting. The Tsurang river of the Mokokchung district is another important name in the state in this regard (Semy and Singh 2021). It is one of the symbols of significance to the Ago Naga tribes and one of the major sources of water supply to the local agrarian community.

Many important rivers start and flow through the state of Manipur also. Some of them include Barak, Imphal, Iril, Khuga, Nambul etc. The Imphal river is one of the most significant rivers in the state, a tributary of the Manipur River and joins it in the Thoubal district. In this state, the rivers play a significant role in providing drinking water as well as water for other activities. In the urban regions, maximum portion of the drinking water is achieved from the three rivers namely Imphal, Iril and Nambul. Besides, the Loktak lake, one of the Ramsar site is the lifeline of the state. Many rivers flow into this lake. However, due to several anthropogenic activities and other natural factors, water of this grand lake is getting deteriorated since ages and is not suitable for drinking purpose as identified in the recent times.

Different rivers with their tributaries flow through the state of Assam. Among them, the mighty Brahmaputra is one of the significant rivers. The main course of the river flows through three countries namely China, India, Bangladesh. As per one reported study, it covers a total length of 2880 km and recognized as the 22nd longest river in the world (Rahaman and Varis 2009). In China, the river is called as the Tsangpo, in India as the Brahmaputra while in Bangladesh it is known as the Jamuna river. Apart from the Brahmaputra, Pagladia, Beki and Kolong (tributaries of the Brahmaputra) are the other important rivers flowing through the state of Assam (Singh et al. 2019). The rivers in this state are also useful for generation of hydropower. The Kopili Hydroelectric project and Karbi Langpi Hydroelectric project located on the Kopili and Borpani rivers are some of the important names in this aspect. Besides these, the Barak River flowing through the south Assam is also a huge source of natural water resource as indicated by many reported studies.

The Teesta River, another major river is called the lifeline of the state of Sikkim. It plays a crucial role in developing agriculture, hydroelectricity as well as local economy. It is also an important transboundary river flowing between India and Bangladesh. Other notable rivers flowing through the state of Sikkim includes Lachen, Rangeet, Lachung, Rangpo etc.

9.3 Water Quality Deterioration Since Times

Water pollution is a pressing issue for a country like India in the contemporary times. Various industrial discharges, vehicular emissions and other anthropogenic activities have exacerbated the situation. The Northeastern region is also not an exception. In a report by the Central Pollution Control Board, 60 river stretches in this region has been demarcated as polluted depending upon the BOD level. One of the major sources of water pollution involves leakage from the municipal landfill leachate. Saha et al. (2023) performed a systematic study for a landfill site nearby Morabharali river at Tezpur town in Assam (map shown in Fig. 9.1). It belongs to the Brahmaputra valley. The authors reported that the waste dumped in the landfill site was not stable. Moreover, a high leachate pollution index of 18.39 confirmed the instability of the dumped waste. So, obviously it indicates a high chance of surrounding groundwater pollution.

In the present times, often water quality index (WQI) is used to signify the overall quality of water from a certain source. In one of the studies, Lkr et al. (2020) reported the WQI from various sampling points from the Doyang river of Nagaland. Various anthropogenic activities such as hydroelectric dam, change in landuse and land cover, enhancement in human settlement etc. resulted in deterioration of the water quality of the river.

Gradual degradation of the water quality of Loktak lake (Manipur), one of the Ramsar sites in India is a major environmental concern over the contemporary times. According to recent studies, Nambul river has been recognized as one of the significant polluters of the water of this lake (Thongam and Meitei 2021). Extremely low dissolved oxygen (DO) (<1 mg/L) accompanied with high concentrations of nitrate and phosphate has turned the river water unfit for drinking. Besides, abnormal sodium percentage, sodium adsorption

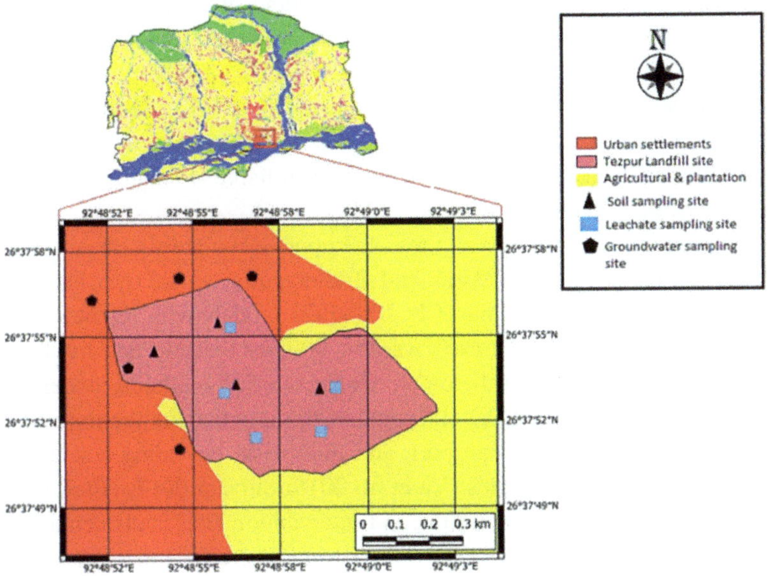

Fig. 9.1 Map of study area showing the location of Tezpur dumping site (Saha et al. 2023)

ratio and Kelly's ratio have made the water unsuitable for irrigation purpose also. The urban stretch of the river receives a vast amount of raw sewage and industrial wastewater from the Imphal municipal region.

Different water bodies of this region are at the risk of eutrophication due to the disposal of sewage as well as industrial effluents. As per the recent study conducted by Laishram et al., some portions of the Loktak lake show eutrophic characteristics at least for some portions of the year (Laishram et al. 2022).

9.4 Integrated Monitoring and Modeling Approaches for River Basin Management

Conventional assessments of water quality and quantity in Northeast Indian rivers, such as those conducted by the Pollution Control Board Assam, provide essential baseline data for surface water monitoring (Biggs 2021; Barua et al. 2018; U. N. World Bank 2023). However, the region's complex hydroclimatic variability-marked by intense monsoon rainfall, diverse topography, and limited in-situ monitoring-necessitates advanced, data-driven, and spatially distributed tools for effective river basin management (Thakur et al. 2020). Recent years have seen the successful integration of satellite-based datasets with hydrological models to address data scarcity. The integration of satellite-based datasets-such as precipitation estimates from GPM and TRMM (Kittel 2020), high-resolution digital elevation models, and land use/land cover maps (Tang et al. 2024; Toosi et al. 2025) from Landsat and Sentinel missions-has greatly enhanced the capabilities of process-based hydrological models like the Soil and Water Assessment Tool (SWAT) and the Hydrologic Engineering Center-Hydrologic Modeling System (HEC-HMS) (Boddepalli et al. 2024). These models can now be calibrated and validated with unprecedented accuracy, even in data-scarce mountainous catchments, enabling more reliable simulation of river discharge and flood forecasting (Chathuranika et al. 2022).

The practical implications of these integrated approaches are significant. For instance, in flood-prone basins such as the Barak and Brahmaputra, where over half the valley is susceptible to flooding, the coupling of hydrological models with remote sensing inputs supports the development of robust early warning systems that can save lives and property (Ghosh and Prakasam 2025). Furthermore, scenario-based modeling under Representative Concentration Pathways (RCPs) allows planners to anticipate the impacts of climate change on river flows, sediment transport, and flood risk, providing a scientific foundation for climate adaptation strategies (Vu et al. 2016; Shivam 2017; Nilawar and Waikar 2019). Spatial prioritization of restoration and intervention sites is also enhanced through the use of satellite imagery, enabling targeted action where it is most needed (Patel and Srivastava 2013; Rahman and Di 2017).

Another critical dimension is the integration of hydrological models with water-energy-food (WEF) nexus frameworks, which is particularly relevant in regions where traditional jhum (shifting cultivation) and intensive agriculture coexist (Jain et al. 2023). By coupling evapotranspiration models such as SEBAL and METRIC with hydrological simulations, water managers can optimize agricultural water use while maintaining ecosystem health (Mondal et al. 2022; Kumar and Hamouda 2025). Moreover, for transboundary rivers like the Brahmaputra, these integrated tools support cross-border data sharing and scenario analysis, fostering cooperative governance and joint management in a region where political boundaries often complicate resource allocation (Yasuda et al. 2017; Stefano et al. 2017; Sarfaraz et al. 2022; Milman et al. 2020; He 2021). In summary, the adoption of integrated monitoring and modeling approaches-leveraging both satellite remote sensing and advanced hydrological models-represents a paradigm shift in river basin management for Northeast India, offering actionable insights for flood risk reduction, climate adaptation, sustainable agriculture, and transboundary cooperation.

9.5 River Basin Management Through Integrated Modeling and Remote Sensing

The sustainable management of river systems in Northeast India is increasingly dependent on the adoption of advanced modeling frameworks and satellite-based earth observation technologies, especially given the region's pronounced climatic variability, data scarcity, and challenging terrain (Ghosh and Prakasam 2025; Schulze 2004; Palash et al. 2018).

Traditional monitoring approaches, while valuable, are often constrained by the inaccessibility of many catchments and the lack of continuous, high-quality in-situ data (Plengsaeng et al. 2014; Kibler et al. 2014; Jahanddideh-Tehrani et al. 2021). In response, process-based hydrological models such as SWAT, the Variable Infiltration Capacity (VIC) model, and HEC-HMS are being widely applied to simulate watershed hydrology, sediment dynamics, and nutrient loading under a variety of land use and climate change scenarios (Chathuranika et al. 2022; Belvederesi et al. 2022; Nesru 2023). These models are now central to projecting river discharge trends under RCP-based climate scenarios, evaluating the hydrological impacts of land use change and deforestation, and designing adaptive infrastructure for flood control and water storage in major basins like the Barak, Subansiri, and Brahmaputra (Dutta et al. 2021; Alam et al. 2021).

Remote sensing platforms have further revolutionized river basin management by providing consistent, time-series data on key hydrological and environmental variables. Sensors such as Landsat-8, Sentinel-2, MODIS, and hyperspectral imagers allow for the monitoring of surface water extent, turbidity, and suspended sediment concentrations using indices like the Normalized Difference Water Index (NDWI) and Suspended Sediment Indices (Sigopi et al. 2024; Pan et al. 2025; Chatrabhuj et al. 2024). These data are crucial for tracking land use and land cover changes associated with shifting cultivation, mining, and urban encroachment-activities that directly impact river health and water quality. Additionally, satellite missions such as SMAP and SMOS, combined with energy balance models like SEBAL and METRIC, enable the assessment of soil moisture and evapotranspiration patterns (Saxena et al. 2024; Huang et al. 2025), which are essential for understanding agricultural water demand and drought vulnerability.

Adopting a water-energy-food (WEF) nexus perspective is particularly important in Northeast India, where livelihoods are closely tied to monsoon-fed rivers, fragile ecosystems, and seasonal water availability (Jain et al. 2023). By integrating hydrological models with crop models such as AquaCrop and CropWat, and incorporating energy balance assessments, planners can identify trade-offs and synergies in water allocation decisions (Yalcin 2024; Zhang et al. 2021; Agrawal et al. 2023). This systems-based approach is vital for districts practicing jhum or facing high seasonal irrigation needs, where water and energy security are interdependent.

The policy and planning relevance of these integrated modeling approaches is profound. Outputs from these models underpin early warning systems for floods and droughts, guide the spatial prioritization of interventions such as pollution control and reforestation, and support the placement of new reservoirs or flood control structures. In transboundary basins like the Brahmaputra and Lohit, integrated modeling provides a scientific basis for cross-border data sharing and cooperative governance, which are essential for managing shared water resources in a politically sensitive region (Yasuda et al. 2017; Stefano et al. 2017; Plengsaeng et al. 2014). Nevertheless, the full potential of these tools can only be realized by addressing persistent challenges, including limited institutional capacity, lack of inter-agency coordination, and the absence of open data platforms (Kibler et al. 2014; Mukuyu et al. 2020; Ibrahim 2020). Strengthening local research infrastructure, investing in the training of water managers, and enhancing citizen participation in water stewardship are critical steps toward more resilient and sustainable river basin management in Northeast India.

9.6 Climate Change Related Impact

Climate change is one of the pressing issues of contemporary times. Rise of temperature, extreme weather conditions along with increase of sea level can be visualized as some of its severity. As per USEPA, increase in concentration of greenhouse gases in the atmosphere has lead to the continuously warming effect considerably since 1990. In this context, water pollution is often referred to as the critical driver behind climate change issues. Pollution level can significantly contribute to the greenhouse emission and therefore change the earth's carbon cycle. Deterioration of water quality can exacerbate the negative impacts of climate change on the ecosystem and human society. It is clear from the recent reports, that polluted rivers are becoming emerging source worldwide for contributing to GHG (Upadhyay et al. 2023).

Several articles of recent times highlighted the risk of methane generation from polluted rivers (Upadhyay et al. 2023). These highly contaminated rivers are often described as hotspots for exchange of GHGs like CH_4 and N_2O. Wang et al. carried out a systematic study on an urban river flowing through Beijing, China (Zhao et al. 2022). The authors concluded that the annual fluxes for CH_4 and N_2O varied in the range of 19.3 to 79.4 kg of C and 17.4 to 44.8 kg of N respectively.

In the recent period, many studies have been performed to analyze the climate change effect in the northeast India. Deb and Sil conducted climate change study in the Barak River basin by considering meteorological parameters such as temperature (minimum and maximum), precipitation, evapotranspiration etc. (Wang et al. 2020). The location of the meteorological sites in the Barak river basin is shown in Fig. 9.2. It was revealed that both the maximum and minimum temperature showed increasing trend while for precipitation a declining trend was observed. However, in case of potential evapotranspiration, a considerable rising trend was found during the monsoon time. In another study, Palash et al. studied the effect of climate change on the hydrology of the Brahmaputra river basin (Deb and Sil 2019). It is projected that there will be an increase in future temperature, which will affect the snowfall and snowmelt process and ultimately will enhance the precipitation in this basin. Further, it is concluded that the upper part of the Brahmaputra will be most impacted by the climate change followed by the middle part.

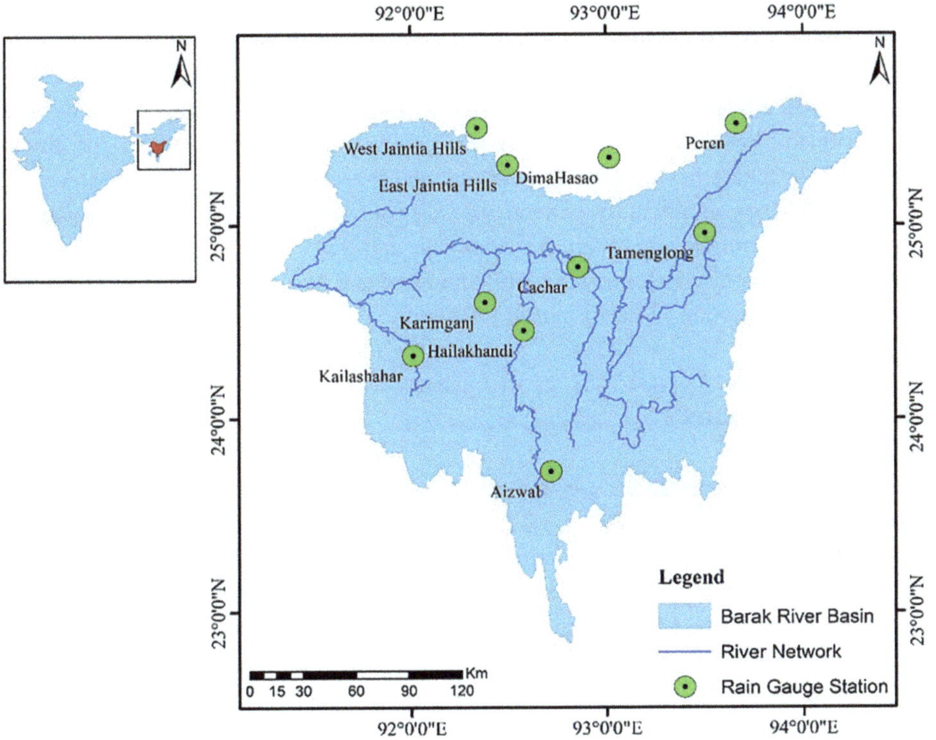

Fig. 9.2 Location of the meteorological sites in the Barak River basin (Wang et al. 2020)

9.7 Sustainable Measures that Can Be Adopted

It is already mentioned in the previous sections, that the lakes, rivers, and natural water resources of Northeast India are continuously being polluted due to different anthropogenic activities. Hence, various actions are required to be adopted immediately to protect this huge natural water resource from further degradation. Among different pollution control measurements, phytoremediation is an effective technique for decontaminating polluted water. Thongam and Meitei (Thongam and Meitei 2021) studied the potential of macrophytes such as *Eichhornia crassipes* and *Pistia stratiotes* as promising phytofilters for treating polluted river water. In another study, Meitei and Prasad (Palash et al. 2023) investigated phoomdi based phytoremediation technique for cleaning of the Loktak lake. Phoomdi is a unique form of floating vegetation available in the Loktak lake region only. The authors reported that these plants were capable of adsorbing nutrients and heavy metals from the water body. Detailed experimental investigation revealed that the maximum uptake capacity reached up to 9.2 mg/kg for Cu, while 9461.4 mg/kg for Fe.

Many rivers in this zone suffer from eutrophication as already highlighted by the recent reports. Among various remediation techniques available for abating eutrophication, adsorption has been recognized as one of the simplest and most versatile methods. Different novel adsorbents have been reported by the researchers for adsorption of nitrate and phosphate. Awual explored amine functionalized composite adsorbent for efficient removal of phosphate from wastewater (Meitei and Prasad 2016). Presence of other interfering ions did not significantly hamper the adsorption process due to the formation of strong hydrogen bonding between the adsorbent and adsorbate. Moreover, thus synthesized novel adsorbent was found reversible in nature and hence found reusable for several cycles without reduction in performance efficiency.

9.8 Discussions

The rivers of Northeast India are not only sources of freshwater but also critical for sustaining regional biodiversity, agriculture, and local livelihoods. The review highlights that most river systems—including the Brahmaputra, Barak, Imphal, Doyang, and Teesta—face increasing pressure from rapid urbanization, unregulated waste disposal, and agricultural runoff. These factors have accelerated the decline in water quality, as reflected in rising concentrations of nutrients, heavy metals, and organic pollutants.

Integrated modeling studies show that combining hydrological tools (SWAT, HEC-HMS, VIC) with remote sensing datasets has improved our understanding of catchment-scale processes. However, these approaches remain underutilized in the Northeast due to sparse ground data and weak institutional coordination. In addition, transboundary data gaps in the Brahmaputra and Barak basins limit basin-wide flood forecasting and climate adaptation planning.

From an ecological standpoint, eutrophication in lakes such as Loktak and the spread of invasive aquatic plants indicate that water quality deterioration is now directly affecting ecosystem services. Phytoremediation studies using local macrophytes, such as *Eichhornia crassipes* and *Pistia stratiotes*, demonstrate the potential for low-cost water purification in rural contexts.

An important observation is the linkage between these river systems and the deltaic environments downstream, especially the Brahmaputra–Ganga–Meghna delta. Sediment and nutrient flows from the upper basins directly influence delta morphology, fisheries, and coastal resilience. Therefore, sustainable river management in the upstream catchments is crucial for maintaining delta health and preventing salinity intrusion and land loss in coastal regions.

9.9 Conclusion

The rivers like Brahmaputra, Barak, Teesta, Doyang, Imphal etc. play a crucial role in maintaining natural water resources and developing economy of the Northeast India. These rivers are basically an integral part of the cultural heritage and biodiversity of this region. Since ages, they are acting as the major sources of drinking water as well as water for other important works. However, during the contemporary times, several anthropogenic activities and other factors deteriorate the water quality as evidenced from the recent studies. Heavy metal pollution and synthetic organic compounds have been detected in surface water along many places in this region. Apart from this severity of climate change is also expected to have a negative impact on this huge natural resource. Hence, sustainable river basin management is a priority of the current times. Besides, environmental remediation techniques involving bioremediation, phytoremediation, adsorption, phosphate and nitrate removal shall be adopted in large scale to save these rivers in the upcoming future.

9.10 Recommendations

Effective management of the rivers in Northeast India requires a coordinated and science-based approach that links upstream catchments with downstream delta systems. Based on the findings discussed in this chapter, the following recommendations are proposed:

1. **Integrated River–Delta Management**: River basin monitoring should be closely linked with deltaic studies to capture the complete pathway of sediment and nutrient movement from the headwaters to the coast. Establishing an integrated management framework between the Brahmaputra and Meghna basins would help in understanding how upstream hydrological changes influence delta morphology, coastal productivity, and the long-term stability of the Brahmaputra–Meghna delta.
2. **Adoption of Nature-based Solutions**: Environmentally sustainable methods such as phytoremediation, wetland restoration, and the creation of vegetative buffer zones along polluted rivers should be promoted to reduce nutrient enrichment and heavy-metal contamination. These approaches not only improve water quality but also enhance biodiversity and ecosystem resilience.
3. **Enhanced Monitoring and Data Sharing**: Hydrological and meteorological monitoring networks in the region should be expanded and modernized to capture real-time changes in flow, sediment load, and water quality. Open-access data platforms and regional data-sharing agreements among India, Bhutan, China, and Bangladesh would strengthen transboundary river management and improve flood forecasting capacity.

4. **Promotion of Climate-resilient Infrastructure**: Future flood control and irrigation projects should be designed using climate projection data to account for changing rainfall intensity, temperature variation, and extreme weather events. Incorporating climate resilience into infrastructure planning will reduce vulnerability to floods, droughts, and erosion in the Brahmaputra and Barak basins.
5. **Capacity Building and Community Participation**: Training programs should be established for local water managers, engineers, and environmental officers to improve their technical and decision-making skills. Indigenous and farming communities should be actively involved in conservation initiatives to ensure that local knowledge complements scientific management. Awareness campaigns and education programs can further promote water stewardship at the community level.
6. **Policy Integration and Governance**: Indicators of river health and ecosystem integrity should be integrated into regional development plans and water governance frameworks. Industrial, agricultural, and urban growth strategies must align with environmental sustainability goals to maintain the ecological balance of river systems and their connected deltaic environments.

Add Acknowledgement (if any).

Na

State whether you used AI or not in preparing the chapter.

Only for increasing resolution of images, it has been done online. Other than that, no AI has been used in preparing the chapter.

References

Agrawal A, Srivastava PK, Tripathi VK, Maurya S, Sharma R, Shrinivasa DJ (2023) Future projections of crop water and irrigation water requirements using a bias-corrected regional climate model coupled with CROPWAT. J Water Clim Change 14(4):1147–1161. https://doi.org/10.2166/wcc.2023.349

Alam S, Ali MM, Rahaman AZ, Islam Z (2021) Multi-model ensemble projection of mean and extreme streamflow of Brahmaputra River Basin under the impact of climate change. J Water Clim Change 12(5):2026–2044. https://doi.org/10.2166/wcc.2021.286

Awual MR (2019) Efficient phosphate removal from water for controlling eutrophication using novel composite adsorbent. J Clean Prod 228:1311–1319. https://doi.org/10.1016/j.jclepro.2019.04.325

Bargohain R, Mokashi-Punekar R (2022) Poetry, identity and the geography of culture: representatives of landscape in poetry in English from Northeast India. Asian Ethn 23(1):47–65. https://doi.org/10.1080/14631369.2020.1727311

Barua A, Vij S, Rahman MZ (2018) Powering or sharing water in the Brahmaputra River basin. Int J Water Resour 34(5):829–843. https://doi.org/10.1080/07900627.2017.1403892

Belvederesi C, Zaghloul MS, Achari G, Gupta A, Hassan QK (2022) Modelling river flow in cold and ungauged regions: a review of the purposes, methods, and challenges. Environ Rev 30:159–173. https://doi.org/10.1139/er-2021-0043

Biggs S (2021) Water management on the brahmaputra and the applicability of the UNECE water convention. Vanderbilt Law Review 555 51(2). https://scholarship.law.vanderbilt.edu/vjtl/vol51/iss2/4

Boddepalli M, Pasupuleti LN, Nalli BR (2024) A comprehensive study of floodplain analysis utilising HEC-HMS, HEC-RAS, and GIS on the Kosasthalaiyar River sub-basin. Water Pract Technol 19(11):4612–4628. https://doi.org/10.2166/wpt.2024.269.

Chathuranika IM, Gunathilake MB, Baddewela PK, Sachinthanie E, Babel MS, Shrestha S, Jha MK, Rathnayake US (2022) Comparison of two hydrological models, HEC-HMS and SWAT in runoff estimation: application to Huai Bang Sai tropical watershed, Thailand. Fluids 7(8):267. https://doi.org/10.3390/fluids7080267

Chatrabhuj, Meshram K, Mishra U, Omar PJ (2024) Integration of remote sensing data and GIS technologies in river management system. Discov Geosci 2(1):67. https://doi.org/10.1007/s44288-024-00080-8

Das T, Das AK (2015) Conservation of plant diversity in rural homegardens with cultural and geographical variation in three districts of Barak valley, Northeast India. Econ Bot 69(1):57–71. https://doi.org/10.1007/s12231-015-9299-6

Deb S, Sil BS (2019) Climate change study for the meteorological variables in the Barak River basin in North-East India. Urban Clim 30:100530. https://doi.org/10.1016/j.uclim.2019.100530

Dutta P, Hinge G, Marak JDK, Sarma AK (2021) Future climate and its impact on streamflow: a case study of the Brahmaputra river basin. Model Earth Syst Environ 7(4):2475–2490. https://doi.org/10.1007/s40808-020-01022-2

Ghosh S, Prakasam C (2025) Geospatial flood modeling and risk assessment of floodplain villages along the Barak River, Northeast India. Asian J Water Environ Pollut 22(2):64–86. https://doi.org/10.36922/AJWEP025040019

He Y (2021) Current and future transboundary water cooperation over the YarlungZangbo/Brahmaputra River basin: from an interdisciplinary perspective. Water Policy 23(5):1107. https://doi.org/10.2166/wp.2021.008

Huang J, Sehgal V, Alvarez LV, Brocca L, Cai S, Cheng R, Cheng X, Du J, Masri BE, Endsley KA, Fang Y, Hu J, Jampani M, Kibria MG, Koren G, Li L, Liu L, Mao J, Moreno HA, Rigden A, Shi MM, Shi X, Wang Y, Zhang X, Fisher JB (2025) Remotely sensed high-resolution soil moisture and evapotranspiration: bridging the gap between science and society. Water Resour Res 61(5) Art. no. e2024WR037929. https://doi.org/10.1029/2024WR037929

Ibrahim IA (2020) Legal implications of the use of big data in the transboundary water context. Water Resour Manage 34(3):1139–1153. https://doi.org/10.1007/s11269-020-02491-x

Jahanddideh-Tehrani M, Bozorg-Haddad O, Daliakopoulos IN (2021) The role of water information and data bases in water resources management. In: Bozorg-Haddad O (eds) Essential tools for water resources analysis, planning, and management. Springer Water, Springer Singapore, https://doi.org/10.1007/978-981-33-4295-8_3

Jain SK, Sikka AK, Alam MF (2023) Water-energy-food-ecosystem nexus in India-a review of relevant studies, policies, and programmes. Front Water. https://doi.org/10.3389/frwa.2023.1128198

Khangembam S, Kshetrimayum KS (2019) Evaluation of hydrogeochemical controlling factors and water quality index of water resources of the Barak valley of Assam, Northeast India. Groundw Sustain Dev 8:541–553. https://doi.org/10.1016/j.gsd.2019.02.001

Kibler KM, Biswas RK, Lucas AMJ (2014) Hydrologic data as a human right? Equitable access to information as a resource for disaster risk reduction in transboundary river basins. Water Policy 16(S2):36–58. https://doi.org/10.2166/wp.201.307

Kittel CMM (2020) Satellite radar observations for hydrologic and hydrodynamic modelling. DTU Environment, Technical University of Denmark, PhD

Kumar N, Hamouda MA (2025) Utility of single-source surface energy balance models in estimation of daily actual evapotranspiration in arid regions. Hydrol Sci J 70(4):664–686. https://doi.org/10.1080/02626667.2024.2446266

Laishram RJ, Yumnam G, Alam W (2022) Assessment of ecohydrogeochemical status of freshwater Loktak Lake of Manipur, India. Environ Monit Assess 194:659. https://doi.org/10.1007/s10661-022-10336-w

Lkr A, Singh MR, Puro N (2020) Assessment of water quality status of Doyang River, Nagaland, India, using water quality index. Appl Water Sci 10:46. https://doi.org/10.1007/s13201-019-1133-3

Meitei MD, Prasad MNV (2016) Bioaccumulation of nutrients and metals in sediment, water, and phoomdi from Loktak Lake (Ramsar site), northeast India: phytoremediation options and risk assessment. Environ Monit Assess 188:329. https://doi.org/10.1007/s10661-016-5339-7

Milman A, Gerlak AK, Albrecht T, Colosimo M, Conca K, Kittikhoun A, Kovacs P, Moy R, Schmeier S, Wentling K, Werick W, Zavadsky I, Ziegler J (2020) Addressing knowledge gaps for transboundary environmental governance. Glob Environ Chang 64:102162. https://doi.org/10.1016/j.gloenvcha.2020.102162

Mondal I, Thakur S, De A, De TK (2022) Application of the METRIC model for mapping evapotranspiration over the Sundarban Biosphere Reserve, India. Ecol Indic 136:108553. https://doi.org/10.1016/j.ecolind.2022.108553

Mukuyu P, Lautze J, Rieu-Clarke A, Saruchera D, McCartney M (2020) The devil's in the details: data exchange in transboundary waters. Water Int 45(7–8):884–900. https://doi.org/10.1080/02508060.2020.1850026

Nath K, Halder S, Gogoi D, Mahanta BN, Goswami TK, Sarmah RK (2023) Morphometric study of selected river basins from the Meghalaya Plateau-implications for the hydrodynamics of the eastern part of Indian subcontinent. Arab J Geosci 16:243. https://doi.org/10.1007/s12517-023-11346-9

Nesru M (2023) A review of model selection for hydrological studies. Arab J Geosci 16(2):102. https://doi.org/10.1007/s12517-023-11194-7

Nilawar AP, Waikar ML (2019) Impacts of climate change on streamflow and sediment concentration under RCP 4.5 and 8.5: a case study in Purna river basin, India. Sci Total Environ 650:2685–2696. https://doi.org/10.1016/j.scitotenv.2018.09.334

Palash W, Jiang Y, Akanda AS, Small DL, Nozari A, Islam S (2018) J Hydrometeorol 19:201–225. https://doi.org/10.1175/JHM-D-16-0202.1

Palash W, Bajracharya SR, Shrestha AB, Wahid S, Hossain MS, Mogumder TK, Mazumder LC (2023) Climate change impacts on the hydrology of the Brahmaputra River Basin. Climate 11(1):18. https://doi.org/10.3390/cli11010018

Pan D, Deng Y, Yang SX, Gharabaghi B (2025) Recent advances in remote sensing and artificial intelligence for river water quality forecasting: a review. Environments 12(5):158. https://doi.org/10.3390/environments12050158

Pandey DK, Momin KC, Dubey SK, Adhiguru P (2022) Biodiversity in agricultural and food systems of jhum landscape in the West Garo Hills, North-eastern India. Food Secur 14:791–804. https://doi.org/10.1007/s12571-021-01251-y

Patel DP, Srivastava PK (2013) Flood hazards mitigation analysis using remote sensing and GIS: correspondence with town planning scheme. Water Resour Manage 27:2353–2368. https://doi.org/10.1007/s11269-013-0291-6

Plengsaeng B, Wehn U, Zaag P (2014) Data-sharing bottlenecks in transboundary integrated water resources management: a case study of the Mekong River Commission's procedures for data sharing in the Thai context. Water Int 39(7):933–951. https://doi.org/10.1080/02508060.2015.981783

Rahaman MM, Varis O (2009) Integrated water management of the Brahmaputra basin: perspectives and hope for regional development. Nat Res Forum 33:60–75. https://doi.org/10.1111/j.1477-8947.2009.01209.x

Rahman MS, Di L (2017) The state of the art of spaceborne remote sensing in flood management. Nat Hazards 85:1223–1248. https://doi.org/10.1007/s11069-016-2601-9

Saha P, Saikia KK, Kumar M, Handique S (2023) Assessment of health risk and pollution load for heavy and toxic metal contamination from leachate in soil and groundwater in the vicinity of dumping site in Mid-Brahmaputra Valley, India. Total Environ Res Themes 8:100076. https://doi.org/10.1016/j.totert.2023.100076

Sarfaraz MU, Hall DM, Rotman RM (2022) Data sharing in transboundary water management. Front Water 4:982605. https://doi.org/10.3389/frwa.2022.982605

Saxena D, Choudhary M, Sharma G (2024) Spatiotemporal trends and evapotranspiration estimation using an improvised SEBAL convergence method for the semi-arid region of Western Rajasthan, India. AQUA Water Infrastruct Ecosyst Soc 73(3):407–423. https://doi.org/10.2166/aqua.2024.220

Schulze RE (2004) Modelling as a tool in integrated water resources management: conceptual issues and case study applications. https://www.wrc.org.za/wp-content/uploads/mdocs/749.pdf

Semy K, Singh MR (2021) Quality assessment of Tsurang River water affected by coal mining along the Tsurangkong Range, Nagaland, India. Appl Water Sci 11:115. https://doi.org/10.1007/s13201-021-01444-y

Shivam (2017) River basin planning of Subansiri river under climate change scenarios: river basin planning of Subansiri river under climate change scenarios. https://gyan.iitg.ac.in/items/05871a34-78de-4326-8911-3f21d5ce31dd/full

Sigopi M, Shoko C, Dube T (2024) Advancements in remote sensing technologies for accurate monitoring and management of surface water resources in Africa: an overview, limitations, and future directions. Geocarto Int 39:5. https://doi.org/10.1080/10106049.2024.2347935

Singh KR, Goswami AK, Kalamdhad AS (2019) Assessment of surface water quality of Pagladia, Beki and Kolong rivers (Assam, India) using multivariate statistical techniques. Int J River Basin Manag 18(4):511–520. https://doi.org/10.1080/15715124.2019.1566236

Stefano LD, Petersen-Perlman JD, Sproles EA, Eynard J, Wolf AT (2017) Assessment of transboundary river basins for potential hydro-political tensions. Glob Environ Chang 45:35–46. https://doi.org/10.1016/j.gloenvcha.2017.04.008

Tang Z, Zhang Y, Tian J, Ma N, Li X, Kong D, Cao Y, Yang X, Wang L, Zhang X, Chen Y (2024) Using hydrological modeling and satellite observations to elucidate subsurface and surface hydrological responses to the extreme drought. J Hydrol 645:132174. https://doi.org/10.1016/j.jhydrol.2024.132174

Thakur PK, Dhole P, Roy A, Aggarwal SP, Nikam BR, Garg V, Chouksey A, Pokhriyal N, Jani M, Chauhan V, Thakur N, Dogra VS, Rao GS, Chauhan P, Kumar AS (2020) Significance of remote sensing based precipitation and terrain information for improved hydrological and hydrodynamic simulation in parts of Himalayan River basins. The international archives of the photogrammetry, remote sensing and spatial information sciences, Volume XLIII-B3–2020, 2020 XXIV ISPRS Congress (2020 edition). https://doi.org/10.5194/isprs-archives-XLIII-B3-2020-911-2020

Thongam N, Meitei MD (2021) Role of dominant macrophytes to treat Nambul river, the main polluter Loktak- a dying Ramsar site in the Indo Burma hot spot (Manipur, India). Int J Phytorem 23(11):1132–1144. https://doi.org/10.1080/15226514.2021.1880367

Toosi AS, Batelaan O, Shanafield M, Guan H (2025) Land use-land cover and hydrological modeling: a review. Wiley Interdiscip Rev Water 12(2):e70013. https://doi.org/10.1002/wat2.70013

U. N. World Bank (2023) Assam integrated river basin management program (P174593): water global practice. South Asia Region

Upadhyay P, Prajapati SK, Kumar A (2023) Impacts of riverine pollution on greenhouse gas emissions: a comprehensive review. Ecol Ind 154:110649. https://doi.org/10.1016/j.ecolind.2023.110649

Vu MT, Raghavan SV, Liong SY (2016) Use of regional climate models for proxy data over transboundary regions. J Hydrol Eng 21(6):05016010. https://doi.org/10.1061/(ASCE)HE.1943-5584.0001342

Wang R, Zhang H, Zhang W, Zheng X, Butterbach-Bahl K, Li S, Han S (2020) An urban polluted river as a significant hotspot for water-atmosphere exchange of CH_4 and N_2O. Environ Pollut 264:114770. https://doi.org/10.1016/j.envpol.2020.114770

Yalcin E (2024) A CMIP6 multi-model ensemble-based analysis of potential climate change impacts on irrigation water demand and supply using SWAT and CROPWAT models: a case study of Akmese Dam, Turkey. Theor Appl Climatol 155(1):679–699. https://doi.org/10.1007/s00704-023-04657-0

Yasuda Y, Aich D, Hill D, Huntjens P, Swain A (2017) Transboundary water transboundary water cooperation over the Brahmaputra River: legal political economy analysis of current and future potential cooperation. The Hague Institute for Global Justice

Zhao F, Zhou Y, Xu H, Zhu G, Zhan X, Zou W, Zhu M, Kang L, Zhao X (2022) Oxic rivers as a potential source of atmospheric methane. Environ Pollut 297:118769. https://doi.org/10.1016/j.envpol.2021.118769

Zhang Y, Wu Z, Singh VP, He H, He J, Yin H, Zhang Y (2021) Coupled hydrology-crop growth model incorporating an improved evapotranspiration module. Agric Water Manag 246:106691. https://doi.org/10.1016/j.agwat.2020.106691. https://www.sciencedirect.com/science/article/pii/S0378377420322356

Part II

Quality Assessment and Modeling

Greywater Recycling and Reuse: Opportunities and Challenges

10

Irshad N. Shaikh and M. Mansoor Ahammed

Abstract

Due to increasing population and economic growth, urban centre, particularly in developing countries, are experiencing water shortages. This has led to increased interest in wastewater reuse. Since greywater is significantly less polluted and contributes to 70% of the total wastewater generated in households, its treatment and reuse is attractive and cost-effective. In delta regions, where populations are highly dependent on both agriculture and fisheries, greywater reuse can help reduce pressure on limited freshwater supplies and prevent untreated discharges into sensitive deltaic waters. The present chapter first discusses the sources and characteristics of greywater reported from around the world. It also covers various reuse applications, along with the relevant standards or guidelines from different countries. Options for its treatment are then outlined along with their suitability for different reuse. The chapter also addresses the problems and challenges involved in greywater recycling and reuse. These challenges are particularly pronounced in deltas, where high population densities, climate risks, and fragile hydrological balance demand robust, locally appropriate systems. Finally, different strategies for the successful implementation of greywater reuse are reported. This study supports the Sustainable Development Goals (SDG), particularly SDG 6, SDG 11 and SDG 12, through sustainable greywater treatment and reuse.

I. N. Shaikh · M. Mansoor Ahammed (✉)
Department of Civil Engineering, Sardar Vallabhbhai National Institute of Technology, Surat, India
e-mail: mansoorahammed@gmail.com

S. Yadav et al. (eds.), *Water Resources in Delta Region in India*, Deltas of the World,
https://doi.org/10.1007/978-3-032-06500-1_10

10.1 Introduction

Water scarcity is a critical global issue, and the quest for alternative water sources is becoming increasingly important as populations grow and urbanization expands. Given these conditions, traditional water management approaches may face challenges in meeting future demands (Shiam Babu et al. 2025). Therefore, exploring reuse options is essential, with greywater being a viable solution. Greywater refers to domestic wastewater typically generated from bathrooms, sinks, kitchens, and laundry (Lanchipa-Ale et al. 2024; Muniz Sacco et al. 2024).

Greywater contains relatively low levels of solids, organic contents, pathogens, and heavy metals, making it simpler to treat than municipal wastewater (Maiga et al. 2024). Repurposing this large volume of minimally contaminated greywater for non-potable uses, such as irrigation of lawns and gardens, and flushing toilets, can significantly decrease potable water demand. This, in turn, lowers energy consumption and the costs associated with water distribution and treatment (Alateeqi et al. 2023). Utilizing treated greywater not only conserves limited water resources but also reduces supply expenses and lessens the burden on centralized wastewater facilities (Busgang et al. 2015), thereby minimizing the environmental and financial costs associated with water extraction and subsequent treatment (Ansari et al. 2022). Studies indicate that reusing greywater can cut potable water consumption by approximately 25–30% (Bidira et al. 2024). According to Haldar et al. (2019), climate change–impacted delta regions such as Khulna City, Bangladesh, face a rising demand for alternative irrigation sources of adequate quality to safeguard both current and future food production.

Use of greywater without any treatment poses health risks to the public, though greywater only holds total organic matter of about 30–40% and nutrients of nearly 10–22% in regular household wastewater (Tusiime et al. 2022), it still poses potential health risks if used without proper treatment. The microbiological dangers linked to using untreated greywater for irrigation and toilet flushing make treatment necessary (Prabha et al. 2021). Various physical, chemical, biological, and natural treatment systems are available, each with its own benefits and limitations (Spychała et al. 2021).

Public acceptance plays a vital role in the success of any strategic water reuse plan (Madzaramba and Zanamwe 2023). The effectiveness of planning, constructing, and operating wastewater reuse facilities relies on public perception, acceptance, and support. Attitude is fundamental to understanding human behaviour (Alateeqi et al. 2023). This underscores the importance of examining individual views on greywater management, as negative perceptions may hinder policy implementation for effective and sustainable greywater solutions (Alateeqi et al. 2023).

In the present study, the quantity and quality characteristics of greywater from various sources are discussed. Key challenges and issues associated with the reuse of treated greywater for non-potable applications are studied. Additionally, opportunities for greywater reuse are analyzed through treatment methods commonly used in over 113 studies, including granular media filtration, coagulation, biological treatment systems, and constructed wetlands (CW). Lastly, the study recommends the most suitable treatment system based on greywater characteristics and intended reuse purposes.

10.2 Greywater Sources and Characteristics

The characteristics of domestic greywater in a household can differ significantly based on various factors, such as the number and age of residents, their daily activities, cultural practices, lifestyle choices, and the use of chemicals, personal care products (PCP) and cleaning agents (Spychala et al. 2019). Greywater composition is significantly influenced by factors such as geographical location and the time of year, with variations in water usage and the discharge of different substances (Eriksson et al. 2002). Greywater primarily arises from activities in bathrooms, kitchens, sinks, and laundry. In existing studies, greywater is often classified into two categories: light greywater (LGW) and dark greywater (DGW), based on the contamination level. Water generated from bathtubs, hand basins, bathroom sinks, and showers is included in LGW, whereas DGW consists of wastewater from washing machines, laundry, dishwashers, and kitchens (Alsulaili and Hamoda 2015). LGW typically has lower pollutant levels than DGW. When both types are mixed, the resulting water is termed mixed greywater (MGW). An overview of various greywater sources and their components is provided in Table 10.1.

Bathroom greywater comes from showers, bathroom sinks, and bathtubs, while hand basin greywater is the wastewater from washbasins or hand basins, whether they're in

Table 10.1 Greywater sources and their composition

Greywater source	Composition
Hand basin/ Wash basin	Toothpaste, body care products, soaps, hairs, shaving waste, and skin cells
Bathroom/ Shower/ Bathtub	Shampoo, toothpaste lint, soap, traces of urine, body care products, skin, body fats, hair oil, hairs, hot water, and sand and clay particles
Kitchen/ Dishwasher	Food residue, dishwashing detergent, raw meat washing, tea or coffee, oil and fats, hot water, traces pf food preservatives, fruit and vegetable peels, sand and clay particles
Laundry/ Washing machine	Solvents, chemicals from detergents, oils, bleaches, hot water, paints, non-biodegradable fibres from clothing

Data source Eriksson et al. (2002), Khajvand et al. (2022), Li and Huang (2009), Mohan (2014), Nolde (2000), Oteng-Peprah et al. (2018a), Spychala et al. (2019)

a bathroom or a toilet. Kitchen greywater is sourced from manual washing of utensils or from dishwashers, while laundry greywater includes wastewater from manual cloth washing or washing machines, encompassing wash, rinse, and spin cycles. Hand basin greywater contains residues such as toothpaste, soap, hair, skin cells, shaving waste, and PCP (Khajvand et al. 2022).

According to Khajvand et al. (2022) soap is one of the major components in hand basin greywater, accounting for approximately 90% of its total mass load. Bathroom greywater typically includes toothpaste, soap, body oils, urine traces, skin particles, shampoo, PCPs, hair, sand or clay and lint (Noah 2002; Shaikh and Ahammed 2024a). The pollutant concentration can rise when activities such as washing pets or disposing of hazardous materials like paint, pesticides, and herbicide residues occur. However, since it lacks urine and faeces, bathroom greywater generally contains lower levels of phosphorus and nitrogen (Noutsopoulos et al. 2018).

Cleaning agents and dishwashing detergents are alkaline in nature and are found in kitchen greywater. Kitchen greywater comprises dishwashing soaps, fats, oils, food preservatives, coffee, food scraps, sand and clay particles, wash water from raw meat, hot water, vegetable and fruit peels, tea, etc. This type of greywater is characterized by high levels of organic matter, TSS, turbidity, and nitrogen. Additional food particles, salinity, odour, oils, grease, soaps, elevated pH, bacteria, TSS, organic content, turbidity, and heat are contributed by automatic dishwashers (Noah 2002). The presence of food residues, abundant fats, and higher temperatures encourages microbial growth, and higher levels of FC have been reported, likely due to organic matter and the washing of raw meat (Blanky et al. 2015).

Laundry greywater contains various chemicals from bleaches, nonbiodegradable fibers shed from clothing, detergents, hot water, solvents, oils, and paints (Shaikh and Ahammed 2024a). Bacteria and viruses may also be present, especially from the washing of dirty clothes and nappies used by babies. Detergents and soiled fabrics are responsible for elevated chemical concentrations in the laundry greywater, which results in elevated levels of TSS, lint, turbidity, and oxygen demand. Laundry greywater tends to be less biodegradable than shower greywater due to the presence of phosphates and heavy metals (Nolde 2000). Furthermore, laundry greywater typically lacks nitrogen (Al-Khatib et al. 2022). In delta regions, where rivers discharge into the sea, untreated greywater from households can directly affect water quality and aquatic life. Greywater reuse is particularly relevant in delta regions (which are extensions of river basins), as untreated discharges from households can directly pollute deltaic water bodies; hence, treatment and reuse can significantly reduce pollution loads in these sensitive areas.

10.3 Greywater Quantity and Quality Characteristics

10.3.1 Greywater Quantity Characteristics

Greywater production is influenced by a variety of factors, including culture, water availability, household size, income, gender, age, country, socioeconomic status and water source (outdoor or indoor) (Al-Khatib et al. 2022). The volume of greywater generation from sinks, bathrooms, and kitchens is typically estimated by multiplying the duration of water use by the flow rate at an individual source (Leong et al. 2018). Greywater volume for manual and machine laundry, dishwashers, and toilets is determined by multiplying usage frequency with water consumption in each cycle (Jamrah et al. 2006).

Several methods have been used to measure greywater volume. To estimate the amount from different greywater sources, some studies used containers that were labelled and premarked. Tally card that could record over 1000 L of flow as an alternative to traditional flow meters was used by Oteng-Peprah et al. (2018a). To estimate greywater production, Leong et al. (2018) multiplied the proportion of total water demand for a specific source, like the bathroom, by the overall annual water demand. Greywater generation is also shaped by factors such as economic status, climate, and others. Accurate estimation of greywater volumes is particularly important in delta regions, where high population densities place additional pressures on limited water resources. Figure 10.1 provides a summary of data on water consumption, greywater generation, and return factors from 82 studies. Daily per-person water use ranged from 35 to 343 L, with an average of 146 ± 63 L (mean ± standard deviation), surpassing the recommended minimum of 30 L/person/day.

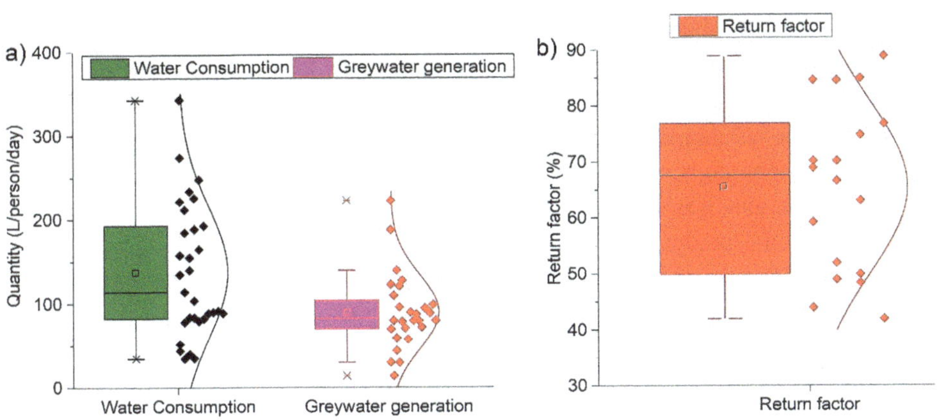

Fig. 10.1 **a** A summary of water consumption and greywater generation, **b** return factor data from 82 studies

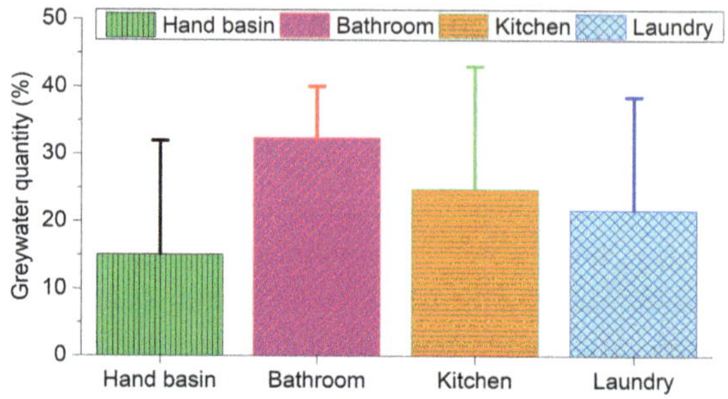

Fig. 10.2 The proportion of GW generated by each source based on 43 studies

Figure 10.1a illustrates daily per-person greywater production, showing significant variation. Reported greywater generation ranged from 14 to 223 L/person/day, with an average of 98 ± 37 L/person/day. While greywater generation is not influenced by floor area, it is affected by household size. Jamrah et al. (2006) found that although total greywater output rises with family size, per capita greywater production decreases as family size grows. Figure 10.1b presents return factor, which shows the amount of greywater generated as a portion of the total water consumption. The return factor varied between 41 and 89%, with a median of 67%, suggesting it is generally independent of water usage. Differences in water consumption and greywater production may be attributed to sanitation facilities, living standards, and household practices.

Greywater is produced from various sources, including bathrooms, sinks, kitchens, and laundry activities. Figure 10.2 illustrates box plots depicting the proportion of greywater generated by each source, based on 43 studies. On average, bathrooms, sinks, kitchens, and laundry contribute 32 ± 16%, 15 ± 16%, 25 ± 13%, and 22 ± 8% of total greywater, respectively. Light greywater originating from bathrooms and sinks, accounts for 48 ± 28% and 49 ± 26%, respectively, while DGW, consisting of kitchen and laundry wastewater, represents 47 ± 22% and 48 ± 23% of total greywater. Kitchen greywater, making up approximately 25% of total greywater, may result from increased use of water-efficient dishwashers (Noah 2002) and lifestyle changes such as consuming more processed foods, less home cooking, and dining out. Laundry greywater generation varies across studies, with higher values observed in cases where washing machines, which use more water than handwashing, are commonly used.

On average, the production of LGW and DGW is nearly balanced. However, LGW usually has a lower pollutant load than DGW, making it easier to treat with simpler systems (Oteng-Peprah et al. 2018a).

10.3.2 Greywater Quality Characteristics

Habits of residents, water source, plumbing systems, geographical location, and source of greywater affects greywater quality (Gani et al. 2015; Winward et al. 2008a). Dishwashing routines, household chemical disposal, bathing habits, choice of cleaning products, and laundry practices are also important in shaping greywater quality (Prathapar et al. 2005). Monitoring the quality of greywater is critical in deltas, where pollutants can rapidly accumulate in low-lying floodplains and affect both agriculture and fisheries. Understanding these factors and their variations is crucial for selecting an appropriate greywater treatment system. This review draws on data from over 169 studies.

Significant variations in greywater quality have been observed across different sources (Bakare et al. 2017; Gupta et al. 2023). Many times, immediate analysis of fresh greywater is not possible, and its quality can be affected by the duration of storage before testing. During storage, solids are more significantly impacted, while organic matter experiences moderate changes (Katukiza et al. 2014a). However, the concentration of microorganisms tends to remain relatively stable, even as greywater settles (Abdel-Shafy et al. 2014a). It was found by Eriksson et al. (2002) that storing untreated greywater for up to 24–48 h can help partially remove solids and organic material. After 48 h, prolonged storage resulted in unpleasant odours due to decomposition. Chin et al. (2009) observed that overnight storage reduced chemical oxygen demand (COD) by 13%, turbidity by 31%, and lowered pH by 0.5 units. Additionally, greywater decomposes faster than blackwater (Prathapar et al. 2005). Figure 10.3 presents the variations in TSS, pH, biological oxygen demand (BOD), and COD across different greywater sources, based on findings from 69 studies.

The variation in TSS concentration is illustrated in Fig. 10.3a, shows kitchen and laundry greywater exhibiting significantly higher TSS levels. In laundry greywater, zeolites and fabric fibers from laundry detergents contribute to elevated TSS, while in kitchen greywater, washing vegetables and fruits adds to the solid content (Shaikh and Ahammed 2024b; Leong et al. 2018). There are notable differences in the reported TSS values

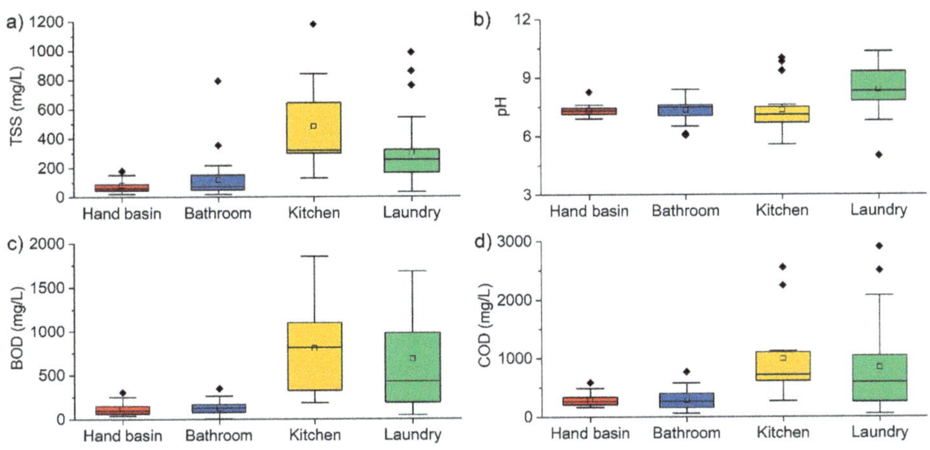

Fig. 10.3 Variations in **a** TSS, **b** pH, **c** BOD and **d** COD across different greywater sources reported in 69 studies

for not only kitchen (483 ± 341 mg/L) greywater but also for laundry (391 ± 377 mg/L) greywater. This variation is influenced by the use of automatic dishwashers and washing machines in some households, while others rely on traditional manual washing methods. Additionally, solids in greywater increase due to the practice in some households of using charcoal ash as well as sand with water to clean utensils, preventing them from blackening (Sall and Takahashi 2006).

Greywater from indoor sources generally has a lower TSS concentration than greywater from outdoor sources (Oteng-Peprah et al. 2018a). A correlation between suspended solids and various microbial groups was found by Winward et al. (2008a; b), including bacteria, protozoa, and bacterial indicators. Lower levels of TSS and turbidity also improve the effectiveness of greywater disinfection.

The variation in pH across different greywater streams is shown in Fig. 10.3b. Most greywater streams exhibit a near-neutral pH, except for laundry greywater, which tends to be slightly alkaline. The pH of greywater is affected by both alkalinity and pH values of the water supply. (Oteng-Peprah et al. 2018b). Laundry greywater's higher pH results from alkaline substances in detergents (Eriksson et al. 2002). The lowest pH was found in kitchen greywater, as oils and food particles break down quickly under anoxic conditions (Bakare et al. 2017).

Some greywater samples exhibited very low pH values, likely due to the formation of organic acids from edible organic matter. On the other hand, highly alkaline pH levels may be a result of sodium hydroxide-based soaps (Noah 2002). Minimal fluctuations in daily greywater pH variation have been observed (Turner et al. 2013). When greywater with a pH above 8.0 is used for irrigation, it can reduce the availability of essential micronutrients like boron for plants (Gani et al. 2015; Turner et al. 2013).

Figure 10.3c and d show the variations reported in the concentrations of BOD and COD across different greywater sources. Both BOD and COD display similar trends, with DGW sources having higher organic content compared to LGW sources. Data from 63 studies showed that the average BOD concentration in kitchen greywater was 802 mg/L, compared to 680 mg/L in laundry greywater. The mean BOD and COD values for bathroom and hand basin greywater were 126 mg/L and 121 mg/L, respectively. Similarly, the average COD concentrations for the hand basin and bathroom were 298 and 281 mg/L, respectively while it was 987 mg/L for the kitchen, and 837 mg/L for the laundry greywater. These values suggest that the average BOD concentration in DGW can be as much as five times greater than in LGW, likely because of the presence of PCPs and XOCs (Bidira et al. 2024). Badr et al. (2024), in their characterization of wastewater from Egypt's western Nile delta, detected Human Adenovirus (HadV) and Human Bocavirus (HBoV) in all 96 samples analyzed, underscoring the necessity of treatment prior to direct discharge.

Household water consumption significantly impacts the BOD and COD levels in generated greywater. Some developing nations have reported exceptionally high organic content values. The elevated organic content in kitchen and laundry greywater may be explained by the extensive use of detergents, along with food and beverage residues containing fats and oils (Ghaitidak and Yadav 2015a). Additionally, chemicals in laundry and dishwashing detergents are responsible for the higher COD levels in DGW sources.

The variations in total nitrogen (TN), total phosphorus (TP), total coliforms (TC), and faecal coliforms (FC) across different greywater sources, based on 69 studies, are presented in Fig. 10.4. Nitrogen and phosphorus concentrations vary significantly among different greywater sources (see Fig. 4a and b). However, the median nitrogen levels in different greywater streams typically range from 7 to 15 mg/L (see Fig. 4a), which is significantly lower than 20 to 80 mg/L commonly found in greywater (Eriksson et al. 2002). Elevated nitrogen levels in some bathroom greywater could be due to the type of washing products employed, while higher nitrogen concentrations in the laundry greywater might be linked to lower water usage. Sources of nitrogen include protein-based shampoos, proteins from vegetables and meat, and cleaning products that contain ammonia or are ammonia-based (Noutsopoulos et al. 2018). Nitrogen content in greywater tends to follow the flow pattern, with peak concentrations aligning with periods of high discharge (Dalahmeh et al. 2014).

Figure 10.4b presents the variation in total phosphorus concentration across greywater sources. Phosphorus levels are mainly influenced by laundry greywater, with kitchen, bathroom, and hand basin greywater contributing to a lesser extent. Soaps, cleaning agents and detergents are major contributors of phosphorus in greywater (Maimon et al. 2014). The average phosphorus concentrations of 10.83 mg/L for the kitchen, 15.51 mg/L for manual laundry and less than 4.6 mg/L for both hand basin and bathroom were present. Elevated phosphorus concentrations in kitchen and laundry greywater lead to higher levels in DGW.

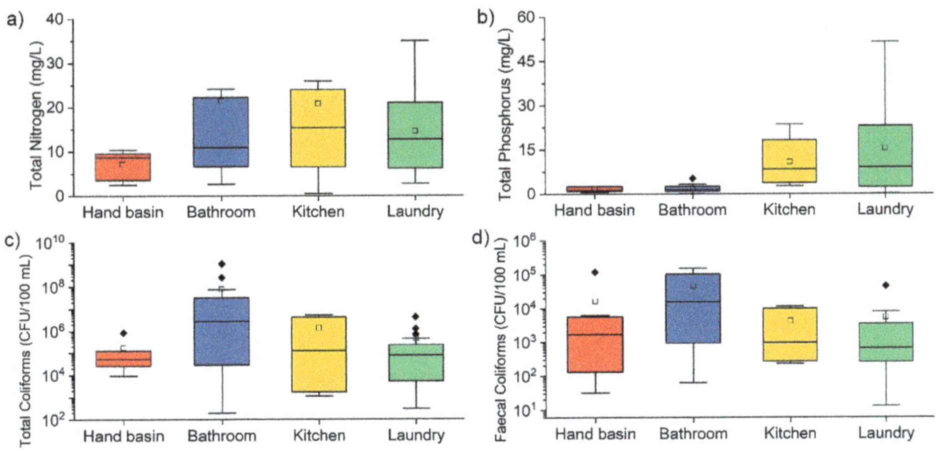

Fig. 10.4 Variations in **a** total nitrogen, **b** total phosphorus, **c** total coliforms and **d** faecal coliforms across different greywater sources reported in 63 studies

Pathogenic organisms represent a significant public health risk in greywater reuse (Maimon et al. 2014), making their removal a crucial objective in greywater treatment. Monitoring the microbiological quality of greywater is therefore essential. Furthermore, Careful attention is needed to monitor the survival and transformation of contaminants in greywater. To assess its microbial quality, common indicators like TC and FC are often used.

Common contributors include washing diapers, use of hand basin after using the toilet, handling of raw food or meat, (O'Toole et al. 2012). Higher faecal contamination was reported in households with young children, and several researchers have shown that the level of contamination in greywater can be influenced by the age of the people living in the household. (Blanky et al. 2015). Lack of proper personal hygiene and improper diaper washing contribute to increased fecal contamination in greywater (Maimon et al. 2014).

Bathrooms and laundry areas are major sources to faecal contamination in greywater (Al-Gheethi et al. 2017; O'Toole et al. 2012). Fecal contamination is primarily caused by activities like showering, washing soiled diapers, and handwashing after handling potentially contaminated items (Busgang et al. 2015). Bathroom greywater, in particular, has been found to contain the highest concentrations of TC compared to other greywater sources (see Fig. 4c), likely due to the presence of dead skin cells, sweat, and traces of urine (O'Toole et al. 2012; Zhu et al. 2018). Additionally, bathing babies and washing diapers in the bathroom can also introduce fecal matter into the greywater (Benami et al. 2016). Fecal coliform concentrations can vary widely across different greywater streams; the levels vary from undetectable to as high as 10^5 CFU/100 mL. In bathroom greywater, the median TC concentration is 7.9×10^7 CFU/100 mL, 1.73×10^5 CFU/100 mL in hand basin water, 1.40×10^6 CFU/100 mL in kitchen greywater, and 4.34×10^5 CFU/100 mL in laundry greywater. Similarly, *E. coli* concentrations exhibit significant variability, from undetectable levels to over 10^6 CFU/100 mL.

Studies from delta region of China, India, Nigeria, Saudi Arabia, Vietnam, Bangaladesh indicated the presence of several parameters above effluent discharge standards values of those countries (Badr et al. 2024; Alomar et al. 2020; Haldar et al. 2019; Nguyen and Tran, 2025; Sheng et al. 2020).

10.4 Problems and Challenges in Greywater Recycling

10.4.1 Quality of Greywater

The types and concentrations of harmful organisms and toxic chemicals in greywater can vary, depending on its source. As such, its reuse presents certain risks that need to be addressed to ensure safety. These risks include potential threats to human health and the environment. The concentration of contaminants that could impact health is a critical factor in health risk assessments. In delta regions, such contaminants pose heightened risks because the water systems are interconnected and support large populations dependent on river delta ecosystems. The concentration of pathogens can vary based on factors such as age, immune system strength, nutrition, existing immunity, and other general host conditions (Maimon et al. 2014). Direct exposure occurs to individuals who handle or clean greywater filters, work in areas irrigated with greywater, or use toilets flushed with greywater. Indirect exposure might happen when pets play on lawns irrigated with greywater.

Beyond human health risks, environmental impacts must also be considered. Studies have shown that greywater irrigation can lead to significantly higher FC levels in soil compared to freshwater irrigation. The number and survival rate of pathogens in soil can vary based on the amount of greywater applied (Dalahmeh et al. 2014). One of the most critical environmental concerns is soil salinization, which can result from the high concentrations of ions such as Na+, Mg^2+, K+, and Ca^2+. Soil salinization leads to diminished soil health, stunted plant growth, and reduced agricultural yields due to ion toxicity and osmotic stress (Katukiza et al. 2015).

Although bacterial indicators such as TC, FC, and *E. coli* are commonly monitored in raw greywater and treated effluent, limited data is available on pathogens like bacteria, viruses, and protozoa due to the high costs and difficulty in measuring them. However, the presence of disease-causing organisms like *Salmonella, Clostridium,* and *Legionella* in greywater highlights the importance of further research in this area (Arden and Ma 2018). A recent comprehensive review revealed the detection of around 280 organic micropollutants in greywater (Etchepare and van der Hoek 2015). Household chemicals, particularly personal care products, were primarily associated with most of these pollutants, while pharmaceutical compounds with potential endocrine-disrupting effects were not detected. Etchepare and van der Hoek (2015) reported that the majority of the chemicals evaluated do not present significant risks to human health.

Although global regulations for treated greywater quality are absent, many countries have established their own developed standards and guidelines for water reuse, based on the reuse objective of the treated greywater. These standards differ across countries, influenced by several factors like water scarcity and varying risk perceptions. Generally, these guidelines focus on reducing organic content and microbial content in treated greywater, and may also address aspects such as turbidity, pH, and residual chlorine. However, these standards are not always based on risk assessments, which leaves the level of protection for human health unclear. This highlights the emergency need for further research on the microbial risk assessments for the specific pathogens.

10.4.2 User Perception

Greywater, which accounts for 50–90% of household wastewater excluding toilet and urinal waste, presents significant environmental challenges and public health. In many semi-urban areas, greywater is often discharged untreated into the environment, contributing to pollution. In peri-urban regions, less polluted greywater is sometimes mixed with toilet waste, unnecessarily increasing the overall contamination load and burdening wastewater treatment plants. Globally, much research has focused on understanding greywater characteristics and developing decentralized treatment solutions. Governments and local authorities are increasingly emphasizing the treatment and reuse of greywater, particularly in large residential areas, as a way to address water scarcity and sanitation issues in developing countries (Madzaramba and Zanamwe 2023). In response to declining rainfall linked to climate change and the rapidly growing water demand in the Pearl River Delta, the Government of China has developed a greywater management strategy (Sheng et al. 2020). Given the ongoing water scarcity and the need to achieve Sustainable Development Goal 6 (SDG6) by 2030, managing and reusing greywater effectively is critical. However, limited research has been conducted on the acceptance of treated greywater by users in developing countries like India, which underscores the importance of studying public perceptions of greywater management. Negative perceptions could hinder the successful implementation of policies aimed at sustainable greywater management (Oteng-Peprah et al. 2020). Enhancing public acceptance of greywater reuse is particularly important in delta regions, where communities rely heavily on local water bodies for daily needs and livelihoods.

Public acceptance is necessary in the success of any water reuse initiative (Chen et al. 2012). The planning and construction followed by operation of wastewater reuse facilities are heavily influenced by public acceptance, attitudes, and support. Attitude is fundamental to understanding human behaviour. The attitude toward wastewater reuse, whether for domestic, agricultural, industrial, or commercial purposes, reflects the collective outcome of individual responses, either in support or opposition (Buyukkamaci and Alkan 2013; Lanchipa-Ale et al. 2024).

Lanchipa-Ale et al. (2024) also reported the increased acceptability to 77% from 34% for treated and untreated greywater reuse, respectively, for reusing greywater in green spaces in the university campus in Peru by students. Newcomer et al. (2017) indicated that health risks, environmental awareness and costs were major factors for resisting the implementation of wastewater treatment systems at decentralized levels. However, a few (21%) of the households were ready to install and pay for the greywater treatment and non-potable reuse facility in the societies once they were informed about the water scarcity and benefits of reuse.

Oteng-Peprah et al. (2018a) also reported a higher willingness of participants to know their own greywater source compared to the acceptability of an unknown source of greywater. Almost 66% of the participants hesitate about indoor reuse of greywater compared to the outdoor reuse (Alsulaili and Hamoda 2015). According to Abdelrahman et al. (2020), approximately 66% of respondents in the United Arab Emirates were hesitant to use treated wastewater for irrigating food crops. About 60% of respondents Akpan et al. (2020), had a favourable attitude towards using treated greywater for garden irrigation.

Treated greywater reuse acceptability of 80 and 62% for irrigating vegetables was reported by Newcomer et al. (2017) and Al-Khatib et al. (2022), respectively. Senthilkumar et al. (2021) reported significantly higher total green and fodder yields when wastewater was used for irrigation in the Cauvery Delta region of Tamil Nadu, India. Portman et al. (2022) showed 90% reuse acceptability of greywater reuse for irrigating food crops while it was 81% in the study conducted by Zhu et al. (2017). Al-Khatib et al. (2022) reported that around 96% of participants in Israel were in favor of using treated wastewater for irrigating public parks. In the Delta region of Bangladesh, 53% of citizens were willing to pay for wastewater treatment, and 66% of farmers were prepared to invest in higher-quality irrigation water, according to Haldar et al. (2019), based on the perceptions of 385 respondents in Khulna City.

Abdelrahman et al. (2020) found that more than 80% of participants from the UAE and Australia were open to reusing treated wastewater for various industrial and commercial purposes. Studies by Buyukkamaci and Alkan (2013), Abdelrahman (2020), and Lanchipa-Ale et al. (2024) highlighted that users generally accept greywater reuse under specific conditions, especially when there is no direct contact with the water. Baghapour et al. (2017) found that the highest level of acceptance for treated wastewater reuse was associated with minimal skin contact. Akpan et al. (2020) found that 80% of participants were open to using treated greywater for toilet flushing, while Buyukkamaci and Alkan (2013) reported a 57% acceptance rate for the same practice in households. Younger individuals were more likely to support greywater reuse for toilet flushing, while those over the age of 51 showed less receptivity. Zhu et al. (2017) also found that in China, individuals over 40 were less accepting of the practice compared to younger participants.

Regarding the reuse acceptance of treated greywater for car washing was 41% in UAE (United Arab Emirates), which is lower than the 70% acceptance in the United States, 73% in Nigeria, and 49% in Palestine (Abdelrahman et al. 2020; Al-Khatib et al. 2022; Akpan et al. 2020; Portman et al. 2022). Sample size, cultural differences, education, awareness campaigns, individual preferences, geographic location, and the availability and reliability of efficient treatment systems are factors governing the acceptance level for car washing reuse of treated greywater.

10.5 Greywater Treatment and Reuse

Greywater can be treated by several means however, major treatment methods adopted worldwide for greywater treated are discussed here. In delta regions, effective treatment and reuse of greywater can help reduce direct discharges into fragile river–delta systems and support water security for local communities. Cities in Southern California draw water from the San Francisco Bay Delta (SFBD), making it necessary to treat greywater before any direct discharge into the delta systems (Dougherty and Murphy 2012).

10.5.1 Granular Media Filtration

The preference for locally available materials is given in the literature which often emphasize reducing material costs. Filter media sizes used in various studies range from 0.09 mm to over 80 mm (Bahrami et al. 2020; Emslie et al. 2021; Prabha et al. 2021). For greywater filtration, the most commonly used effective sizes range from 0.12 mm to 2.85 mm, with numerous studies utilizing media finer than 4.75 mm sieve and having an 0.65 mm or less as an effective diameter. In most studies, filter media typically have a coefficient of uniformity (UC) under 4. The UC of materials is ideally between 1.5 and 2.5, for optimal filtration (Zipf et al. 2016). A lower UC enhances total pore space and infiltration rates, thereby improving oxygenation.

Loading rate, influent composition, media depth, operational mode, and media properties govern the potential of filters to treat greywater. Studies have shown significant variations in filtrate quality in response to changes in HLR and OLR. HLR between 3.2 cm/day and 42.1 m/h has been adopted in the literature, while OLR was between 7.83 and 496.00 gBOD/m^2/day (Abdel-Shafy et al. 2014a; Emslie et al. 2021). Choosing the right HLR and OLR is essential since they influence both clogging and the quality of the effluent. Elevated HLR and OLR values typically decrease the filters' effectiveness in eliminating organic matter, nutrients, and microorganisms. Raising the OLR to 35 from 4 g COD/m^2/day had no significant impact on turbidity removal in sand filters (Patil and Munavalli 2016). Changes in HLR and OLR had little effect on the values of pH and EC of the filtered greywater (Zipf et al. 2016).

The potential of filters in the removal of nutrients, organic matter, and microorganisms decreases with an increase in HLR and OLR. For instance, in a study on MGW from a Norwegian household, it was observed that increasing the HLR by 2.5 times led to a decrease in BOD removal from 85 to 78% (Karabelnik et al. 2012). In Uganda, while HLR was doubled to 40 from 20 cm/day, TC and *E. coli* removal declined from 1.83 to 1.62 log and from 2.52 to 2.31 log, respectively, in a crushed lava rock filter treating DGW (Katukiza et al. 2014b).

Increasing media size negatively impacts filter efficiency in removing solids, organic matter, nutrients, and microorganisms. In a comparison of ceramic filters with particle sizes of 0.25, 0.72, and 1.18 mm, Khalfan Al Jahmani et al. (2021) found that the 0.25 mm filter provided the best removal of turbidity, total organic carbon (TOC), TSS, and COD in kitchen greywater. Greywater pH is unaffected due to media size during filtration (Bahrami et al. 2020). The TN removal increased to 75% from 53% when the size of filter media decreased from 0.6 to 0.3 mm in the treatment of LGW from a Japanese household Ochoa et al. (2015). Abdel-Shafy et al. (2014a) in the treatment of MGW observed NH_4-N removal rate of 40% using gravel, which increased to 55% when sand media was used. The TC removal was reduced to below 1.60 log from 1.93 log while filter media size was increased (to 0.80–1.18 mm from 0.40–0.80 mm) (Singh et al. 2021).

The depth of media is important in the performance of filters. Studies commonly use media depths between 30 and 100 cm, as deeper layers can lead to anaerobic conditions (Ochoa et al. 2015; Shaikh and Ahammed 2023a). Bacterial removal is highest in the upper layers and decreases with depth due to reduced oxygen and organic matter (Ochoa et al. 2015). The development of a biolayer on the surface of the media improves turbidity removal in the upper layers (Raj et al. 2023). The depth of the sand bed is crucial for the removal of nutrients, as nitrification and denitrification processes occur simultaneously throughout the filter depth (Gupta et al. 2023). The top 10 cm removed most NH_4-N, while deeper layers further reduced NH_4–N levels. Although PO_4–P removal was highest in the upper layers, deeper media enhanced overall removal. According to Verma et al. (2017), a 67% removal of TC was observed within the top 10 cm of the media, while Emslie et al. (2021) observed the formation of a biolayer on the media surface after the first few days of filter operation.

The quality of effluent from greywater treatment filters depends on the operating mode, which can be either intermittent (processing specific volumes in batches) or continuous, with vertical or horizontal flow, up-flow or downflow, and either saturated or unsaturated conditions. Sabogal-Paz et al. (2020) noted that continuous filters performed better in turbidity removal than intermittent ones. Similarly, Verma et al. (2017) found higher *E. coli* removal in continuously operated filters compared to intermittent ones. In intermittent filtration, dosing frequency is a key factor. Dalahmeh et al. (2014) distributed 70% of the total daily greywater volume during the first 9 h, while the remaining 10%, and 20% were distributed after 16 and 20 h, respectively. A few studies administered filter doses 8 to 11 times per day (Assayed et al. 2014; Friedler et al. 2005).

The intact biolayer in saturated filters aids in the removal of PO_4–P through biodegradation and adsorption, as noted by Abdel-Shafy et al. (2014b). Conversely, the aerobic conditions in unsaturated filters enhance NH_4–N removal, as deeper layers in saturated filters tend to be anoxic (Achak et al. 2009). Adsorption and biodegradation play key roles in reducing PO_4–P concentrations, and their effectiveness depends on the filter's saturation state (Assayed et al. 2015). Karabelnik et al. (2012) documented that the vertical filters achieved better pollutant removal than horizontal ones. Downflow systems in vertical filters are more energy-efficient and provide better removal of a broad range of contaminants than upflow systems (Abdel-Shafy et al. 2014a). In deltas, granular media filtration offers a low-cost and locally adaptable solution for treating greywater before it reaches floodplains and coastal wetlands (Nguyen and Tran 2025). In the delta region of Vietnam, Nguyen and Tran (2025) used locally available materials in a filter for on-site greywater treatment, reporting about 75% removal of BOD and COD and 80% removal of TSS.

In conclusion, granular filters are more effective in removing turbidity and TSS, but their efficiency in BOD and COD removal is moderate, averaging around 50–60%. These filters perform well in phosphorus removal but show lower efficiency in nitrogen removal. The removal of bacteria varied during filtration, ranging between 0.1 and 4.3 log, based on the operational conditions. Particle size variations, HLR, operation mode, and OLR have minimal impact on turbidity removal; they significantly influence the removal of organic contents. The effectiveness of granular filters in removing surfactants, bacterial pathogens, organic micropollutants, and viruses is still not well understood and requires additional research. When appropriately designed with proper pre-treatment, media selection, depth, and operational mode suited to the greywater source, granular filters can be a viable option for on-site greywater treatment.

10.5.2 Chemical Treatment System

Coagulants consist of highly charged molecules that neutralize the electrical charges on colloidal and fine particles, leading to destabilization. In wastewater treatment, a coagulant facilitates the aggregation of colloidal particles into flocs, allowing them to settle. Additionally, organic chemicals known as flocculants are added to enhance the flocculation process, improving particle aggregation and removal (Tchobanoglous et al. 2011). Chemical treatment systems can be particularly useful in densely populated deltas, where rapid removal of colloids and pathogens is needed to prevent pollution of shallow surface waters.

In the realm of greywater treatment, both natural synthetic and organic polymers, as well as pre-hydrolyzed metal salts and metal, are extensively employed as flocculants and coagulants. Research has extensively documented the use of various natural and chemical coagulants for this purpose. Among the chemical coagulants, aluminium sulphate (alum, $Al_2(SO_4)_3$), polyaluminium chloride (PAC) ferrous sulphate ($FeSO_4$), and ferric chloride ($FeCl_3$), are frequently utilized (Ghaitidak and Yadav 2015a). Additionally, other chemical agents such as aluminium chloride, aluminium tripolyphosphate, hydrogen peroxide (H_2O_2), calcium hydroxide (lime, $Ca(OH)_2$), and magnetic ion exchange resin (MIEX) have been effectively employed in the treatment of greywater. On the natural front, *Moringa oleifera* stands out as a widely used coagulant. Furthermore, studies have delved into the potential of banana peels, sapodilla seeds, mango peels (from Mangifera indica), Citrus sinensis (sweet orange)orange peels, tamarind seeds, and Nirmali seeds (Strychnos potatorum), while treating greywater. These natural alternatives have shown promise, offering sustainable and eco-friendly solutions for water purification.

The task of removing colloidal particles becomes particularly challenging when a low dose of coagulant is employed. According to Tchobanoglous et al. (2011), cationic polyelectrolytes are primarily used for charge neutralization, whereas non-ionic polymers and anionic and play a key role in the bridging processes of chemical coagulation. The removal process is mainly influenced by the type of coagulant applied. Polyelectrolytes eliminate colloidal particles by neutralizing charges and forming polymer bridges, while hydrolyzed metal ions aid in removal through adsorption, charge neutralization, and bridging between particles. At higher doses, sweep coagulation also important in the removal of solids

(Kaewsarn and Yu 2001; Tchobanoglous et al. 2011). Mononuclear and polynuclear metal hydrolysis species are primarily involved in adsorption and charge neutralization, whereas polynuclear metal hydrolysis species and polymers are linked to adsorption and interparticle bridging. When an adequate quantity of metal salts is introduced, macroflocculation occurs, leading to the formation of metal hydroxide flocs. These flocs aid in the settling and subsequent removal of colloidal particles, ensuring effective treatment.

10.5.2.1 Aluminium-Based Coagulants

Alum coagulation could reduce the average turbidity of 88% and TSS of 78% from raw greywater. However, the dosage of PAC tends to increase with the rise in turbidity levels of the raw greywater, as highlighted by Ghaitidak and Yadav (2015b). The decrease in greywater pH is proportionate to the amount of alum adopted. Performance of different coagulants in selected studies is shown in Table 10.2. According to Antonopoulou et al. (2013), raising the alum dose from 100 to 800 mg/L led to a rise in TSS removal, which increased to 79% from 61% with an increase in alum dose to 800 mg/L from 100 mg/L, COD removal from 38 to 81%. However, beyond this dosage range, further increases in alum did not yield any substantial improvement in the removal efficiencies of TSS and COD. This suggests an optimal dosage range for achieving effective treatment outcomes.

In various studies, the dosage of alum employed for greywater coagulation has been observed to range between 5 and 1200 mg/L. This wide variation is primarily attributed to differences in the pH of greywater, which typically falls between 6.60 and 9.34. Alum is effective across a broad pH range, but it works most efficiently in removing TSS and turbidity, and when the pH is between 5 and 9. In this optimal range, the formation of $Al(OH)_3$ and positively charged polymeric species is enhanced, aiding pollutant removal through mechanisms like precipitation, charge neutralization, and sweep coagulation. Remarkably, alum coagulation achieves removal efficiencies exceeding 75% for turbidity, 50% for BOD, and 42% for COD, even without the need for pH adjustment. Furthermore, significant reductions in TC and FC have been recorded, with average removals exceeding 2 log, specifically 1.35 log for TC and 1.37 log for FC. These results underscore alum's effectiveness as a dependable coagulant for treating greywater under diverse conditions.

Proper rapid mixing can significantly improve the potential of alum as a coagulant, as highlighted by Shaikh and Ahammed (2023b). However, its performance tends to decline at lower temperatures, and its limited capacity to remove organic suspended solids further limits its application (Pidou et al. 2008). On average, alum achieves a moderate removal of BOD and COD with an efficiency of around 65%. Antonopoulou et al. (2013) reported a remarkable 80% COD removal with an alum dose of 800 mg/L.

As the alum dosage increases, the effectiveness of removing solid and organic matter improves, but this improvement becomes marginal beyond a certain threshold (Antonopoulou et al. 2013). Šostar-Turk et al. (2005) in Slovenia while treating laundry greywater found that alum had a minimal effect on removing anionic surfactants.

Table 10.2 Performance of chemical and natural coagulants in greywater treatment

Source of greywater	Coagulant	Coagulant dose range$ in mg/L	pH range#	Removal (%)					Refs.
				Turbidity*	BOD*	COD*	TC*	FC*	
S, B, HB, L	Alum	140–260 (190)	7.32–8.36	85.7 (12.0)	64.5 (39)	65.2 (80)	95.53 (2.1 × 10^5)	95.75 (1.7 × 10^4)	Singh et al. (2021)
HB, B, K, WM, L, S	Ferric chloride	(40)	7.42–8.78 (7.9)	51.9 (3.8)		88.0 (75)	80.63 (2.5X10^2)		Ahmad et al. (2020)
MGW	Alum	200–1200 (1000)	2.00–12.00	93.0 (6.2)	–	96.1 (24)	–	–	Chitra and Muruganandam (2019)
HB, S, B	PAC			89.5 (3.6)	85.2 (15)			82.86 (2.4 × 10^4)	Ghaitidak and Yadav (2016)
L	Ferrous sulphate	30–120 (120)	7.96–8.37	59.0 (25.8)		38.4 (269)			Al-Gheethi et al. (2017)
B, S, HB, WM, K, DW	Ferric chloride	(100)	6.05–7.96	93.4 (7.3)	87.7 (38)	81.7 (80)			Abdel-Shafy and Al-Sulaiman (2014)
S, B, HB, L	PAC	70–400 (199)	6.54–9.30	93.4 (4.0)	–	74.2 (42)	–	–	Vinitha et al. (2018)
S, B, HB	Ferrous sulphate	5–50	4.50–7.00 (4.50)	88.8 (5.2)	85.4 (30)	63.6 (288)	99.99 (<1)		Pidou et al. (2008)

B: bathroom; HB: hand basin; L: laundry; S: shower; DW: dishwasher; K: kitchen; WM: washing machine; PAC: poly aluminium chloride; MGW: mixed greywater; $the values in parentheses represent the optimal coagulant dose; #The values in the parenthesis represent the optimal pH; *The concentrations of treated greywater are indicated in parentheses, with NTU for turbidity, mg/L for BOD and COD, and CFU/100 mL for TC and FC

Moreover, Alum coagulation demonstrated limited effectiveness in decreasing dissolved organic carbon and TOC, with removal rates staying under 15% and 46%, respectively (Thompson et al. 2020, 2017). In contrast, PAC proves to be highly effective in removing TSS and turbidity, achieving removal rates of over 73% and 83%, respectively. In contrast to alum coagulation, the turbidity removal efficiency of PAC remained consistent regardless of the pH value of the greywater (Pidou et al. 2008). However, to remove TSS, higher PAC doses were necessary at increased pH levels. Furthermore, as turbidity in the raw greywater increased, the necessary PAC dose also rose (Vinitha et al. 2018). Overall, PAC remains effective across a wide pH range, from 4.5 to 9.5.

The introduction of a coagulant into greywater significantly impacts its pH and alkalinity, with the extent of alkalinity reduction being closely linked to the dosage of PAC used. The success of coagulation and hydrolysis in treating greywater is influenced by the initial alkalinity of the water, as the chemical coagulation process tends to deplete alkalinity (Pidou et al. 2008). Vinitha et al. (2018) showed that a PAC dosage of 50 and 250 mg/L resulted in a 70 and 84% reduction in COD, respectively. Much like alum, PAC coagulation also lowers the pH of greywater due to the release of H + ions during hydrolysis. In delta environments, aluminium-based coagulants can provide a quick means to improve water clarity, which is crucial where greywater often enters fisheries and irrigation channels.

10.5.2.2 Iron-Based Coagulants

Iron-based coagulants such as ferric chloride and ferrous sulfate have gained prominence for their effectiveness. Ferric chloride, in particular, stands out for its remarkable ability to eliminate solids from greywater. Studies have shown that it can reduce turbidity levels in raw greywater by an impressive 55 to 98%, with an average removal rate hovering around 87%. Notably, the removal of TSS often surpasses 73%, irrespective of any pH adjustments made. To illustrate the performance of these coagulants, Table 10.2 provides a comparative analysis of ferrous sulfate and ferric chloride. Friedler et al. (2008) showed a 55% reduction in turbidity was achieved with an optimal ferric chloride dosage of 50 mg/L. Another investigation by Friedler (2008) highlighted an even more remarkable 94% reduction in turbidity at dosages ranging from 40 to 50 mg/L. This highlights ferric chloride's potential as an effective coagulant in the effective treatment of greywater, making it a valuable asset in water management practices.

One interesting aspect of using ferric chloride for coagulation is its impact on the pH of greywater. Moreover, the carbon dioxide gas produced after adding ferric chloride reacts with water to form carbonic acid, further contributing to the decrease in pH. It has been observed that ferric chloride works best in more acidic conditions; for instance, a drop in pH from 6 to 5 can enhance the charge neutralization potential by as much as 2.81 times compared to a pH of 6 (Ghaitidak and Yadav 2015b). This highlights the importance of pH in optimizing the coagulation process and underscores the effectiveness of ferric chloride in treating greywater. Friedler and Alfiya (2010), who found that a slightly alkaline environment was optimal for achieving a maximum BOD reduction of 51% in the treatment of greywater from an office. Iron-based coagulants may be highly beneficial in deltas, as they not only reduce turbidity but also help control nutrient loads that otherwise lead to algal blooms in deltaic waters.

Research on the nutrient removal efficiency of ferric chloride remains quite limited, and there is a notable lack of data regarding its ability to eliminate microbial contaminants. The sole significant study conducted by Ghaitidak and Yadav (2015b) demonstrated that ferric chloride could reduce TC by 2.02 to 3.23 logs and FC by 2.03 to 2.44 logs. This points to a gap in knowledge, indicating the need for further research to fully explore the microbial treatment capabilities of ferric chloride at different dosages, both with and without pH adjustments.

Interestingly, while many studies report a decrease in pH during coagulation processes, Almeida et al. (2013) found that the pH of greywater actually increased when treated with ferrous sulfate. They attributed this phenomenon to the dissolution of electrolytes in the solution. In contrast, it is generally observed that higher doses of ferrous sulfate lead to a corresponding decrease in greywater pH (Al-Gheethi et al. 2017).

The mechanisms of pollutant removal also vary with pH levels. In acidic conditions, charge neutralization is crucial for the effectiveness of iron salts, while in alkaline environments, adsorption and sweep coagulation become the dominant processes (Dong et al. 2015). Specifically, under alkaline conditions, iron salts primarily remove pollutants through sweep coagulation (Noutsopoulos et al. 2018). For ferric chloride, the formation of Fe(OH)2+ and Fe(OH)3 occurs at pH levels of 4.6 and 8.3, respectively, indicating its dual effectiveness in both acidic and alkaline conditions (El-Gohary et al. 2010). This duality underscores the importance of pH in optimizing the use of ferric chloride for effective greywater treatment (Dong et al. 2015).

Ferrous sulfate has shown a moderate level of effectiveness in reducing organic content in greywater, achieving an average reduction of about 50% in COD. However, its performance can vary significantly depending on the source of the greywater. For instance, it tends to be more effective in removing BOD from MGW compared to that from bathrooms. Interestingly, while increasing the dosage of ferrous sulfate initially leads to a decrease in effluent BOD concentration, there comes a point where adding more coagulant does not yield significant improvements in removal efficiency, as the BOD levels tend to stabilize (El-Gohary et al. 2010). The findings of Pidou et al. (2008) reveal that ferrous sulphate can achieve a modest reduction of 1% in total nitrogen, while it excels in phosphate removal, achieving an impressive 96% reduction from greywater. Despite these promising results, there remains a significant gap in research regarding its effectiveness against microbial contaminants.

10.5.2.3 Natural Coagulants

Natural coagulants, as the name suggests, are primarily sourced from plants. Among them, *Moringa oleifera* stands out as the most commonly used, but there are several other noteworthy options as well. These include tamarind seeds, banana peels, *Citrus sinensis* (orange peels), *Mangifera indica* (mango peels) and *Strychnos potatorum* (Nirmali seeds). These plant-based coagulants are rich in soluble proteins that, when dissolved in water, acquire a positive charge. This allows them to effectively bind with negatively charged colloidal particles, facilitating their removal from water (Nautiyal et al. 2017).

While it is true that natural coagulants often require larger quantities compared to their chemical counterparts to achieve effective coagulation, they offer significant advantages. One of the most notable benefits is that they produce considerably less sludge. Furthermore, the sludge generated from natural coagulants is biodegradable, in stark contrast to the potentially harmful waste associated with chemical coagulants. This makes natural coagulants a more environmentally friendly option for water treatment, aligning well with sustainable practices. In deltas, natural coagulants like moringa oleifera or agricultural by-products can be promoted as community-level solutions, leveraging local biomaterials to treat greywater sustainably.

Natural coagulants have demonstrated a commendable turbidity removal efficiency of around 63%, with Total Suspended Solids (TSS) removal varying between 11 and 51%. Among these, *Moringa oleifera* has consistently proven to be effective, achieving an impressive average turbidity reduction of $82 \pm 11\%$ and a TSS removal rate of $66 \pm 30\%$.

In a similar vein, Chitra and Muruganandam (2019) while treating MGW from a hostel in India reported an 84% turbidity removal with a *Moringa oleifera* dosage of 600 mg/L. It is important to note that the efficiency of solid removal tends to improve with increasing doses of natural coagulants, but this improvement plateaus after reaching a certain threshold, beyond which additional coagulant does not yield further benefits (Nautiyal et al. 2017). However, among the various natural coagulants studied, *Strychnos potatorum* (Nirmali seeds) and orange peels exhibited the lowest efficiency in solid removal (Mohan 2014). This highlights the varying effectiveness of different natural coagulants in treating greywater.

Natural coagulants also contribute to a decrease in the pH of treated greywater, and this reduction tends to be proportional to the amount of coagulant used (Mohan 2014). While natural coagulants can moderately reduce BOD and COD, they generally do not match the efficiency of chemical coagulants. Interestingly, the effectiveness of organic content removal tends to improve with longer contact times between the coagulant and the greywater. For example, Rathore et al. (2019) found that when mango peels were allowed to interact with raw greywater for 60 min instead of just 20 min, COD removal increased significantly from 15 to 53% while treating greywater from an Indian residential society.

On the other hand, orange peels and nirmali seeds, whether used individually or in combination, showed only slight improvements in the removal of organic matter, even at higher dosages (Mohan 2014). Many studies focusing on natural coagulants have largely overlooked their effectiveness in removing nutrients and microorganisms, with the exception of the work by Kwabena Ntibrey et al. (2020). In their research in the treatment of bathroom greywater using *Moringa oleifera*, they achieved a phosphate removal rate of 45%, 1.06 log removal of TC and 3.11 log removal of *E. coli*.

This highlights a significant gap in the existing literature, indicating a pressing need for more comprehensive research to explore the potential of various natural coagulants. Future studies should aim to assess their effectiveness in removing not only nutrients but also heavy metals, microbiological contaminants, and organic micropollutants from greywater. Such investigations could provide valuable insights and pave the way for more sustainable and effective water treatment solutions.

10.5.3 Biological Treatment Systems

A range of biological treatment systems, including advanced technologies such as biological aerated filters (BAF), anaerobic fluidized bed reactors (AFBR), moving bed membrane bioreactors (MBBMR), membrane bioreactors (MBR), sequential batch reactors (SBR), upflow anaerobic sludge blankets (UASB), rotating biological contactors (RBC), integrated fixed-film activated sludge (IFAS), and moving bed biofilm reactors (MBBR).

The use of biological treatment systems in isolation is quite rare. More often, these systems are integrated with physical pre-treatment methods such as storage, equalization, sedimentation tanks, or screening, either individually or in combination. After the biological treatment process, additional post-treatment options are typically employed, which may include sedimentation (Abdel-Kader 2013) and disinfection methods like chlorine treatment or UV light (Couto et al. 2015; Fountoulakis et al. 2016).

Biological treatment systems have found applications in a variety of settings, effectively treating greywater from single households (Fountoulakis et al. 2016), apartment complexes (Abdel-Shafy et al. 2014a), student hostels (Shaikh and Ahammed 2024c), sports facilities, guest accommodations (Abdel-Kader 2013), and even airports (Couto et al. 2015). This versatility highlights the potential of biological systems in diverse environments, contributing to more sustainable water management practices. In delta regions, biological systems are especially important because untreated greywater often discharges directly into river channels and coastal zones, impacting agriculture, fisheries, and public health.

The HRT varied widely across different treatment systems, ranging from 1.5 to 24.0 h, while the OLR fluctuated between 0.09 and 1.93 kg COD/m^3/day. Notably, higher HRTs and OLRs were commonly used for treating high-strength greywater (Kraume et al. 2010; Hocaoglu and Orhon 2010).

However, it is worth mentioning that SBR systems often fell short of meeting the standards for unrestricted reuse concerning BOD levels (USEPA 2012). Among the various methods, anaerobic systems proved to be the least effective in removing BOD. Overall, it was observed that as the HRT increased, the effluent COD concentrations decreased across all biological systems, indicating a positive correlation between retention time and treatment efficiency.

The MBR system has proven to be quite effective in treating MGW, as highlighted by the work of Hocaoglu and Orhon (2010). Their study involved greywater sourced from various facilities, including kitchens in apartments, lodging establishments, and guest-houses. Coarse screens of 6 mm and fine screens of 3 mm were employed so as to eliminate larger and smaller particles before the greywater was stored in a holding tank for optimal treatment. It was then fed into a pilot-scale MBR system, which utilized a 0.4-μm membrane with a surface area of 5 m^2.

The results were impressive, with the MBR system significantly reducing TSS, BOD, and COD concentrations from initial levels of 51, 117, 310 mg/L to below 2, 5, 10 mg/L, respectively. The membrane's cut-off size effectively acted as a barrier against pathogens, resulting in TC and FC levels dropping to undetectable limits (Liberman et al. 2016). Furthermore, the MBR facilitated the complete breakdown of biodegradable COD, leaving only residual COD in a soluble form. Despite variations in influent concentration and the challenges posed by local conditions, the MBR system consistently produced effluent that met the standards for greywater reuse in outdoor applications (Atanasova et al. 2017). This underscores the reliability and efficiency of MBR technology in managing greywater effectively. Such high-quality effluent is particularly relevant for deltas, where even small improvements in water quality can reduce the cumulative pollution load entering sensitive wetlands and estuaries.

MBR technology was applied by Liberman et al. (2016) to treat shower greywater with a low organic load, maintaining an MLSS of 2000 mg/L. This comparatively lower MLSS concentration helps prolong membrane performance without the need for backwashing, chemical cleaning, or significant sludge disposal. However, membrane fouling remains a significant challenge for MBR systems, mainly caused by the formation of a thin biological layer, a gel layer on the outer membrane surface, and microbial buildup on the inner surface (Bani-Melhem et al. 2014). In space-limited settings like apartments, hotels, leisure clubs, and airports, the compact design of MBR systems effectively alleviates these challenges.

Kraume et al. (2010) studied the use of an SBR to treat shower greywater, comparing two HRTs of 13 and 16.5 h. Their results showed that the SBR system achieved COD removal efficiency similar to that of an MBR. However, the SBR effluent had significantly TSS levels than the MBR effluent, and the system also failed to produce fecal contamination-free effluent. Performance improved with an increase in HRT, but even with an extended HRT of 36 h, the SBR system was unable to meet Malaysian water quality standards for agricultural use when treating MGW (Krishnan et al. 2008). The lack of nutrients in MGW was identified as a key limiting factor, while nutrient-enriched greywater met the standards under the same HRT conditions.

Upflow anaerobic sludge blanket (UASB) reactors and anaerobic filters have also been explored for greywater treatment, particularly for MGW. UASB systems showed moderate organic removal efficiency, ranging from 40 to 64%, but were less effective in removing total nitrogen (22% to 30%) and total phosphorus (15% to 21%) (Elmitwalli and Otterpohl 2007). Additionally, UASB reactors demonstrated limited efficiency in TSS removal (Abdel-Shafy et al. 2014b). When compared to aerobic systems, anaerobic systems were less effective in removing organic matter and nutrients, with BOD and COD removal rates of 37 and 45% while TN and TP removal rates were 17%, and 9%, respectively, while similar-sized aerobic systems achieved 40%, 53%, 20%, and 30% removal (Abu Ghunmi et al. 2010). Abu Ghunmi et al. (2010) recommended anaerobic treatment as an initial step, followed by aerobic treatment and a disinfection stage to enhance overall performance.

The RBC has been recognized as an effective attached-growth biological treatment for greywater, with most studies focusing on their use for LGW. Friedler et al. (2005) assessed an RBC system with two basins in series, each having an HRT of 2 h, for LGW treatment. The system successfully reduced BOD and COD concentrations from 59 and 110 mg/L to 6.6 and 46 mg/L, respectively, while TSS was reduced from 43 to 16 mg/L. A combination of low BOD and nutrient content reduced the potential for fouling and regrowth in the reuse system. Additionally, FC levels were reduced by over five orders of magnitude (Friedler et al. 2005). Notably, the RBC system required less energy (1.2 kWh/m^3) for greywater treatment than both MBR (1.7 kWh/m^3) and SBR (3.6 kWh/m^3) systems (Friedler et al. 2005; Liberman et al. 2016).

10.5.4 Constructed Wetlands

Constructed wetlands (CW) are systems that combine physical, chemical, and biological processes to effectively remove a variety of pollutants. Several types of CW systems studied for greywater treatment, such as horizontal flow CW (HFCW), reedbeds, floating treatment wetlands (FTW), vertical flow CW (VFCW), recycling vertical flow CW (RVFCW) and green rooftop water recycling (GROW), and. Table 10.3 provides an overview of how these systems perform in removing chemical, physical, and microbial contaminants.

In natural systems, larger particles are removed using sedimentation (Patil and Munavalli 2016), followed by sand filtration as a post-treatment to capture any remaining media or particles carried by treated water (Patil and Munavalli 2016). Depending on the reuse purpose of the treated water, disinfection may be applied as an optional final step. Overall, wetlands are found to be effective in removing organic matter and solids, but their ability to remove microorganisms is more limited, even when HFCW and VFCW systems are used together (Muniz Sacco et al. 2024; Paulo et al. 2009). The HRT in these systems ranges from 0.5 day to 1 year, with 5.1 days as a median. On average, HLR was 28 cm/day and OLR was 39 kg $COD/m^2/day$, respectively.

Table 10.3 Summary of performance of selected natural treatment systems

Source of greywater	Treatment system*	HRT (h)	Removal (%)							Reference
			BOD	COD	TSS	TN	TP	TC	FC	
S	HSSFCW	8	97.49 (71.83)	92.82 (306)	97.91 (47.5)	64.04 (158)	68 (6.91)		99.99 (136,000.00)	Maiga et al. (2024)
SLGW	VFCW		10 (90)	96 (13)	89 (5.0)	81 (2)	24 (3.10)			Stefanatou et al. (2023)
MGW	HFWSFCW			71.52 (171)			34.26 (1.78)	48.54 (308.75)	61.06 (50.63)	Chikogu Ameso et al. (2023)
	VSSFCW			52.23 (287)			1.43 (2.69)	41.71 (349.75)	28.17 (93.375)	
	VFSFCW			70.23 (179)			0.23 (2.69)	47.21 (316.75)	55.96 (57.25)	
	HSSFCW			51.42 (292)			1.71 (2.65)	38.23 (370.63)	45.67 (70.63)	
SGW	VSSFCW	5		58 (84)		20 (15)				Alateeqi et al. (2023)
		4		92 (16)		94.44 (1)				
		3		93 (14)		68.75 (5)				

(continued)

Table 10.3 (continued)

Source of greywater	Treatment system*	HRT (h)	Removal (%)							Reference
			BOD	COD	TSS	TN	TP	TC	FC	
SGW	FTW		57.6 (15)	27 (94)						Abed et al. (2017)
B	HSSFCW	4–10	64 (59)	63 (71)			65 (0.47)			Patil and Munavalli (2016)

SGW: synthetic graywater; SLGW: synthetic light greywater; LGW: light graywater; MGW: mixed graywater; S: shower; B: bathroom; WB: wash basin; WM: washing machine; GROW: green roof top water recycling; HFWSFCW: horizontal flow free water surface constructed wetland; VSSF: vertical subsurface flow constructed wetland; VFSF: vertical free surface flow constructed wetland; HSSF: horizontal subsurface flow constructed wetland; HSSFCW: horizontal subsurface flow constructed wetland; PET: polyethylene terephthalate; FTW: floating treatment wetland

The values in parentheses represent effluent concentration, with all measurements in mg/L, except for TC and FC, which are expressed in CFU/100 mL or MPN/100 mL

*The removal efficiencies shown in the table apply solely to the constructed wetlands and do not represent the complete treatment system

Phragmites communis and *Phragmites australis* are frequently and commonly used plants in wetlands (Harikumar and Mol 2012). Arunbabu et al. (2015) used *Axonopus compressus* to enhace the visual aesthetics of treatment systems. Other frequently chosen plants include *Scirpus lacustris*, pennywort (*Hydrocotyle leucocephala*), dwarf papyrus (*Cyperus haspan*), *Lagenandra meeboldii*, and *Acorus calamus*. Many researchers opt for locally available plant species for wetland applications.

The choice of media in CW systems has a significant impact on their performance. Coarse gravel, crushed rock, limestone aggregates, charnockite rock chips, and coarse sand are also used as media. Several studies have tested various materials for the inlet, main basin, and outlet. For instance, Ramprasad and Philip (2016) utilized gravel for the inlet and outlet, while the main basin was filled with an equal mix of brick bats and sand. A few studies have explored multimedia combinations. Gross et al. (2007) designed a VFCW with three layers: planted organic soil on top, tuff or plastic media in the middle, and limestone pebbles at the bottom. Typically, extended retention times result in reduced effluent concentrations and enhanced removal efficiency. The enhanced pollutant removal at higher HRT is attributed to the increased contact time, which promotes better adsorption and biodegradation. RVFCW systems operate efficiently even at significantly lower HRT values.

Interestingly, BOD removal potential of wetlands was better than the COD removal potential. Most systems, except for FTW, demonstrated moderate COD removal performance. Planted CW achieved a 95% COD removal, while unplanted systems achieved 85% under comparable conditions (Arunbabu et al. 2015). In a study by Ramprasad and Philip (2016) on HFCW and VFCW with *Phragmites australis* for treating hotel wastewater, VFCW outperformed HFCW. Maiga et al. (2024) showed BOD removal of 97.49% and COD removal of 92.82%, respectively, in treating shower greywater using HSSFCW planted with *Chrysopogon zizanioides*. The removal of pollutants was more efficient in the summer than in the other seasons. Furthermore, an increase in the OLR initially improved pollutant removal efficiency; however, beyond a certain threshold, further increases in OLR caused a decline in efficiency.

Bidira et al. (2024) explored the treatment of wash basin greywater in Ethiopia using an HSSFCW system planted with Canna flowers. They achieved PO_4-P removal rates of 92.58% and 96.27% with brick and gravel rock media, respectively. Jokerst (2011) studied greywater treatment with an FWS CW containing *Typha latifolia*, followed by an SSF HFCW planted with *Scirpus acutus*, reporting removal rates of 92.8% for surfactants, 94.1% for TN, and 80% for TP. Muniz Sacco et al. (2024) evaluated a VFCW system with *Phragmites australis* in Luxembourg, achieving more than 90% micropollutant removal in light SGW at an HLR of 100 L/m^2/d. In a Yangtze Delta city in China, Sheng et al. (2020) achieved removal efficiencies of 80% for NH_4–N, 19% for total phosphorus, and 52% for COD using VFCW.

Constructed wetland systems are free from chemicals, economical, and energy-efficient; however, they demand a large surface area and are prone to clogging. Their performance can also be affected by climate, making them less suitable for hot regions where plant evapotranspiration may exceed the inflow (Arunbabu et al. 2015; Bidira et al. 2024). Arden and Ma (2018) conducted a thorough review and concluded that wetlands, by themselves, are not enough to meet all reuse standards. Generally, they achieve a 1–2 log reduction in bacterial indicators when the HRT is optimized. With an HRT of 3–5 days, they are capable of meeting TSS and BOD standards of 30 mg/L. In deltas, constructed wetlands can be integrated into community-level infrastructure to filter greywater before it reaches main river channels, thereby preserving deltaic biodiversity and reducing health risks.

10.5.5 Optimum Treatment System

Chemical and energy requirements, scale, the quality and quantity of greywater, the need for skilled supervision, intended reuse, land availability, and geographic location determine the appropriate greywater treatment system. Since no single system is universally applicable, each system has unique strengths and limitations. A detailed review of existing literature suggests that use of biological systems is more frequent than other treatment methods. The LGW and MGW contain different levels of contaminants and should be treated separately. This section explores treatment options based on the type of greywater source. For LGW, physicochemical treatment could result in effluent that meets unrestricted reuse standards (USEPA 2012). This method is particularly suited for LGW sources that lack the nutrients needed for biological treatment.

After sedimentation, biological systems such as UASB, SBR, or RBC can be used for the treatment of LGW, depending on greywater quality characteristics and cost considerations. RBC systems, while effective, have high initial costs, making them economically viable for apartments, societies or clusters of buildings. These systems are effective at removing organics and solids. To make biologically treated greywater suitable for restricted reuse in non-potable applications, it should undergo unsaturated continuous sand filtration and disinfection.

Wetland systems provide another option for treating LGW. These natural treatment methods are especially beneficial in developing countries due to their low construction, maintenance, and operational costs, although they require long HRT. Treated water should undergo polishing via continuous saturated sand or GAC filtration before being reused for restricted purposes. Additionally, since natural systems have limited pathogen removal capabilities, a disinfection stage is necessary for unrestricted reuse.

Membrane filtration, while more expensive, is another alternative. It produces high-quality effluent and serves as a barrier against viruses and bacteria, negating the need for disinfection to meet unrestricted reuse standards. However, membrane systems require large surface areas, significant energy, and skilled personnel for efficient operation of the system. Their main limitation is membrane fouling, which negatively impacts performance and increases operational costs. For MGW, biological treatment systems are effective after pre-treatment, as they provide shorter HRTs and efficiently remove solids and organic matter. Selecting the right treatment method should consider the energy needs, characteristics of the settled greywater, and total costs. In delta contexts, choosing the optimum system is crucial because high population densities, agricultural demands, and climate vulnerabilities create greater pressures on water quality. An optimal mix of low-cost natural systems and high-efficiency technologies can help preserve delta ecosystems while supporting local communities.

Wetland systems can also be used for MGW treatment, but their effluent must undergo further processing through GAC filtration to meet restricted reuse standards. Membrane bioreactors (MBRs) offer the most effective treatment for MGW, as the effluent from MBR systems can be reused directly without additional disinfection. However, MBR systems' high initial costs and the need for skilled personnel make them less viable for widespread use, particularly in developing countries.

10.6 Discussions

The analysis presented in this chapter demonstrates that greywater represents a valuable but underutilized resource in the context of growing global water stress. The technical review highlights that the characteristics of greywater vary significantly depending on its source, household practices, and socio-economic factors. Light greywater (from bathrooms and wash basins) is generally more suitable for low-cost and simple treatment systems, while dark greywater (from kitchens and laundry) contains higher loads of organic matter, detergents, oils, and microbes, making its management more complex. These variations emphasize the importance of source-separation strategies and tailored treatment approaches to ensure both technical and economic feasibility.

A wide spectrum of treatment technologies is available, ranging from simple natural systems (such as constructed wetlands, sand filtration, or natural coagulants) to advanced engineered solutions (including membrane bioreactors, sequencing batch reactors, and hybrid systems). Natural and low-cost systems offer affordability, low energy demand, and operational simplicity, making them attractive in developing regions where financial and technical capacity is limited. However, they often require large land areas and may show limited pathogen removal, necessitating supplementary disinfection. Conversely, advanced treatment systems produce effluent of consistently high quality suitable for unrestricted reuse, but their high capital, operation, and maintenance costs may restrict applicability in resource-constrained settings. Thus, a context-specific selection of treatment technologies based on water quality targets, intended reuse application, land availability, and cost considerations becomes essential.

An equally critical dimension is the role of public perception and acceptance. Evidence indicates that although there is increasing awareness of water scarcity, hesitation remains regarding the use of treated greywater, especially in applications that involve direct human contact such as food crop irrigation. Acceptance levels are significantly higher for uses such as toilet flushing, landscape irrigation, or industrial cooling. Therefore, education, awareness campaigns, and transparent communication about health and environmental safeguards are central to mainstreaming greywater reuse. Without public trust, even technically sound solutions may fail in real-world implementation.

From an environmental standpoint, greywater reuse reduces dependence on freshwater supplies, decreases the burden on centralized wastewater treatment plants, and mitigates the discharge of untreated wastewater into the environment. Additionally, reuse can lower energy consumption and greenhouse gas emissions associated with pumping, transporting, and treating freshwater. At the household and community levels, decentralized greywater treatment promotes resource efficiency and reduces utility costs. These benefits extend beyond the immediate technical and economic gains, supporting larger urban resilience strategies and sustainable development planning.

Beyond the technical considerations, the findings of this chapter have strong relevance to the United Nations Sustainable Development Goals (SDGs). Greywater treatment and reuse directly support SDG 6 (Clean Water and Sanitation) by providing an alternative water resource, thereby reducing stress on freshwater supplies. Indirectly, reuse contributes to SDG 11 (Sustainable Cities and Communities) by promoting resilient urban water management, and to SDG 12 (Responsible Consumption and Production) by encouraging resource recovery and efficient use of water. By integrating greywater reuse into urban planning and household practices, water scarcity can be alleviated while advancing sustainability at local, regional, and global scales. In delta regions, where water scarcity, pollution, and rapid urbanization converge, making greywater reuse a strategic tool for both environmental sustainability and community resilience.

10.7 Concluding Remarks

The literature on greywater's quantitative characteristics shows considerable variation in water consumption, greywater generation, and return factors. Among different sources, laundry greywater exhibited the highest quantity variation, while hand basin greywater had the lowest. Dark greywater sources, particularly kitchen and laundry, were the primary contributors to turbidity, organic content, and nutrients, whereas hand basin greywater had the highest bacteriological load.

Variability in greywater quantity and quality across different sources poses a key challenge in implementing on-site greywater treatment technologies. Another major hurdle is public perception and acceptance of treated greywater for non-potable uses, including toilet flushing, irrigation of food and non-food crops, car washing, groundwater recharge, and industrial applications. Public acceptance plays a crucial role in the success of water reuse strategies. The effective planning, construction, and operation of wastewater reuse facilities rely heavily on public attitude, support, and willingness to adopt such systems.

The quantity and quality of greywater in households, apartments, and communities can fluctuate significantly over time, necessitating treatment methods that can effectively accommodate these variations. The advantages of greywater reuse extend from direct benefits for users to broader positive impacts on entire cities or regions. Although advanced technologies are capable of treating greywater to high standards for safe reuse, selecting the most appropriate system dependent on various factors. A stand-alone treatment method is insufficient to meet reuse standards, necessitating a combination of approaches. Since most on-site greywater reuse systems are small-scale and operated by non-experts, they must be robust, sustainable, and capable of adapting to significant variations in water characteristics. In delta regions, where populations rely heavily on agriculture and fisheries, greywater reuse can help reduce pressure on limited freshwater resources while minimizing pollution to fragile deltaic ecosystems. The chapter findings contribute to SDG 6, SDG 11 and SDG 12 by promoting sustainable greywater management and reuse. This is particularly crucial for deltas, where water scarcity and pollution risks directly threaten livelihoods and biodiversity.

10.8 Recommendations

Readers are advised to recognize that greywater varies greatly in both quantity and quality depending on its source, with light greywater from bathrooms and sinks being easier to treat than dark greywater from kitchens and laundry. Source segregation is therefore encouraged, as it simplifies treatment and reduces costs. Since greywater quality deteriorates quickly during storage, holding times should be minimized, and systems should include equalization tanks and monitoring of key parameters such as BOD, COD, TSS, nutrients, and microbial indicators. The challenges of greywater reuse lie not only in its quality such as pathogens, surfactants, and chemical residues but also in user perception, as acceptance is often higher for outdoor uses like gardening than for indoor uses such as toilet flushing. Awareness and engagement programs can help overcome these barriers and enhance public confidence in reuse initiatives.

In terms of treatment, no single method is sufficient for all situations. Granular media filtration is effective for turbidity and solids removal but requires disinfection for safe reuse. Chemical treatments using alum, PAC, ferric salts, or natural coagulants like *Moringa oleifera* can be effective for solids and turbidity removal, though dosage optimization and sludge management are critical. Biological systems, including MBRs, SBRs, and RBCs, can provide high-quality effluent but may be limited by fouling, energy needs, or costs, while constructed wetlands offer low-cost, eco-friendly solutions but require larger land areas and seasonal management. Hybrid or integrated systems that combine physical, chemical, and biological methods, followed by disinfection, are generally the most reliable option. For delta regions, the adoption of hybrid systems and constructed wetlands can be especially advantageous, as they align with local hydrological conditions and help protect downstream water bodies.

For practical reuse, treated greywater is best applied in non-potable uses such as toilet flushing, irrigation of non-food crops, landscaping, and car washing, with careful soil monitoring where irrigation is involved. Decentralized systems in housing societies, institutions, and apartment complexes can help reduce the burden on municipal wastewater facilities and support water security. Looking forward, research and practice should focus on hybrid systems, natural coagulants, removal of emerging contaminants, and smart monitoring technologies. At the same time, user awareness and education must be strengthened to build trust and ensure widespread adoption of greywater reuse. In deltas, such measures can simultaneously address water shortages, safeguard agricultural productivity, and reduce contamination of sensitive aquatic habitats. Such efforts not only address water scarcity but also directly support the achievement of Sustainable Development Goals, particularly SDG 6 (Clean Water and Sanitation), SDG 11 (Sustainable Cities and Communities), and SDG 12 (Responsible Consumption and Production), by promoting resource efficiency, resilience, and sustainable urban living. This recommendation is highly relevant to deltas, where the nexus of water, food, and livelihoods makes greywater reuse a strategic priority.

Statement
The authors acknowledge that Artificial Intelligence (AI) tools were employed exclusively for language rephrasing and refinement of the manuscript text. The scientific content, interpretations, and conclusions remain entirely the responsibility of the authors.

References

Abdelrahman RM, Khamis SE, Rizk ZE (2020) Public attitude towards expanding the reuse of treated wastewater in the United Arab Emirates. Environ Dev Sustain 22:7887–7908. https://doi.org/10.1007/s10668-019-00551-w

Abdel-Kader AM (2013) Studying the efficiency of grey water treatment by using rotating biological contactors system. J King Saud Univ Eng Sci 25:89–95. https://doi.org/10.1016/j.jksues.2012.05.003

Abdel-Shafy H, Al-Sulaiman AM (2014) Assessment of physico-chemical processes for treatment and reuse of greywater. Egypt J Chem 57:215–231. https://doi.org/10.21608/EJCHEM.2014.1042

Abdel-Shafy HI, Al-Sulaiman AM, Mansour MSM (2014a) Greywater treatment via hybrid integrated systems for unrestricted reuse in Egypt. J Water Process Eng 1:101–107. https://doi.org/10.1016/j.jwpe.2014.04.001

Abdel-Shafy HI, El-Khateeb MA, Shehata M (2014b) Greywater treatment using different designs of sand filters. Desalination Water Treat 52:5237–5242. https://doi.org/10.1080/19443994.2013.813007

Abed SN, Almuktar S, Scholz M (2017) Remediation of synthetic greywater in mesocosm—scale floating treatment wetlands. Ecol Eng 102:303–319. https://doi.org/10.1016/j.ecoleng.2017.01.043

Abu Ghunmi L, Zeeman G, Fayyad M, van Lier JB (2010) Grey water treatment in a series anaerobic-Aerobic system for irrigation. Bioresour Technol 101:41–50. https://doi.org/10.1016/j.biortech.2009.07.056

Achak M, Mandi L, Ouazzani N (2009) Removal of organic pollutants and nutrients from olive mill wastewater by a sand filter. J Environ Manage 90:2771–2779. https://doi.org/10.1016/j.jenvman.2009.03.012

Ahmad A, Al Ghadeer S, Al-Hosni TK (2020) Study on gray water treatment using cement kiln dust and chemical coagulants. J Appl Water Eng Res 8:161–170. https://doi.org/10.1080/23249676.2020.1761900

Al-Khatib IA, Al Shami AH, Garcia GR, Celik I (2022) Social acceptance of greywater reuse in rural areas. J Environ Public Health 2022:1–11. https://doi.org/10.1155/2022/6603348

Alateeqi M, Khajah M, Abu-Arabi M, AlMansour H (2023) Greywater treatment using vertical subsurface flow constructed wetland system. J Eng Res 11:7–12. https://doi.org/10.1016/j.jer.2023.100052

Al-Gheethi AA, Mohamed RMSR, Wurochekke AA, Nurulainee NR, Mas Rahayu J, Amir Hashim MK (2017) Efficiency of moringa oleifera seeds for treatment of laundry wastewater. In MATEC web of conferences. EDP Sciences. https://doi.org/10.1051/matecconf/201710306001

Almeida JS, Rocha NRAF, Franco Junior MR (2013) Treating domestic greywater and expectations to be reused. Am J Environ Eng 2013:195–198. https://doi.org/10.5923/j.ajee.20130304.05

Alsulaili AD, Hamoda MF (2015) Quantification and characterization of greywater from schools. Water Sci Technol 72:1973–1980. https://doi.org/10.2166/wst.2015.408

Akpan VE, Omole DO, Bassey DE (2020) Assessing the public perceptions of treated wastewater reuse: opportunities and implications for urban communities in developing countries. Heliyon 6:e05246. https://doi.org/10.1016/j.heliyon.2020.e05246

Ansari K, Shrikhande A, Malik MA, Alahmadi AA, Alwetaishi M, Alzaed AN, Elbeltagi A (2022) Optimization and operational analysis of domestic greywater treatment by electrocoagulation filtration using response surface methodology. Sustainability 14:1–18. https://doi.org/10.3390/su142215230

Antonopoulou G, Kirkou A, Stasinakis AS (2013) Quantitative and qualitative greywater characterization in Greek households and investigation of their treatment using physicochemical methods. Sci Total Environ 454–455C:426–432. https://doi.org/10.1016/j.scitotenv.2013.03.045

Arden S, Ma X (2018) Constructed wetlands for greywater recycle and reuse: a review. Sci Total Environ 630:587–599. https://doi.org/10.1016/j.scitotenv.2018.02.218

Arunbabu V, Sruthy S, Antony I, Ramasamy EV (2015) Sustainable greywater management with Axonopus compressus (broadleaf carpet grass) planted in sub surface flow constructed wetlands. J Water Process Eng 7:153–160. https://doi.org/10.1016/j.jwpe.2015.06.004

Assayed A, Chenoweth J, Pedley S (2015) Assessing the efficiency of an innovative method for onsite greywater treatment: drawer compacted sand filter-a case study in Jordan. Ecol Eng 81:525–533. https://doi.org/10.1016/j.ecoleng.2015.04.042

Assayed A, Chenoweth J, Pedley S (2014) Drawer compacted sand filter: a new and innovative method for on-site grey water treatment. Environ Technol (United Kingdom) 35:2435–2446. https://doi.org/10.1080/09593330.2014.909886

Atanasova N, Dalmau M, Comas J, Poch M, Rodriguez-Roda I, Buttiglieri G (2017) Optimized MBR for greywater reuse systems in hotel facilities. J Environ Manage 193:503–511. https://doi.org/10.1016/j.jenvman.2017.02.041

Badr KR, Elmahdy EM, Shaheen MNF (2024) Molecular detection of Human Adenovirus and Human Bocavirus in wastewater and drain water in the western Nile Delta of Egypt. Water Supply 24:453–464. https://doi.org/10.2166/ws.2024.001

Baghapour MA, Shooshtarian MR, Djahed B (2017) A survey of attitudes and acceptance of wastewater reuse in Iran: Shiraz city as a case study. J Water Reuse Desalination 7:511–519. https://doi.org/10.2166/wrd.2016.117

Bahrami M, Amiri MJ, Badkubi M (2020) Application of horizontal series filtration in greywater treatment: a semi-industrial study. Aust J Water Resour 24:236–247. https://doi.org/10.1080/13241583.2020.1824610

Bakare BF, Mtsweni S, Rathilal S (2017) Characteristics of greywater from different sources within households in a community in Durban, South Africa. J Water Reuse Desalination 7:520–528. https://doi.org/10.2166/wrd.2016.092

Bani-Melhem K, Al-Qodah Z, Al-Shannag M, Qasaimeh A, Qtaishat MR, Alkasrawi M (2014) On the performance of real grey water treatment using a submerged membrane bioreactor system. J Membr Sci 476:40–49. https://doi.org/10.1016/j.memsci.2014.11.010

Benami M, Gillor O, Gross A (2016) Potential microbial hazards from graywater reuse and associated matrices: a review. Water Res 106:183–195. https://doi.org/10.1016/j.watres.2016.09.058

Bidira F, Muleta WS, Hailu FG, Asaithambi P (2024) Nutrient removal from domestic greywater using constructed wetlands made of gravel, bricks, and the canna flowers plant. Total Environ Eng 1:100002. https://doi.org/10.1016/j.teengi.2024.100002

Blanky M, Rodríguez-Martínez S, Halpern M, Friedler E (2015) Legionella pneumophila: from potable water to treated greywater; quantification and removal during treatment. Sci Total Environ 533:557–565. https://doi.org/10.1016/j.scitotenv.2015.06.121

Busgang A, Friedler E, Ovadia O, Gross A (2015) Epidemiological study for the assessment of health risks associated with graywater reuse for irrigation in arid regions. Sci Total Environ 538:230–239. https://doi.org/10.1016/j.scitotenv.2015.08.009

Buyukkamaci N, Alkan HS (2013) Public acceptance potential for reuse applications in Turkey. Resour Conserv Recycl 80:32–35. https://doi.org/10.1016/j.resconrec.2013.08.001

Chen Z, Ngo HH, Guo W (2012) A critical review on sustainability assessment of recycled water schemes. Sci Total Environ 426:13–31. https://doi.org/10.1016/j.scitotenv.2012.03.055

Chikogu Ameso V, Essandoh HMK, Donkor EA, Nwude MO (2023) Comparative analysis of greywater pollutant removal efficiency with horizontal free water surface flow wetland with other wetland technologies. Heliyon 9:e17637. https://doi.org/10.1016/j.heliyon.2023.e17637

Chin WH, Roddick FA, Harris JL (2009) Greywater treatment by UVC/H2O2. Water Res 43:3940–3947. https://doi.org/10.1016/j.watres.2009.06.050

Chitra D, Muruganandam L (2019) Performance of natural coagulants on greywater treatment. Recent Innov Chem Eng (Former Recent Patents Chem Eng) 13:81–92. https://doi.org/10.2174/2405520412666190911142553

Couto EDAD, Calijuri ML, Assemany PP, Santiago ADF, Lopes LS (2015) Greywater treatment in airports using anaerobic filter followed by UV disinfection: an efficient and low cost alternative. J Clean Prod 106:372–379. https://doi.org/10.1016/j.jclepro.2014.07.065

Dalahmeh SS, Jönsson H, Hylander LD, Hui N, Yu D, Pell M (2014) Dynamics and functions of bacterial communities in bark, charcoal and sand filters treating greywater. Water Res 54:21–32. https://doi.org/10.1016/j.watres.2014.01.019

Dong H, Gao B, Yue Q, Wang Y, Li Q (2015) Effect of pH on floc properties and membrane fouling in coagulation–ultrafiltration process with ferric chloride and polyferric chloride. Chemosphere 130:90–97. https://doi.org/10.1016/j.chemosphere.2015.03.049

Dougherty E, Murphy M (2012) Graywater Use in California Single and Multi-Residential Units: potential Best Management Practices. Carter H, Oropeza M, Pimental L, Ponce Otoya R (Eds), California Urban Water Conservation Council, pp 1–77

El-Gohary F, Tawfik A, Mahmoud U (2010) Comparative study between chemical coagulation/precipitation (C/P) versus coagulation/dissolved air flotation (C/DAF) for pre-treatment of personal care products (PCPs) wastewater. Desalination 252:106–112. https://doi.org/10.1016/j.desal.2009.10.016

Elmitwalli TA, Otterpohl R (2007) Anaerobic biodegradability and treatment of grey water in upflow anaerobic sludge blanket (UASB) reactor. Water Res 41:1379–1387. https://doi.org/10.1016/j.watres.2006.12.016

Emslie D, Siddiqua S, Crawford B, Teece W (2021) Biofilm formation and effectiveness of biosand filtration systems with typical and innovative filter media. Geotech Geol Eng 40:1251–1266. https://doi.org/10.1007/s10706-021-01960-3

Eriksson E, Auffarth K, Henze M, Ledin A (2002) Characteristics of grey wastewater. Urban Water 4:85–104. https://doi.org/10.1016/S1462-0758(01)00064-4

Etchepare R, van der Hoek JP (2015) Health risk assessment of organic micropollutants in greywater for potable reuse. Water Res 72:186–198. https://doi.org/10.1016/j.watres.2014.10.048

Fountoulakis MS, Markakis N, Petousi I, Manios T (2016) Single house on-site grey water treatment using a submerged membrane bioreactor for toilet flushing. Sci Total Environ 551–552:706–711. https://doi.org/10.1016/j.scitotenv.2016.02.057

Friedler E (2008) The water saving potential and the socio-economic feasibility of greywater reuse within the urban sector-Israel as a case study. Int J Environ Stud 65:57–69. https://doi.org/10.1080/00207230701846697

Friedler E, Alfiya Y (2010) Physicochemical treatment of office and public buildings greywater. Water Sci Technol 62:2357–2363. https://doi.org/10.2166/wst.2010.499

Friedler E, Katz I, Dosoretz CG (2008) Chlorination and coagulation as pretreatments for greywater desalination. Desalination 222:38–49. https://doi.org/10.1016/j.desal.2007.01.130

Friedler E, Kovalio R, Galil NI (2005) On-site greywater treatment and reuse in multi-storey buildings. Water Sci Technol 51:187–194. https://doi.org/10.2166/wst.2005.0366

Gani P, Sunar NM, Matias-Peralta HM, Abdul Latiff A, Kamaludin NS, Parjo UK, Emparan Q, Er CM (2015) Experimental study for phycoremediation of <i>Botryococcus</i> Sp. on greywater. Appl Mech Mater 773–774:1312–1317. https://doi.org/10.4028/www.scientific.net/AMM.773-774.1312

Ghaitidak DM, Yadav KD (2016) Greywater treatment for reuse: comparison of reuse options using analytic hierarchy process. J Water Reuse Desalination 6:108–124. https://doi.org/10.2166/wrd.2015.177

Ghaitidak DM, Yadav KD (2015a) Reuse of greywater: effect of coagulant. Desalination Water Treat 54:2410–2421. https://doi.org/10.1080/19443994.2014.902335

Ghaitidak DM, Yadav KD (2015b) Effect of coagulant in greywater treatment for reuse: selection of optimal coagulation condition using analytic hierarchy process. Desalination Water Treat 55:913–925. https://doi.org/10.1080/19443994.2014.924036

Gross A, Kaplan D, Baker K (2007) Removal of chemical and microbiological contaminants from domestic greywater using a recycled vertical flow bioreactor (RVFB). Ecol Eng 31:107–114. https://doi.org/10.1016/j.ecoleng.2007.06.006

Gupta G, Ahammed MM, Shaikh IN (2023) Greywater treatment by zero-valent iron-modified sand filters: performance and modelling using artificial neural network. Mater Today Proceed 83:24–32. https://doi.org/10.1016/j.matpr.2023.01.021

Haldar K, Kujawa-Roeleveld K, Dey P, Bosu S, Datta DK, Rijnaarts HHM (2019) Spatio-temporal variations in chemical-physical water quality parameters influencing water reuse for irrigated agriculture in tropical urbanized deltas. Sci Total Environ 708:134559. https://doi.org/10.1016/j.scitotenv.2019.134559

Harikumar PS, Mol B (2012) A synoptic study on the preparation of a liquid waste management plan for Kerala State, India. Environ Nat Resour Res 2:74–83. https://doi.org/10.5539/enrr.v2n2p74

Hocaoglu SM, Orhon D (2010) Fate of soluble residual organics in membrane bioreactor. J Memb Sci. 364:65–74. https://doi.org/10.1016/j.memsci.2010.07.050

Jamrah A, Al-Omari A, Al-Qasem L, Ghani NA (2006) Assessment of availability and characteristics of greywater in Amman. Water Int 31:210–220. https://doi.org/10.1080/02508060.2006.9709671

Jokerst (2011) Seasonal performance of an outdoor constructed wetland for graywater treatment in a temperate climate. Water Environ Res 83:2187–2198. https://doi.org/10.2175/106143011X12989211841412

Kaewsarn P, Yu Q (2001) Cadmium(II) removal from aqueous solutions by pre-treated biomass of marine alga Padina sp. Environ Pollut 112:209–213. https://doi.org/10.1016/S0269-7491(00)00114-7

Karabelnik K, Kõiv M, Kasak K, Jenssen PD, Mander Ü (2012) High-strength greywater treatment in compact hybrid filter systems with alternative substrates. Ecol Eng 49:84–92. https://doi.org/10.1016/j.ecoleng.2012.08.035

Katukiza AY, Ronteltap M, Niwagaba CB, Kansiime F, Lens PNL (2015) Grey water characterisation and pollutant loads in an urban slum. Int J Environ Sci Technol 12:423–436. https://doi.org/10.1007/s13762-013-0451-5

Katukiza AY, Ronteltap M, Niwagaba CB, Kansiime F, Lens PNL (2014a) Grey water treatment in urban slums by a filtration system: optimisation of the filtration medium. J Environ Manage 146:131–141. https://doi.org/10.1016/j.jenvman.2014.07.033

Katukiza AY, Ronteltap M, Niwagaba CB, Kansiime F, Lens PNL (2014b) A two-step crushed lava rock filter unit for grey water treatment at household level in an urban slum. J Environ Manage 133:258–267. https://doi.org/10.1016/j.jenvman.2013.12.003

Khajvand M, Mostafazadeh AK, Drogui P, Tyagi RD (2022) Management of greywater: environmental impact, treatment, resource recovery, water recycling, and decentralization. Water Sci Technol 86:909–937. https://doi.org/10.2166/wst.2022.226

Khalfan Al Jahmani A, Nageswara Rao L, M.J V, Lakavat M, Shaik F (2021) Treatment of kitchen grey water by ceramic filter and optimization using response surface methodology. Mater Today Proc 80:2366–2374. https://doi.org/10.1016/j.matpr.2021.06.350

Kraume M, Scheumann R, Baban A, El Hamouri B (2010) Performance of a compact submerged membrane sequencing batch reactor (SM-SBR) for greywater treatment. Desalination 250:1011–1013. https://doi.org/10.1016/j.desal.2009.09.093

Krishnan V, Ahmad D, Jeru JB (2008) Technical Note Influence of COD:N:P ratio on dark greywater treatment using a sequencing batch reactor. J Chem Technol Biotechnol 83:756–762. https://doi.org/10.1002/jctb.1842

Kwabena Ntibrey RA, Kuranchie FA, Gyasi SF (2020) Antimicrobial and coagulation potential of Moringa oleifera seed powder coupled with sand filtration for treatment of bath wastewater from public senior high schools in Ghana. Heliyon 6:e04627. https://doi.org/10.1016/j.heliyon.2020.e04627

Lanchipa-Ale T, Cruz-Baltuano A, Molero-Yañez N, Chucuya S, Vera-Barrios B, Pino-Vargas E (2024) Assessment of greywater reuse in a university building in a hyper-arid region: quantity, quality, and social acceptance. Sustainability 16:1–25. https://doi.org/10.3390/su16073088

Leong JYC, Chong MN, Poh PE (2018) Assessment of greywater quality and performance of a pilot-scale decentralised hybrid rainwater-greywater system. J Clean Prod 172:81–91. https://doi.org/10.1016/j.jclepro.2017.10.172

Liberman N, Shandalov S, Forgacs C, Oron G, Brenner A (2016) Use of MBR to sustain active biomass for treatment of low organic load grey water. Clean Techn Environ Policy 18:1219–1224. https://doi.org/10.1007/s10098-016-1112-4

Li Y, Huang G (2009) Dynamic analysis for solid waste management systems: an inexact multistage integer programming approach. J Air Waste Manage Assoc 59:279–292. https://doi.org/10.3155/1047-3289.59.3.279

Madzaramba TH, Zanamwe P (2023) User perceptions and acceptance of treated greywater reuse in low-income communities: a narrative review. J Water Clim Change 14:4236–4244. https://doi.org/10.2166/wcc.2023.414

Maiga Y, Compaoré COT, Diallo/Koné M, Sossou SK, YempalaSomé H, Sawadogo M, Nagalo I, Mihelcic JR, Ouattara AS (2024) Development of a constructed wetland for greywater treatment for reuse in arid regions: case study in rural Burkina Faso. Water 16:1–16. https://doi.org/10.3390/w16131927

Maimon A, Friedler E, Gross A (2014) Parameters affecting greywater quality and its safety for reuse. Sci Total Environ 487:20–25. https://doi.org/10.1016/j.scitotenv.2014.03.133

Mohan SM (2014) Use of naturalized coagulants in removing laundry waste surfactant using various unit processes in lab-scale. J Environ Manage 136:103–111. https://doi.org/10.1016/j.jenvman.2014.02.004

Muniz Sacco FC, Venditti S, Wilmes P, Steinmetz H, Hansen J (2024) Vertical-flow constructed wetlands as a sustainable on-site greywater treatment process for the decrease of micropollutant concentration in urban wastewater and integration to households' water services. Sci Total Environ 946:174310. https://doi.org/10.1016/j.scitotenv.2024.174310

Nautiyal R, Uliana S, Raj I, Shah B, Rathore K, Singh A (2017) Decentralized treatment of grey water by natural coagulants in the presence of coagulation aid. In: World congress on civil, structural, and environmental engineering. Avestia Publishing 158, pp 1–5. https://doi.org/10.11159/awspt17.158

Newcomer E, Boyd C, Nyirenda L, Opong E, Marquez S, Holm R (2017) Reducing the burden of rural water supply through greywater reuse: a case study from northern Malawi. Water Sci Technol Water Supply 17:1088–1096. https://doi.org/10.2166/ws.2017.004

Noah M (2002) Graywater use still a gray area. J Environ Health 64:22–25

Nolde E (2000) Erratum to "greywater reuse systems for toilet flushing in multi-storey buildings–over ten years experience in Berlin" [Urban Water 1 (1999) 275–284]. Urban Water 2:260. https://doi.org/10.1016/s1462-0758(01)00002-4

Noutsopoulos C, Andreadakis A, Kouris N, Charchousi D, Mendrinou P, Galani A, Mantziaras I, Koumaki E (2018) Greywater characterization and loadings–physicochemical treatment to promote onsite reuse. J Environ Manage 216:337–346. https://doi.org/10.1016/j.jenvman.2017.05.094

Ochoa SIC, Ushijima K, Hijikata N, Funamizu N (2015) Treatment of domestic greywater by geotextile filter and intermittent sand filtration bioreactor. J Water Reuse Desalination 5:39–49. https://doi.org/10.2166/wrd.2014.042

Oteng-Peprah M, Acheampong MA, deVries NK (2018a) Greywater characteristics, treatment systems, reuse strategies and user perception—a review. Water Air Soil Pollut 229.https://doi.org/10.1007/s11270-018-3909-8

Oteng-Peprah M, de Vries NK, Acheampong MA (2018b) Greywater characterization and generation rates in a peri urban municipality of a developing country. J Environ Manage 206:498–506. https://doi.org/10.1016/j.jenvman.2017.10.068

Oteng-Peprah M, Vries ND, Acheampong MA (2020) Households ' willingness to adopt greywater treatment technologies in a developing country–exploring a modified theory of planned behaviour (TPB) model including personal norm. J Environ Manage 254:109807. https://doi.org/10.1016/j.jenvman.2019.109807

O'Toole J, Sinclair M, Malawaraarachchi M, Hamilton A, Barker SF, Leder K (2012) Microbial quality assessment of household greywater. Water Res 46:4301–4313. https://doi.org/10.1016/j.watres.2012.05.001

Patil YM, Munavalli GR (2016) Performance evaluation of an integrated on-site greywater treatment system in a tropical region. Ecol Eng 95:492–500. https://doi.org/10.1016/j.ecoleng.2016.06.078

Paulo PL, Begosso L, Pansonato N, Shrestha RR, Boncz MA (2009) Design and configuration criteria for wetland systems treating greywater. Water Sci Technol 60:2001–2007. https://doi.org/10.2166/wst.2009.542

Pidou M, Avery L, Stephenson T, Jeffrey P, Parsons SA, Liu S, Memon FA, Jefferson B (2008) Chemical solutions for greywater recycling. Chemosphere 71:147–155. https://doi.org/10.1016/j.chemosphere.2007.10.046

Portman ME, Vdov O, Schuetze M, Gilboa Y, Friedler E (2022) Public perception and perspectives on alternative sources of water for reuse generated at the household level. Water Reuse 12:157–174. https://doi.org/10.2166/wrd.2022.002

Prabha D, Laxmipriya G, Sudha S, Sujatha K, Sivakumar S (2021) Evaluation of the efficiency of native plant species integrated with sanfilters for greywater reclamation: a pilot study. Desalination Water Treat 209:182–196. https://doi.org/10.5004/dwt.2021.26519

Prathapar SA, Jamrah A, Ahmed M, Al Adawi S, Al Sidairi S, Al Harassi A (2005) Overcoming constraints in treated greywater reuse in Oman. Desalination 186:177–186. https://doi.org/10.1016/j.desal.2005.01.018

Raj DA, Ahammed MM, Shaikh IN (2023) Use of zero-valent iron-modified sand filters for greywater treatment: performance evaluation and modelling using response surface methodology. Environ Sci Pollut Res 31:1–16. https://doi.org/10.1007/s11356-023-31182-4

Ramprasad C, Philip L (2016) Surfactants and personal care products removal in pilot scale horizontal and vertical flow constructed wetlands while treating greywater. Chem Eng J 284:458–468. https://doi.org/10.1016/j.cej.2015.08.092

Rathore K, Raj I, Nautiyal R, Uliana S, Shah B, Singh A (2019) Effect of settling time on the treatment of domestic grey water using mango seeds as coagulant. In: Urbanization challenges in emerging economies, pp 112–121. https://doi.org/10.1061/9780784482025.012

Sabogal-Paz LP, Campos LC, Bogush A, Canales M (2020) Household slow sand filters in intermittent and continuous flows to treat water containing low mineral ion concentrations and Bisphenol A. Sci Total Environ 702:135078. https://doi.org/10.1016/j.scitotenv.2019.135078

Sall O, Takahashi Y (2006) Physical, chemical and biological characteristics of stored greywater from unsewered suburban Dakar in Senegal. Urban Water J 3:153–164. https://doi.org/10.1080/15730620600961387

Senthilkumar (2021) Effect of treated wastewater irrigation combined with manure and inorganic nutrients on fodder yield and proximate parameters of Bajra Napier hybrid grass in Cauvery Delta Region of Tamil Nadu. Intern J Environ Climate Change 11: 595–604. https://doi.org/10.9734/ijecc/2021/v11i1230664

Shaikh IN, Ahammed MM (2024a) Effect of washing method and detergent type on laundry greywater characteristics. J Water Process Eng 66:106103. https://doi.org/10.1016/j.jwpe.2024.106103

Shaikh IN, Ahammed MM (2024b) Comparative evaluation of different pre-treatment alternatives for granular media filters treating greywater and their ranking using analytical hierarchy process. Water Sci Technol 89:2625–2645. https://doi.org/10.2166/wst.2024.155

Shaikh IN, Ahammed MM (2024c) Comparative analysis of greywater quantity and quality characteristics in household and hostels. Technologies for sustainable transportation infrastructures.

Lecture notes in civil engineering 529. Springer, Singapore, pp 751–769. https://doi.org/10.1007/978-981-97-4852-5_61

Shaikh IN, Ahammed MM (2023a) Sand filtration for greywater treatment: long-term performance evaluation and optimization by response surface methodology. Urban Water J 20:450–464. https://doi.org/10.1080/1573062X.2023.2179928

Shaikh IN, Ahammed MM (2023b) Use of coagulation for on-site greywater reuse. In: Resource recovery in municipal waste waters. Elsevier, pp 291–317. https://doi.org/10.1016/B978-0-323-99348-7.00013-8

Sheng X, Qiu S, Xu F, Shi J, Song X, Yu Q, Liu R, Chen L (2020) Management of rural domestic wastewater in a city of Yangtze delta region: performance and remaining challenges. Bioresour Technol Rep 11:100507. https://doi.org/10.1016/j.biteb.2020.100507

Shiam Babu R, Prasanna K, Senthil Kumar P (2025) Assessing the suitability of agro-waste consortium in greywater treatment: a sustainable strategic exploration. Desalination Water Treat 321:100896. https://doi.org/10.1016/j.dwt.2024.100896

Singh S, Ahammed MM, Shaikh IN (2021) Combined coagulation and intermittent sand filtration for on-site treatment of greywater. IOP Conf Ser Mater Sci Eng 1114:012031. https://doi.org/10.1088/1757-899x/1114/1/012031

Šostar-Turk S, Petrinić I, Simonič M (2005) Laundry wastewater treatment using coagulation and membrane filtration. Resour Conserv Recycl 44:185–196. https://doi.org/10.1016/j.resconrec.2004.11.002

Spychała M, Nawrot T, Matz R (2021) A preliminary study on the use of xylit as filter material for domestic wastewater treatment. Appl Sci 11:1–13. https://doi.org/10.3390/app11115281

Spychala M, Nieć J, Zawadzki P, Matz R, Nguyen TH (2019) Removal of volatile solids from greywater using sand filters. Appl Sci 9:1–15. https://doi.org/10.3390/app9040770

Stefanatou A, Schiza S, Petousi I, Rizzo A, Masi F, Stasinakis AS, Fyllas N, Fountoulakis MS (2023) Use of climbing and ornamental plants in vertical flow constructed wetlands treating greywater. J Water Process Eng 53:103832. https://doi.org/10.1016/j.jwpe.2023.103832

Tchobanoglous G, Franklin L, Burton EC, Stensel HD (2011) Wastewater engineering treatment and reuse, Fourth Edi. McGraw Hill, New York, USA

Thompson KA, Summers RS, Cook SM (2017) Development and experimental validation of the composition and treatability of a new synthetic bathroom greywater (SynGrey). Environ Sci 3:1120–1131. https://doi.org/10.1039/c7ew00304h

Thompson KA, Valencia EW, Scott Summers R, Cook SM (2020) Sorption, coagulation, and biodegradation for graywater treatment. Water Sci Technol 81:2152–2162. https://doi.org/10.2166/wst.2020.273

Turner RDR, Will GD, Dawes LA, Gardner EA, Lyons DJ (2013) Phosphorus as a limiting factor on sustainable greywater irrigation. Sci Total Environ 456–457:287–298. https://doi.org/10.1016/j.scitotenv.2013.02.061

Tusiime A, Solihu H, Sekasi J, Mutanda HE (2022) Performance of lab-scale filtration system for grey water treatment and reuse. Environ Challenges 9:100641. https://doi.org/10.1016/j.envc.2022.100641

USEPA (United States Environmental Protection Agency) (2012) Guidelines for water reuse. Environmental Protection Agency, Municipal Support Division Office of Wastewater Management Office of Water Washington, DC. EPA/625/R- 92/004.

Verma S, Daverey A, Sharma A (2017) Slow sand filtration for water and wastewater treatment–a review. Environ Technol Rev 6:47–58. https://doi.org/10.1080/21622515.2016.1278278

Vinitha EV, Mansoor Ahammed M, Gadekar MR (2018) Chemical coagulation of greywater: modelling using artificial neural networks. Water Sci Technol 2017:869–877. https://doi.org/10.2166/WST.2018.263

Winward GP, Avery LM, Frazer-Williams R, Pidou M, Jeffrey P, Stephenson T, Jefferson B (2008a) A study of the microbial quality of grey water and an evaluation of treatment technologies for reuse. Ecol Eng 32:187–197. https://doi.org/10.1016/j.ecoleng.2007.11.001

Winward GP, Avery LM, Stephenson T, Jefferson B (2008b) Ultraviolet (UV) disinfection of grey water: particle size effects. Environ Technol 29:235–244. https://doi.org/10.1080/09593330802030069

Zhu J, Wagner M, Cornel P, Chen H, Dai X (2018) Feasibility of on-site grey-water reuse for toilet flushing in China. J Water Reuse Desalination 8:1–13. https://doi.org/10.2166/wrd.2016.086

Zhu Z, Li A, Wang H (2017) Public perception and acceptability of reclaimed water: the case of Shandong province, China. J Water Reuse Desalination 8:308–330. https://doi.org/10.2166/wrd.2017.022

Zipf MS, Pinheiro IG, Conegero MG (2016) Simplified greywater treatment systems: slow filters of sand and slate waste followed by granular activated carbon. J Environ Manage 176:119–127. https://doi.org/10.1016/j.jenvman.2016.03.035

11 Recent Development in Wastewater Recycling and Future Perspective

Pooja Maibam and Subhadeep Biswas

Abstract

Water and wastewater recycling is one of the pressing topics of contemporary times. It is one of the sustainable water management options for providing an adequate water supply to water-scarce regions. The countries in the Middle East and various African nations fall into the category of arid regions, where water recycling can be a promising technology. Apart from meeting water demand, recycling and reuse of wastewater also help in wastewater management and resource recovery, maintaining health of river and deltas and attaining several goals of UNSDGs. Various articles of the present time are available in the literature regarding the importance and utility of wastewater recycling. These prompted us to write the current chapter focusing on the recent developments that took place in this field globally. After a brief introduction, the next section highlights the importance of water and wastewater recycling in the present-day world by citing different articles from recent times. Some of the commonly adopted techniques often practiced are presented with their suitable applications. Different interesting and important mechanisms, as reported by researchers, are also covered under this section. The next section focuses on the global status of water recycling across different countries of the world. It also critically evaluates the feasibility of the recycling process for large-scale and in a real field scenario. Although recycling technologies are one of the main pillars of sustainability, several challenges and hurdles are often encountered. They are presented with a future perspective before concluding the chapter.

P. Maibam · S. Biswas (✉)
Civil Engineering Department, National Institute of Technology Silchar, Assam, India
e-mail: sdeep0819@gmail.com; subhadeep@civil.nits.ac.in

S. Yadav et al. (eds.), *Water Resources in Delta Region in India*, Deltas of the World,
https://doi.org/10.1007/978-3-032-06500-1_11

11.1 Introduction

Water is one of the indispensable elements required to support life on this earth. The contribution of water to the growth of human civilization is beyond doubt. The development of any industrial sector without the aid of water is unthinkable. On average, a healthy adult person consumes 2–3 L of water per day through drinking. However, with the rapid increase in population and the emergence of urbanization across various parts of the globe, a huge stress on the existing water resources is experienced in the present times. With the increase in the human population, water demand is also rising at a faster pace. Hence, the supply of adequate water and maintaining its quality for the specific task is a great challenge in the current world. Besides, industrial waste, agricultural runoff, domestic sewage, etc., have contributed to the contamination of various fresh water resources like lakes, rivers and consequently their deltas. Moreover, effluents from different categories of industries have made the situation worse by deteriorating the existing water quality. For example, effluent generated by the petrochemical industries is severely contaminated with heavy metals, organic compounds, surfactants, and different kinds of salt (Mansour et al. 2024). Wastewater generated from pharmaceutical units often contains high concentrations of recalcitrant organic compounds, inorganic phosphate, nitrate, etc. These components are often carried away by sewage to the aquatic ecosystem. A study showed that China's Yangtze River carries 172 tons of pharmaceutical products per year (Kayode-Afolayan et al. 2022). Consequently, rivers' delta are also badly affected. The Yangtze River Delta, which is a significant food grain production in China, has been facing increasing challenges from agricultural non-point source pollution (Liang et al. 2021). The sediments of the Nile River are moderately to highly polluted by Ni, Cd, Cr, Cu, Pb, and Zn, which pollute the Nile River Delta (Abotalib et al. 2023). Some industrial wastewater, e.g. tannery effluents cause adverse effects on the aquatic ecosystem if not treated properly. They can enhance the biochemical oxygen demand (BOD), chemical oxygen demand(COD), total suspended solids(TSS), Cr(VI), and sulphide levels of the receiving water bodies to a great extent (Bhardwaj et al. 2023). Hence, these high-strength industrial wastewaters are creating a tremendous impact by polluting the ecosystem. Industrial pollution often degrades the quality of natural water resources, thus producing a freshwater crisis globally.

Apart from industrial wastewater discharge, there are some climatic factors that also generate serious water stress. Frequent droughts and other climatic adversities have created water scarcity across different regions of the world. Several nations are already suffering a severe water crisis. Recent studies showed that climate change and abrupt changes in rainfall patterns are also some of the crucial factors behind water shortage in the African continent (Filho et al. 2022). Gulf countries like Yemen, Qatar, Kuwait, and Oman are some of the well-known water-stressed regions of the world. However, in the present times, climate change has made the situation more serious in these areas (Aljawzi et al. 2022). Hence, finding a sustainable solution to this critical problem is a major challenge for environmental scientists and researchers across the globe. Among various policies for sustainable development, it is important to mention that wastewater recycling and reuse are undoubtedly one of the promising approaches to mitigate the existing problem.

Water reuse and wastewater recycling are currently being practiced across various nations throughout the world. The Gulf Cooperation Countries (GCC) explore non-conventional water sources like desalinated seawater and treated wastewater for meeting various demands. As per the report of Qureshi, 4 billion m^3 of wastewater is collected yearly, and out of which approximately 73% is treated in the wastewater treatment plants. However, only 39% is reused for several activities (Qureshi 2020). Like the Gulf nations, Australia also receives low rainfall throughout the year. For years, Western Australia has been experiencing a declining rainfall pattern. Moreover, by 2019, the condition of eastern Australia had deteriorated due to the drought. Hence, wastewater recycling is reviewed as an alternate water source for the region (Radcliffe and Page 2020). According to the report of the Ministry of Environment (2012), annual discharge from wastewater treatment plants in South Korea was approximately 7.1 billion m^3. The government has initiated the reutilization of treated wastewater for agricultural purposes, especially for paddy, since 2005. Moreover, in 2011, the amount of recycled water used for paddy cultivation reached up to 45 million m^3 (Jeong et al. 2016). Egypt, which is witnessing one of the highest water budget deficits in Africa, is offset by intensive and vast-scale reuse of wastewater in the Nile Delta (Abotalib et al. 2023). Apart from these, water recycling and reuse prevail widely in the countries, including the US (Hopson and Fowler 2022), many European nations, Israel (Kfir et al. 2012), etc. The recycling process goes hand in hand with the implementation of the circular economy (CE) in the water and wastewater sector. Several reports of contemporary times describe various model frameworks in connection with wastewater recycling and CE as an integral part of sustainable development. Smol et al. mentioned the six actions, such as reduction, reclamation, reuse, recycling, recovery, and rethink, in connection with sustainable wastewater management (Smol et al. 2020). Various studies report the reclamation of useful materials from wastewater through proper treatment. Recently, we also reported a novel technique for the selective recovery of anionic dye from a binary dye mixture following a technique resembling solid phase extraction (SPE) (Biswas and Pal 2022). In the first stage, cationic surfactant-modified alginate-xanthan hydrogel beads have been explored for capturing both cationic and anionic dyes from a binary dye mixture. After successful adsorption, the dyes-loaded beads were immersed in a 1-butanol solution for the selective extraction of anionic dye. Robles et al., in one of their latest articles, describe the significance of nutrient recovery from wastewater as an integral part of the upliftment of CE (Robles et al. 2020). The authors described the pros and cons of different technologies for nutrient recovery, including membrane separation, photosynthesis, crystallization, and biological processes. Vinayagam et al., in their article, discussed several innovative approaches in the water and wastewater domain for achieving successful implementation of CE (Vinayagam et al.

2024). The authors critically assessed the principles of reclaim, restore, recycle, reduce, recover, and reuse. Mannina et al. argued justifiably towards reusing reclaimed water for irrigation purposes in order to strengthen the CE concept (Mannina et al. 2022). Various existing hurdles have been identified by the authors in relation to this practice. Further, several practical solutions have also been put forward with reference to some adopted roadmaps.

Different articles from recent times highlight several novel advancements in the wastewater recycling sector. Various innovative technologies, the significance of wastewater reuse, and different environmental-related policies are discussed in various articles. The current chapter focuses on providing insights regarding the latest developments that have been carried out in this domain. Firstly, after the introduction, the significance of water and wastewater recycling from a global perspective is highlighted. The importance of this concept in mitigating water shortage in water-stressed regions, reducing environmental impact, recovering useful materials, and generating electricity is discussed briefly. The next section deals with the description of various technologies commonly adopted for the purpose. Membrane-based processes, adsorption, photocatalytic conversion, etc., are highlighted. The status of wastewater recycling across different countries around the globe is presented. Finally, existing hindrances and future perspectives are discussed before concluding the chapter.

11.2 Importance of Water and Wastewater Recycling in the Present-Day World

Wastewater recycling is very significant in the present-day world. Consequently, it has become a part and parcel of the water demand management process, finds useful in conservation of fresh water as well as keeping the environment pollution-free (Jefferson et al. 1999). Surface and groundwater resources are continuously being polluted in many places across the globe. In this respect, recycling has become one of the most sustainable options to mitigate water scarcity (Gupta et al. 2012). Moreover, climate change has also become a serious issue in several places in contemporary times. For example, Australia received an annual rainfall of 277.6 mm in 2019, which is nearly 40% lower in comparison to the average rainfall received during the period 1961–1990 (465.2 mm) (Radcliffe and Page 2020). A study of 270 WWTPs across the Yangtze River Delta conducted by Huang et al. showed that although most plants were not operating at full energy efficiency, significant potential exists for improvement (Huang et al. 2021). In the Niger Delta, refinery wastewaters from oil have become a neglected yet significant environmental contaminant. Osin et al. discussed the potential upgrade of conventional treatment technologies including the use of nanotechnology (Osin et al. 2017). Melbourne, one of the most important cities of the Australian continent, practices wastewater reuse and recycling to a great extent for different purposes. Besides combating water shortage crises, recycling is also helpful in reducing environmental pollution to a great extent (Lou and Lin 2008). Nutrient recovery (Robles et al. 2020) as well as energy production (Nguyen et al. 2021) are the other important aspects of wastewater recycling. Various such significances are discussed as follows:

11.2.1 Mitigating Water Scarcity

Rapid urbanization and an increase in population have increased the water demand globally to a great extent. Further, in a small nation like the Maldives, there is a lack of fresh water resources. It belongs to small island developing nations (SIDS) and faces different economic and environmental challenges, including water shortage. The only dependable water source of the region is the rainwater and groundwater. Moreover, due to several existing problems including elevated sea-level, higher extraction and lower recharge rate has resulted in salinization of the available ground water (Jaleel et al. 2020). In addition to it, the tsunami of 2004 caused great damage to the natural freshwater reservoir of the Maldives (Acciarri et al. 2021). The nation is currently practicing different techniques, including rainwater harvesting, desalination, and groundwater recharge, to address the issue. Water scarcity is also one of the most common features in countries like Yemen, Qatar, and several areas in the Middle East. Many countries in the Arabian Peninsula are highly dependent on the desalination plants, which is undoubtedly a costly affair. On the other hand, greywater recycling may be a very prominent solution to this existing crisis. In fact, greywater recycling can be achieved at a cost that is 2.2 times less than that of the desalination plants (Roshan and Kumar 2020). In Egypt's Nile Delta, agricultural drainage water reuse partly mitigates water shortages but is considered a source of pollution (heavy metals) because it is practiced without treatment (Abotalib et al. 2023). In the Yangtze River Delta's decentralized systems, constructed wetlands showed strong ammonia and organic carbon removal, while hybrid systems like Anoxic–Oxic-Constructed wetlands (AO–CWs) and Anaerobic Anoxic–Oxic-Constructed wetlands (AAO–CWs) achieved higher total nitrogen and phosphorus reduction (Sheng et al. 2020). It also offers wastewater recycling option. In one of the recent studies, Madhuranthakam et al. investigated the economic feasibility of a decentralized wastewater treatment plant for providing freshwater from greywater located in Abu Dhabi University campus in the United Arab Emirates (Madhuranthakam et al. 2023). Facilities for filtration, chemical treatment, and disinfection were provided for the treatment purpose. The treated water thus produced was found suitable for horticultural purposes. Moreover, provision for handling 90,000 tons of water can be achieved by investing USD 8 M with 34% discounted rate of return. Thus, it proved that a greywater recycling facility is not only economical but also sustainable in nature.

Adewumi et al. reviewed the opportunities and potential of treated wastewater reuse in parts of South Africa (Adewumi et al. 2010). Some parts of South Africa are facing severe water shortages, especially the Western Cape. The authors cited suitable examples of household wastewater reuse in places like Carnarvon, the Northern Province, and Hull Street, Kimberly. Besides, in this article, the cases of district wastewater reuse in Lynedoch, Ecovillage, Western Cape, as well as industrial wastewater reuse in the mining areas of Driefontein have been highlighted. Among various recommendations, the authors mentioned there is an urgent necessity for DWAF to provide a proper guideline regarding wastewater reuse. Secondly, life cycle cost–benefit analysis is important to understand the economic aspects. Suitable tariffs may also be a proper option for implementing a recycling scheme. One of the notable dry regions of the world, the Kingdom of Saudi Arabia has a large area of 2.25 million km^2 with no perennial rivers or permanent natural water source. Dawoud et al., in their article, mentioned different prospects and key challenges for implementing a wastewater recycling project successfully in this arid region (Dawoud et al. 2022). Practice of wastewater reuse will offer several benefits, including a sustainable solution to the water scarcity problems, development of new infrastructures, reduction of net carbon footprint, etc.

11.2.2 Reduction in the Environmental Pollutant

Recycling of wastewater is obviously a sustainable treatment technology for reducing environmental impacts. A huge quantity of wastewater is generated from various industrial sectors, including tannery, electroplating, paper, winery, dairy, textile, etc. Reuse of the supernatant in the wastewater treatment plant is helpful for reducing the effluent volume and results in the ultimate diminishing of total wastewater. Many of the developing and developed countries explore this option. In February 2017, China introduced the 13th five-year plan for national urban sewage treatment and recycling facilities construction plan (Liu et al. 2021). On the other hand, data from another report showed that annual urban sewage discharge in China increased from 378.83×10^6 m^3 in 2010 to 492.895×10^6 m^3 (approximately 1.3 times) in 2017 (Li et al. 2021). Moreover, the recycling capacity of the country also increased to 713.21×10^6 m^3 as reported in the same year. According to the provision of the 13th five-year plan, it was aimed that by 2020, reuse of reclaimed wastewater would be increased throughout the country, and it would not be less than 20% in water shortage regions. Saravanan et al. in another study reported that Beijing has adopted the principle of recycling in wastewater treatment, which has enhanced the treatment efficiency up to 85% (Saravanan et al. 2021). Reuse of wastewater can be practiced either by means of direct reuse or process reuse. Lou and Lin discussed about the Cheng-Ching water treatment plant in South Taiwan (Lou and Lin 2008). It is one of the largest drinking water suppliers in Kaohsiung City. The wastewater generated is transferred to the raw water basin for recycling purposes. Wyczarska-Kokot and Dudziak described the possibility of discharging washing generated from backwashing of filter beds in swimming pools into the surface stream, as well as their reuse in the pool (Wyczarska-Kokot and Dudziak 2022). It was found that without further treatment, these residues cannot be discharged directly into the surface waters, due to the violation of the discharge limits. However, the authors explored remediation techniques in three stages (namely: sedimentation, coagulation, and ultrafiltration in stages 1, 2, and 3). It was revealed that after three stages of treatment, the permeate can be used for loss replenishment purposes.

A wastewater treatment facility requires a good amount of investment for infrastructure purposes. Hence, the installation of a recycling facility should be economically and environmentally viable. In one of the studies, Savun-Hekimoglu et al. reported that various economic and technical factors need to be analyzed for efficient wastewater treatment (Savun-Hekimoglu et al. 2023). The authors reported that a considerable amount of water reuse can reduce the cost of wastewater treatment significantly. Further, it was analyzed that 5% wastewater reuse can reduce the total cost by 1%, where 20% recycling can lead to a deduction of more than 5.74% cost. The results can be clearly visualized as shown in Fig. 11.1. In another study, Hamad et al. analyzed a fabric plant for wastewater treatment at Dubai, UAE (Hamad et al. 2003). The authors concluded that a combination of treatment technologies, including segregation, recycling, and mixing, would lead to a reduction in the overall cost of wastewater treatment.

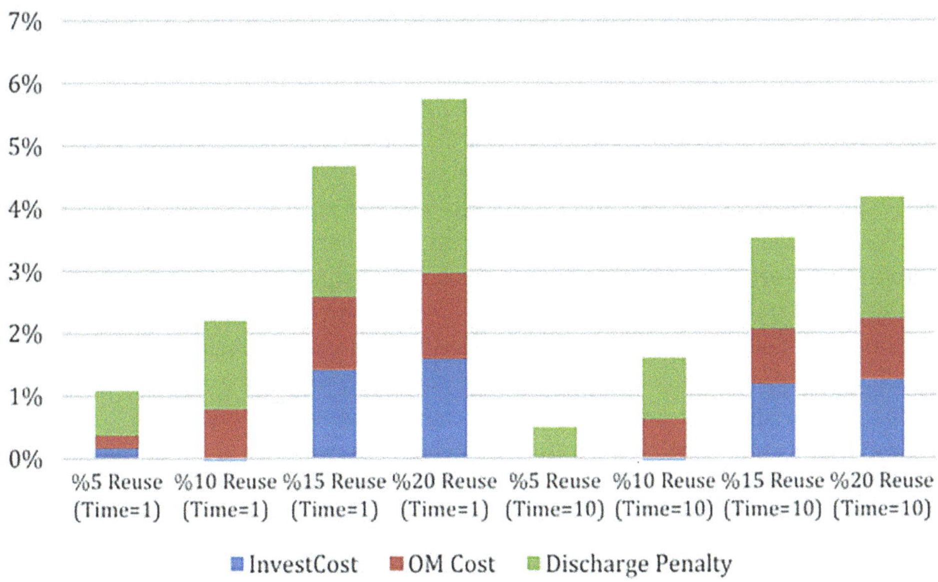

Fig. 11.1 Distribution of cost savings in wastewater system due to water recycling (Savun-Hekimoglu et al. 2023)

11.2.3 Recovery of Useful Materials

Several recently reported studies have been published in the literature describing the potential of wastewater treatment technologies in generating value-added products. Cao et al., in their recent review article, described the significance of algae-based methodologies regarding the recovery of rare earth elements (REE) from wastewater (Cao et al. 2021). REEs are the wonder materials that possess distinguished physical and chemical properties and are found useful for several purposes, including low power consumption, high performance, and thermal stability. One of the attractive features of the dried/living algae is that their cell walls are composed of cellulose, hemicellulose, starch, glycogen, etc. These components possess an abundant amount of hydroxyl, amino, and carboxyl functional groups, which facilitate REE trapping. Sanchez-Contreras et al. developed a photo-biological technique for simultaneous treatment of industrial wastewater along with production of carbohydrate-rich biomass (Sanchez-Contreras et al. 2021).

Ye et al. reviewed the latest developments that took place in the domain of various technologies, such as different biological and chemical methods, membrane techniques, etc., along with their mechanisms with respect to nutrient recovery from wastewater (Ye et al. 2020). Among various chemical methods, the authors mainly discussed the chemical precipitation and adsorption techniques. On the other hand, biological phosphate recovery is one of the famous techniques for the extraction of nutrients. Apart from these, other advanced methods such as membrane-based systems, osmotic membrane bioreactors, bioelectrochemical systems, and membrane photobioreactors are also briefly discussed.

Life cycle assessment (LCA) study constitutes one of the novel approaches for determining the environmental sustainability of any adapted process. Lam et al. reviewed 65 LCA studies related to nutrient recycling during wastewater treatment (Lam et al. 2020). Generally, in most cases, positive outcomes are obtained from an environmental point of view, especially when fewer chemicals are used and there is a separate source separation for human excreta. However, the authors suggested more improvement of the LCA study by including several multidisciplinary subjects such as agricultural science, environmental toxicology, and involving multifunctionality, contaminant accounting, etc.

Apart from nutrient recovery, obtaining precious metals like gold, silver, etc., also constitutes an important commercial benefit. In one of the studies, Li et al. explored nano zerovalent iron (ZVI) for the recovery of gold from hydrometallurgical smelting wastewater (Li et al. 2019). A huge quantity of gold (more than 5 kg) was recovered while treating a huge volume of wastewater (120,000 m^3) within a period of 1 year. High recovery efficiency resulted from the large driving force generated due to the reaction between Au(III) and nano ZVI, as well as owing to the core–shell structure and unique surface chemistry of nanoparticles. Moreira et al. applied an integrated system comprising a membrane distillation unit, contactors, followed by solvent extraction for treatment of gold mining wastewater as well as for recovery of sulfuric acid and copper (Moreira

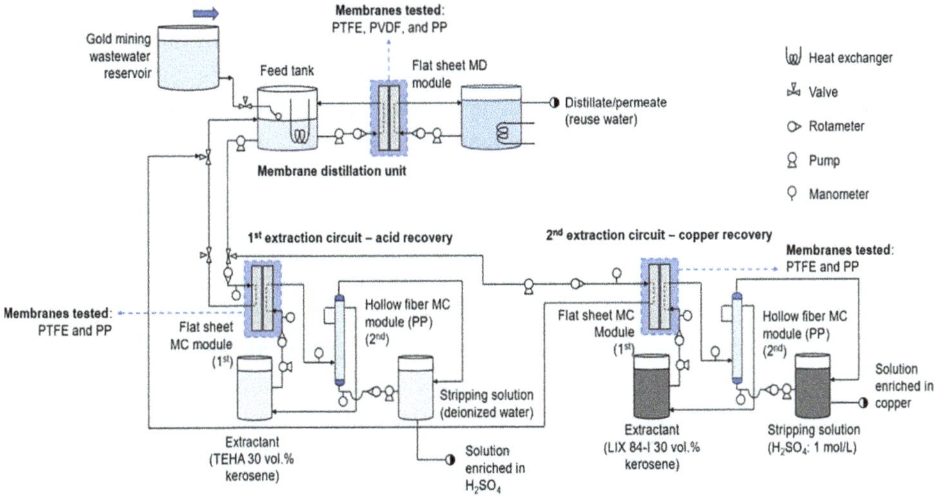

Fig. 11.2 Integrated system of membrane distillation and membrane contactors for simultaneous water, sulfuric acid, and copper recovery from gold mining wastewater. MC: membrane contactor, PP: polypropylene, PTFE: polytetrafluorethylene, PVDF: polyvinylidene fluoride, MD: membrane distillation (Moreira et al. 2023)

et al. 2023). Out of various membrane materials used, the authors reported that the best result was yielded with the polypropylene membrane. 94% acid and 97.3% copper were extracted during the process. The integrated system for wastewater treatment and resource recovery is shown in Fig. 11.2. Nawaz and Sengupta deployed commercially available resins containing a thiol functional group for the recovery of silver from synthetically prepared laundry greywater (Nawaz and Sengupta 2017). The system showed promising performance by recovering more than 90% silver from its trace concentration. Further, the system has been found sustainable due to easy regeneration. Regeneration experiments were performed using thiourea (0.5 M) and dilute HNO_3 (0.1 M) as the regenerant.

11.2.4 Energy Production

Global energy demand is continuously rising due to rapid population growth. On the other hand, energy production through wastewater treatment may provide a sustainable solution to this crisis, and it also helps to reduce the impact of wastewater treatment. Anaerobic digestion and microbial fuel cell (MFC) based technologies are commonly adopted for energy generation purposes. MFC is a wonderful idea of the current times, where wastewater can be treated with the simultaneous production of bioelectricity. It can convert chemical energy into electrical energy with the help of microorganisms acting as the biocatalysts. Nawaz et al., in their recent review article, discussed the power generation through the action of MFC while utilizing different types of wastewaters as the substrate (Nawaz et al. 2022). The authors provided a useful discussion regarding electricity production by comparing various categories of wastewater, including the chocolate industry, the brewer, the petroleum refinery, the dairy industry, etc. Moreover, the review focused on several types of MFC design, such as double chamber MFC, single chamber MFC, flat plate MFC, cylindrical MFC, etc. Khandaker et al. reviewed the latest advances that took place in utilizing MFCs for sustainable wastewater treatment as well as bioelectricity, bio-hydrogen production, and carbon sequestration (Khandaker et al. 2021).

Apart from electricity generation, many research articles of contemporary times report on methane production during anaerobic wastewater treatment. Tang et al. utilized an anaerobic membrane bioreactor for swine wastewater treatment and simultaneous methane generation (Tang et al. 2022). The as-explored technique was found promising in converting waste to energy. Among various parameters, hydraulic retention time (HRT) was one of the major ones. At an HRT of 15 days, the highest methane yield of 0.24 L/g-COD removed was attained. In another study, Zhou et al. found that using polyaniline hydrogel as the conducting medium of direct interspecies electron transfer can improve the methane yield during anaerobic wastewater treatment (Zhou et al. 2021). It provided a good biocompatibility with the microbes and accelerated the methane generation. From experimental investigation, it was revealed that at an optimal dose of 3000 mg/L, 28.77% methane production was enhanced. Some of the studies also describe the formation of hydrogen gas (H_2) during wastewater treatment. Tak et al. reviewed several new-age novel technologies often practiced for H_2 evolution during wastewater treatment (Tak et al. 2022). The authors divided the technologies into three categories, namely: light-dependent, light-independent, and others. Among the light-dependent processes, the major ones are the photocatalysis, photo fermentation, photo electrochemical cell, and microbial photoelectrochemical cell. On the other hand, in the section on light-independent techniques, the authors discussed dark fermentation and microbial electrolysis cell (MEC). Supercritical water gasification and reverse electrodialysis are some of the other advanced techniques for H_2 evolution. Khan et al. evaluated the potential of MEC regarding the urban wastewater treatment as well as hydrogen production in the Kingdom of Saudi Arabia (Khan et al. 2017). The authors concluded that MEC is undoubtedly a promising technology for energy recovery in contemporary times. Further, it is also reported that a surplus amount of 508 Mwe and 637 Mwe of electricity can be generated for the nation in the years 2025 and 2035.

11.3 Various Techniques Adopted for Recycling of Water and Wastewater

Wastewater treatment and reuse are an emerging and very crucial issue across the world. Several research groups practice advanced technologies for recycling purposes. The conventional method of wastewater treatment and recycling consists of three processes, namely primary, secondary, and tertiary treatment. These conventional wastewater treatment processes have succeeded to some extent, but there's a need for improvement in wastewater treatment processes to make treated wastewater reusable for agricultural, industrial, and domestic purposes (Ezugbe and Rathilal 2020). Wastewater can be recycled using various mechanisms such as adsorption, membrane separation, microbial fuel cells (MFC), photocatalytic conversion, phytoremediation, etc., as advanced methods.

Coagulation and flocculation, which are part of the processes of the conventional wastewater treatment plant, have also been reported in some studies for recycling purposes. In the study carried out by Lee and Chang, non-ionic polycrylamide (NPAM) has been used as a flocculant for treating swine wastewater (Lee and Chang 2022). Detailed experimental analysis showed that at a solution pH of 11, the coagulation-flocculation was obtained best. Under suitable conditions, copper ions, zinc ions, total nitrogen, etc., were eliminated at a percentage higher than 90%. In another study, Wang et al. reported the application of aluminium chloride as the coagulant and a modified natural polymer as a suitable flocculating agent for paper-recycling wastewater treatment (Wang et al. 2007). Further, the authors utilized response surface methodology for optimizing the treatment process. With a coagulant dose of 759 mg/L, a flocculant dose of 22.3 mg/L, and a solution pH of 5.4, the optimal condition was achieved. In this section, various such techniques often explored by the researchers have been described in brief as follows.

11.3.1 Recycling of Wastewater by the Process of Adsorption

Adsorption is one of the simplest and versatile techniques to recycle wastewater. Traditionally, silica, alumina, and activated carbon were the commonly explored adsorbents for wastewater purification purposes. However, in contemporary times, these adsorbents are greatly replaced by naturally occurring biopolymers for adsorption purposes. Various biopolymers such as chitosan, alginate, carrageenan, tannin, xanthan gum, guar gum, pectin, starch, and cellulose are widely explored by researchers for pollutant removal due to the presence of various functional groups (Biswas and Pal 2021). Apart from these, various novel adsorbents are also synthesized from different agricultural, domestic, and industrial waste possessing high adsorption and recycling capacity materials, offering a sustainable waste management option (Dawn and Vishwakarma 2021). Various nanomaterials-based adsorbents are utilised by several research groups for pollutant abatement purposes.

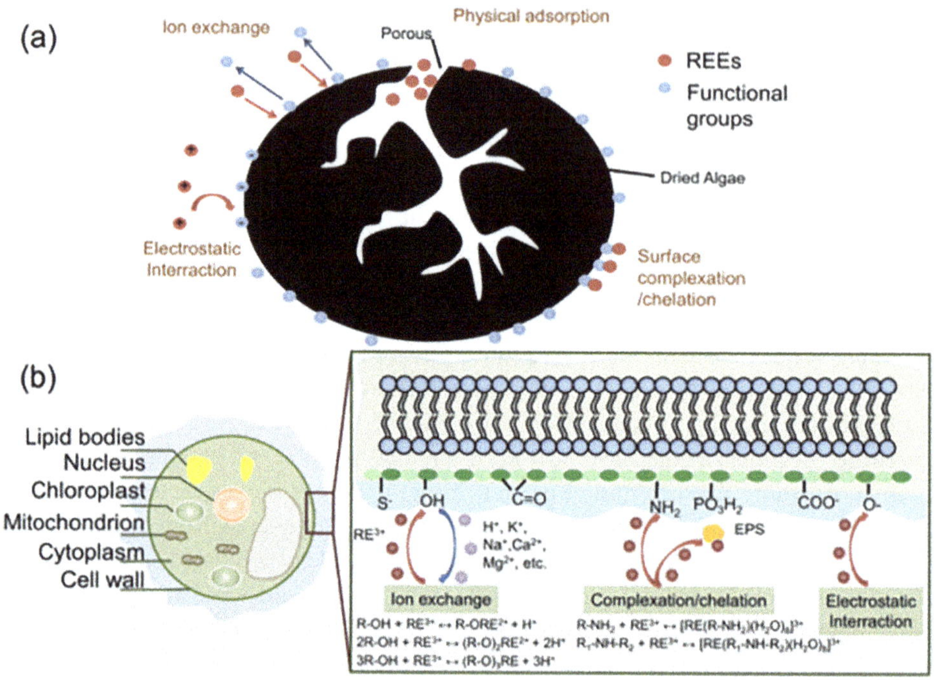

Fig. 11.3 Schematic diagram of the possible adsorption mechanisms of REEs by algae: **a** dried algae and **b** living algae (Cao et al. 2021)

Cao et al. reviewed the use of algae for the recovery of REEs from wastewater. Both dried and living algae are capable of removing REEs from wastewater by the mechanism of adsorption. The REEs removal capacity of dried/living algae is enhanced by adopting physical and chemical modification methods such as grinding, pre-treatment, and grafting for dried algae (Cao et al. 2021). REE uptake by algae mass has been schematically shown in Fig. 11.3. Biomass-derived adsorbents are beneficial in removing and recycling phosphate from wastewater. In order to enhance the adsorption capacity, the biomass adsorbents are generally incorporated with metals. It has been found that rare metals such as La, Zr, and Ce have more rapid phosphate adsorption kinetics as compared to Mg-, Al-, Fe-, or Ca- modified adsorbents (Jiao et al. 2021).

11.3.2 Recycling of Wastewater by the Process of MembraneTechnology

Membrane technology has emerged as one of the promising techniques used to recycle wastewater from different streams. Membrane technology is a pressure-driven process

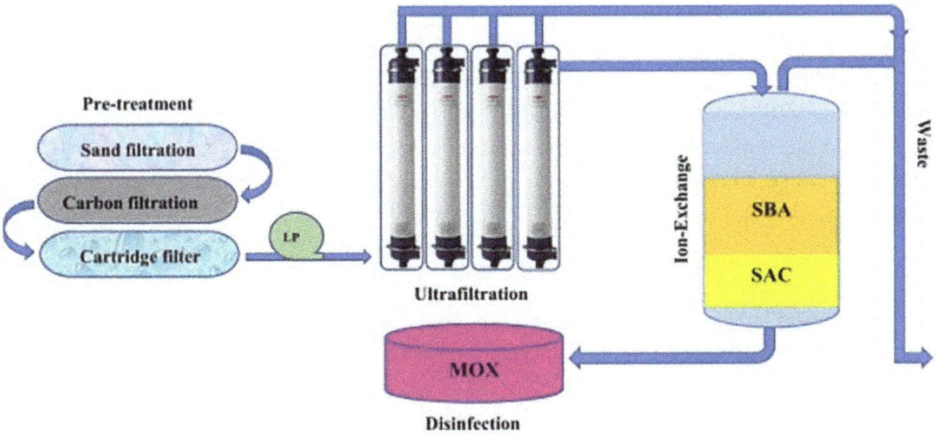

Fig. 11.4 Schematic diagram of UF-IXMB/MOX hybrid system operation (Hashemi et al. 2022)

that utilizes a semi-permeable membrane to split contaminants from the water based on size, charge, or other properties. The several types of membrane technologies applied to wastewater recycling are as follows: Microfiltration (MF), which removes suspended solids, bacteria and larger pathogens; Ultrafiltration (UF) which proves effective in eliminating macromolecules, proteins and colloidal particles; Nanofiltration (NF) which targets divalent and larger monovalent ions, organic compounds and hardness components; Reverse Osmosis (RO): It provides the highest separation factor and removes dissolved salts, organic matter and even pathogens (Ikhsan et al. 2017). Hashemi et al. used a combination of wastewater treatment technology for the petroleum industry, including membrane resin and multioxidant system (Hashemi et al. 2022). Thus, treated effluent was used for makeup water in the cooling tower in the Kermanshah oil refinery, Iran, during the period April to December 2018. The schematic of this hybrid membrane-based wastewater recycling system is shown in Fig. 11.4.

Membrane technology is often integrated with biological processes to produce membrane bioreactors (MBR) for the treatment of wastewater. MBRs have various advantages, such as high efficiency, rapid setup, and a small footprint. Despite these several advantages, membrane technology also poses a problem, which is fouling, high energy consumption, and maintenance costs. Fouling, originating from organic matter, biofouling, effectively diminishes the lifespan as well as efficiency of membranes (Zandi et al. 2019). Solutions to these challenges include: Pre-treatment techniques, which involve implementing filtration and coagulation before membrane processes; Membrane cleaning, which is a regular chemical and physical cleaning method to remove accumulated fouling; and the development of advanced membranes, such as the use of antifouling coatings and materials like mixed matrix membranes to enhance performance (Ikhsan et al. 2017).

Membrane technology has been successfully applied in various sectors of wastewater recycling. Recycled water from membrane processes is used for irrigation, thereby reducing the demand for freshwater resources (Ezugbe and Rathilal 2020). Industries such as oil and gas, textiles, and pharmaceuticals utilize membrane processes to recover water and meet discharge regulations (Ikhsan et al. 2017). Most cities are using MBRs for treating domestic sewage so that treated water can be reused for irrigation and industrial purposes (Zandi et al. 2019).

11.3.3 Recycling of Wastewater by the Process of Solid Phase Extraction

Solid phase extraction (SPE) is a novel analytical technique often practised for the separation, purification, and pre-concentration of analytes prior to chromatographic analysis. Various research articles of the current times report their applicability towards the detection and quantification of organic pollutants from wastewater. SPE is considered to be one of the sustainable physico-chemical techniques of the present era, useful for the recovery of precious materials from wastewater as well as for recycling purposes. Brewer et al., in their recent article, describe the potential of the SPE technique regarding the recovery of critical materials from industrial effluents. For this purpose, firstly, a stationary phase, usually a cartridge material, is selected for passing the wastewater through it. The target organic compound is retained in the stationary phase, which is further recovered using a suitable solvent (Brewer et al. 2022). Recently, we also reported a study resembling the SPE technique for the treatment of mixed dye wastewater, followed by the selective recovery of anionic dye (Biswas and Pal 2022). Firstly, a mixture of cationic (methylene blue) and anionic (methyl orange) was allowed to get adsorbed on the surface of cationic surfactant-modified alginate-xanthan hydrogel beads. Methyl orange got attached to the bead surface by adsolubilization while methylene blue dye got adsorbed due to electrostatic attraction. In the next step, these dye-loaded beads were immersed in a 1-butanol solution for the selective reclamation of methyl orange dye. The schematic is shown in Fig. 11.5. In another study, Sengupta and Pandit explored iron oxide nanoparticles loaded anion exchanger as a suitable adsorbent material for trapping phosphorus from wastewater. Consequently, the loaded phosphorus was extracted using an eluent solution containing NaOH (Sengupta and Pandit 2011).

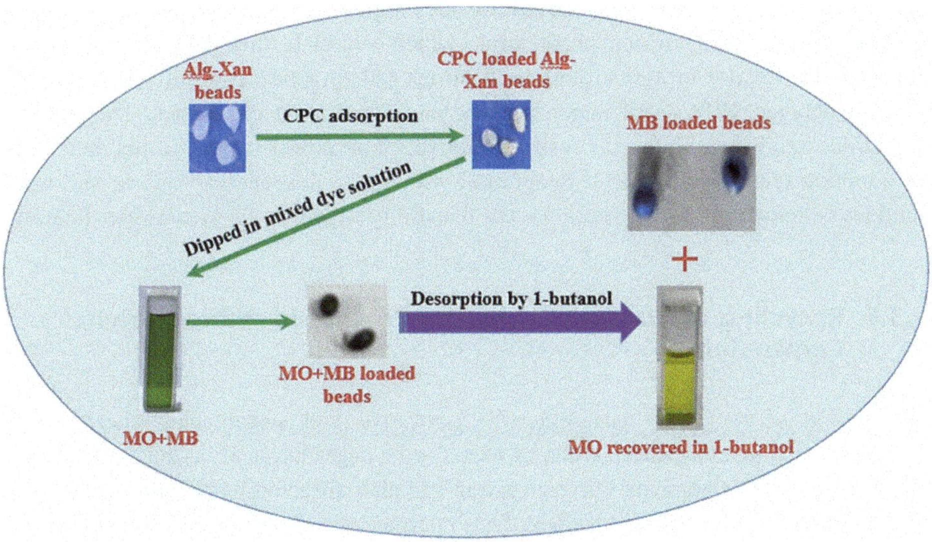

Fig. 11.5 Schematic showing the selective reclamation of methyl orange from a mixture of methyl orange and methylene blue by application of cationic surfactant modified alginate–xanthan hydrogel beads (Biswas and Pal 2022)

11.3.4 Recycling of Wastewater by the Process of Microbial Fuel Cell (MFC)

Microbial Fuel Cells (MFCs) is a new technology for wastewater treatment that produces electricity while leaving low concentrations of nutrients in the effluent. It is a promising technology that seems to be economically viable and can reduce water and energy shortages (Abourached et al. 2016).

Typically, the core working mechanism of an MFC involves a two-chamber setup, an anode chamber and a cathode chamber, divided by a proton exchange membrane. In the chamber with the bacteria, they break down organic substrates and release electrons and protons. Electrons then travel out to an external circuit to the cathode chamber, where they generate electricity; meanwhile, protons migrate through the membrane and react with oxygen to form water (Jatoi et al. 2021).

MFCs use electroactive bacteria to oxidise organic matter, capturing the electrons and allowing them to flow out to an external circuit to produce power. MFCs have shown effective efficiency in treating wastewater from diverse sources. Removal efficiencies of COD as high as more than 90% are reported in the literature, depending on the wastewater type, such as agricultural, dairy, and paper industry effluents. In one of the studies, Choudhury et al. utilized a single-chamber-based MFC for the treatment of dairy wastewater (Choudhury et al. 2021). Under the optimal reaction condition, a promising removal efficiency of 92.21% COD was achieved. Besides organic matters, MFCs can also remove nutrients such as nitrogen and phosphorus, which would further enhance the potential reuse of treated water in agricultural applications (Abourached et al. 2016). Despite the above advantages, MFCs still face economic and operational challenges. The high cost of electrodes and membranes, as well as relatively low power output, limits their extensive adoption (Jatoi et al. 2021). Integration with renewable energy sources and hybrid treatment systems may further improve the feasibility of MFCs for wastewater treatment.

11.3.5 Recycling of Wastewater by the Process of Photocatalytic Conversion

This method of recycling wastewater is innovative and sustainable, whereby light-activated semiconductor materials can be used for the degradation of organic and inorganic pollutants in water. Photocatalytic conversion has also attracted much interest because it decomposes complex pollutants such as dyes, pharmaceuticals, and heavy metals into non-toxic products such as carbon dioxide and water (Ren et al. 2021). Several reported studies of the present times reveal the promising nature of different novel photocatalyst materials. Velempini et al. reviewed the latest developments that took place in the domain of metal oxides regarding photocatalytic mineralisation of pharmaceutical compounds present in wastewater (Velempini et al. 2021). Due to various distinguishing properties such as simplicity in the synthesis process, high surface area to mass ratio, inexpensive, etc., they are widely explored for environmental remediation purposes.

Some innovative studies reported the conversion of heavy metal pollution to sustainable photocatalysts as well as wastewater treatment. Chen et al. explored calcium silicate hydrate nanosheet for the elimination of heavy metals such as Cu^{2+}, Zn^{2+}, and Ni^{2+}, followed by utilizing the metal-loaded adsorbent as a promising catalyst for CO_2 reduction (Chen et al. 2020). In another study, Wang et al. utilised zeolite materials for the adsorption of heavy metals Cd^{2+}, Zn^{2+}, and further reused metal-loaded zeolite for the hydrogen evolution reaction (Wang et al. 2020). Wang et al. prepared a novel CaCr-layered double hydroxide (CaCr-LDH) by utilizing Cr^{3+} from the industrial wastewater (Wang et al. 2023). Further, CaCr-LDH showed excellent performance in the adsorption of heavy metals such as Ni, Cu, Co, etc. NiCaCr-LDH thus produced was found to be an active photocatalyst for CO_2 reduction purposes.

11.3.6 Recycling of Wastewater by the Process of Phytoremediation

Phytoremediation is the process in which plants are used for the recovery of nutrients and other useful materials while simultaneously decontaminating the environment. It has emerged as a low-cost, eco-friendly, and sustainable approach as compared to traditional biological and physicochemical processes (Hu et al. 2020). There have been studies showing how different aquatic plants can be useful in treating industrial and livestock waste. Water hyacinth *Eichhornia crassipes* is widely used in treating wastewater due to its rapid growth and high capacity for pollutant uptake. A study showed that water hyacinths could remove up to 94.67% of cadmium, 65.4% of dissolved oxygen, and 64.41% of BOD in wastewater (Ajayi and Ogunbayo 2012). Similarly, hydrophytes like duckweed, water lettuce, and constructed wetlands have been effective in nutrient recovery and heavy metal removal from livestock wastewater (Hu et al. 2020). These systems can significantly reduce the BOD and TSS, making the treated water suitable for reuse in agricultural and industrial applications.

Buslima et al. reviewed the latest developments of the application of biofilm-integrated phytoremediation technique for the treatment of domestic wastewater following its further reuse (Buslima et al. 2024). Tan et al. provided details of the technical perspective of the phytoremediation process with respect to heavy metals recovery from wastewater stream (Tan et al. 2023). The authors described the process as a highly efficient and sustainable methodology for metal decontamination from wastewater through a plethora of processes, including phytoextraction, rhizofiltration, phytostabilization, etc. However, often there is a chance of secondary pollutant release from the plants in the case of the phytoremediation technique. In another article, Mustafa et al. discussed the influence of phytoremediation in the promotion of the circular economy in the water reuse sector (Mustafa et al. 2022).

The green walls technique can be considered as one of the novel advancements in the field of phytoremediation for treating wastewater. They are also termed as vegetated walls (Yadav et al. 2023). These are basically vertical structures covered in green plants and useful for improving existing air quality, reducing energy costs, wastewater treatment, and reuse. Some research articles of the present period describe promising characteristics of green wall-based techniques for grey water recycling. The vertECO®green wall-based remediation technique has been investigated by Vobruba et al. in a full-scale study for the successful treatment and recycling of grey water generated from an Austrian ecovillage (Vobruba et al. 2025). Masi et al. explored green walls for onsite treating and reusing grey water from an office complex in Pune, India (Masi et al. 2016). Lightweight expanded clay aggregates, along with coconut fibres as the materials inside green walls, offered a highly efficient treatment to the incoming wastewater.

11.4 Water and Wastewater Recycling Status Across Different Nations

In recent times, wastewater recycling is actively practised across several nations. Among various countries, the US is one of the leading nations in regard to the reuse of wastewater for different purposes. There has been a growing interest in recycling grey water from domestic sources such as hand basins, kitchen sinks, and washing machines. Maier et al.

studied how much potable water could be saved by recycling wastewater. The study was conducted in a park in California, which was connected to the recycled water supply. The study concluded that the unit cost of recycled wastewater is less than that of potable water. The estimated rate of displacement in the study period was found to be 81.7% which means that each unit of recycled water used decreased 0.817 units of potable water usage (Maier et al. 2022). Jeong et al. in one of their works studied the feasibility of a small-scale hybrid system for reclamation of grey water generated from laundry, showers, dishwashing, etc. through submerged membrane bioreactors in Atlanta, USA (Jeong et al. 2018). Wastewater thus collected was subjected to treatment via submerged membrane bioreactors, and the reclaimed water can be used for non-potable purposes, including irrigation and toilet flushing. Further, the authors evaluated the LCA of the hybrid system as well as the conventional centralized system. It was found that the LCA score of the conventional system was much lower than that of the newly incorporated hybrid one.

Reuse of treated wastewater is highly encouraged and prevalent across different European nations. Arena et al. in one of their recent studies, nicely discussed the cost–benefit and other economic aspects of using treated wastewater for irrigation purposes in Puglia, southern Italy (Arena et al. 2020). Wastewater from the coastal region was aimed for reuse in agricultural and recreational activities. Among various beneficial aspects of the project, the authors identified the most prominent ones as the reduced irrigation-related expenses, provision of better water quality to the farmers, and cessation of effluent discharge into the seawater. Operational data, statistics, and various other relevant documents revealed the environmental benefits of the wastewater reclamation project. Being located on the Betic orography, southeastern Spain is one of the driest zones of the European continent. Unconventional sources of water, including municipal sewage reuse, are undoubtedly a useful solution in such a dry environment. Gil-Meseguer et al. reported that 110×10^3 m^3 purified sewage is produced yearly in this region (Gil-Meseguer et al. 2019). Interestingly, during the time span of 2003–2016, 92.7% of the sewage got regenerated, and 89.4% was judiciously utilized for agricultural purposes in southeastern Spain. Pronk et al. evaluated reuse of municipal and industrial wastewater for irrigation purposes, for industrial applications, and groundwater recharge in the Netherlands (Pronk et al. 2021). In all the cases, it was seen that through this practice, dependency on the groundwater extraction can be greatly reduced. From the Sankey diagrams, it was concluded that such a sustainable approach can effectively improve the water system robustness for the Dutch nation. The green wall system is one of the novel techniques of the current period, where plants are allowed to grow on a vertical wall and their roots are effectively utilized for wastewater treatment. Vobruba et al., in one of the articles, reported the successful implementation of the vertECO® green wall-based treatment system for wastewater decontamination as well as its reuse for agricultural activities (Vobruba et al. 2025). The system thus designed received 500 L of treated wastewater per day from an ecovillage in Austria. Further, the effluent released from vertECO® green wall complied with the norms of the EU water reuse regulation (class C: for drip irrigation purposes). Various schemes have been tested across the UK, demonstrating the feasibility of grey water reuse (Jefferson et al. 1999). Reuse of treated rainwater, grey water, and black water has become one of the most important elements of the urban lifestyle in the UK. Diaper et al. described the instance of many single-house greywater recycling sites across the country (Diaper et al. 2001). Many of these belong to the Water Company or Environmental Agency monitoring programme.

Saudi Arabia is one of the notable arid countries in the world, having minimal rainfall and high water demand. The discovery of oil has led to a rapid increase in population in the region during the last few decades. Aquifers and desalination are the major water sources for the country, which seems to be inadequate in comparison to the growing demand. Hence, the policy makers of the country are of the opinion that recycled wastewater may be one of the viable sustainable options for this serious problem. It was estimated that by 2025, full utilization of recycled wastewater can be done on a large scale. Abubakar and Muazu investigated the willingness of people residing in various parts of Saudi Arabia to utilize treated greywater for domestic applications. It is mentioned that the government took an ambitious initiative to recycle treated greywater in major cities by 2040. On the basis of a questionnaire survey of a population of 625 households, the authors concluded that among people, there is more willingness towards the reuse of the treated greywater for laundering purposes rather than for bathing purposes (Abubakar and Mu'azu 2022).

There are some encouraging reported studies available in the literature describing the sustainable wastewater reuse and reclamation in the Eurasian country of Turkey in contemporary times. Alkaya et al. investigated the water recycling opportunity for a soft drink manufacturing company (Alkaya and Demirer 2015). Based on several evaluations, water reusing practice was applied in the cooling system due to its large water-saving potential. Interestingly, it reduced the water demand for cooling purposes from 14.4 m^3/m^3 to 1.2 m^3/m^3. Moreover, this step reduced the annual water consumption by 503,893 m^3.

A study conducted in Iran, where water scarcity is revealed, stated the need for an integrated Water Supply and Wastewater Collection System (WSWCS) (scheme shown in Fig. 11.6) (Fathollahi-Fard et al. 2020). In one of the latest studies, Naghedi et al. described the installation of a framework for an integrated industrial wastewater recycling system in Toos Industrial Park of Khorasan Rezavi Province, Iran (Naghedi et al. 2020). The system is composed of three aspects, namely the collection system, the treatment process, and the reuse target, which are collectively termed as "functional group." The study suggests future research should concentrate on the development of tools for assessing the sustainability of alternatives for wastewater recycling purposes. Like many places in Iran, Qom is a dry province receiving very little rainfall and is located adjacent to two deserts. Shamabadi et al. reported an innovative onsite grey water recycling system initiated by Hazrat-e-Masoumeh University (Shamabadi et al. 2015). It is primarily made up of suspended solid media acting as the trickling filter. Further, the processed water is stored in a settling tank followed by chlorination.

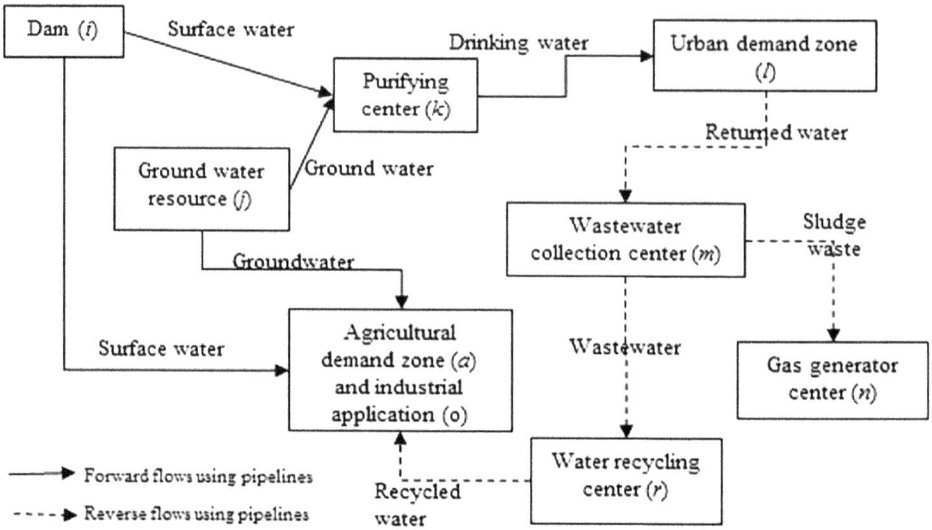

Fig. 11.6 Proposed integrated WSWCS (Fathollahi-Fard et al. 2020)

Like many nations across the globe, Australia is also practising and encouraging wastewater reclamation extensively. It is a dry continent and experienced a prolonged drought period (from 1997 to 2010) known as the millennium drought. This harsh climatic condition forced the government to adopt the water reuse policy more strongly (Radcliffe 2015). Radcliffe and Page, in their article, highlighted the state of the art of treated wastewater reuse in Australia from the past to recent times (Radcliffe and Page 2020). Throughout the country, recycled water has been accepted among the citizens as one of the legitimate water sustainable sources of water for non-drinking purposes (Apostolidis et al. 2011). For decades, Australia has emerged as one of the leading countries across the globe in the water recycling sector. It has set a wonderful example for the whole world by developing its own national guidelines regarding the usage of this reclaimed water for several non-drinking purposes.

Recently, in the Indian scenario, several water reuse innovation schemes have come up. Unplanned industrialization, rapid urbanization, and a drastic increase in population have created tremendous pressure on natural water resources across several Indian cities. In many places, the groundwater table has lowered alarmingly. In one of the latest studies, Sudarsan et al. highlighted the significance of water reuse in Tiruvananthapuram city, Kerala, India (Sudarsan et al. 2021). The authors quantified the potential of greywater recycling in two places, namely Palayam and Nanthancode in the city, and justified the greywater reusing strategy across residential areas. It is reported that, on average, about 700 m^3 of recyclable wastewater is generated from these two places. This wastewater is collected in a large tank and subjected to sedimentation, filtration, and phytoremediation treatment. Thus, treated wastewater was found suitable for toilet flushing and gardening purposes. In another study, Sarkar evaluated life cycle and benefit cost analysis of greywater reuse in commercial and residential sectors in Ahmedabad and Mumbai (Sarkar 2023). Both the commercial and residential complexes possess grey water recycling systems. However, in the case of the commercial complex in Ahmedabad, the system was imported from a German manufacturer. On the other hand, in the case of the residential one, the recycling system was manufactured in India. On detailed cost calculation, it was revealed that in the case of the Indian system, the annual cost incurred was about Rs. 187/m^2, while it was Rs. 327/m^2 for the imported one. Moreover, the benefit–cost ratio for the residential complex was 2.09, while it was 1.59 for the commercial complex at Ahmedabad. Subramanian et al., in an article, discussed the feasibility of a decentralised greywater recycling system located at a school in the rural region of Karnataka (Subramanian et al. 2020). Greywater in this study was obtained from handwash and kitchen wash sinks. Due to a high difference in composition, these streams were separated. Various treatments, including screens, grease traps, slow sand biofiltration, anaerobic reactor, disinfection (ozone-based), etc., were provided to the generated greywater. Several parameters, including TSS, BOD, COD, fecal coliform, etc., were drastically reduced after the treatment. Thus, reclaimed wastewater was utilized for the toilet flushing activity. Greenwall technology has also been explored in India. Masi et al. reported the successful installation and running of a green wall-based project for greywater treatment at pilot scale in an urban area of Pune, Maharashtra (Masi et al. 2016). In the first phase, green walls were filled only with lightweight expanded clay aggregate for the treatment purpose. However, due to poor treatment performance, coconut fibres were incorporated along with aggregates in the second phase. It significantly enhanced the treatment efficiency. Yadav et al. explored a treatment system comprising green walls coupled with a slow sand filtration system for the treatment of greywater generated from household activities (Yadav et al. 2023). The system thus designed produced treated effluent of high quality (~85% COD removal, ~72% total nitrogen removal) for wastewater samples of varying strength. These green walls are not only environmental remediation tools but also helpful for maintaining environmental sustainability by fixing CO_2 in plant biomass.

11.5 Current Challenges and Future Outlook

It is widely accepted that wastewater recycling is a novel initiative of the present-day world. The recycling and reuse approach helps to reduce the environmental pollution level significantly. It helps in providing a sustainable solution to the water-scarce region. Further, nutrients, heavy metals, and other useful materials recovery during wastewater treatment is an economically attractive option. Hydrogen evolution, electricity generation, and methane production as part of wastewater decontamination schemes are helpful in energy supply. However, despite offering several advantages, there still exist quite a considerable number of challenging issues and hindrances that need to be overcome in the upcoming days to make wastewater recycling and reuse schemes more acceptable globally.

Oh et al. in their review article discussed some of the major constraints behind the implementation of grey water recycling projects in Malaysia (Oh et al. 2018). Greywater typically includes domestic wastewater coming out from the kitchen sinks or bathrooms, excluding toilet inputs. It comprises nearly half of the total wastewater generated and can be reasonably reused for non-potable purposes such as toilet flushing, irrigation, due to low pathogen and organic fraction. However, long-term application of this type of water for irrigation purposes may affect the soil quality. Therefore, disinfection is required. Among several current hindrances, the authors justifiably identified that legal issues are one of the most significant ones. Although the government is interested in wastewater recycling projects, due to inexperience regarding management and maintenance, it cannot be approved.

Undoubtedly, wastewater recycling contributes significantly to the growth of the circular economy. However, there are several existing drawbacks as identified by Kakwani and Kalbar in one of their recent articles while discussing the Indian scenario (Kakwani and Kalbar 2020). In India to date, there is an inadequate wastewater collection system, a significant gap between the wastewater actually generated and that which gets sufficient treatment. Further linear approach towards water management is also a severe drawback to wastewater reuse. The authors divided the existing challenges into technical, economic, institutional, and social challenges. It is true that there are a number of technologies available for the conversion of wastewater into potable water. However, the selection of a suitable one, depending upon several conditions, is a challenging task for the concerned authority. Besides, conventional wastewater treatment plants are initially designed for tackling BOD, N, and P removal. However, with urbanization and a changed lifestyle, emerging pollutants are occupying a significant place in the wastewater streams. As per previous reports (Tare 2009), the operation and maintenance cost for treating per m^3 of wastewater annually was around Rs. 3.5. However, due to recycling, the cost may rise up to Rs. 6.5–9/m^3. Hence, economically, the option is not attractive from the Indian perspective. Additionally, intensive application of recycled water for irrigation purposes may reduce the market value of the crops significantly.

Vaidya et al. discussed several challenging issues in the Indian scenario regarding the nationwide implementation of wastewater recycling and reuse policy (Vaidya et al. 2024). However, the authors also suggested some valuable solutions to overcome these issues. Many places in the country have inadequate sewage treatment provision, while in many states, there are no sewage treatment plants at all. Under these deteriorating conditions, some useful intervention in the form of the construction of a small reactor-type anaerobic treatment system can upgrade the existing sanitation system. Previously conducted studies revealed that within a hydraulic retention time period of 24 h, BOD and COD can be reduced in the range of 20–30%. As per the reports of 2021, the compliance rate of the operating sewage treatment plants in the Indian scenario was only 23% in comparison to the standards provided by the National Green Tribunal. Moreover, discharge standards are getting stricter day by day for maintaining a safe environment and preventing further pollution. In this condition, a more detailed analysis of water quality (TSS, BOD, COD) at the outlet of the reactors is necessary. Many modifications can be practiced to obtain the optimal condition. The operators of the treatment plant should be properly trained to have a thorough idea of various physical and chemical parameters.

Many novel technologies are emerging in modern times with respect to sustainable wastewater treatment. However, application of these innovative processes in the real field situation from the laboratory scale is a challenging issue. Setting up a recycling facility at a pilot scale may offer several hurdles. A rigorous investigation of these problems needs to be carried out. Further biological treatment processes are found to be robust under a set of certain environmental conditions. A proper knowledge of these environmental parameters is required. Detailed benefit–cost analysis, LCA, materials flow analysis, etc., are essential for determining the feasibility of wastewater recycling projects. Several studies have been reported in the recent past. However, more such detailed study is expected from the scientific community.

11.6 Discussions

Water and wastewater recycling are among the most prioritized environmental activities of the present-day world. It plays a pivotal role in water scarcity mitigation, wastewater management, recovery of useful materials, energy production, etc. For this purpose, various techniques such as adsorption, photocatalytic oxidation, phytoremediation, SPE, and membrane technology are often explored among researchers. Adsorption is one of the most versatile and simplest techniques often adopted by engineers. In recent times, the scientific community has been focusing on low-cost, highly efficient, bio-based adsorbents for achieving sustainable wastewater treatment (Biswas et al. 2025). Phytoremediation is a nature-based solution, and photocatalytic conversion is a promising sustainable technology of the current era. SPE is useful on many occasions for extracting valuable components from the wastewater stream and is hence considered one of the important recycling techniques (Brewer et al. 2022).

Recycling of wastewater also aligns with many sustainable development goals (SDGs), as proposed by the United Nations, to be achieved by 2030. It directly correlates with SDG 6: Clean water and sanitation, SDG 12: Responsible consumption and production, and SDG 14: Life below water. Besides, water and wastewater recycling also supports the principles of SDG 13: Climate action, SDG 7: Affordable and clean energy, and SDG 3: Good health and well-being.

11.7 Conclusions

Increase in environmental pollution, generation of water scarcity across several places, and rapid depletion of natural water resources are some of the hot environmental issues of contemporary times. Scientific and engineering communities are continuously working to find sustainable solutions. However, in this regard, it is true that wastewater recycling and reuse may provide answers to several environmental problems of the present era. The importance of the reuse of treated wastewater is immense. Various techniques, including adsorption, SPE, photocatalysis, MFC, green wall, etc., are explored by researchers across the globe for reclamation purposes. Various nations, including India, the USA, Turkey, Italy, Australia, Saudi Arabia, etc., are practicing wastewater recycling extensively. Several benefits are highlighted by the researchers. Recent developments in this field and the status of reuse practice across various countries have been elaborated in the present article. Although reuse and recycling have become an integral part of present-day waste management policies, there are several existing challenges associated with them. In the near future, more thorough analysis of these issues and obtaining prominent solutions may be obtained from the upcoming works.

Recommendations

A few valuable recommendations can be drafted for the readers. They are as follows:

- Water and wastewater recycling are an integral part of the water management scheme. They need to be practiced more in the upcoming days to ameliorate several environmental problems as well as to meet several SDGs.
- Recycling techniques need to be environmentally sustainable. Hence, an environmental impact assessment study and a life cycle analysis of several techniques are recommended.
- Undoubtedly, recycling processes should be robust and most efficient in nature in order to meet stringent environmental guidelines.
- Future recycling technologies should focus more on biobased materials as well as utilize indigenous technology in a more effective way.

Add Acknowledgement (if any).

Not applicable.

State whether you used AI or not in preparing the chapter.

Only for increasing resolution of images, it has been done online. Other than that, no AI has been used in preparing the chapter.

References

Abotalib AZ, Abdelhady AA, Heggy E, Salem SG, Ismail E, Ali A, Khalil MM (2023). Irreversible and large-scale heavy metal pollution arising from increased damming and untreated water reuse in the Nile Delta. Earth's Future, 11, e2022EF002987, https://doi.org/10.1029/2022EF002987.

Abourached C, English MJ, Liu H (2016) Wastewater treatment by Microbial Fuel Cell (MFC) prior irrigation water reuse. J Clean Prod 137:144–149. https://doi.org/10.1016/j.jclepro.2016.07.048

Abubakar IR, Mu'azu ND (2022) Household attitudes toward wastewater recycling in Saudi Arabia. Utilities Policy 76, 101372, https://doi.org/10.1016/j.jup.2022.101372

Acciarri MF, Checola S, Galli P, Magatti G, Stefani S (2021) Water resource management and sustainability: A case study in the Republic of Maldives. Sustain 13:3484. https://doi.org/10.3390/su13063484

Adewumi JR, Ilemobade AA, Zyl JEV (2010) Treated wastewater reuse in South Africa: Overview, potential and challenges. Resour Conserv Recycl 55:221–231. https://doi.org/10.1016/j.resconrec.2010.09.012

Ajayi TO, Ogunbayo AO (2012) Achieving environmental sustainability in wastewater treatment by phytoremediation with water hyacinth (*Eichhornia Crassipes*). J Sustain Dev. 5(7). https://https://doi.org/10.5539/jsd.v5n7p80

Aljawzi AA, Fang H, Abbas AA, Khailah EY (2022) Assessment of water resources in Sana's region, Yemen Republic (Case study). Water 14(7):1039. https://doi.org/10.3390/w14071039

Alkaya E, Demirer GN (2015) Water recycling and reuse in soft drink/beverage industry: A case study for sustainable industrial water management in Turkey. Resour Conserv Recycl 104:172–180. https://doi.org/10.1016/j.resconrec.2015.08.011

Apostolidis N, Hertle C, Young R (2011) Water recycling in Australia. Water 3:869–881. https://doi.org/10.3390/w3030869

Arena C, Genco M, Mazzola MR (2020) Environmental benefits and economical sustainability of urban wastewater reuse for irrigation- A cost-benefit analysis of an existing reuse project in Puglia. Italy. Water 12:2926. https://doi.org/10.3390/w12102926

Bhardwaj A, Kumar S, Singh D (2023) Tannery effluent treatment and its environmental impact: a review of current practices and emerging technologies. Water Qual Res J 58(2):128–152. https://doi.org/10.2166/wqrj.2023.002

Biswas S, Pal A (2021) Application of biopolymers as a new age sustainable material for surfactant adsorption: A brief review. Carbohydr Polym Technol Appl 2:100145. https://doi.org/10.1016/j.carpta.2021.100145

Biswas S, Pal A (2022) Fabrication of a novel composite gel bead to reclaim methyl orange from a binary dye mixture: An active role of adsolubilization phenomenon. Mater Today Commun 33:104386. https://doi.org/10.1016/j.mtcomm.2022.104386

Biswas S, Nayak AK, Pal A (2025) Surfactant-influenced biosorption as a sustainable and effective way for the eradication of environmental pollutants: a review. RSC Sustain 3:112–133. https://doi.org/10.1039/D4SU00574K

Brewer A, Florek J, Kleitz F (2022) A perspective on developing solid-phase extraction technologies for industrial-scale critical materials recovery. Green Chem 24:2752–2765. https://doi.org/10.1039/D2GC00347C

Buslima FA, Hasan HA, Abdullah SRS, Othman AR (2024) Water recovery from domestic wastewater using integrated biofilm-phytoremediation technology: A review. J Water Process Eng 65:105875. https://doi.org/10.1016/j.jwpe.2024.105875

Cao Y, Shao P, Chen Y, Zhou X, Yang L, Shi H, Yu K, Luo X, Luo X (2021) A critical review of the recovery of rare earth elements from wastewater by algae for resources recycling technologies. Resour Conserv Recycl 169:105519. https://doi.org/10.1016/j.resconrec.2021.105519

Chen L, Wang X, Chen Y, Zhuang Z, Chen F, Zhu Y, Yu Y (2020) Recycling heavy metals from wastewater for photocatalytic CO_2 reduction. Chem Eng J 402:125922. https://doi.org/10.1016/j.cej.2020.125922

Choudhury P, Ray RN, Bandyopadhyay TK, Basak B, Muthuraj M, Bhunia B (2021) Process engineering for stable power recovery from dairy wastewater using microbial fuel cell. Int J Hydrogen Energy 46:3171–3182. https://doi.org/10.1016/j.ijhydene.2020.06.152

Dawn SS, Vishwakarma V (2021) Recovery and recycle of wastewater contaminated with heavy metals using adsorbents incorporated from waste resources and nanomaterials-A review. Chemosphere 273:129677. https://doi.org/10.1016/j.chemosphere.2021.129677

Dawoud MA, Ewea HA, Alaswad SO (2022) The future of wastewater treatment and reuse in Kingdom of Saudi Arabia. Desalin Water Treat 263:127–138. https://doi.org/10.5004/dwt.2022.28217

Diaper C, Jefferson B, Parsons SA, Judd SJ (2001) Water-recycling technologies in the UK. Water Environ J 15(4):282–286. https://doi.org/10.1111/j.1747-6593.2001.tb00355.x

Ezugbe EO, Rathilal S (2020) Membrane technologies in wastewater treatment: A review, Membranes 10, 89; https://https://doi.org/10.3390/membranes10050089

Fathollahi-Fard AM, Ahmadi A, Al-e-Hashem SMJM (2020) Sustainable closed-loop supply chain network for an integrated water supply and wastewater collection system under uncertainty. J Environ Manage 275:111277. https://doi.org/10.1016/j.jenvman.2020.111277

Filho, W L, Totin, E, Franke, J A, Andrew, S M, Abubakar, I R, Azadi, H, Nunn, P D, Ouweneel, B, Williams, P A, Simpson, N P & The Global Adaptation Mapping Initiative Team (2022) Understanding responses to climate-related water scarcity in Africa. Sci Total Environ 806:150420. https://doi.org/10.1016/j.scitotenv.2021.150420

Gil-Meseguer E, Bernabe-Crespo MB, Gomez-Espin JM (2019) Recycled sewage- A water resource for dry regions of Southeastern Spain. Water Resour Manage 33:725–737. https://doi.org/10.1007/s11269-018-2136-9

Gupta VK, Ali I, Saleh TA, Nayak A, Agarwal S (2012) Chemical treatment technologies for wastewater recycling-an overview. RSC Adv 2:6380–6388. https://doi.org/10.1039/C2RA20340E

Hamad A, Aidan A, Douboni M (2003) Cost-effective wastewater treatment and recycling in mini-plants using mass integration. Clean Technol Environ Policy 4:246–256. https://doi.org/10.1007/s10098-002-0166-7

Hashemi F, Hashemi H, Abbasi A, Schreiber ME (2022) Life cycle and economic assessments of petroleum refineries wastewater recycling using membrane, resin and on site disinfection (UF-IXMB-MOX) processes. Process Saf Environ Prot 162:419–425. https://doi.org/10.1016/j.psep.2022.04.027

Hopson MN, Fowler L (2022). An analysis of and recommendations for comprehensive state water recycling policy strategies in the U.S. Resources, Conservation and Recycling, 183, 106356, https://doi.org/10.1016/j.resconrec.2022.106356

Hu H, Li X, Wu S, Yang C (2020) Sustainable livestock wastewater treatment via phytoremediation: Current status and future perspectives. Biores Technol 315:123809. https://doi.org/10.1016/j.biortech.2020.123809

Huang R, Shen Z, Wang H, Xu J, Ai Z, Zheng H, Liu R (2021) Evaluating the energy efficiency of wastewater treatment plants in the Yangtze River Delta: Perspectives on regional discrepancies. Appl Energy 297:117087. https://doi.org/10.1016/j.apenergy.2021.117087

Ikhsan SNW, Yusof N, Aziz F, Misdan N (2017) A review of oilfield wastewater treatment using membrane filtration over conventional technology. Malays J Anal Sci, 21(3), 643 – 658, https://doi.org/10.17576/mjas-2017-2103-14

Jaleel MI, Ibrahim SA, Hussain A, Mustafa M, Pathirana A (2020) A screening approach for assessing groundwater quality for consumption in small islands: Case study of 45 inhabited islands in the Maldives. Water 12:2209. https://doi.org/10.3390/w12082209

Jatoi AS, Akhter F, Mazari SA, Sabzoi N, Aziz S, Soomro SA, Mubarak NM, Baloch H, Memon AQ, Ahmed S (2021) Advanced microbial fuel cell for waste water treatment—a review. Environ Sci Pollut Res 28:5005–5019. https://doi.org/10.1007/s11356-020-11691-2

Jefferson B, Laine A, Parsons S, Stephenson T, Judd S (1999) Technologies for domestic wastewater recycling. Urban Water 1:285–292. https://doi.org/10.1016/S1462-0758(00)00030-3

Jeong H, Seong C, Jang T, Park S (2016) Classification of wastewater reuse for agriculture: A case study in South Korea. Irrig Drain 65:76–85. https://doi.org/10.1002/ird.2053

Jeong H, Broesicke OA, Drew B, Crittenden JC (2018) Life cycle assessment of small-scale greywater reclamation systems combined with conventional centralized water systems for the City of Atlanta, Georgia. J Clean Prod 174:333–342. https://doi.org/10.1016/j.jclepro.2017.10.193

Jiao G, Ma J, Li Y, Jin D, Ali Z, Zhou J, Sun R (2021) Recent advances and challenges on removal and recycling of phosphate from wastewater using biomass-derived adsorbents. Chemosphere 278:130377. https://doi.org/10.1016/j.chemosphere.2021.130377

Kakwani NS, Kalbar PP (2020) Review of Circular Economy in urban water sector: Challenges and opportunities in India. J Environ Manage 271:111010. https://doi.org/10.1016/j.jenvman.2020.111010

Kayode-Afolayan SD, Ahuekwe EF, Nwinyi OC (2022) Impacts of pharmaceutical effluents on aquatic ecosystems. Sci Afr 17:e01288. https://doi.org/10.1016/j.sciaf.2022.e01288

Kfir O, Tal A, Gross A, Adar E (2012) The effect of reservoir operational features on recycled wastewater quality. Resour Conserv Recycl 68:76–87. https://doi.org/10.1016/j.resconrec.2012.08.002

Khan MZ, Nizami AS, Rehan M, Ouda OKM, Sultana S, Ismail IM, Shahzad K (2017) Microbial electrolysis cells for hydrogen production and urban wastewater treatment: A case study of Saudi Arabia. Appl Energy 185:410–420. https://doi.org/10.1016/j.apenergy.2016.11.005

Khandaker S, Das S, Hossain MT, Islam A, Miah MR, Awual MR (2021) Sustainable approach for wastewater treatment using microbial fuel cells and green energy generation- A comprehensive review. J Mol Liq 344:117795. https://doi.org/10.1016/j.molliq.2021.117795

Lam KL, Zlatanovic L, Hoek JP (2020) Life cycle assessment of nutrient recycling from wastewater: A critical review. Water Res 173:115519. https://doi.org/10.1016/j.watres.2020.115519

Lee W, Chang C (2022) Effectively recycling swine wastewater by coagulation–flocculation of nonionic polyacrylamide. Sustain 14:1742. https://doi.org/10.3390/su14031742

Li S, Li J, Wang W, Zhang W (2019) Recovery of gold from wastewater using nanoscale zero-valent iron. Environ Sci Nano 6:519–527. https://doi.org/10.1039/c8en01018h

Li X, Feng Y, Chiu Y, Lin T, Chiu S (2021) Recycling water and sludge disposal efficiency in China's sewage treatment industry. Manag Decis Econ 42:1703–1717. https://doi.org/10.1002/mde.3338

Liang, X, Yang, J, He, S, Liu, B, Li, J, Xu, X, Fu, B (2021). Promoting construction of zero-direct discharge engineering for agricultural wastewater from paddy fields in Yangtze River delta region. Bull Chin Acad Sci (Chinese Version): 36:7, Article 12, https://doi.org/10.16418/j.issn.1000-3045.20210429001

Liu C, Cai W, Zhai M, Zhu G, Zhang C, Jiang Z (2021) Decoupling of wastewater eco-environmental damage and China's economic development. Sci Total Environ 789:147980. https://doi.org/10.1016/j.scitotenv.2021.147980

Lou J, Lin Y (2008) Assessing the feasibility of wastewater recycling and treatment efficiency of wastewater treatment units. Environ Monit Assess 137:471–479. https://doi.org/10.1007/s10661-007-9782-3

Madhuranthakam CMR, AbuZaid M, Chaahal O, Ghannam T (2023) Sustainable water management with design and economic evaluation of recycling greywater at Abu Dhabi University-A case study on decentralization. Sustain 15(3):16208. https://doi.org/10.3390/su152316208

Maier J, Palazzo J, Geyer R, Steigerwald DG (2022) How much potable water is saved by wastewater recycling? Quasi-experimental evidence from California. Resour Conserv Recycl 176:105948. https://doi.org/10.1016/j.resconrec.2021.105948

Mannina G, Gulhan H, Ni B (2022) Water reuse from wastewater treatment: The transition towards circular economy in the water sector. Biores Technol 363:127951. https://doi.org/10.1016/j.biortech.2022.127951

Mansour MSM, Abdel-shafy HI, Ibrahim AM (2024) Petroleum wastewater: Environmental protection, treatment, and safe reuse: An overview. J Environ Manage 351:119827. https://doi.org/10.1016/j.jenvman.2023.119827

Masi F, Bresciani R, Rizzo A, Edathoot A, Patwardhan N, Panse D, Langergraber G (2016) Green walls for greywater treatment and recycling in dense urban areas: a case-study in Pune. J Water, SanitIon Hyg Dev 6(2):342–347. https://doi.org/10.2166/washdev.2016.019

Moreira VR, Torres EA, Balarini JC, Amaral MCS (2023) Rethinking gold mining wastewater treatment with an integrated process of membrane distillation and membrane contactors for minimal waste discard and resource recovery. Chem Eng J 476:146906. https://doi.org/10.1016/j.cej.2023.146906

Mustafa HM, Hayder G, Mustapa SI (2022) Circular economy framework for energy recovery in phytoremediation of domestic wastewater. Energies 15:3075. https://doi.org/10.3390/en15093075

Naghedi R, Moghaddam MRA, Piadeh F (2020) Creating functional group alternatives in integrated industrial wastewater recycling system: A case study of Toos Industrial Park (Iran). J Clean Prod 257:120464. https://doi.org/10.1016/j.jclepro.2020.120464

Nawaz T, Sengupta S (2017) Silver recovery from greywater: Role of competing cations and regeneration. Sep Purif Technol 176:145–158. https://doi.org/10.1016/j.seppur.2016.11.076

Nawaz A, Haq I, Qaisar K, Gunes B, Raja SI, Mohyuddin K, Amin H (2022) Microbial fuel cells: Insight into wastewater treatment and bioelectricity generation. Process Saf Environ Prot 161:357–373. https://doi.org/10.1016/j.psep.2022.03.039

Nguyen TKL, Ngo HH, Guo W, Nguyen TLH, Chang SW, Nguyen DD, Varjani S, Lei Z, Deng L (2021) Environmental impacts and greenhouse gas emissions assessment for energy recovery and material recycle of the wastewater treatment plant. Sci Total Environ 784:147135. https://doi.org/10.1016/j.scitotenv.2021.147135

Oh KS, Leong JYC, Poh PE, Chong MN, Lau EV (2018) A review on greywater recycling related issues: Challenges and future prospects in Malaysia. J Clean Prod 171:17–29. https://doi.org/10.1016/j.jclepro.2017.09.267

Osin OA, Yu T, Lin S (2017) Oil refinery wastewater treatment in the Niger Delta, Nigeria: current practices, challenges, and recommendations. Environ Sci Pollut Res 24:22730–22740. https://doi.org/10.1007/s11356-017-0009-z

Pronk GJ, Stofberg SF, Dooren TCGWV, Dingemans MML, Frijns J, Koeman-Stein NE, Smeets PWMH, Bartholomeus RP (2021) Increasing water system robustness in the Netherlands: Potential of cross-sectoral water reuse. Water Resour Manag 35:3721–3735. https://doi.org/10.1007/s11269-021-02912-5

Qureshi AS (2020) Challenges and prospects of using treated wastewater to manage water scarcity crises in the Gulf Cooperation Council (GCC) countries. Water 12(7):1971. https://doi.org/10.3390/w12071971

Radcliffe JC (2015) Water recycling Australia-during and after the drought. Environ Sci: Water Res Technol 1:554. https://doi.org/10.1039/C5EW00048C

Radcliffe JC, Page D (2020) Water reuse and recycling in Australia- history, current situation and future perspectives. Water Cycle 1:19–40. https://doi.org/10.1016/j.watcyc.2020.05.005

Ren G, Han H, Wang Y, Liu S, Zhao J, Meng X, Li Z (2021) Recent advances of photocatalytic application in water treatment: A review. Nanomaterials 11:1804. https://doi.org/10.3390/nano11071804

Robles A, Aguado D, Barat R, Borras L, Bouzas A, Gimenez JB, Marti N, Ribes J, Ruano MV, Serralta J, Ferrer J, Seco A (2020) New frontiers from removal to recycling of nitrogen and phosphorus from wastewater in the Circular Economy. Biores Technol 300:122673. https://doi.org/10.1016/j.biortech.2019.122673

Roshan A, Kumar M (2020) Water end-use estimation can support the urban water crisis management: A critical review. J Environ Manage 268:110663. https://doi.org/10.1016/j.jenvman.2020.110663

Sanchez-Contreras MI, Morales-Arrieta S, Okoye PU, Guillen-Garces RA, Sebastian PJ, Arias DM (2021) Recycling industrial wastewater for improved carbohydrate-rich biomass production in a semi-continuous photobioreactor: Effect of hydraulic retention time. J Environ Manage 284:112065. https://doi.org/10.1016/j.jenvman.2021.112065

Saravanan A, Kumar PS, Jeevanantham S, Karishma S, Tajsabreen B, Yaashikaa PR, Reshma B (2021) Effective water/wastewater treatment methodologies for toxic pollutants removal: Processes and applications towards sustainable development. Chemosphere 280:130595. https://doi.org/10.1016/j.chemosphere.2021.130595

Sarkar D (2023) Life cycle costing analysis of grey water recycling systems for commercial and residential projects of Ahmedabad, India. materials today: proceedings 77:254–259. https://doi.org/10.1016/j.matpr.2022.11.298

Savun-Hekimoglu B, Isler Z, Hekimoglu M, Burak S, Karli D, Yucekaya A, Akpinar E, Ediger VS (2023) Optimization of wastewater treatment systems for growing industrial parks. Sci Total Environ 905:167223. https://doi.org/10.1016/j.scitotenv.2023.167223

Sengupta S, Pandit A (2011) Selective removal of phosphorus from wastewater combined with its recovery as a solid-phase fertilizer. Water Res 11:3318–3330. https://doi.org/10.1016/j.watres.2011.03.044

Shamabadi N, Bakhtiari H, Kochakian N, Farahani M (2015) The investigation and designing of an onsite greywater treatment systems at Hazrat-e-Masoumeh University, Qom, IRAN. Energy Procedia 74:1337–1346. https://doi.org/10.1016/j.egypro.2015.07.780

Sheng X, Qiu S, Xu F, Shi J, Song X, Yu Q, Liu R, Chen L (2020) Management of rural domestic wastewater in a city of Yangtze delta region: Performance and remaining challenges. Bioresour Technol Rep 11:100507. https://doi.org/10.1016/j.biteb.2020.100507

Smol M, Adam C, Preisner M (2020) Circular economy model framework in the European water and wastewater sector. J Mater Cycles Waste Manage 22:682–697. https://doi.org/10.1007/s10163-019-00960-z

Subramanian PSG, Raj AV, Jamwal P, Connelly S, Yeluripati J, Richards S, Ellis R, Rao L (2020) Decentralized treatment and recycling of greywater from a school in rural India. J Water Process Eng 38:101695. https://doi.org/10.1016/j.jwpe.2020.101695

Sudarsan JS, Vaishampayan S, Srihari V (2021) Greywater recycling as a tangible solution to the water crisis: A case study in Tiruvananthapuram, India. Int J Energy Water Resour 5:441–445. https://doi.org/10.1007/s42108-021-00116-6

Tak SS, Shetye O, Muley O, Jaiswal H, Malik SN (2022) Emerging technologies for hydrogen production from wastewater. Int J Hydrogen Energy 47:37282–37301. https://doi.org/10.1016/j.ijhydene.2022.06.225

Tan HW, Pang YL, Lim S, Chong WC (2023) A state-of-the art of phytoremediation approach for sustainable management of metals recovery. Environ Technol Innov 30:103043. https://doi.org/10.1016/j.eti.2023.103043

Tang J, Pu Y, Zeng T, Hu Y, Huang J, Pan S, Wang XC, Li Y, Abomohra AE (2022) Enhanced methane production coupled with livestock wastewater treatment using anaerobic membrane bioreactor: Performance and membrane filtration properties. Biores Technol 345:126470. https://doi.org/10.1016/j.biortech.2021.126470

Tare V, Bose P (2009) Compendium of sewage treatment technologies national river conservation directorate. Minist Environ For, GovMent India

Vaidya R, Verma K, Kumar M, Hoysall C, Rao L (2024) Assessing wastewater management challenges in developing countries: a case study of India, current status and future scope. Environ Dev Sustain 26:19369–19396. https://doi.org/10.1007/s10668-023-03540-2

Velempini T, Prabakaran E, Pillay K (2021) Recent developments in the use of metal oxides for photocatalytic degradation of pharmaceutical pollutants in water- a review. Mater Today Chem 19:100380. https://doi.org/10.1016/j.mtchem.2020.100380

Vinayagam V, Sikarwar D, Das S, Pugazhendhi A (2024) Envisioning the innovative approaches to achieve circular economy in the water and wastewater sector. Environ Res 241:117663. https://doi.org/10.1016/j.envres.2023.117663

Vobruba T, Hartl M, Langergraber G, Pucher B, Gattringer H, Bertino G, Panzenbock F, Kisser J (2025) VertECO® vertical green wall system demonstration for domestic wastewater treatment and on-site reuse in an Austrian eco-village. Ecol Eng 211:107460. https://doi.org/10.1016/j.ecoleng.2024.107460

Wang J, Chen Y, Ge X, Yu H (2007) Optimization of coagulation-flocculation process for a paper-recycling wastewater treatment using response surface methodology. Colloids Surf, A 302:204–210. https://doi.org/10.1016/j.colsurfa.2007.02.023

Wang M, Yao H, Zhang L, Zhou X (2020) Synthesis of highly-efficient photocatalyst for visible-light-driven hydrogen evolution by recycling of heavy metal ions in wastewater. J Hazard Mater 383:121149. https://doi.org/10.1016/j.jhazmat.2019.121149

Wang J, Wang Y, Xiong W, Li Z, Kong X, Yan H, Lin Y, Duan H, Zhao Y (2023) Super-stable mineralization of multiple heavy metal ions from wastewater for utilization in photocatalytic CO_2 reduction and trace metal recovery. Chem Eng Sci 271:118583. https://doi.org/10.1016/j.ces.2023.118583

Wyczarska-Kokot J, Dudziak M (2022) Reuse-Reduce-Recycle: water and wastewater management in swimming pool facilities. Desalin Water Treat 275:69–80. https://doi.org/10.5004/dwt.2022.28756

Yadav RK, Sahoo S, Yadav AK, Patil SA (2023) Green wall system coupled with slow sand filtration for efficient greywater management at households. Npj. Clean Water 6:73. https://doi.org/10.1038/s41545-023-00285-3

Ye Y, Ngo HH, Guo W, Chang SW, Nguyen DD, Zhang X, Zhang J, Liang S (2020) Nutrient recovery from wastewater: From technology to economy. Bioresour Technol Rep 11:100425. https://doi.org/10.1016/j.biteb.2020.100425

Zandi S, Nemati B, Jahanianfard D, Davarazar M, Sheikhnejad Y, Mostafaie A, Kamali M, Aminabhavi TM (2019) Industrial biowastes treatment using membrane bioreactors (MBRs) -a scientometric study. J Environ Manage 247:462–473. https://doi.org/10.1016/j.jenvman.2019.06.066

Zhou N, Wang T, Chen S, Hu Q, Cheng X, Sun D, Vupputuri S, Qiu B, Liu H, Guo Z (2021) Conductive polyaniline hydrogel enhanced methane production from anaerobic wastewater treatment. J Colloid Interface Sci 581:314–322. https://doi.org/10.1016/j.jcis.2020.07.075

Water Quality Assessment and Management in Rural India: Challenges, Contaminants, Health Risks, and Sustainable Solutions

12

Aditi Sharan, Abhradeep Majumder, and Manoj Kumar Yadav

Abstract

Water assessment and management remain one of the major global challenges, with rural livelihoods and public health distinctively affected by growing degraded water quality, increasing the risk of diseases. Rural India faces a severe drinking water crisis, worsened by rapid urbanization, lifestyle changes along with the industries and agriculture explosion. Access to clean water remains difficult placing rural communities under persistent water stress. This chapter discusses the current water status in the country and the significance of water quality assessment tools, with a brief on physicochemical water parameters analysis, biological monitoring, and Water Quality Index (WQI), Nemerow's Pollution Index (NPI), and other groundwater and surface water indices. In addition, the chapter discusses recent water quality advancements in relation to monitoring techniques, such as AI models, GIS-based techniques, Internet of Things (IoT) tools, and hydro-chemical facies analysis. This chapter contributes to achieve Sustainable Development Goals (SDGs) directly related to water, sanitation, agriculture, ecosystem, and health by promoting sustainable water management

A. Sharan · M. K. Yadav (✉)
Department of Civil and Environmental Engineering, Indian Institute of Technology Patna, Bihta, India
e-mail: mkyadav@iitp.ac.in

A. Sharan
e-mail: aditi_2421ce15@iitp.ac.in

A. Majumder
Department of Civil Engineering, Birla Institute of Technology and Science, Pilani-Hyderabad Campus, Hyderabad, India
e-mail: abhradeep.majumder@hyderabad.bits-pilani.ac.in

S. Yadav et al. (eds.), *Water Resources in Delta Region in India*, Deltas of the World,
https://doi.org/10.1007/978-3-032-06500-1_12

and universal access to safe drinking water. The chapter throws light on emerging contaminants and other long-known toxics, associated health risks, primarily affecting economically and socially disadvantaged populations. This is further discussed in state level case studies such as West Bengal, Bihar, and Rajasthan, demonstrating the regional water quality variability and the water safety challenges in rural India. To identify and sort water related challenges, the chapter further discusses sustainable management field-scale techniques, such as affordable water purification technologies, water conservation and community-based management models, particularly relevant to rural settings. The chapter completes itself by showing the major involvement and key roles of government policy frameworks, capacity building, and community engagement to ensure the long-term sustainability of water quality interventions in rural India.

12.1 Introduction

Water is one of the most significant natural resources for living beings on earth, often considered as an abundant and renewable resource with both ecological and economic values (Musie and Gonfa 2023; Tyagi et al. 2013). However, under current usage trends driven by population growth, industrialization, urbanization, climate change, and agriculture, the long-held perception of water as a renewable resource is increasingly challenged (Ingrao et al. 2023). In addition, the available water resources are becoming contaminated due to both natural and human activities (Lin et al. 2022). Natural processes contributing to this are weathering and leaching from minerals from subsurface rocks, greater quantities of trace metals in groundwater, high salinity and sodium-related issues, and others. Human activities, such as untreated industrial effluents, hazardous chemicals, and wastewater from hospitals and households, further exacerbate pollution (Lin et al. 2022). Surface and groundwater in deltaic regions and coastal landscapes are also facing water quality challenges, not only due to contamination from anthropogenic and natural geological sources but also due to saline intrusion, hydrogeological dynamic factors like floods, storm surges, tidal inundations (Quamar et al. 2025; Tuan et al. 2024; Kannan et al. 2023; Rao et al. 2022; Jothivel et al. 2023). Additionally, the chapter also discusses the recent advanced water assessment approaches being conducted by researchers, giving the main pollution drivers and key pollutants of freshwater resources available in the Indian delta landscapes. Unfortunately, contaminated water, along with poor sanitation, has contributed to a rise in water-related illnesses such as typhoid, diarrhoea, poliomyelitis, hepatitis A, cholera, trachoma, gastrointestinal diseases, dysentery, and schistosomiasis (Lin et al. 2022; WHO 2023). As per the 2018 NITI Aayog report, nearly 2 lakh people in India die each year because of exposure to contaminated water (MoJS 2022a). In India, per capita water availability was over 5300 m^3 annually in 1951 but dropped to 1905 m^3 by 1999 and further to 1486 m^3 in 2021, and is likely to decline further to nearly 1341 m^3 in 2025 and 1140 m^3 or less in 2050 (Nath et al. 2024; Panda 2019; MoJS 2024a). Regions where per capita water availability falls below 1700 m^3 each year are classified as water-stressed, as this threshold reflects the inability to meet basic water needs (MoJS 2024a). India is facing water scarcity, and one of the key factors is the irrigation-focused agricultural sector, as it is the largest groundwater consumer and is responsible for approximately 87% of total groundwater extraction annually (CGWB 2024). Agronomic practices, encompassing improved cropping patterns and water-efficient irrigation methods, are essential for addressing this issue and are discussed in this chapter. Water resource management is primarily observed in rural areas, in the smallest effective social units of the communities (Singh 2004). Integrated water management (IWM) promises sufficient water supply during dry seasons or droughts, promotes resilience, local solutions, sustainability, less dependence on conventional water sources, and reduces energy consumption following the path to Sustainable Development Goal (SDG) 6 by balancing social, economic, and environmental water needs (Ssekyanzi et al. 2024). The components of IWM are water quality monitoring, wastewater treatment units, water supply systems with smart sensors, and participatory water management (MoRD 2011). Other advancements in water monitoring techniques include Internet of Things (IoT) enabled sensors, Artificial Intelligence (AI), Machine Learning (ML) algorithms, and Geographic Information Systems (GIS) often as integrated approaches, for achieving streamlined water monitoring, especially in remote locations (Abdulwahid 2020; Mishra et al. 2024; Sharma et al. 2024). In India, the critical challenges for implementation of the rural water management framework, particularly focusing on the groundwater depletion issues, are also discussed in the chapter. Focus is more on groundwater development than on management and proper utilization, resulting in a growing groundwater crisis, particularly in northwest India, with Punjab facing many cases of over-exploitation due to its intensive agriculture practices (Srivastava et al. 2015). The chapter also discusses solutions for water resource management, offering a comprehensive overview of historical and modern approaches and innovations. The chapter also covers the community-based initiatives, highlighting how local involvement and collaboration can be pivotal in sustainable water management. Through the field applications and case studies, the chapter provides a holistic understanding of solutions to be implemented for the long-term availability of groundwater, especially in rural areas.

The chapter further captures the recent state-wise performance in groundwater management, including recharge trends, extraction, along with contaminant profiles, successful prevention and mitigation steps implemented in affected rural regions, and the government initiatives for rural water resource management. The chapter extends to India's major deltaic, riverine, and coastal regions, demonstrating the recent research work in the field of integrated water assessment strategies, applied in Indian delta systems. These regions are of immense ecological and socioeconomic importance (Foufoula-Georgiou 2013; Bhadra et al. 2020). Overall, the chapter discusses the water quality status, contaminants, their associated health risks, challenges, and recent management strategies in addition to government initiatives, and sustainable pathways for safeguarding India's water resources for future generations.

12.2 Water Quality Status in Rural Regions of India

Water resources play a crucial role in a nation's overall development as it produces impacts on various sectors such as agriculture, energy production, industrial activities (textiles, tanneries, paper, and pulp production), and domestic water supply, thus playing a vital role in a nation's overall development (Lin et al. 2022). In India, only 47% of the total population consumes safe drinking water. If we talk in context to the drinking water sources of rural India, almost 44% village people rely on (handpump) followed by 18% and 13% people relying on tap water (treated source and untreated source), and less than 10% people rely on open wells (12%), closed wells (2%), tube-wells (8%), local water bodies, and other sources (Naika 2022). Sustainable water resource managment is crucial to ensure long-term financial growth, ecosystem balance, and social well-being. In this context, the water availability and distribution is critical for effective planning and utilization. Table 12.1 presents a detailed overview of India's available water resources.

Currently, the total available water is about 1139 billion cubic meters (BCM). However, by 2050, the projected water demand is expected to rise to 1180 BCM, which will exceed the available water resources (MoJS 2025a). Since, the annual average per capita water demand is 1486 m^3, less than 1700 m^3, which is the threshold for water-stressed condition, thus, India is stated as a water-stressed country (MoJS 2025a; Kulinkina et al. 2016). This highlights the problem of enormous pressure on water resources driven by increasing population density, global warming, inadequate conservation, huge water wastage, and the impact of climate change (IDSA 2010). Despite receiving nearly 4000 BCM of annual

Table 12.1 Water resources of India. *Data source* (MoJS 2025a, 2025b)

Availability of water resources data	Quantity (BCM/year)
Annual precipitation	3880
Total water available for utilization	1139
Available surface water	690
Available groundwater	449
Annual extractable groundwater	406.19
Total water resource's potential utilization	691
Annual surface water utilization	450
Annual groundwater extraction	245.64
Annual groundwater recharge	446.90
Increase in annual groundwater recharge (compared to 2017)	15
Decrease in groundwater extraction (compared to 2017)	3
Annual average per capita water availability (2021)	1486 m^3
Recent stage of groundwater extraction	60.47%

precipitation, many rural areas, even high-rainfall regions, face severe water stress during non-monsoon seasons (Nath et al. 2024). As a result, more than half of India, nearly 54% of India experiences high to extreme water stress, with economically weaker rural regions being the most affected (Dutta et al. 2018; Chaturvedi 1985). There have been decreasing trends in water levels in recent years. In India, recent studies indicate a higher percentage of wells, which shows a declining trend in groundwater levels (CGWB 2024, 2023a). The percentage of wells with falling water levels has increased, while those with rising levels have decreased over the years. In November 2021, 54.8% of wells that showed a rise in water levels decreased to 42.81% in November 2022 and further decreased to 38.01% in late 2023 (CGWB 2024; MoJS 2019; Dangar et al. 2021). However, during the same period, that is, in November 2021, 44.1% of wells that showed a decline in water levels increased to 55.67% in November 2022 and rose to 61.12% in late 2023 (CGWB 2024, 2023a, 2022).

According to the 2018 NITI Aayog report, 70% of India's water resources are contaminated, placing India at 120th rank out of 122 countries in the context of the Water Quality Index (WQI) (MoJS 2019). The groundwater resources available for extraction are 406.19 BCM, of which 241 BCM extraction potential is utilized annually (MoJS 2025a). For a deeper analysis of the groundwater scenario across India, mapping and categorizing of groundwater extraction units is a crucial step. The Central Groundwater Board (CGWB) takes this responsibility under the initiative named National Aquifer Mapping and Management Program (NAQUIM). A total of 6746 groundwater source units were identified (CGWB 2024), mapped, and categorized in the country. The categorization depends on the severity of groundwater extraction and recharge levels. It includes the safe category (the extraction stage is $\leq$ 70% of available extractable groundwater), semi-critical (the extraction stage is 70 to 90%), critical (the extraction stage is 90% to 100%), and overexploited (extraction levels exceed annual recharge levels, which is > 100%). Approximately 73.39% (4951) are safe, and 11.13% (751) of groundwater extraction units are overexploited. The overexploited category needs prime attention, followed by 3.05% (206) critical, 10.54% (711) semi-critical, and 1.88% (127) saline all over India (CGWB 2024).

As per the latest data, a total of 751 groundwater extraction units were observed in an overexploited category. Greater than 50 assessed groundwater sources are under the overexploited category for five states: Rajasthan, Uttar Pradesh, Punjab, Tamil Nadu, and Haryana. The state with the most overexploited units is Rajasthan, with 214 units, followed by Punjab in the second and Tamil Nadu in the third, with 115 and 106, respectively. Dadra Nagar Haveli and Daman, and Diu (DNH and DD) have all accessed groundwater sources under the overexploited category (CGWB 2024).

In a populous country like India, rising groundwater demand for irrigation is the key driver of groundwater level changes. The other key factors are climate change-induced rainfall variations (increased precipitation in the north and drought conditions in the south) and free electricity schemes. Particularly in northwestern India like Punjab and

Haryana, farmers are pushed towards unsustainable water-intensive irrigation practices due to other factors, including subsidized electricity schemes, leading to a shift from canal-fed irrigation to private groundwater extraction units. This signifies the groundwater's role in agricultural enhancement (Omarova et al. 2019; Singh and Goyal 2025). Sadly, this transition points to a decline in resources and adverse environmental impacts, with 839 out of 5723 blocks facing overexploitation, mainly in Punjab and Haryana, where groundwater extraction rate is more than 100% (MoJS 2019). In addition to groundwater, other freshwater sources include surface water and rainwater whose sustainable usage and management are crucial for rural and remote areas.

12.3 Socioeconomic Impacts and Health Consequences of Water Contamination in Rural Communities

Being the epicentre of human activities, water should be free from contaminants to avoid disease outbreaks. Basic facilities such as potable water access vary largely within the wide margin of socioeconomic inequality within rich and poor. In many developing countries, contaminated water usage has been reported to spread enteric diseases and high mortality, with a larger population mass from rural regions (Adelodun et al. 2021).

12.3.1 Socioeconomic Aspects and Health-Related Expenditures in the Context of Water-Borne Diseases

One of the significant health and socioeconomic challenges in India is groundwater contamination. Globally, 844 million people lacks adequate access to safe drinking water, and 79% of them are rural residents (Omarova et al. 2019; Singh and Goyal 2025). In India, waterborne diseases create a significant annual economic burden of $600 million due to the yearly loss of approximately 73 million working days (Khurana et al. 2008). Arsenic (As) contamination in groundwater is still a serious concern across the Ganga–Brahmaputra–Meghna floodplains and shallow aquifers of the Bengal Delta (Ivy 2025; Kumar and Nayak 2025; Marghade et al. 2023; Sarkar et al. 2022). Nearly 63% of residents adopt a preventive approach (alternative water sources), while only 23% population adopted remediation strategies in As-affected Bihar, highlighting increased awareness but limited access to affordable treatment technologies (Kumar and Nayak 2025). Average monthly losses higher than ₹200 ($\approx$2.25 USD) are reported for these affected individuals, each in medical bills and wage reductions, due to As-related ailments only (Kumar and Nayak 2025). The menace of As and fluoride (F) in rural areas causes the suffering of affected rural populations from physical and psychological trauma due to social ostracism created by society (Chakraborti et al. 2016). They are abandoned by society, face marital breakdowns, lose jobs, and are forced to live a sub-standard lifestyle. There is a high number of As-related cases, with people in such rural areas being stuck in a vicious cycle of poverty and diseases (Chakraborti et al. 2016). In India, hospitals are responsible for 60 – 80% of healthcare fund expenses, with a smaller amount of contribution towards basic medical services (Dar 2023). Primarily, the focus is on cities and curative treatments, focusing less on preventive measures, benefiting urban areas over vulnerable rural communities (Khurana et al. 2008; Dar 2023). Between the years of 2004 and 2005, 29.2% of government-funded health expenses went to allopathic medical services in urban areas. However, the share was only 11.8% for rural clinical services (Balarajan et al. 2011). Rural populations face a significant financial burden due to deteriorating water quality, which accounts for nearly 5.28% of their average annual income (Khurana et al. 2008). Based on WHO estimates, over 1 billion people risk falling into poverty when 10% or more of their household budget is spent on health-related expenses (WHO 2025). An overall societal approach is required to overcome health risks and related expenditures. Primary Health Care (PHC) is one such approach that ensures the highest achievable level of mental and physical wellness, strengthens the accountability of healthcare systems, enhances equity and accessibility to disease prevention and cure measures, centres for rehabilitation, health services, and palliative care located close to people's daily surroundings (WHO 2025). An overview of states affected by significant groundwater contaminants is displayed in Fig. 12.1 and Table 12.2. These contaminants are pharmaceuticals, like analgesics and antibiotics; geogenic pollutants like iron (Fe), F, As, and uranium (U); heavy metals and pesticides containing persistent organic pollutants (POPs) like atrazine, malathion, lindane, endosulfan, chlorpyrifos, and alachlor.

12.3.2 Socioeconomic, Environmental, and Health Dimensions of Water in Rural Deltaic and Coastal Regions

Deltas are among the most productive and densely populated global ecosystems, being a home to nearly half a billion people through agriculture, and fisheries serving as national food baskets (Foufoula-Georgiou 2013). Deltaic regions are a hotspot of rich biodiversity and encompass complex networks of rivers, wetlands, estuaries, and coastal barrier landforms (Foufoula-Georgiou 2013). In rural deltaic and coastal areas, groundwater is a crucial source of drinking and for agriculture; however, its quality is deteriorated by saline water intrusion, which undermines its suitability for agriculture, aquaculture, and domestic use (Tuan et al. 2024; Bhadra et al. 2020). Alongside groundwater, rainwater, and surface water are other freshwater resources in these areas (Hoque et al. 2019). Yet, both surface water bodies and groundwater aquifers are increasingly at risk from salinity and contamination, primarily resulting from storm surge–induced inundation events (Feist et al. 2023).

The Ganga–Brahmaputra delta currently faces rising pollution risks from geogenic contaminants and microplastics, mainly due to sediment deposition and increasing

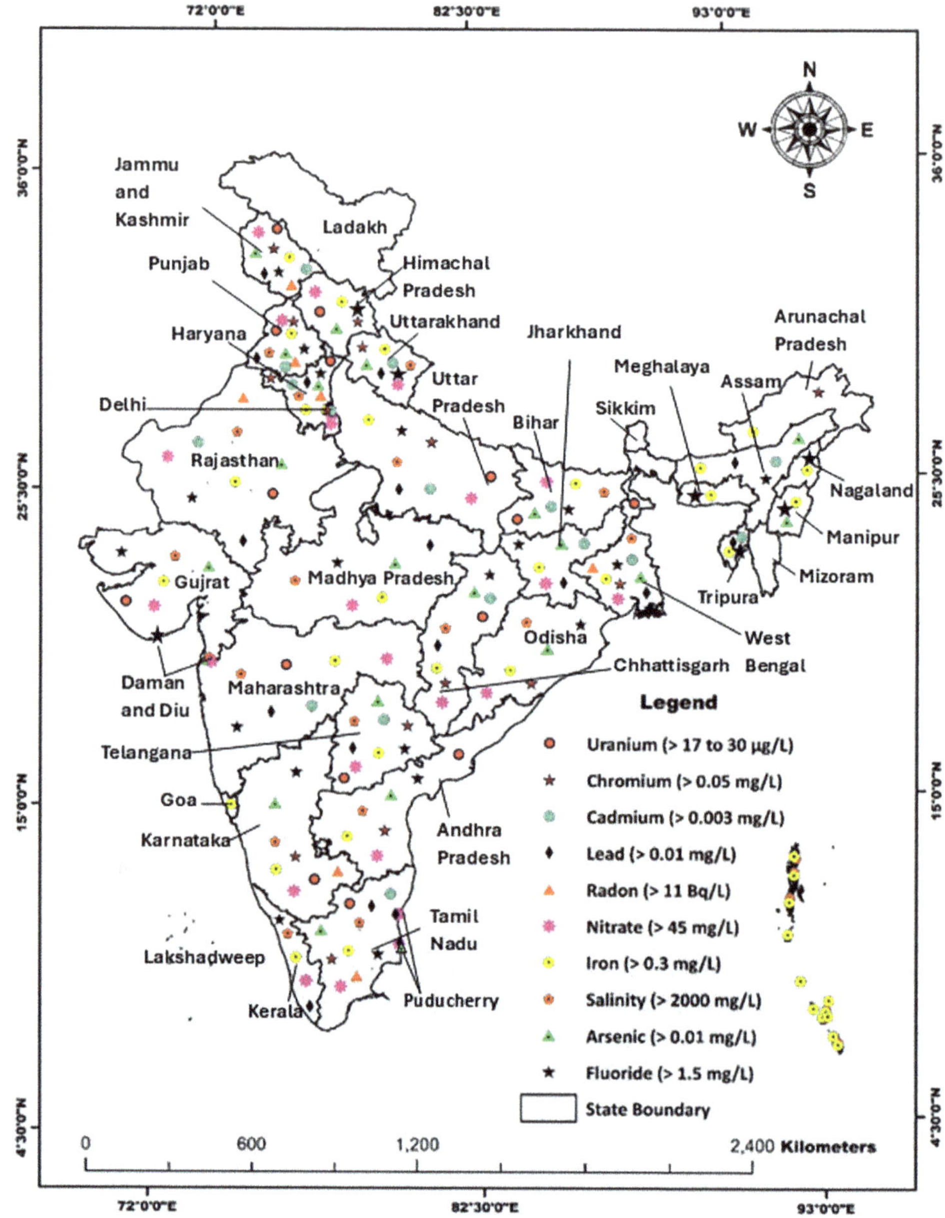

Fig. 12.1 Occurrence of major groundwater contamination in different parts of India. *Data source* (NGT 2024; Brindha et al. 2020; Kanmani and Gandhimathi 2013; Malyan et al. 2019; Bhutiani et al. 2016; Goswami et al. 2023; Kumar et al. 2018, 2019; Ratnalu and Dhakate 2021; Sujathan and Singh 2023; Singh and Sharma 2020)

Table 12.2 States with different contaminants in groundwater

Water contaminants	Permissible limits	Regulatory Standards	Concentrations in affected areas	Affected areas (states and UTs)	Sources
Fluoride	1500 µg/L	BIS	> 1500 µg/L	27 States/UTs, 469 districts	(BIS 2012; NGT 2024)
Arsenic	10 µg/L	BIS	> 10 µg/L	25 states/UTs, 230 districts	
Salinity	2000 (EC in micromhos/ cm	BIS	> 3000 micromhos/ cm	20 states/UTs, 275 districts	(BIS 2012; MoJS 2022b)
Nitrate	45,000 µg/L	BIS	> 45,000 µg/L	23 states/UTs, 507 districts	(BIS 2012; MoJS 2022b)
Iron	300 µg/L	BIS	> 1000 µg/L	516 districts in 29 states/ UTs	
Malathion	290 µg/L	BIS, (GCDWQ)	150–7215 µg/L	Rural (G.W. of West Bengal)	(BIS 2012; Duttagupta et al. 2020)
Atrazine	2 µg/L	BIS	2–2.9 µg/L	(G.W. of West Bengal)	(BIS 2012; Duttagupta et al. 2020)
Alachlor	20 µg/L	BIS	0.56–3.23 µg/L	(G.W. of West Bengal)	(BIS 2012; Duttagupta et al. 2020)
Lindane	2 µg/L	BIS	0.21–2.96 µg/L	(G.W. of West Bengal)	(BIS 2012; Duttagupta et al. 2020)
Insecticides	–	–	0.0014–0.01 µg/L	(G.W. of Patna)	(Richards et al. 2021a)
Carbamazepine (Anticonvulsant)	–	–	0.067 µg/L		
Artificial sweeteners	–	–	0.003–1.2 µg/L		
Fluorinated surfactants	–	–	0.0008–0.35 µg/L		

(continued)

Table 12.2 (continued)

Water contaminants	Permissible limits	Regulatory Standards	Concentrations in affected areas	Affected areas (states and UTs)	Sources
Antibiotics	_	_	0.034–0.12 µg/L	(G.W. of Uttar Pradesh)	(Lapworth et al. 2018)
Phenoxyacetic acid	_	_	0.02–0.21 µg/L		
Ofloxacin	–	–	0.48 µg/L	Wells, Hyderabad	(Fick et al. 2009)
Microplastics	_	_	3–80 items/L	(G.W. of Kodungaiyur, Perungudi) Chennai	(Kannaiyan and Sekhar 2023)
Analgesics/ NSAIDS	–	–	0.00192–0.0234 µg/L	(G.W. of Patna)	(Sharma et al. 2019)
Cadmium	3 µg/L	BIS	> 3 µg/L	G.W. of more than14 states/UTs	(BIS 2012; Brindha et al. 2020; Kanmani and Gandhimathi 2013; Kerketta et al. 2013; Malyan et al. 2019; MoJS 2021a; Mahajan et al. 2022)
Lead	10.µg/L	BIS	> 10 µg/L	G.W. of 18 states/UTs	(BIS 2012; Brindha et al. 2020; Malyan et al. 2019; CGWB 2025; Gaur et al. 2014; Sabarathinam et al. 2011)
Chromium	50.µg/L	BIS	> 50 µg/L	G.W. of 15 states/UTs	(BIS 2012; CGWB 2025; Bhutiani et al. 2016; Goswami et al. 2023; Kumar et al. 2018; 2020; Ratnalu and Dhakate 2021)
Radon	11 Bq/L	_	> 11 Bq/L	G.W. of 7states/UTs	(Malyan et al. 2019; Manawi et al. 2024)

(continued)

Table 12.2 (continued)

Water contaminants	Permissible limits	Regulatory Standards	Concentrations in affected areas	Affected areas (states and UTs)	Sources
Uranium	30 µg/L	WHO	> 30 µg/L	12 states and two UTs	(Sujathan and Singh 2023; WHO 2022)
Pesticides	0.1 µg/L	European Union (EU) drinking water limits	Up to 1.4 µg/L	In-situ pumps, boreholes, Bengaluru	(Brauns et al. 2024; EU 2020)

Note G.W. = Groundwater

groundwater vulnerability (Pandey et al. 2021). Increased salinity and contaminants loading of F, As, and heavy metals have been widely detected in Indian coastal aquifers (CGWB 2023b). Organic and inorganic pollutants, sourced from natural processes and anthropogenic dumping (like municipal solid waste), enter deltaic river systems, with recent evidence highlighting microplastics carrying toxic metals such as cadmium (Cd), mercury (Hg), chromium (Cr), and lead (Pb) across northeastern Indian coasts (Anandavelu et al. 2024). In the southeastern coast, microplastics (MPs) are widely detected, responsible for the disruption of marine food chains and ecosystems, extending from shorelines to deeper oceanic layers (Kannan et al. 2023). Consumption of water with high sodium concentrations contributes to increased infant mortality, hypertension, kidney and heart illness (Feist et al. 2023). In the Bengal Delta, dependence on saline groundwater has been associated with a higher incidence of sodium-related diseases among rural populations (Rahman et al. 2019). A recent study on the Hooghly River reported a temporal increase in potentially toxic elements (PTEs) such as Cd, cobalt (Co), and zinc (Zn), with children facing higher non-carcinogenic risks than adults. PTEs, encompassing both metals and non-metals, pose carcinogenic and non-carcinogenic threats even at concentrations below permissible limits (Ghosh et al. 2021).

Saline intrusion and frequent flooding are reported to cause a reduction in agricultural and aquaculture yields, lowered household incomes, and migration toward urban centres (Hoque et al. 2019; Feist et al. 2023). Ineffective and costly water supply interventions, such as hand pumps, yield poor-quality saline water, wasting financial resources and limiting potable water access (Hoque et al. 2019; Rivett et al. 2019). Gender inequalities are another consequence of water scarcity and poor water quality, where women are more involved in water collection. Other challenges observed include rising irrigation costs and heightened food insecurity, ultimately deepening economic vulnerability among rural communities. Salinization-driven migration and food insecurity have been reported in the Indus, Ganga–Brahmaputra–Meghna, including other global deltas such as the Mekong, Niger, Orinoco, Rhine, and Sacramento–San Joaquin (Rahman et al. 2019).

These interlinked aspects require effective water management and resilient livelihood strategies in the deltaic and coastal plain for long-term sustainability. Each delta faces distinct hydrogeological pressures and contamination patterns, emphasizing the importance of identifying dominant pollutants and their source in the target locations (Table 12.3) (Scown et al. 2023).

12.3.3 Rural Health Deterioration: Groundwater Contaminants and Waterborne Diseases

A large part of India, which is 37.7 million, suffers from water-related diseases annually, resulting in nearly 1.5 million child deaths due to diarrhoea alone (Dar 2023). Gallbladder

Table 12.3 An overview of key contamination patterns, their pollution drivers, impacts, and health risks of major Indian deltas

Major Indian delta regions	Groundwater-dominant contaminants	Dominant surface water contaminants	Causes	Socioeconomic impacts	Health issues	Sources
Ganges–Brahmaputra Delta (Sundarbans)	Strontium (Sr), calcium (Ca), barium (Ba), high salinity	Large MPs (white, transparent), organic/inorganic pollutants, agrochemicals, As, heavy metals, COVID- virus, high salinity, microbial load (Fecal coliform, E. coli, Enterococcus)	Industrial, fishing, tourism, oil spills, agriculture, poor sanitation; groundwater–marine water mixing	Ecological shifts, water scarcity, loss of sanitation access, dissolved oxygen (DO) reduction by MPs, weed overgrowth, loss in crop yield	COVID -19, gastrointestinal illness, high risk of waterborne diseases for farmers	(Chakrabarti et al. 2018; Hassan and Westerhoff 2024; Chakraborty et al. 2022; Singh et al. 2024; Ghosh and Mistri 2021) (Chakrabarti et al. 2018; Hassan and Westerhoff 2024; Chakraborty et al. 2022; Singh et al. 2024; Ghosh and Mistri 2021)
Krishna delta	Magnesium (Mg), sodium (Na) ions, high salinity, fluoride, deep aquifer nitrate	Hexachlorocyclohexane (HCH), Dichlorodiphenyltrichloroethane (DDT), Endosulfan	Fluoride minerals, pesticide-intensive agriculture, nitrate leaching, monsoon runoff contamination	Water unsuitability for drinking/ irrigation, crop yield loss, soil cementation, alkalinity rise from Mg water	Fluorosis, diabetes, methemoglobin-emia, GI cancer, thyroid hypertrophy	(Gandla et al. 2023; Sathe et al. 2025; Vaiphei and Kurakalva 2021)

(continued)

Table 12.3 (continued)

Major Indian delta regions	Groundwater-dominant contaminants	Dominant surface water contaminants	Causes	Socioeconomic impacts	Health issues	Sources
Godavari delta	High salinity, Mg hazard, organochlorines	MPs, Fe, Zn (abundant), Ni, Cu (excess)	Domestic sewage, industrial and agricultural discharges, plastic dumping, GW overuse, waterlogging	Fish contamination (91% with MPs), crop damage, saline GW unfit for irrigation	Neurological issues, seizures, unconsciousness, cholinergic effects	(Rao et al. 2022; Choudhary et al. 2025; Hussain et al. 2017; Sharma et al. 2023) Rao et al. 2022; Choudhary et al. 2025; Hussain et al. 2017; Sharma et al. 2023)
Mahanadi delta	Chloride, carbonate, nitrate, phosphate	MPs, PAHs, silicates, mild alkaline	Sea–aquifer mixing, fertilizers, waste discharge, industrial and vehicular pollution	Contaminated food chains, eutrophication, low yields, aquatic species loss, financial strain	Cancer risk, algal toxin effects, poor sanitation diseases, MPs in blood	(Acharya et al. 2022; Das 2025a; Das 2025b; Patidar et al. 2023)
Cauvery (Kaveri) Delta	Cd, Pb, phosphate, sulfate, nitrate, PFAS, salinity	PFAS, emerging organic contaminants, antimicrobials, plastics, heavy metals (Fe, Cd, Pb, phosphate, chloride, sulfate, As, selenium (Se), colored MPs	Rock weathering, seawater intrusion, over-pumping, industrial effluents poor sanitation, fishing	Ecosystem and food chain toxicity, social inequality, poor water access, 40% water species loss	Cancer, organ failure, skin/neurological disorders, bone softening, arthritis	(Brauns et al. 2024; Singh et al. 2024; Earth5R 2025; Keerthan et al. 2023; Myvizhi and Devi 2020)

Note Sr = Strontium; Ca = Calcium; Ba = Barium; DO = Dissolved oxygen; Mg = Magnesium; Na = Sodium; HCH = Hexachlorocyclohexane; DDT = Dichlorodiphenyl-trichloroethane; Se = Selenium; PAHs = Polyaromatic hydrocarbons cancers have increased manifold in Bihar, and a recent study has reported a strong positive correlation between gallbladder cancer patients and "As" concentrations reported in their gallbladder tissue, blood, bile, and hair samples (Kumar et al. 2023). According to recent drinking water analyses, the incidence of arsenicosis in Bihar is alarmingly high—approximately 72 cases per 1,000 people—with males being more affected than females, at nearly 83 cases per 1,000 individuals (Kumar and Nayak 2025). A study conducted in 21 villages of southwest Punjab found groundwater contaminated with arsenic, lead, nitrate, fluoride, and uranium, all exceeding WHO/BIS drinking water limits. Higher cancer rates have been observed among the elderly, greater than 45 years, along with other health issues, who were under prolonged exposure to carcinogenic water contaminants like arsenic and uranium, contributing to a mortality rate of 235 per lakh due to cancer in the last 5 years (Kaur et al. 2021). Major sources of groundwater contamination, various contaminants, and their exposure to humans are represented in Fig. 12.2.

Groundwater contamination in rural India is a pressing concern, primarily influenced by geogenic and anthropogenic factors. The main sources of contaminants, such as arsenic, fluoride, iron, radon, uranium, strontium, and salinity, are naturally occurring geochemical processes involving the dissolution of minerals from rocks (CGWB 2014). Human activities, such as excessive use of synthetic fertilizers and pesticides, solid waste dumping in unlined landfills, and improper waste management, exacerbate the issue. Leachate forms when rainwater seeps through wastes, dissolving soluble solids

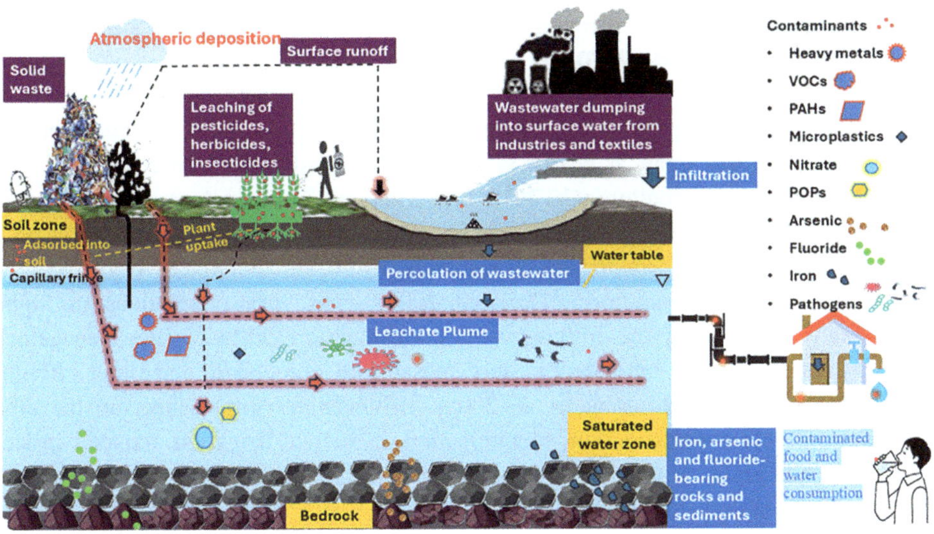

Fig. 12.2 Schematic diagram depicting sources of water pollution from large-scale anthropogenic activities and pathways of emerging groundwater contaminants

and contaminants, infiltrating and percolating through soil layers, resulting in groundwater pollution. Agricultural runoff adds heavy metals, persistent organic pollutants (POPs), and nutrients like nitrate and phosphorus to waters (Xia et al. 2020). Globally, fluoride contamination impacts around 240 million people, while arsenic severely affects nearly 220 million individuals (Biswas et al. 2023), out of which more than 50 million people suffer from arsenic poisoning, primarily in the Ganga and Brahmaputra basins in India (Marghade et al. 2023), often leading to social discrimination. In 2018, the global use of 4.12 million pesticides was reported to seep into surface and groundwater through runoff, infiltration, and geochemical pathways (FAO 2021). Pesticides like malathion, atrazine, DDTs, and endosulfan come into the category of Persistent Organic Pollutants (POPs) (Kyriakopoulos and Sebos 2024) and degrade the quality of surface waters, making it a more hostile environment for aquatic species. *Leptospira interrogans*, a pathogen found in surface water contaminated with pathogen-laden animal tissues and urine, causes waterborne leptospirosis. It enters the human body through cuts, wounds, or mucous membranes such as the mouth, nose, or eyes. This pathogen is significant in drinking water safety as it cannot be detected using thermotolerant coliform (E. coli) tests (WHO 2022). Contaminated water and its long exposure result in severe chronic health issues, as shown in Fig. 12.3. These interconnected problems highlight the significance of addressing the water contamination and its far-reaching socioeconomic and health impacts on the rural population.

12.4 Significant Challenges and Decentralized Approaches to Water Management

To maintain agricultural productivity (Nath et al. 2024), sustainable water management is the key factor in rural areas, as nearly 87% of the extracted groundwater (nearly 213.29 BCM) is required for agriculture (CGWB 2024). This remains a challenge considering the spatial distribution of Indian villages, including transportation and communication problem\s, failing to receive drinking water resources from remote villages from receiving consistent water quality monitoring (Srikanth 2009). Thus, rural populations are highly susceptible to suffering by the drinking water contamination. Surface water management that faces challenges like untreated waste disposal, idol immersion, and ritual offerings in open water systems (Khurana et al. 2008). Groundwater governance, on the other hand, must address over-extraction, contamination, inefficient irrigation, reduced agricultural yields, and equity issues. Critical issues and their implications in sustainable water management is tabulated in Table 12.4.

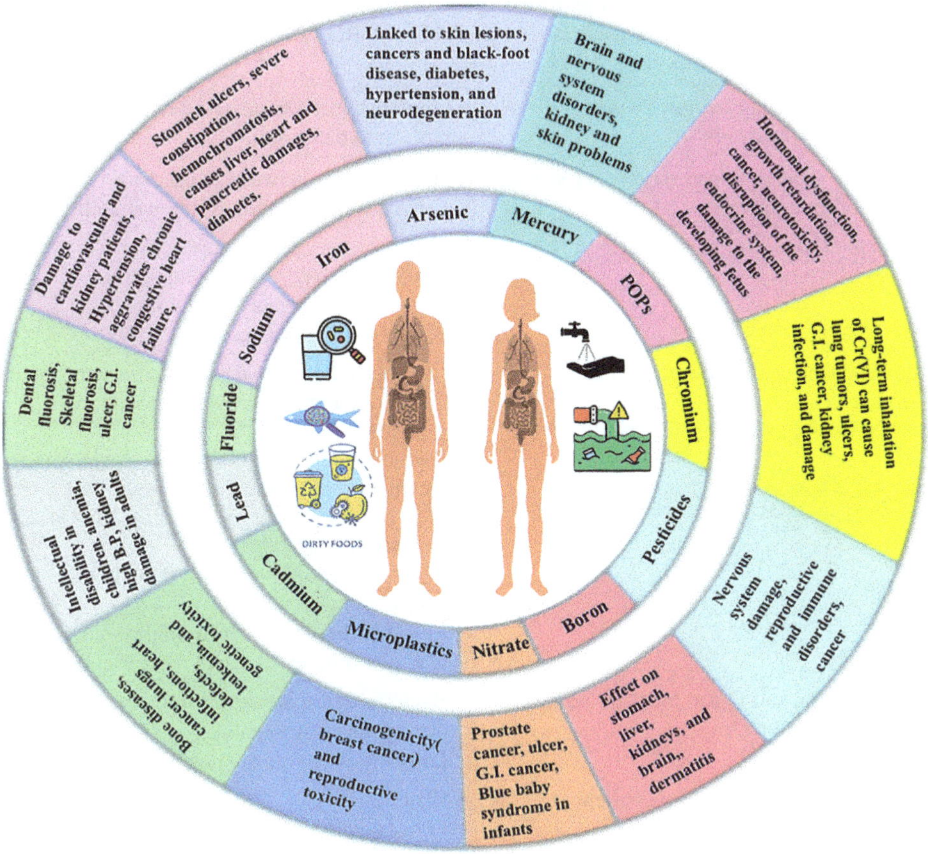

Fig. 12.3 Schematic diagram reflecting long-term health effects on the human body because of prolonged consumption of pollutant-laden water

12.4.1 Water Resources Related Challenges and Their Management Strategies in Deltaic and Coastal Regions

If we talk about deltaic regions, sustainable water management and regulated monitoring are crucial to overcome water storage challenges, such as reducing risks of geomorphic hazards, enhancing the water resource management framework, climate change impacts, and human disturbances in deltas (Quamar et al. 2025). Sustainable groundwater management is of high priority for socioeconomic growth and long-term sustainability of a deltaic region (Tuan et al. 2024). For instance, groundwater and resource exploitation can exacerbate land subsidence, thereby increasing the vulnerability of deltaic habitats to sea-level rise, storm surges, and wave-induced erosion, with floods and inundation usually occurring due to extreme weather events (Elneel et al. 2024). In addition, rural

Table 12.4 Significant challenges and their consequences in the context of water management. Data source (Nath et al. 2024; Ssekyanzi et al. 2024; MoRD 2011; Srikanth 2009; Chopra and Ramachandran 2021; Mehta et al. 2024)

Challenges affecting the rural water management framework	Cumulative impacts observed due to the significant challenges
Increasing population and rural water source sustainability issues	• Increased groundwater dependence • Decrease in per capita water availability
Overexploitation of groundwater	• Groundwater table decline • Entry of minerals into the aquifer • Geogenic pollutants contamination • Saline water intrusion
Climate changes and destructive natural phenomena	• Unexpected monsoon rainfall affects rain-fed agroecosystems, decreasing agricultural produce • Decline in rainfall patterns and rise in extreme weather events • Rise in sea levels and salinization of coastal groundwater systems • Annual declining trends in rainfall in different states • External shocks (floods, drought) • Global warming increases mass evaporation
Poor operation and O and M, poor sanitation, and hygiene challenges	• Bacteriological contamination, especially during the rainy season • Drinking water contamination
Institutional and policy-level challenges	• Water crisis • Water conflict cases and inter-state disputes • Weak enforcement of regulatory guidelines • Inefficient stakeholder participation, particularly in marginalized communities • Lack of multi and interdisciplinary strategies in water governance
Technical and infrastructure challenges	• Poor operation and maintenance • Inconsistent availability of reagents.in remote villages hinders periodic water quality monitoring • Absence of a single state-level agency serving as a water quality data repository for improved coordination

(continued)

Table 12.4 (continued)

Challenges affecting the rural water management framework	Cumulative impacts observed due to the significant challenges
Water scarcity and pollution	• Poor water quality issues • Water insecurity due to inadequate access to potable water, especially during scarcity and drought • Disputes over water sharing between states and within water-dependent sectors
Regional variation in water quality and quantity	• Failure of centralized approaches, need for decentralization • Less water availability in high population density • Need for rural community participation

deltaic regions often face maintenance issues in community-managed systems, emphasizing the need for professional and sustained post-construction support to ensure reliable water supply services (Hoque et al. 2019). The key challenges in deltaic regions, along with their associated impacts and corresponding management strategies, are summarized in Table 12.5.

The delta ecosystems are unable to absorb excess nutrient loads, and they end up reaching coastal waters, leading to eutrophication and water quality degradation. Moreover, the salinity dynamics are one of the regulators of the coastal ecosystems, which increases in groundwater aquifers with increased groundwater abstraction, past transgression, mainly in dry seasons and high sea levels (Tuan et al. 2024). In the delta region, agriculture is one of the dominant utilities of groundwater, but high salinity concentrations limit the type of crop (salinity higher than 0.4% is unsuitable for crops, if they are not salt-tolerant) (Tuan et al. 2024).

In deltaic landscapes, rural regions serve as active zones of decentralized groundwater extraction, presenting challenges for large-scale monitoring in the absence of centralized assessment systems (Tuan et al. 2024). Regional variations in water quality need adaptive management approaches and localized solutions. Local resources, economic and cultural aspects, play a vital role in shaping effective water management strategies. Several concerns with water contamination and management across various Indian states are summarized as follows:

12.4.2 West Bengal and Bihar Arsenic Crisis

In groundwater, high arsenic levels are a major threat in West Bengal and Bihar (Srikanth 2009). In Bihar, 18 of 38 districts and 19 of 23 districts in West Bengal exceed the

Table 12.5 Challenges, impacts, and management strategies in deltaic regions

Deltaic region challenges	Impacts observed	Management strategies in deltaic regions	References
Geomorphic hazards and hydrodynamic variations (siltation, channel shifts, bank erosion, sediment redistribution patterns, and variable surface water flows)	Affects/disrupts water availability; increased risk of floods (due to reduced river flow)	Monitoring land use changes, river water flow dynamics, and erosion mitigation techniques. Riverbank stabilization methods	(Quamar et al. 2025)
Overexploitation of groundwater; rising dependence from inland to coastal areas; intensive agricultural use	Flood risks, groundwater depletion; high risk of salinity intrusion in aquifers of coastal areas; land subsidence	Management of groundwater use; agricultural development; treatment of saline sea waters and groundwaters; promoting alternative sources	(Tuan et al. 2024)
Imbalance in regional development and water sectoral water utilities	More groundwater vulnerability in coastal areas; costly affair for water services for financially challenged communities	Avoidance of non-essential groundwater use; promotion of RWH to augment existing water sources; ensure timely maintenance services	(Tuan et al. 2024)
Ineffective community-based water management (demand driven); common resource issues	Non-functional water points; groundwater depletion; increased vulnerability	Regulate withdrawal; enhance recharge; strengthen maintenance service; post-construction support from local authorities and NGOs	(Tuan et al. 2024; Ghosh et al. 2021)
Infrastructure, technological, and financial constraints; unplanned investments; lack of regulation and accountability in the rural water sector	Improper monitoring and data collection; inequitable water security among people	Strengthen capacity training activities; stakeholder participation; encourage NGO/private involvement	(Tuan et al. 2024; Hoque et al. 2019)

(continued)

Table 12.5 (continued)

Deltaic region challenges	Impacts observed	Management strategies in deltaic regions	References
Unregulated Fisheries and tourism, poor municipal garbage collection, illegal dumbing	Rise of plastics pollution in surface waters of coastal and deltaic areas	Regular beach cleaning activities like beach garbage collection	(Kannan et al. 2023)
Sea level rise, cyclones, storm surges, and extreme precipitation; global climate change	Floods and inundations; livelihood loss; agricultural damage; severe cyclone impacts (on kutcha houses); high risks to coastal and low-lying areas; embankment (dykes) breaching	Flood and cyclone hazard mapping; projection of future SLR impacts; evaluation of mitigation measures; salinity safety plan; tidal river management	(Feist et al. 2023; Elneel et al. 2024; Barui et al. 2024) (Feist et al. 2023; Elneel et al. 2024; Barui et al. 2024)

WHO drinking water standard of 10 μg/L (ADB 2020; Kumar et al. 2021). Now, this data has further increased for Bihar from 18 to 31 districts affected by arsenic, iron, and fluoride (Ivy 2025). West Bengal is severely affected, with certain regions exhibiting concentrations as high as 4622 μg/L (Marghade et al. 2023). In Bihar, studies have documented cases of arsenic-induced cancers such as GBC and skin diseases among rural populations (Kumar et al. 2023). A recent drinking water analysis report of Bihar highlighted arsenicosis disease was higher in males than in females, with a monthly economic loss of up to ₹462.64 or 6.21 USD (Kumar and Nayak 2025). Cost-effective techniques tailored for rural areas differ from those in developed regions (Obayomi et al. 2024). CSIR Kolkata has implemented efficient ceramic membrane technology in West Bengal to remove arsenic and iron (NJJM 2023a). Collaborative efforts at international, national, and institutional levels have developed cost-effective technologies maintained through community engagement (NJJM 2023a).

12.4.3 Rajasthan Water Scarcity and Fluoride Mitigation

In Rajasthan, only 12% of the 299 blocks have adequate groundwater levels, while the remaining 88% face serious issues of excessive groundwater withdrawal, reaching overexploited, semi-critical, or critical zones (Sharma 2024). In the Balotra district of Rajasthan, water scarcity is driven by inadequate rainwater harvesting (RWH) and parallel illegal water supply activities. Rajasthan accounts for about 10% of the world's fluoride-affected population, with 4,177 contaminated habitations and 16,560 affected villages, representing nearly 51% of India's 33,211 fluoride-affected villages (MoJS 2025c; Rani 2015).

Although significant progress has been achieved under the Jal Jeevan Mission (JJM), 80 rural households of the state, from the overall country's 250 fluoride-affected rural households, remain in Rajasthan as water-challenged, though they are provided with safe drinking water through community water purification plants (CWPPs) and Individual Household Purifiers (IHP) supply systems (MoJS 2025c). Rajasthan's water scarcity, one of the characteristics of arid regions, is worsened by droughts and low rainfall, which pressurize already limited water resources. This imbalance in water resources due to temporary events affects the ecosystem and poses severe threats to agriculture, economic stability, and livelihoods. For strengthening of drought monitoring and effective mitigation measures in arid zones, the Discrepancy Precipitation Index (DPI) serves an easy metric for drought severity assessment (Swain et al. 2024).

Sustainable solutions are implemented by Indian government in collaborations with UNICEF, NGOs, and private companies. For instance, for fluoride mitigation, these solutions include solar based defluoridation, the Nalgonda technique, and advanced defluoridation technologies developed by the Council of Scientific and Industrial Research (CSIR)-NEERI, Nagpur with The Energy and Resources Institute (TERI), New Delhi (NJJM 2023a).

12.5 Rural Water Quality Management in India: Possibilities and Approaches

India being a densely populated and predominantly rural nation is dependent on monsoon rainfall and faces significant water quality management challenges (Singh and Goyal 2025). Effective water management includes monitoring, identifying pollutants and their sources, and then selecting treatment options based on water quality designated use (CPCB 2008).

12.5.1 Water Quality Assessment Tools: WQI and Pollution Indices

Water quality is assessed through biological, chemical, and physical parameters, which can be harmful if they exceed safe limits. The WQI provides a convenient metric to simplify large multivariate data sets into a single unitless value, serving as a unified water quality indicator. WQIs are useful for measuring the aggregate impact of land use on water quality and are significantly considered in integrated catchment management (Locke 2025). Several indexing systems utilized in India to monitor the quality of the water are given in Table 12.6. The most frequently used water quality metrics for WQI include temperature, dissolved oxygen (DO), pH, ammonia, turbidity, suspended particles, total dissolved solids (TDS), and biochemical oxygen demand (BOD) (Uddin et al. 2021).

Table 12.6 Water quality indices and their types

Several types of WQI	Standard Indices adopted and used in India	References
Surface water quality indices	Canadian Council of Ministers of the Environment (CCME) WQI; Horton Index (HI); Environmental Quality Index (EQI); Smith index; Fuzzy interface system (FIS); Bascaron Index (BI); National Sanitation Foundation (NSF) WQI, Bhargava WQI, Overall index of pollution; Drinking water quality index (DWQI)	(Das 2025b; Uddin et al. 2021; Tirkey et al. 2013; Kumar et al. 2024)
Groundwater quality indices	Irrigation WQI (Sodium absorption ratio, Residual sodium carbonate, Kelly Ratio, Sodium percentage, Magnesium hazard, and Permeability index), Drinking Water Quality Index, Hybrid Fuzzy-GIS-based WQI (FG) WQI,	(Acharya et al. 2018; Jha et al. 2020; Ramesh et al. 2010)
Macroinvertebrate-based biological surface water quality indices	Invertebrate Community Index (ICI); Biological Monitoring Working Party (BMWP) Score; Family Biotic Index (FBI); Macroinvertebrate WQI (MWQI); EPT (Ephemeroptera, Plecoptera, Trichoptera) richness index, Entropy Water Quality Index (EWQI), Bhargava method (May not be applicable for all kinds of surface waters)	(Das 2025a; Bhatt and Pandit 2010; Kripa et al. 2013)
Pollution indices for groundwater as well as surface water	Heavy Metal Pollution Index (HMPI), Nemerow Pollution Index, Pollution load Index, and Metal Contamination Index (MCI), Modified heavy metal pollution index (m – HPI), Synthetic Pollution Index (SPI), Overall Index of Pollution (OIP)	(MoJS 2024a; Das 2025c; Dash et al. 2019; Ephsy and Raja 2023; Swati and Umesh 2015; Das 2025d)

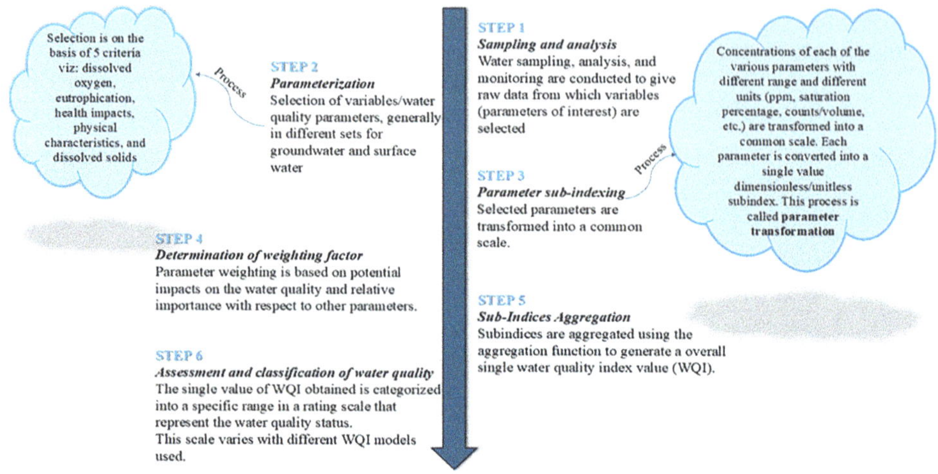

Fig. 12.4 Pictorial representation of a general WQI framework. *Data Source* (Tyagi et al. 2013; Uddin et al. 2021; Tirkey et al. 2013)

The basic framework of the WQI consists of four essential components (Uddin et al. 2021). The first step is to select the parameters that must be analysed. The second step involves parameter transformation, creating sub-indices for each chosen parameter. Then, the weight is assigned to each parameter based on its significance, followed by the aggregation function. The subindices are merged using the aggregation function to calculate the overall water quality index, indicating the water quality of the chosen location according to the rating scale specific to each WQI. The steps involved in the WQI framework are shown in Fig. 12.4.

In Rajasthan's Kot Dam, water quality was assessed by analysing 13 parameters (pH, TDS, hardness, alkalinity, DO, BOD, chloride, sulfate, nitrate, Ca, Mg, electrical conductivity (EC), and fluoride, followed by calculating WQI using the Weighted arithmetic index method (Singodia et al. 2024). The results pointed out that dam water has organic pollutants, pesticides from agricultural runoff, and other toxics because of waste dumping by locals (WQI value was near 85). According to this method, WQI in the range of 76 to 100 indicate severe water pollution (Singodia et al. 2024).

The Nemerow's Pollution Index (NPI) is referred to as the ratio of concentration of a water parameter to its corresponding permissible limit. The NPI value for each water quality indicates contamination contributed by that specific pollutant or water parameter. The NPI value exceeding one indicates that the pollutant/water parameter is present at higher levels than the permissible limits, whereas a value less than or equal to one means compliance with the water regulatory standards. The NPI measure for a region helps to identify the potential pollutants and improve the studied water quality (Swati and Umesh 2015). Recent research studies used NPI for a comprehensive heavy metal pollution level in the study sites along the Halda River, Bangladesh, revealing that all water samples are highly polluted with six heavy metals (Cd, Cr, Fe, Pb, Cu, and As), largely crossing their drinking water limits (Dey et al. 2021).

12.5.1.1 Biological Test Methods

Biomonitoring and bioanalytics are two biological testing techniques essential for evaluating and forecasting water quality (Szczerbińska and Gałczyńska 2015). Bioanalytics is the analysis of water quality to identify contaminants by use of biosensors or biological water quality parameters such as total and faecal coliforms (Kulinkina et al. 2016; Szczerbińska and Gałczyńska 2015). Biomonitoring monitors water conditions by examining biosensors or bioindicators (living organisms or any biologically active material, especially sensitive to specific pollutants). Some of these bioindicator species include bacteria (*E. Coli* as pathogenic indicators in water), larval invertebrates, planktons, algae (cyanobacteria), fungi (Daphnia used for toxicity tests in water), macroinvertebrates (like Coleoptera, Trichoptera, and Diptera utilized for the detection of heavy metals and pesticides in water), and fishes (Kulinkina et al. 2016; Szczerbińska and Gałczyńska 2015). Another macroinvertebrate, like molluscs, including snails, serves as bio-monitors for PAHs, Polychlorinated Biphenyls (PCBs), and inorganic substances such as Cd, Cu, Hg, tin (Sn), and other pollutants as they accumulate in larger amounts, which is observed by their behaviour change in the presence of the above toxic chemicals (Salánki et al. 2003). An herbaceous species, such as Tradescantia pallida, along with planktonic crustaceans like Daphnia magna, has recently been utilized for online water quality testing to detect urine levels and assess genotoxicity in freshwater (Júnior et al. 2024).

12.5.1.2 Integrating AI and GIS for Effective Water Resource Management and Quality Monitoring

Traditional water quality assessment techniques often encounter difficulties in remote regions. It fails to observe how local water resources quickly change with time due to floods, hurricanes, or human-caused pollution and cannot prepare a plan for future droughts, underscoring the necessity for real-time monitoring by including advanced technologies such as AI, GIS, and remote sensing (Yang et al. 2022). GIS gathers spatial information and analyses phenomena that vary across distinct locations (Bari et al. 2023). The data collection, storage, database management, spatial analysis, spatial browsing, analysis of spatial data, reports, charts, and visualizations using maps and other representation tools that communicate the analysed information are provided by the help of GIS outputs (Ali 2020). Map-based visualization includes automated mapping, thematic mapping, composite mapping, or overlays (Ali 2020). Integrating different water quality data into practical insights remains a significant challenge. This challenge is solved by automated sensor technologies such as smart meters, which streamline complex data and help customers understand water consumption patterns (Gupta et al. 2020). Water demand prediction and leak detection signals can also be integrated along with the water consumption-based automated sensors to minimize wastage (Jayakumar et al. 2024). AI-driven smart sensors track water pollution, spread awareness for excessive water use patterns, toxic components, and DO levels, providing notifications for these unusual observations or leaks (Gupta et al. 2020). Smart water systems (SWS) minimize the human capital investment as they possess self-learning capabilities, address problems such as inconsistent data, inadequate sampling, and human errors by enabling ongoing, real-time observation and prediction. Thus, SWS is increasingly being implemented for local water resource assessment without any discontinuities and redundancies in time and space (Gupta et al. 2020).

In rural settings, SWS has enhanced water safety and mitigated contamination. The rise of smart farming in villages employs sensors to boost crop yields, conserve water, and optimize irrigation schedules, especially in regions facing water shortages (Gupta et al. 2020).

Groundwater quality forecasting is examined to be successful with different AI approaches, such as Support Vector Machines (SVM), Adaptive Neuro-Fuzzy Inference Systems (ANFIS), and Artificial Neural Networks (ANN) (Hanoon et al. 2021). The combination of these AI-related models with GIS and other approaches significantly improves groundwater management. For instance, a recent study merged GIS with survey data, Participatory Rural Appraisals (PRA), and water testing to create a water quality map and collect community feedback (Mishra et al. 2024). This illustrates the promise of AI and GIS in enhancing water quality management in rural localities.

12.5.1.3 Integrated Water Assessment Approaches in Recent Studies Conducted in Indian Deltas

Integrated water assessment approaches including WQI and advanced analytical tools such as principal component analysis, partial least square regression, and GIS provides more accurate, site-specific evaluation, interpretations, water quality changes and key pollutant sources (Das 2025c; Isaac et al. 2024).

These integrated approaches (Table 12.7) involve tools for higher precision, easier for the public to understand the water health (WQI), and other tools used to detect the hidden trends in the data by reducing dimensionality and noise, ultimately simplifying the intricate datasets (principal component analysis, or PCA) along with GIS which offers a spatiotemporal analysis and visual representation of continuous variables for interpretations, and decision-making in water quality assessment. Other tools involved in integrated water assessment strategies are multivariate statistical analysis, partial least square regression (PLSR), GIS mapping and spatial reflectance index (SPI) (Das 2025d; Isaac et al. 2024).

Table 12.7 Integrated approaches for surface and groundwater quality assessment across major Indian river deltas and their surroundings, highlighting the application of water assessment strategies

River/delta region	Location/state	Water source assessed	WQI applied	Other integrated approaches	Key findings	References
Mahanadi River Basin (Deltaic zone)	Odisha	Surface water	Entropy weighted WQI (EWQI), pollution index of surface water (PIS), nitrate pollution index (NPI)	GIS, multivariate statistical analysis (PCA), bland–altman validation	High turbidity, TDS, and coliform beyond WHO limits; turbidity, sulfate, nitrate, and coliform as key pollutants	(Das 2025a)
Karamana River Estuary	Kerala	Surface water	Weighted reckoning WQI	Multivariate statistical analysis, GIS mapping, trace metals identification	Fair (in midstream) to good water quality, chromium, nickel (above the WHO limits), 95% water fit for drinking	(Shanmugasundharam et al. 2023)
Yamuna river	Prayagraj, Uttar Pradesh	Surface water (river)	Weighted arithmetic WQI	Multivariate statistics (PCA), ICP-MS elemental analysis	High WQ in summer, rain, winter, poor WQ in spring, PCA result (seasonal variation and untreated sewage dumping point in the river, negatively affects WQI), low-level heavy metals	(Isaac et al. 2024)

(continued)

Table 12.7 (continued)

River/delta region	Location/state	Water source assessed	WQI applied	Other integrated approaches	Key findings	References
Odisha surface water tract	Odisha	Surface water	Drinking water quality index (DWQI)	Partial Least Squares Regression (PLSR), Spatial Reflectance Indices (SRI), GIS mapping	44% assessed locations – excellent to good, nitrate, phosphate (excess than WHO limits), key drivers: fertilizer, organic waste, soil leaching	(Das 2025b)
Godavari delta region	Coastal Andhra Pradesh	Groundwater	Irrigation water quality index (IRWQI)	Hydro-chemical analysis	High EC, sodium, magnesium, over-pumping, evaporation deteriorated GW quality, and highly saline water shows saline intrusion	(Rao et al. 2022)
Cauvery river delta region	Tamil Nadu	Groundwater and drinking water	User specific water quality index (USWQI), parameters quality index (PQI)	Principal Component modelling (PCA), hydro-chemical analysis	Drinking water fit for consumption, GW is suitable for irrigation	(Jothivel et al. 2023)

12.5.2 Rural Water Resources Management Framework

12.5.2.1 Role of Strategic Government Planning for Water Resources

Central and state governments have undertaken significant national and state-level projects supporting international organizations like the World Bank to solve rural water issues (Nath et al. 2024). The central government provides funding and encourages research in relation to water quality analysis and treatment methods. State governments work on water supply schemes in districts through departments like Public Health Engineering Departments (PHEDs) and Panchayati Raj Institutions (PRIs) (MoRD 2011). The PHEDs/ Rural Water Supply and Sanitation (RWS) department responsible for implementing the Jal Jeevan Mission (JJM) in the country, allows the testing of water sources at least twice a year (MoJS 2021b).

There are five strategies to overcome the challenges associated with the rural water sector (MoRD 2011). These are effective community engagement, sustainable sourcing, infrastructure modernization, decentralized governance, and capacity building. The challenges faced in rural water sector are addressed by National Rural Drinking Water Programme (NRDWP) by promoting conjunctive water use and supporting convergence with other water-related initiatives, such as the Mahatma Gandhi National Rural Employment Guarantee Scheme (MNREGs) and watershed development programs (Nath et al. 2024; MoRD 2011). In 2019, NRDWP was restructured and subsumed into the JJM 2019, whose focus is to supply every rural household with operational tap-water connections by 2024 (MoJS 2023a). In 2024, out of 19.27 crore rural Indians, nearly 73.93% (14.24 crore) had tap water facilities (MoJS 2024b). The MNREGs support water conservation projects financially, providing 75% of their funds for water conservation, enhancing rural water infrastructure, and generating employment. Some of the water management achievements in this scheme are drought-proofing, water conservation units, and irrigation facilities (Bassi and Kumar 2010; MoRD 2019). This effort complements the Jal Shakti Abhiyan (JSA) 2019, prioritizing RWH and river rejuvenation for sustainable water conservation. Mainly for effective irrigation and water usage, the Pradhan Mantri Krishi Sinchayee Yojana (PMKSY) scheme with the timeline of 2021–22 to 2025–26 has been introduced with a total expense of Rs. 93,068.56 crore, to target 13.88 lakh hectares of additional irrigation potential, benefiting 22 lakh farmers (MoJS 2023a, 2023b). Another groundwater related government scheme is Atal Bhujal Yojana, with the key aim of introducing sustainable groundwater management in 8,220 water-stressed gram panchayats in 80 districts across seven states (MoJS 2023c). Over the previous 20 years, the World Bank has introduced 9 rural water supply and sanitation projects, reducing the burden on rural women and saving an estimated 5.5 crore hours daily by improving access to water (Nath et al. 2024; MoJS 2025d).

To ensure water quality, over 2,119 laboratories have been established nationwide, guided by the drinking water quality monitoring and surveillance framework, which emphasizes communication, coordination, and collaboration—the three Cs—for effective monitoring (NJJM 2023b).

12.5.2.2 Participatory Water Management

Planning, building, and maintaining a holistic rural water management framework requires the active participation of communities and PRIs. Studies in Dharwad, Karnataka, showed that community-driven initiatives had positive impacts on local engagement in improved water management. In participatory groundwater management, local communities and NGOs manage water extraction, use, and the initiative to replenish and recharge the resource, such as villagers, along with UTTHAN NGO, solved the fluoride water crisis in Gujarat (Nath et al. 2024; Khurana et al. 2008). Another example is of APFAMGS groundwater management system installed by NGOs in Andhra Pradesh (Gulati et al. 2019). Private company's participation aids in overcoming the challenges related to heavy initial investment and has introduced technologies like the water ATM system, solar-driven water purification, and water delivery schemes in rural areas (Nath et al. 2024). An overview of Indian government programs for rural water resource management is presented in Table 12.8.

12.5.2.3 Low-Cost Water Treatment Technologies in Rural India

Given the financial constraints to combat water issues in rural regions, it is essential to ensure sustainable sources, cost-effective operation, and management of the system, along with social and environmental sustainability, while maintaining high efficiency (MoDWS 2013). Water purification technologies must be economically sound and be constructed using locally available materials to minimize expenses. For instance, lanthanum-impregnated silica gel and freshly burned bricks are sustainable methods for fluoride removal (Wasay et al. 1996). Sustainable removal techniques are currently installed in rural areas to address widespread contaminants such as arsenic, fluoride, nitrate, and iron, which affect more than 20 states and are listed in Table 12.9. Technical efficiency has been significantly enhanced through the joint efforts of leading Indian institutions, international partners, and national agencies, in collaboration with the central and state governments. Innovations like the arsenic filters developed by IIT Bombay and the AMRIT filters developed by IIT Madras are prime examples of how research and collaboration can efficiently address water quality issues (NJJM 2023a).

Table 12.8 Indian government initiatives for rural water management

Ministry/ department	Schemes/Policies	Main objectives	References
MoJS, DoDWS	JJM Har Ghar Jal on 15 August 2019	• To provide safe and assured water through tap-water networks to rural residents by 2024	(MoJS 2024b; MoJS 2021c)
MoJS, department of water resources, river development and Ganga Rejuvenation (DoWR, RD and GR)	Jal Shakti Abhiyan: Catch the Rain (JSA: CTR) 2024 (fifth in the series of JSA)	• Spreading awareness • Construction, repair, and maintenance of rooftop and other RWH structures	(MoJS 2024c)
	New National Water Policy, 2012	• Laws and regulations for managing, conserving, and securing the water supply and water availability with a unified national perspective • Enabling water footprints and water audit	(MoJS 2021c; MoWR 2012)
Department of agriculture and farmers welfare, in coordination with the respective ministries responsible for agriculture, water resources, and rural development	Prime Minister's Agricultural Irrigation Scheme /Pradhan Mantri Krishi Sinchayee Yojana (PMKSY) to 2025 -26) 1. PMKY-Accelerated Irrigation Benefit (AIBP) 2. PMKY- Har Khet ko Pani (HKKP)	• Water-efficient irrigation methods under per drop more crop project • Surface minor irrigation scheme (SMI) • Restoration of surface water bodies under the Repair, Renovation, and Restoration (RRR) scheme • The target of 34.63 lakh Ha. irrigation potential is completed • AIBP have fifty prioritized projects completed out of ninety-nine projects	(MoJS 2023a; MoJS 2023b)
MoJS, DoWR, RD and GR	Support for irrigation modernization program (SIMP) (2021–26)	• To improve irrigation potential • Enhance crop water productivity	(MoJS 2023b)

(continued)

Table 12.8 (continued)

Ministry/ department	Schemes/Policies	Main objectives	References
MoJS, DoWR, RD, and GR	Atal Bhujal Yojana; 2020 to 2024–25	• Participatory groundwater management • Artificial groundwater recharges	(MoJS 2023a; RUPIB 2022)
	National hydrology project	• To establish a hydrological information system • Improvisation of the targeted water resources, groundwater management, and regulation	(MoJS 2023b)
CGWB, MoJS, DoWR, RD, and GR,	Master plan for artificial recharge to groundwater-2020	• Construction of nearly 1.42 crore RWH structures Conservation of 185 BCM of rainfall water	(MoJS 2024c)
	Ground water management and regulation (GWMR) Scheme (2021–2026)	• Regulation and control of groundwater extraction, • Construction of Piezometers and aquifer mapping under the NAQUIM program	(MoJS 2023a; MoJS 2023b; MoJS 2023d)
	MGNREGS	• The construction of water conservation and watershed • Includes management structures	(MoJS 2024c)
MoJS (national water mission)	Bureau of water use efficiency (BWUE) scheme	• Promote improved water utilization efficiency for different sectors	(MoJS 2025c)
DoWR	National water awards (2018)	• To appreciate and facilitate the efforts towards water conservation • To raise public awareness towards water sustainable management	(MoJS 2025b)

(continued)

Table 12.8 (continued)

Ministry/ department	Schemes/Policies	Main objectives	References
MoJS in collaboration with government of Gujarat	Jal Sanchay Jan Bhagidari initiative	• Focus on groundwater recharge • Construction of 1 million recharge units across India • Aims for 24,800 RWH units in Gujrat state • To increase stakeholders and public participation	(MoJS 2025e; PIBRU 2024)
MoJS, (Department of DoWR, RD and GR)	Namami Gange programme	• To reduce pollution, improve conservation and rejuvenation of River Ganges • Installation of 36 real time and 97 manual water quality monitoring • Forestry interventions • Rural sanitation • Construction and renovation of kunds/ ponds	(NMCG 2020)
Ministry of water resources, river development and Ganga Rejuvenation	Flood management and border areas programme (2021–26)	• Flood control and drainage development • Flood plain zoning, flood forecasting	(MoWR and GR, 2024)

12.6 Sustainable Solutions for Rural Water Resources Management

Sustainable water management involves RWH and other traditional water harvesting methods, efficient irrigation (smart/AI-based irrigation, drip irrigation), brackish water irrigation, and reuse of wastewater (as energy, water, and fertilizer source), water conservation as a priority, and water demand as a control. (Sharma et al. 2024; Levidow et al. 2014; Richards et al. 2021b; Wang and Su 2020).

Table 12.9 Sustainable and cost-effective treatment methods for major groundwater contaminants installed in rural areas

Low-cost, tested and installed technologies	Household-level	Community level	Removal of contaminants	Working principle and locally available materials used instead of high-cost commercial materials	Limitation	Capital cost/ operation costs	Source
Kanchan arsenic filter	Yes	No	Arsenic	Slow sand filter working mechanisms, concrete containers, gravels, sand, and iron fillings	For low turbid waters	The unit cost Rs 750	(NJJM 2023a; Ngai et al. 2005)
GEH 102 Granular Ferric Hydroxide	No	Yes	As, Cu, Cr, PO_4^{3-}, molybdenum (Mo), Se, antimony (Sb), dissolved organic matter	Granular ferric hydroxide, Adsorption	Raw water must be free from iron, turbidity, and calcium precipitates	Cost of Water Per Liter (Paise per liter): 4.00	(NJJM 2023a; GEH 2021)

(continued)

Table 12.9 (continued)

Low-cost, tested and installed technologies	Household-level	Community level	Removal of contaminants	Working principle and locally available materials used instead of high-cost commercial materials	Limitation	Capital cost/ operation costs	Source
Community-scale IITB Arsenic filter	No	Yes	(As, Fe, PO_4^{3-})	Oxidation, adsorption, and iron fillings	Requires frequent cleaning of nails and gravel to maintain efficiency	Water cost per 1000 L: Rs. 0.1 to 1; for filter maintenance, Rs. 10/month per family	(NJJM 2023a; Banerji and Chaudhari 2017)
Rama Krishna Mission (RKM) filter	Yes	No	Arsenic	Oxidation, coagulation, precipitation. Low-cost earthen pots instead of ceramic	Requires chemicals	Water costs per 10 L: Rs. 0.75–1.00; the unit cost is Rs. 250 per unit	(NJJM 2023a)
Filter tablet system	Yes	No	Arsenic	Coagulation and precipitation. Low-cost earthen pots, activated carbon media	Chemicals required, cost varies with construction material	The cost is Rs 100–1500	(NJJM 2023a; Das et al. 2001)

(continued)

Table 12.9 (continued)

Low-cost, tested and installed technologies	Household-level	Community level	Removal of contaminants	Working principle and locally available materials used instead of high-cost commercial materials	Limitation	Capital cost/ operation costs	Source
Electrolytic De-Fluoridation Technology (CSIR-NEERI)	No	Yes	Fluoride, nitrate	Electrolysis, aluminum electrodes, sedimentation/ filtration	Electrode corrosion, Replacement, and cleaning of electrodes	Water costs 1000 L Rs.12	(NJJM 2023a; CSIR-NEERI 2023)
Nalgonda Technique	Yes	Yes	Fluoride	Coagulation, flocculation, sedimentation, filtration	Chemicals involved; high sludge produced	Extremely low cost	(NJJM 2023a; Bulusu et al. 1979; Nawlakhe and Paramasivam 1993)
Bamboo charcoal filters	Yes	Yes	Iron	Filtration, Pebbles, gravels, bamboo charcoal	No chemicals required	Extremely low cost	(NJJM 2023a; Baruah et al. 2011)

12.6.1 Structural Water Harvesting and Groundwater Recharge Management

India's total water availability is primarily derived from nearly 4000 BCM of annual precipitation (Nath et al. 2024). This water, received in rainfed areas, is lost as runoff, often leading to water scarcity in a protracted dry spell. To prevent the loss of rainwater and storm runoff, which are crucial for water balance, they should be collected, used, and recycled through structural conservation methods. For instance, the stored water can be used for dry season irrigation practices in rural areas. Farm ponds are created by digging pits for flat plains, whereas, in undulatory terrain, they are of surface impoundment type (Nath et al. 2024). Different lining materials such as bentonite clays and plastic liners, reduce percolation losses for proper storage (Nath et al. 2024; Msangi and Cline 2016). For groundwater recharge, to amplify the percolation and increase soil moisture, particularly in watershed areas, nala bunds, percolation tanks, and johads are built (Nath et al. 2024; Mehta et al. 2024). The central government launched the Jal Shakti Abhiyan project in 2019 to conserve rainwater. This project includes *Jal Shakti Abhiyan:* Catch the Rain (JSA-CTR) 2021, which focused on RWH and modifying the traditional methods like jhalaras, zings, kuhls, johads, recharge pits, zabo, etc., to conserve water (Sharma et al. 2024). These methods are eco-friendly and unique to a specific region. An overview of traditional rainwater harvesting practices in rural India and in other regions worldwide is shown in Fig. 12.5.

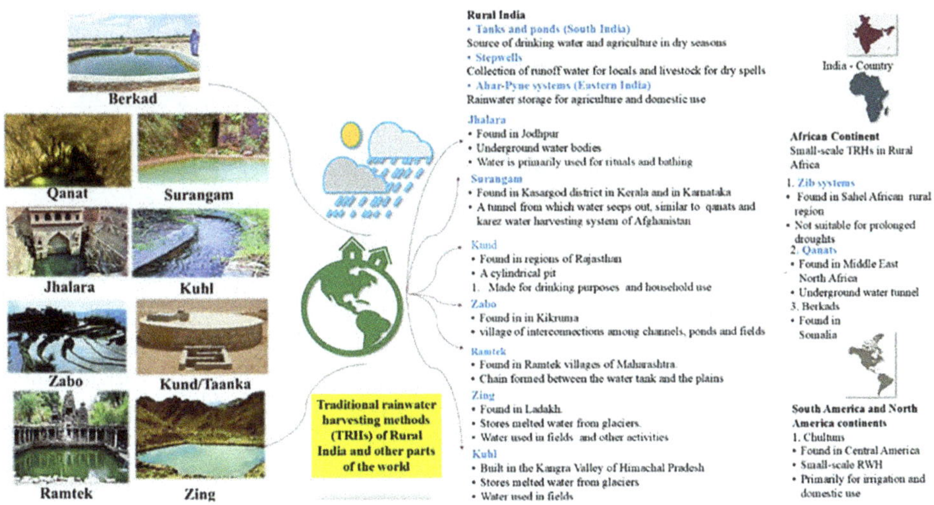

Fig. 12.5 Traditional rainwater harvesting systems practiced in rural India and around the world. *Source* (Ssekyanzi et al. 2024; Sharma et al. 2024, 2011; Contreras et al. 2013; Ghai 2021; Hein 2019; Life 2017; Mahmoudian 2024; Siphambe et al. 2024)

12.6.2 Agronomic Practices

In India, with a population of 1.4 billion, more than one-third of the people rely on agriculture for their living (Kapoor and Anand 2024). Agricultural practices form one of the largest water sectors and must be monitored and modified to cut heavy dependence on groundwater and use alternative water sources. Sustainable agricultural measures, along with efficiently managing the water utilized, along the enhancement of crop production (Nath et al. 2024).

12.6.2.1 In-Situ Conservation

Water is conserved in the form of soil moisture during in situ conservation. This method uses several techniques for growing crops in the off-season. These are deep tilling of the soil (deep plowing), creating dead furrows, naturally covering and protecting the soil from direct sun's heat (mulching), tillage patterns following natural land contours (keyline cultivation), square-shaped lowered earthen bunds to capture rainwater (compartmental bunding), and planting cover crops in and around main crops (Nath et al. 2024; Pathak et al. 1989). These practices increase the rainwater seepage rate into the soil, conserve soil moisture, increase water retention capacity, and reduce vulnerability to insects and pests (Nath et al. 2024; Prihar and Sandhu 1987).

12.6.2.2 Smart Farming and Efficient Irrigation

Smart farming involves the use of AI, in which automated smart sensors analyse various factors like weather, soil moisture, and soil nutrition, and suggest suitable sowing and harvesting times of crops, which is responsible for better crop yield (Sharma et al. 2024). With the help of this data, farmers can record their crop water requirements per day. This saves up to 20% of water in field irrigation (Sharma et al. 2024; Gupta et al. 2020).

12.6.2.3 Advancement in Irrigation Techniques

In India, the most considerable portion of the agricultural area, nearly 63%, is rainfed and dependent on rainfall for irrigation, and the rest is under manual irrigation (Sharma et al. 2024). Groundwater is depleting rapidly due to poor irrigation efficiency, changes in cropping patterns (paddy fields), shifts to groundwater irrigation, and energy sources, mainly in the northwestern region of the country, in Punjab (Srivastava et al. 2015). In Punjab, groundwater levels have dropped from 8 m in 1999 to 15 m in 2014, mainly due to agriculture-related factors like changes in cropping patterns and subsidies that encourage a shift to groundwater (Sharma et al. 2024; Srivastava et al. 2015). Most villages in the northern plains follow cost-effective well irrigation but require complete dependency and fail in dry seasons. Farmers in these irrigation practices either overwater or underwater the crops, struggling with lower yield and low income (Sharma et al. 2024). Canal irrigation is effective in floodplains and low-lying areas with perennial water sources, such as the Ganga–Brahmaputra plains. It is widely practiced in the northern plains of Uttar Pradesh,

Haryana, Punjab, Rajasthan, and Bihar (Pandey 2013). It encompasses nearly 44.28% of Haryana's irrigated land, 64.7% of Odisha's, and 91.72% of Jammu and Kashmir's agricultural fields (Sharma et al. 2024). Inundation canals divert excess river water to fields but are functional only during floods or the rainy season, while perennial canals provide a year-round water supply for irrigation. While conventional canals are beneficial, they also pose challenges such as waterlogging, soil salinity, insect breeding, and waste accumulation (Pandey 2013).

Drip irrigation is micro irrigation in which the water, without any wastage, directly reaches the roots of plants in the droplet form, which is much more efficient than surface irrigation (Yadav et al. 2019). India faced agricultural problems, and to overcome this, drip irrigation was adopted, which showed an increase in crop yields and reduced the cultivation cost, reduced atmospheric humidity (to achieve lower disease risk in plants), and minimized the overexploitation of water (Yadav et al. 2019).

12.6.2.4 Autonomous IoT-Based Irrigation: A Boon to Agriculture

Irrigation with the help of pressure pumps and motors, with erratic power supply and labor charges, has proven inefficient, leading to reduced crop productivity (Sharma et al. 2024). For instance, in Andhra Pradesh, Balayapalli village, the farmers suffered from constant trouble due to the irregular water supply to fields, resulting in crop damage and huge water loss, until a device named Kisanraja was introduced, benefiting over 34,200 farmers (Sharma et al. 2024; Gogoi 2019).

This device is developed using advanced technologies like AI and IoT, which form the framework, making the device include many features like wireless sensors, wireless valve controllers (Sharma et al. 2024). Wireless sensors analyse the parameters, and, on the cloud, these data are uploaded and sent to the receiver's mobile phones. Accordingly, the farmers can modify specific irrigation practices.

12.6.3 Groundwater Sustainable Management

Water of 790 BCM is estimated to enter the soil, of which 430 BCM will contribute to soil moisture. An estimated 360 BCM of water percolates into the soil, leading to water changes inside the aquifers and contributing to groundwater enrichment, and nearly 255 BCM is commercially extractable (Nath et al. 2024). Groundwater, besides being a safe freshwater source, also protects against droughts, is internally connected to surface water sources, and is a significant component of the hydrological cycle (Saha and Ray 2019). However, groundwater contamination and scarcity have been observed in recent years due to increasing population demands, urbanization, industrialization, and other factors (Datta 2019). Groundwater extraction and consumption often reflect atomistic individualism, where individual use is assumed to have little or no implications for other users. However, the conservation approach requires transitioning to an aquifer community where every user is mutually vulnerable and dependent, as it is a central resource for livelihoods (Shah 2012). Additionally, groundwater resources are often developed without a thorough understanding of their spatial and temporal occurrence. Therefore, scientific management is essential to ensure long-term sustainability (Nath et al. 2024; Kumar and Ballabh 2000).

12.6.4 Aquifer-Centric Groundwater Resource Development and Management

Groundwater and aquifer development strategies comprise delineation and characterization, aquifer recharge, deep aquifer development, and management (MoJS 2021c, 2023e). Four situations that evidence distinctive systematic patterns of coping and adaptation for aquifer development and recharge are the Ganga–Meghana–Brahmaputra (GBM) basin, arid alluvial aquifers, salinity-prone aquifers, and hard rock aquifers (Shah 2012). Due to the underutilization of deep aquifers in areas of unleashed groundwater potential, there are problems of rapid water quality deterioration and stagnant water conditions at subsurface depths in several states like Haryana, Punjab, and Uttar Pradesh (Nath et al. 2024). Effective mining, proper development, and management of deep aquifers are required to utilize accessible groundwater in the long term. The common methods are drilling horizontal bores inside the open well (enhancing connectivity between pores and wells), borewell blasting, and aquifer blasting (increasing well yield and rainwater storage (Shah 2012). Northeast India has a greater scope for aquifer development, requiring flood control and groundwater table management to enable canal-based recharge and irrigation use (Nath et al. 2024). These strategies benefit subsurface drainage, groundwater recharge, and better water resource use.

12.7 Discussions

This chapter has discussed the significant water management-related challenges in India, such as rising water demand, salinity, flood and cyclone hazards, groundwater overextraction, including widespread geogenic contamination, and insufficient water treatment and monitoring infrastructures. These challenges increase due to population surge, rapid urbanization, climate change, improper waste disposal, and industrial and chemical-based agricultural activities (Balkhi et al. 2023; Karthik et al. 2025). Pollutants, primarily of geogenic nature, continue to exceed safe limits in groundwater across most Indian states, with iron (29 states) > fluoride (27 states) > arsenic (25 states) > uranium (12 states) > radon (7 states), followed by heavy metals such as Cr, Cd, and Pb, exceeding the regulatory limits in more than 14 states. Emerging contaminants (investigated widely, only in the last 20 years) are now increasingly detected in Indian waters, and most lack regulatory standards or routine monitoring (Noguera-Oviedo and Aga 2016). MPs contamination was reported in all major Indian deltas, as discussed in the chapter. This is in support with a recent study which reported the highest MPs contamination in the Ganga-Yamuna River basins followed by Godavari and Mahanadi basins with sediments having higher MPs levels than waters (Ahmed et al. 2024). Addressing these concerns help to achieve SDGs, calling for sustainable water management and universal access to safe drinking water (Das 2025d). Integrated approaches are essential for effective and long-term sustainable water management which focuses on affordable, effective assessment and treatment approaches, with reuse of reclaimed wastewater for industrial cooling, irrigation, municipal cleaning, and other sustainable applications (Qian 2016). Integrated water assessment approaches, including seasonal water monitoring and WQI, address pollution across biological, physical, and chemical aspects enhancing the water quality monitoring and aligning with the SDGs (Kumar et al. 2024; Das 2025c, 2025d). For instance, in deltas and riverine regions, which are highly exposed to risks from relative sea-level rise, flooding, and salinisation require integrated strategies combining WQI with advanced monitoring tools such as GIS mapping, AI, sensors, other multivariate and hydro-chemical analyses along with flood/cyclone hazard mapping, SSP, riverbank stabilization, and marine pollution control measures to significantly improve water management (Scown et al. 2023; Das 2025a; Shanmugasundharam et al. 2023). UNEP also supported the idea of locally adapted solutions based on existing water quality, contamination type, and cultural acceptance to ensure long-term success. The study highlights both household-level and community-level treatment options as practical and cost-effective measures for rural and remote areas. However, recent studies show that point-of-use treatment is often more feasible in rural and remote areas with low population masses, where centralized treatment cannot be effective. Decentralized systems also reduce contamination due to faults in distribution systems and house storage, thereby preventing the burden of water-borne disease (Malan and Sharma 2023; CTCN 2016).

12.8 Conclusion

In recent years, around 2022, India experienced a decline in groundwater levels, posing water stress for half of the population. Agriculture has a crucial role in groundwater management, with the potential to avert the groundwater crisis. In coastal and riverine delta plains, key challenges include water and soil salinity challenges, and food insecurity. Addressing these challenges require integrated strategies with predictive modelling for better understanding of salinity dynamics and their related hazards within deltas. The current management strategies need upgrading in terms of technical, financial, social, and political aspects. Additionally, in India, water planning is often limited to state borders rather than considering river basins, leading to water disputes. Integrated and decentralized sustainable management strategies with stakeholder's involvement, should be implemented at the local and regional levels for policy making, to detect the water resources variability, and formulate comprehensive solutions at a larger level. Finally, considering water quality, sanitation, and hygiene problems highlighted in SDGs, it is the responsibility of communities, the private sector and government to work together for sustainable and equitable water resource utilization.

12.9 Recommendations

In the context of the chapter's findings, effective rural water management in India require a shift from development of new water sources to development of existing sources. Criteria for successful treatment techniques are site-specific and community-acceptable treatment techniques, according to the local water quality and community acceptable treatment techniques, making them more economically feasible and suitable for decentralised techniques, especially in rural, coastal, and deltaic regions. Adoption of smart water monitoring and irrigation practices can enable more targeted and efficient treatment methods with sustainable water usage. These smart monitoring techniques, along with water quality indices for simplifying the interpretations of the water quality data for the public, will encourage stakeholder's involvement, reduce the risk of waterborne outbreaks, and move closer towards socioeconomic development. Collectively, these approaches can support long-term resilience in vulnerable populations and can ensure sustainable rural water management in countries, like India.

No AI tools were not used in preparing the chapter; all content was developed and written by the authors.

References

Abdulwahid AH (2020) IoT based water quality monitoring system for rural areas. In: 9th international conference on renewable energy research and applications, ICRERA 2020. pp 279–282

Acharya S, Sharma SK, Khandegar V (2018) Assessment of groundwater quality by water quality indices for irrigation and drinking in South West Delhi, India. Data Brief 18:2019–2028. https://doi.org/10.1016/j.dib.2018.04.120

Acharya SS, Hishamunda V, Chakrabarti R (2022) Natural sources and anthropogenic influences on the river water and groundwater chemistry of the lower Mahanadi Basin: Insights from radiogenic Sr isotopes and major ion chemistry. Frontiers in Water 4:846438. https://doi.org/10.3389/frwa.2022.846438

ADB (2020) Arsenic and fluoride in drinking water in West Bengal - characteristics, implications, and mitigation

Adelodun B, Ajibade FO, Ighalo JO, Odey G, Ibrahim RG, Kareem KY, Bakare HO, Tiamiyu AGO, Ajibade TF, Abdulkadir TS, Adeniran KA, Choi KS (2021) Assessment of socioeconomic inequality based on virus -contaminated water usage in developing countries: A review. Environ Res 192:110309. https://doi.org/10.1016/j.envres.2020.110309

Ahmed F, Hussain ER, Rasid SS, Hussain SA, Pathak D (2024) Investigating microplastic pollution in freshwater fish populations: An Indian perspective. Int J Entomol Res 9:50–58

Ali E (2020) Geographic Information System (GIS): Definition, development, applications & components. Department of Geography, Ananda Chandra College, India 1–12

Anandavelu I, Karthik R, Robin RS, Hariharan G, Mugilarasan M, Ramesh R, Purvaja R (2024) Morphometric characteristics and spatiotemporal heterogeneity of microplastics on the north-east coast of India. J Hazard Mater 480:136180. https://doi.org/10.1016/j.jhazmat.2024.136180

Balarajan Y, Selvaraj S, Subramanian SV (2011) Health care and equity in India. Lancet 377:505–515. https://doi.org/10.1016/S0140-6736(10)61894-6

Balkhi SAA, Karki KB, Philip L, Maliyekkal SM (2023) Water quality status and challenges in India and Nepal. In: Philip L, Pradeep T, Bhallamudi SM (eds) Technological solutions for water sustainability—towards a water-secure India. IWA Publishing, pp 13–23

Banerji T, Chaudhari S (2017) A Cost-effective technology for arsenic removal: Case study of Zerovalent Iron- Based IIT Bombay Arsenic Filter in West Bengal. Water SanitIon New Millenn 127–137. https://doi.org/10.1007/978-81-322-3745-7

Bari JBA, Tiwari SP, Mitra B, Elhaj AI, Siddiqui A, Reshi OR, Rahman SM (2023) GIS-based modelling for estimation of water quality parameters: A review. In: Mushtaq F, Farooq M, Mukherjee AB, Ghosh Nee Lala M (eds) Geospatial Analytics for Environmental Pollution Modeling. Springer, Cham, pp 57–89

Baruah BK, Das B, Haque A, Misra K, Mishra AK (2011) Iron removal efficiency of different bamboo charcoals: A study on modified indigenous Water filtration technique in rural areas of Assam. J Chem Pharm Res 3:454–459

Barui I, Bhakta S, Ghosh K, Shaw R (2024) Assessment of household vulnerability to embankment breaching in the coastal area of the Indian Sundarban. Int J Disaster Risk Reduct 110:104632. https://doi.org/10.1016/j.ijdrr.2024.104632

Bassi N, Kumar MD (2010) NREGA and rural water management in India: Improving the welfare effects. Hyderabad

Bhadra T, Hazra S, Sinha Ray SP, Barman BC (2020) Assessing the groundwater quality of the coastal aquifers of a vulnerable delta: A case study of the Sundarban Biosphere Reserve. India. Groundw Sustain Dev 11:100438. https://doi.org/10.1016/j.gsd.2020.100438

Bhatt JP, Pandit MK (2010) A macro-invertebrate based new biotic index to monitor river water quality. Curr Sci 99:196–203

Bhutiani R, Kulkarni DB, Khanna DR, Gautam A (2016) Water quality, pollution source apportionment and health risk assessment of heavy metals in groundwater of an industrial area in North India. Expo Health 8:3–18. https://doi.org/10.1007/s12403-015-0178-2

BIS (2012) Indian Standard drinking water specification (second revision). New Delhi

Biswas T, Chandra Pal S, Saha A, Ruidas D (2023) Arsenic and fluoride exposure in drinking water caused human health risk in coastal groundwater aquifers. Environ Res 238:117257. https://doi.org/10.1016/j.envres.2023.117257

Brauns B, Chandra S, Civil W, Lapworth DJ, MacDonald AM, McKenzie AA, Read DS, Sekhar M, Singer AC, Thankachan A, Tipper HJ (2024) Presence of emerging organic contaminants and microbial indicators in surface water and groundwater in urban India. Environ Pollut 362:124983. https://doi.org/10.1016/j.envpol.2024.124983

Brindha K, Paul R, Walter J, Tan ML, Singh MK (2020) Trace metals contamination in groundwater and implications on human health: Comprehensive assessment using hydrogeochemical and geostatistical methods. Environ Geochem Health 42:3819–3839. https://doi.org/10.1007/s10653-020-00637-9

Bulusu K, Pathak B, Nawlakhe W, Kulkarni N, Thergaonkar V (1979) Fluorides in water, defluoridation methods and their limitations. Natl Environ Eng Res Inst, NEERI 60:83

CGWB (2014) Concept note on geogenic contamination of groundwater in India with a special note on nitrate. Faridabad

CGWB (2022) National compilation on dynamic groundwater resources of India, 2022. Department of water resources, river development & Ganga Rejuvenation, Ministry of Jal Shakti, Government of India

CGWB (2023) National compilation on dynamic groundwater resources of India, 2023. Department of water resources, river development & Ganga Rejuvenation, ministry of Jal Shakti, Government of India

CGWB (2023) Overview of ground water quaility. https://cgwb.gov.in/en/overview-ground-water-quaility. Accessed 11 Oct 2025

CGWB (2024) National compilation on dynamic groundwater resources of India, 2024. Department of water resources, river development & Ganga Rejuvenation ministry of Jal Shakti Government of India

CGWB (2025) States wise details of partly affected districts with selected contaminants in groundwater of India

Chakrabarti R, Mondal S, Acharya SS, Lekha JS, Sengupta D (2018) Submarine groundwater discharge derived strontium from the Bengal Basin traced in Bay of Bengal water samples. Sci Rep 8:4383. https://doi.org/10.1038/s41598-018-22299-5

Chakraborti D, Rahman MM, Chatterjee A, Das D, Das B, Nayak B, Pal A, Chowdhury UK, Ahmed S, Biswas BK, Sengupta MK, Lodh D, Samanta G, Chakraborty S, Roy MM, Dutta RN, Saha KC, Mukherjee SC, Pati S, Kar PB (2016) Fate of over 480 million inhabitants living in arsenic and fluoride endemic Indian districts: Magnitude, health, socio-economic effects and mitigation approaches. J Trace Elem Med Biol 38:33–45. https://doi.org/10.1016/j.jtemb.2016.05.001

Chakraborty P, Vinod PG, Syed JH, Pokhrel B, Bharat GK, Basu AR, Fouzder T, Pasupuleti M, Urbaniak M, Beskoski VP (2022) Water-sanitation-health nexus in the Indus-Ganga-Brahmaputra River Basin: Need for wastwater surveillance of SARS-CoV-2 for preparedness during the future waves of pandemic. Ecohydrol Hydrobiol 22:283–294. https://doi.org/10.1016/j.ecohyd.2021.11.001

Chaturvedi MC (1985) Water resources in India: An overview. Sadhana 8:13–38. https://doi.org/10.1007/BF02811270

Chopra A, Ramachandran P (2021) Understanding water institutions and their impact on the performance of the water sector in India. Water Policy 23:466–486. https://doi.org/10.2166/wp.2021.207

Choudhary A, George L, Mandal A, Biswas A, Ganie ZA, Darbha GK (2025) Assessment of microplastics and associated ecological risk in the longest river (Godavari) of peninsular India: A comprehensive source-to-sink analysis in water, sediment and fish. Mar Pollut Bull 212:117560. https://doi.org/10.1016/j.marpolbul.2025.117560

Contreras SM, Sandoval TS, Tejada SQ (2013) Rainwater harvesting, its prospects and challenges in the uplands of Talugtog, Nueva Ecija, Philippines. Int Soil Water Conserv Res 1:56–67. https://doi.org/10.1016/S2095-6339(15)30031-9

CPCB (2008) Guidelines for water quality management. Ministry of Environment and Forests. Delhi

CSIR-NEERI (2023) Technologies—Electrolytic defluoridation EDF Technique. In: CSIR-National environmental engineering research institute (NEERI). https://www.neeri.res.in/abouts/technology_detail/electrolytic-defluoridation-edf-technique/5b1f95ae232a3#googtrans(en%7Cen). Accessed 31 Dec 2025

CTCN (2016) Point of use water treatment. In: United Nation Clim Technol Cent & Netw. https://www.ctc-n.org/technologies/point-use-water-treatment. Accessed 25 Oct 2025

Dangar S, Asoka A, Mishra V (2021) Causes and implications of groundwater depletion in India: A review. J Hydrol (Amst) 596:126103. https://doi.org/10.1016/j.jhydrol.2021.126103

Dar JA (2023) Health status issues and strategies of portable water in Indian villages: A review report. Environ Science: Ind J 19:266. https://doi.org/10.37532/0974-7451.2023

Das A (2025) Comprehensive analysis, hydrogeochemical characterization and processes controlling surface water quality: Entropy-based WQI, geospatial assessment, PIS, NPI, and multivariate approaches in Mahanadi Basin, Odisha (India). Water-Energy Nexus. https://doi.org/10.1016/j.wen.2025.07.003

Das A (2025b) Surface water quality evaluation of Mahanadi and its tributary Katha Jodi River, Cuttack District, Odisha, using WQI, PLSR, SRI, and geospatial techniques. Appl Water Sci 15:26. https://doi.org/10.1007/s13201-024-02357-2

Das A (2025c) A data-driven approach utilizing Machine Learning (ML) and Geographical Information System (GIS)-based time series analysis with data augmentation for water quality assessment in Mahanadi River Basin, Odisha. India. Discov Sustain 6:555. https://doi.org/10.1007/s43621-025-01464-7

Das A (2025d) Drinking water resources suitability assessment in Brahmani River Odisha based on pollution index of surface water utilizing advanced water quality methods. Sci Rep 15:34101. https://doi.org/10.1038/s41598-025-19539-w

Das D, Chatterjee A, Samanta G, Chowdhary TR, Mandal BK, Dhar R, Chanda CR, Lodh D, Chowdhary PP, Basu GK, Biswas BK, Uttam Kumar Chowdhary, Rahman MM, Paul K, Chakraborti D (2001) A simple household device to remove arsenic from groundwater and two years performance report of arsenic removal plant for treating groundwater with community participation. BUETUNU Int Work Technol Arsen Remov Drink Water

Dash S, Borah SS, Kalamdhad A (2019) A modified indexing approach for assessment of heavy metal contamination in Deepor Beel. India. Ecol Indic 106:105444. https://doi.org/10.1016/j.ecolind.2019.105444

Datta PS (2019) Water harvesting for groundwater management: Issues, perspectives, scopes, and challenges. John Wiley and Sons Ltd., Delhi, India

Dey M, Akter A, Islam S, Chandra Dey S, Choudhury TR, Fatema KJ, Begum BA (2021) Assessment of contamination level, pollution risk and source apportionment of heavy metals in the Halda River water. Bangladesh. Heliyon 7:e08625. https://doi.org/10.1016/j.heliyon.2021.e08625

Dutta V, Shukla KK, Chander S, Sobti RC (2018) Water policy in India: Building blocks for synergy with Science, Technology and Innovation (STI) policy for inclusive growth. SSRN Electronic Journal. https://doi.org/10.2139/ssrn.2873478

Duttagupta S, Mukherjee A, Bhattacharya A, Bhattacharya J (2020) Wide exposure of persistent organic pollutants (PoPs) in natural waters and sediments of the densely populated Western Bengal basin. India. Sci Total Environ 717:137187. https://doi.org/10.1016/j.scitotenv.2020.137187

Earth5R (2025) The Kaveri River: From sacred waters to urban crisis – a call for restoration. https://earth5r.org/kaveri-river-restoration-call/. Accessed 14 Oct 2025

Elneel L, Zitouni MS, Mukhtar H, Galli P, Al-Ahmad H (2024) Exploring Key Aspects of Sea Level Rise and Their Implications: An Overview. Water (Basel) 16:388. https://doi.org/10.3390/w16030388

Ephsy D, Raja S (2023) Characterization of microplastics and its pollution load index in freshwater Kumaraswamy Lake of Coimbatore. India. Environ Toxicol Pharmacol 101:104207. https://doi.org/10.1016/j.etap.2023.104207

EU (2020) Directive (EU) 2020/2184 of the European Parliament and of the Council of 16 December 2020 on the Quality of Water Intended for Human Consumption

FAO (2021) Pesticides use. Global, regional and country trends 1990–2018. FAOSTAT Analytical Brief 16. Rome

Feist SE, Hoque MA, Ahmed KM (2023) Coastal salinity and water management practices in the Bengal Delta: A critical analysis to inform salinisation risk management strategies in Asian deltas. Earth Syst Environ 7:171–187. https://doi.org/10.1007/s41748-022-00335-9

Fick J, Söderström H, Lindberg RH, Phan C, Tysklind M, Larsson DGJ (2009) Contamination of surface, ground, and drinking water from pharmaceutical production. Environ Toxicol Chem/ SETAC 28:2522–2527. https://doi.org/10.1897/09-073.1

Foufoula-Georgiou E (2013) A vision for a coordinated international effort on delta sustainability. In: Deltas: Landforms, ecosystems and human activities. pp 3–11

Gandla VK, Chiluka M, Gupta H, Sinha SN, Chakraborty P (2023) Sediment-water partitioning and risk assessment of organochlorine pesticides along the urban, peri-urban and rural transects of Krishna River Basin, Peninsular India. Sci Total Environ 874:162360. https://doi.org/10.1016/j.scitotenv.2023.162360

Gaur S, Singh N, Saxena S (2014) Status of lead present in ground drinking water samples of Uttarakhand (Garhwal Region) in India. Asian J Biomed Pharm Sci 1:32–38. https://doi.org/10.15272/AJBPS.V1I1.473

GEH (2021) Granular ferric hydroxide for drinking water treatment. In: GEH Wasserchemie GmbH & Co. KG. https://www.geh-wasserchemie.com/assets/application-downloads/41b96aa99e/GEH102_EN.pdf. Accessed 29 Oct 2025

Ghai R (2021) Fate of Afghanistan's karez system uncertain, but south India's surangam thriving. Down Earth

Ghosh S, Mistri B (2021) Assessing coastal vulnerability to environmental hazards of Indian Sundarban delta using multi-criteria decision making approaches. Ocean Coast Manag 209:105641. https://doi.org/10.1016/j.ocecoaman.2021.105641

Ghosh S, Bakshi M, Mahanty S, Gaine T, Bhattacharyya S, Biswas JK, Chaudhuri P (2021) Spatiotemporal distribution of potentially toxic elements in the lower Gangetic delta and their implications for non-carcinogenic health risk management. Geosci Lett 8:19–24. https://doi.org/10.1186/s40562-021-00189-5

Gogoi A (2019) Meet the man whose "KisanRaja" smart irrigation device helps over 34200 farmers. The Better India. https://thebetterindia.com/204393/india-water-pumps-innovation-agritech-invention-irrigation-agriculture-farmers-kisanraja/. Accessed 14 Dec 2024

Goswami R, Bhagat C, Lollen I, Neog N, Barache UB, Thakur R, Mahlknecht J, Kumar M (2023) Potential arsenic–chromium–lead co-contamination in the hilly terrain of Arunachal Pradesh, north-eastern India: Genesis and health perspective. Chemosphere 323:138067. https://doi.org/10.1016/j.chemosphere.2023.138067

Gulati A, Sharma B, Banerjee P, Mohan G (2019) Getting more from less: Story of India's shrinking water resources. New Delhi

Gupta AD, Pandey P, Feijóo A, Yaseen ZM, Bokde ND (2020) Smart water technology for efficient water resource management: A review. Energies (Basel) 13:6268. https://doi.org/10.3390/en13236268

Hanoon MS, Ahmed AN, Fai CM, Birima AH, Razzaq A, Sherif M, Sefelnasr A, El-Shafie A (2021) Application of artificial intelligence models for modeling water quality in groundwater: Comprehensive review, evaluation and future trends. Water Air Soil Pollut 232:411. https://doi.org/10.1007/s11270-021-05311-z

Hassan Z, Westerhoff H V. (2024) Arsenic contamination of groundwater is determined by complex interactions between various chemical and biological processes. Toxics 12

Hein C (2019) Adaptive strategies for water heritage: Past, present and future, 1st ed. Springer Nature

Hoque SF, Hope R, Arif ST, Akhter T, Naz M, Salehin M (2019) A social-ecological analysis of drinking water risks in coastal Bangladesh. Sci Total Environ 679:23–34. https://doi.org/10.1016/j.scitotenv.2019.04.359

Hussain J, Husain I, Arif M, Gupta N (2017) Studies on heavy metal contamination in Godavari river basin. Appl Water Sci 7:4539–4548. https://doi.org/10.1007/s13201-017-0607-4

IDSA (2010) Water security for India: The external dynamics. New Delhi

Ingrao C, Strippoli R, Lagioia G, Huisingh D (2023) Water scarcity in agriculture: An overview of causes, impacts and approaches for reducing the risks. Heliyon 9:e18507. https://doi.org/10.1016/j.heliyon.2023.e18507

Isaac R, Siddiqui S, Higgins P, Sahil Paul A, Abhick Lawrence N, Samson Lall A, Khatoon A, Singh A, Andrew Majeed P, Massey S, Prasad A (2024) Assessment of seasonal impacts on Water Quality in Yamuna river using Water Quality Index and Multivariate Statistical approaches. Waste Manag Bull 2:145–153. https://doi.org/10.1016/j.wmb.2024.07.006

Ivy N (2025) Arsenic contamination in GBM plains with a focus on Bihar, India: Risks and remediation. Discov Environ 3:5. https://doi.org/10.1007/s44274-024-00179-2

Jayakumar D, Bouhoula A, Al-Zubari WK (2024) Unlocking the potential of artificial intelligence for sustainable water management focusing operational applications. Water (Switzerland) 16:1–31. https://doi.org/10.3390/w16223328

Jha MK, Shekhar A, Jenifer MA (2020) Assessing groundwater quality for drinking water supply using hybrid fuzzy-GIS-based water quality index. Water Res 179:115867. https://doi.org/10.1016/j.watres.2020.115867

Jothivel S, Fu D, Mary S, Rene ER, Singh RP (2023) Potable and irrigation suitability of different water supply systems in riverine communities at River Cauvery distributaries. India. Environ Res 236:116809. https://doi.org/10.1016/j.envres.2023.116809

Júnior EO de C, Campos JMS de, Dias RJP, Barros NO (2024) Novelties on Tradescantia: Perspectives on water quality monitoring. Chemosphere 368:143732. https://doi.org/10.1016/j.chemosphere.2024.143732

Kanmani S, Gandhimathi R (2013) Investigation of physicochemical characteristics and heavy metal distribution profile in groundwater system around the open dump site. Appl Water Sci 3:387–399. https://doi.org/10.1007/s13201-013-0089-y

Kannaiyan N, Sekhar SI (2023) Microplastics in freshwater ecosystems of India: Current trends and future perspectives. ACS Omega 8:34235–34248. https://doi.org/10.1021/acsomega.3c01214ACS

Kannan G, Kolandhasamy P, Anbukkarasu S, Sigamani S, Ayyappan S, Rajendran R (2023) Marine plastics on the beaches of Cuddalore coast, southeast coast of India: A assessment of their abundance during Covid lockdown and post lockdown. Reg Stud Mar Sci 65:103051. https://doi.org/10.1016/j.rsma.2023.103051

Kapoor A, Anand M (2024) Addressing groundwater depletion crisis in India: Institutionalizing rights and technological innovations

Karthik KC, Naz M, Chaitanya K, Manasa KR (2025) Assessment of drinking water quality in various parts of India—a review. AJ J Med Sci 2:1–7. https://doi.org/10.71325/ajjms.v2i1.24.16

Kaur G, Kumar R, Mittal S, Sahoo PK, Vaid U (2021) Ground/drinking water contaminants and cancer incidence: A case study of rural areas of South West Punjab, India. Hum Ecol Risk Assess 27:205–226. https://doi.org/10.1080/10807039.2019.1705145

Keerthan L, RamyaPriya R, Elango L (2023) Geogenic and anthropogenic contamination in river water and groundwater of the lower Cauvery Basin, India. Front Environ Sci 11:1–17. https://doi.org/10.3389/fenvs.2023.1001052

Kerketta P, Baxla SL, Gora RH, Kumari S, Roushan RK (2013) Analysis of physico-chemical properties and heavy metals in drinking water from different sources in and around Ranchi, Jharkhand, India. Vet World 6:370–375. https://doi.org/10.5455/vetworld.2013.370-375

Khurana I, Sen R, India W (2008) Drinking water quality in rural India: Issues and approaches

Kripa PK, Prasanth KM, Sreejesh KK, Thomas TP (2013) Aquatic Macroinvertebrates as Bioindicators of Stream Water Quality-A Case Study in Koratty, Kerala, India. Res J Recent Sci 2:217–222

Kulinkina AV, Mohan VR, Francis MR, Kattula D, Sarkar R, Plummer JD, Ward H, Kang G, Balraj V, Naumova EN (2016) Seasonality of water quality and diarrheal disease counts in urban and rural settings in south India. Sci Rep 6:20521. https://doi.org/10.1038/srep20521

Kumar S, Nayak D (2025) Assessing the health costs of arsenic-contaminated drinking water in Bihar, India: A bivariate probit analysis. Groundw Sustain Dev 29:101429. https://doi.org/10.1016/j.gsd.2025.101429

Kumar A, Sharma S, Mehra R, Kanwar P, Mishra R, Kaur I (2018) Assessment of radon concentration and heavy metal contamination in groundwater of Udhampur district, Jammu & Kashmir, India. Environ Geochem Health 40:815–831. https://doi.org/10.1007/s10653-017-0027-2

Kumar M, Nagdev R, Tripathi R, Singh VB, Ranjan P, Soheb M, Ramanathan AL (2019) Geospatial and multivariate analysis of trace metals in tubewell water using for drinking purpose in the upper Gangetic basin, India: Heavy metal pollution index. Groundw Sustain Dev 8:122–133. https://doi.org/10.1016/j.gsd.2018.10.001

Kumar S, Toppo S, Kumar A, Tewari G, Beck A, Bachan V, Singh TB (2020) Assessment of heavy metal pollution in groundwater of an industrial area: A case study from Ramgarh, Jharkhand, India. Int J Environ Anal Chem 102:7290–7312. https://doi.org/10.1080/03067319.2020.1828391

Kumar A, Ali M, Kumar R, Kumar M, Sagar P, Pandey RK, Akhouri V, Kumar V, Anand G, Niraj PK, Rani R, Kumar S, Kumar D, Bishwapriya A, Ghosh AK (2021) Arsenic exposure in Indo Gangetic plains of Bihar causing increased cancer risk. Sci Rep 11:2376. https://doi.org/10.1038/s41598-021-81579-9

Kumar A, Ali M, Raj V, Kumari A, Rachamalla M, Niyogi S, Kumar D, Sharma A, Saxena A, Panjawani G, Jain P, Vidyarthi A, Kumar N, Kumar M, Niraj PK, Rahman MS, Bishwapriya A, Kumar R, Sakamoto M, Kumar S, Singh M, Ghosh AK (2023) Arsenic causing gallbladder cancer disease in Bihar. Sci Rep 13:4259. https://doi.org/10.1038/s41598-023-30898-0

Kumar D, Kumar R, Sharma M, Awasthi A, Kumar M (2024) Global water quality indices: Development, implications, and limitations. Total Environ Adv 9:200095. https://doi.org/10.1016/j.teadva.2023.200095

Kumar MD, Ballabh V (2000) Water management problems and challenges in India: An analytical review

Kyriakopoulos GL, Sebos I (2024) Marketable and banned pesticides in agriculture: Categorization, simulation, and crystallography review. Int J Mol Sci 25:11885. https://doi.org/10.3390/ijms252211885

Lapworth DJ, Das P, Shaw A, Mukherjee A, Civil W, Petersen JO, Gooddy DC, Wakefield O, Finlayson A, Krishan G, Sengupta P, MacDonald AM (2018) Deep urban groundwater vulnerability in India revealed through the use of emerging organic contaminants and residence time tracers. Environ Pollut 240:938–949. https://doi.org/10.1016/j.envpol.2018.04.053

Levidow L, Zaccaria D, Maia R, Vivas E, Todorovic M, Scardigno A (2014) Improving water-efficient irrigation: Prospects and difficulties of innovative practices. Agric Water Manag 146:84–94. https://doi.org/10.1016/j.agwat.2014.07.012

Life (2017) Life works with local village to build berkad. https://localinitiativesforeducation.us/news/2017/11/16/life-works-with-local-village-to-build-berkad. Accessed 8 Jan 2025

Lin L, Yang H, Xu X (2022) Effects of water pollution on human health and disease heterogeneity: A review. Front Environ Sci 10:880246. https://doi.org/10.3389/fenvs.2022.880246

Locke K (2025) Using Nemerow's pollution index to investigate correlations between land use and water quality across a sample of South African catchments. Water Pract Technol 20:91–107. https://doi.org/10.2166/wpt.2024.307

Mahajan M, Gupta PK, Singh A, Vaish B, Singh P, Kothari R, Singh RP (2022) A comprehensive study on aquatic chemistry, health risk and remediation techniques of cadmium in groundwater. Sci Total Environ 818:151784. https://doi.org/10.1016/j.scitotenv.2021.151784

Mahmoudian M (2024) What is Qanat? | Iran's ancient qanat system. In: Eavar. https://www.eavartravel.com/blog/2023/5/8/130631/qanat/. Accessed 8 Jan 2024

Malan A, Sharma HR (2023) Assessment of drinking water quality and various household water treatment practices in rural areas of Northern India. Arab J Geosci 16:96. https://doi.org/10.1007/s12517-022-11153-8

Malyan SK, Singh R, Rawat M, Kumar M, Pugazhendhi A, Kumar A, Kumar V, Kumar SS (2019) An overview of carcinogenic pollutants in groundwater of India. Biocatal Agric Biotechnol 21:101288. https://doi.org/10.1016/j.bcab.2019.101288

Manawi Y, Hassan A, Atieh MA, Lawler J (2024) Overview of radon gas in groundwater around the world: Health effects and treatment technologies. J Environ Manage 368:122176. https://doi.org/10.1016/j.jenvman.2024.122176

Marghade D, Mehta G, Shelare S, Jadhav G, Nikam KC (2023) Arsenic contamination in Indian groundwater: From origin to mitigation approaches for a sustainable future. Water (Switzerland) 15:4125. https://doi.org/10.3390/w15234125

Mehta P, Jangra MS, Baweja PK, Srivastav AL (2024) Impact of climate change on rural water resources and its management strategies. In: Srivastav AL, Izah SC, Van HE (eds) Madhav S. Challenges and Mitigation. Elsevier Ltd., Water Resources Management for Rural Development, pp 45–54

Mishra AP, Singh S, Sarkar MS, Singh R, Chandra N, Đurin B, Sahu H, Ahmed MFM, Abdo HG (2024) Integrating community perceptions, scientific data and geospatial tools for sustainable water quality management. Results Eng 23:102563. https://doi.org/10.1016/j.rineng.2024.102563

MoDWS (2013) National rural drinking water programme: Movements towards ensuring people's drinking water security in rural India. Guidelines-2013. New Delhi, India

MoJS (2019) NITI aayog report on water crisis. Press Information Bureau, Government of India

MoJS (2021) Groundwater pollution. Press Information Bureau, Government of India. Delhi

MoJS (2021) Drinking water quality monitoring & surveillance framework. Government of India, Ministry of Jal Shakti, Department of Drinking Water & Sanitation, National Jal Jeevan Mission. New Delhi

MoJS (2021) New national water policy. Delhi

MoJS (2022) Deaths due to lack of clean water. Press Inf Bur, Govment India

MoJS (2022) Measures for curtailing contamination of groundwater. Press Information Bureau, Government of India

MoJS (2023) Year-end review 2022. Press Information Bureau, Government of India.

MoJS (2023) Targets achieved under Pradhan Mantri Krishi Sinchayee Yojana. Press Information Bureau, Government of India.

MoJS (2023) Atal Bhuljal Yojna is being implemented in seven states for a period of 5 years from 01.04.2020. Press Information Bureau, Government of India

MoJS (2023) Ground water management and regulation (GWMR). department of water resources, river development and Ganga Rejuvenation, government of India. Ground Water Manag Regul (GWMR). Accessed on: 24/12/2024. Accessed 24 Dec 2024

MoJS (2023) National aquifer mapping program. press information bureau, government of India. https://pib.gov.in/PressReleaseIframePage.aspx?PRID=1944419. Accessed 17 Dec 2025
MoJS (2024) Annual report 2023–2024
MoJS (2024) Over 14.24 crore (73.93 %) rural households now have tap water supply in their homes. Delhi
MoJS (2024) Major steps taken by government for water conservation and rainwater harvesting to reduce water stress in the country. Delhi
MoJS (2025) Annual report 2024–2025. Department of water resources, river development and Ganga Rejuvenation, Government of India
MoJS (2025) India's groundwater revival making the invisible visible. Press Inf Bur, Govment India. Delhi
MoJS (2025) Fluoride and arsenic free drinking water under JJM. Delhi
MoJS (2025) Jal Jeevan Mission: Ensuring tap water for 15 crore rural families. Press Information Bureau. Government of India. New Delhi
MoJS (2025) Year end review 2024: department of water resources, river development and Ganga Rejuvenation. Delhi
MoRD (2011) Ensuring drinking water security in rural India - Strategic plan 2011–2022
MoRD (2019) Water conservation under MGNREGS. Press Information Bureau, Government of India
MoWR (2012) National water policy (2012)
MoWR RD and GR (2024) Cabinet approves Flood Management and Border Areas Programme (FMBAP) for the period 2021–26. Press Information Bureau, Government of India
Msangi S, Cline SA (2016) Improving groundwater management for Indian agriculture: Assessing tradeoffs across policy instruments. Water Econ Policy 2:1650027. https://doi.org/10.1142/S2382624X16500272
Musie W, Gonfa G (2023) Fresh water resource, scarcity, water salinity challenges and possible remedies: A review. Heliyon 9:e18685. https://doi.org/10.1016/j.heliyon.2023.e18685
Myvizhi P, Devi PA (2020) Heavy metal contamination in water of the River Cauvery—a case study of Erode, Salem and Namakkal districts, Tamil Nadu. J Himalayan Ecol Sustain Dev 15:190–202
Naika R (2022) Water management in rural communities: A step towards achieving sustainability by 2030. Sustain Water Resour Manag 8:115. https://doi.org/10.1007/s40899-022-00702-4
Nath D, Peramaiyan P, Kumari V, Laik R, Dotaniya ML (2024) Water resources of rural India: Challenges and management strategies for sustainable development. Elsevier Ltd.
Nawlakhe WG, Paramasivam R (1993) Defluoridation of potable water by Nalgonda technique. Curr Sci 65:743–748. https://www.jstor.org/stable/24095996
Ngai T, Dangol B, Murcott S, Shrestha RR (2005) Kanchan Arsenic Filter. Mass Inst Technol (MIT) Environ Public Health Organ (ENPHO). https://web.mit.edu/watsan/Docs/Other%20Documents/KAF/KAF%20booklet%20final%20Jun05.pdf
NGT (2024) Report by state of Uttarakhand in OA No. 728 of 2023 (In re: News item appearing in Hindustan dated 30.11.2023 titled arsenic found in groundwater in 25 States, fluoride in 27).
NJJM (2023) Handbook on drinking water treatment technologies. Delhi
NJJM (2023) Concise handbook for monitoring water quality of piped drinking water supply to rural households. New Delhi
NMCG (2020) Namami Gange Programme. https://nmcg.nic.in/NamamiGanga.aspx. Accessed 1 Nov 2025
Noguera-Oviedo K, Aga DS (2016) Lessons learned from more than two decades of research on emerging contaminants in the environment. J Hazard Mater 316:242–251. https://doi.org/10.1016/j.jhazmat.2016.04.058

Obayomi OV, Olawoyin DC, Oguntimehin O, Mustapha LS, Kolade SO, Oladoye PO, Oh S, Obayomi KS (2024) Exploring emerging water treatment technologies for the removal of microbial pathogens. Curr Res Biotechnol 8:100252. https://doi.org/10.1016/j.crbiot.2024.100252

Omarova A, Tussupova K, Hjorth P, Kalishev M, Dosmagambetova R (2019) Water supply challenges in rural areas: A case study from central Kazakhstan. Int J Environ Res Public Health 16:688. https://doi.org/10.3390/ijerph16050688

Panda P (2019) A study on water scarcity and security in India. Indian J Appl Econ Bus 1:73–90

Pandey A (2013) Environmental impacts of canal irrigation in India. Mediterr J Soc Sci 4:138–143. https://doi.org/10.5901/mjss.2013.v4n11p138

Pandey D, Singh A, Ramanathan A, Kumar M (2021) The combined exposure of microplastics and toxic contaminants in the floodplains of North India: A review. J Environ Manage 279:111557. https://doi.org/10.1016/j.jenvman.2020.111557

Pathak P, Laryea KB, Singh S (1989) A modified contour bunding system for Alfisols of the semi-arid tropics. Agric Water Manag 16:187–199. https://doi.org/10.1016/0378-3774(89)90002-4

Patidar K, Ambade B, Verma SK, Mohammad F (2023) Microplastic contamination in water and sediments of Mahanadi River, India: An assessment of ecological risk along rural-urban area. J Environ Manage 348:119363. https://doi.org/10.1016/j.jenvman.2023.119363

PIBRU (2024) Jal Sanchay Jan Bhagidari: A community-driven path to water sustainability in India. New Delhi

Prihar S, Sandhu B (1987) Irrigation of field crops: principles and practices. Indian Counc Agric Res (ICAR) 3:1–2. https://www.semanticscholar.org/paper/Irrigation-of-field-crops%3A-principles-and-practices-Prihar-Sandhu/51185ab5a07a5f864f6fa73020434fcd10111368

Qian Y (2016) Sustainable management of water resources. Eng 2:23–25

Quamar S, Parhi PK, Dwivedi AK, Singh N, Singh D, Singh H (2025) Geomorphic hazards and riverine dynamics in the deltaic region of Eastern India. Discov Appl Sci 7:670. https://doi.org/10.1007/s42452-025-07050-w

Rahman MM, Penny G, Mondal MS, Zaman MH, Kryston A, Salehin M, Nahar Q, Islam MS, Bolster D, Tank JL, Müller MF (2019) Salinization in large river deltas: Drivers, impacts and socio-hydrological feedbacks. Water Secur 6:100024. https://doi.org/10.1016/j.wasec.2019.100024

Ramesh S, Sukumaran N, Murugesan AG, Rajan MP (2010) An innovative approach of Drinking Water Quality Index-A case study from southern Tamil Nadu, India. Ecol Indic 10:857–868. https://doi.org/10.1016/j.ecolind.2010.01.007

Rani B (2015) Hazardous fluoride ion on groundwater of Rajasthan state: Its irrefutable manifestations, precautionary measures & attenuation methodologies. 3:1–13. https://doi.org/10.17577/IJERTCONV3IS03031

Rao KN, Latha PS, Kumar PVR (2022) Groundwater quality assessment for irrigation use in the Godavari Delta region of east coast India using IRWQI and GIS. Water Supply 22:2612–2629. https://doi.org/10.2166/WS.2021.454

Ratnalu GV, Dhakate R (2021) Human health hazard evaluation with reference to chromium (Cr^{+3} and Cr^{+6}) in groundwater of Bengaluru metropolitan city. South India. Arab J Geosci 14:24272. https://doi.org/10.1007/s12517-021-08671-2

Richards LA, Kumari R, White D, Parashar N, Kumar A, Ghosh A, Kumar S, Chakravorty B, Lu C, Civil W, Lapworth DJ, Krause S, Polya DA, Gooddy DC (2021a) Emerging organic contaminants in groundwater under a rapidly developing city (Patna) in northern India dominated by high concentrations of lifestyle chemicals. Environ Pollut 268:115765. https://doi.org/10.1016/j.envpol.2020.115765

Richards S, Rao L, Connelly S, Raj A, Raveendran L, Shirin S, Jamwal P, Helliwell R (2021b) Sustainable water resources through harvesting rainwater and the effectiveness of a low-cost water treatment. J Environ Manage 286:112223. https://doi.org/10.1016/j.jenvman.2021.112223

Rivett MO, Budimir L, Mannix N, Miller AVM, Addison MJ, Moyo P, Wanangwa GJ, Phiri OL, Songola CE, Nhlema M, Thomas MAS, Polmanteer RT, Borge A, Kalin RM (2019) Responding to salinity in a rural African alluvial valley aquifer system: To boldly go beyond the world of hand-pumped groundwater supply? Sci Total Environ 653:1005–1024. https://doi.org/10.1016/j.scitotenv.2018.10.337

RUPIB (2022) Atal bhujal yojana (Atal Jal)—A central sector scheme to improve groundwater management through community participation

Sabarathinam C, Karmegam U, Prasanna MV, Sasidhar P, Murugesan V (2011) A study on hydrochemical elucidation of coastal groundwater in and around Kalpakkam region, Southern India. Environ Earth Sci 64:1419–1431. https://doi.org/10.1007/s12665-011-0966-3

Saha D, Ray RK (2019) Groundwater resources of India: Potential, challenges and management. Groundwater Development and Management. Springer, Cham, pp 19–42

Salánki J, Farkas A, Kamardina T, Rózsa KS (2003) Molluscs in biological monitoring of water quality. Toxicol Lett 140–141:403–410. https://doi.org/10.1016/S0378-4274(03)00036-5

Sarkar A, Paul B, Darbha GK (2022) The groundwater arsenic contamination in the Bengal Basin—a review in brief. Chemosphere 299:134369. https://doi.org/10.1016/j.chemosphere.2022.134369

Sathe SS, Bhan U, Kushwaha A, Shabbir M, Patil K, Kamboj V, Goswami L, Oh S, Bahukhandi KD, Bind A (2025) Fluoride distribution, groundwater quality and health risk assessment for contaminated region near Krishna River (Maharashtra) India. Environ Nanotechnol Monit Manag 23:101033. https://doi.org/10.1016/j.enmm.2024.101033

Scown MW, Dunn FE, Dekker SC, van Vuuren DP, Karabil S, Sutanudjaja EH, Santos MJ, Minderhoud PSJ, Garmestani AS, Middelkoop H (2023) Global change scenarios in coastal river deltas and their sustainable development implications. Glob Environ Chang 82:102736. https://doi.org/10.1016/j.gloenvcha.2023.102736

Shah T (2012) Community response to aquifer development: Distinct patterns in India's alluvial and hard rock aquifer areas. Irrig Drain 61:14–25. https://doi.org/10.1002/ird.1656

Shanmugasundharam A, Akhina SN, Adhithya RP, Singh DSH, Krishnakumar S (2023) Water quality index (WQI), multivariate statistical and GIS for assessment of surface water quality of Karamana River estuary, west coast of India. Total Environ Res Themes 6:100031. https://doi.org/10.1016/j.totert.2023.100031

Sharma JR, Pearlman J, Sharma C (2011) Developing earth observation based end user technology for making sustainable development a living reality in semi arid areas - Nurturing through convergence of technologies at grass root level. Proceedings - 2011 IEEE Global Humanitarian Technology Conference. GHTC 2011:177–188. https://doi.org/10.1109/GHTC.2011.67

Sharma BM, Bečanová J, Scheringer M, Sharma A, Bharat GK, Whitehead PG, Klánová J, Nizzetto L (2019) Health and ecological risk assessment of emerging contaminants (pharmaceuticals, personal care products, and artificial sweeteners) in surface and groundwater (drinking water) in the Ganges River Basin, India. Sci Total Environ 646:1459–1467. https://doi.org/10.1016/j.scitotenv.2018.07.235

Sharma S, Patel P, Kulkarni SV, Deoshatwar A, Yadav R, Tanwar S, Manohar K, Dolla JR, Jain SK, Singh SK, Dikid T (2023) An outbreak of acute neurological illness associated with drinking water source following a cyclone in Eluru, West Godavari district, Andhra Pradesh, India, December 2020. Clin Epidemiol Glob Health 20:101261. https://doi.org/10.1016/j.cegh.2023.101261

Sharma M (2024) Tanker economy revealed: Rajasthan's groundwater level takes a dip as water mafias emerge. Down Earth

Sharma M, Gupta V, Gupta D (2024) Advancement in technologies of water management in rural areas. In: Water Resour Manag Rural Dev: ChallS Mitig. Elsevier, pp 117–126

Singh N (2004) Water management traditions in rural India: Valuing the unvalued. In: 18th European conference on modern south asian studies 9. https://www.diva-portal.org/smash/get/diva2%3A481013/FULLTEXT01.pdf

Singh S, Goyal MK (2025) A review of India's water policy and implementation toward a sustainable future. J Water Clim Chang 16:493–510. https://doi.org/10.2166/wcc.2025.560

Singh N, Sharma M (2020) Assessment of the quality of drinking water sources and human health in a rural area of Solan, north India. Mapan 35:301–308. https://doi.org/10.1007/s12647-019-00354-4

Singh J, Yadav BK, Schneidewind U, Krause S (2024) Microplastics pollution in inland aquatic ecosystems of India with a global perspective on sources, composition, and spatial distribution. J Hydrol Reg Stud 53:101798. https://doi.org/10.1016/j.ejrh.2024.101798

Singodia RR, Nirmal NK, John PJ (2024) Assessment of water quality of Kot Dam, Rajasthan, India, using water quality index. J Environ Biol 45:171–181. https://doi.org/10.22438/jeb/45/2/MRN-5180

Siphambe TV, Aliyu A, Souadji K, Bayongwa SA, Amans T, Fomena-Tchinda H, Banaon PY, Gina CS, Vuai HA, Farah AM, Niang AB, Taicha A, Ahmed S, Bashir A, Abdelbaki C, Mwamila TB, Gwenzi W, Nya EL, Noubactep C (2024) Mitigating flash flooding in the city: Drain or harvest? Water Supply 24:812–834. https://doi.org/10.2166/ws.2024.023

Srikanth R (2009) Challenges of sustainable water quality management in rural India. Curr Sci 97:317–325

Srivastava SK, Chand R, S.S. Raju RJ, I. K, Sachdeva J, And JS, Kaur AP (2015) Unsustainable groundwater use in Punjab agriculture: Insights from cost of cultivation survey. Indian J Agric Econ 70:367–378. https://doi.org/10.22004/ag.econ.230215

Ssekyanzi G, Ahmad MJ, Choi K-S (2024) Sustainable solutions for mitigating water scarcity in developing countries: A comprehensive review of innovative rainwater storage systems. Water (Switzerland) 16:0–21. https://doi.org/10.3390/w16172394

Sujathan S, Singh A (2023) Investigation of potential drivers of elevated uranium prevalence in Indian groundwaters with a unified speciation model. Environ Sci Technol 57:1970–1986. https://doi.org/10.1021/acs.est.2c08524

Swain S, Mishra PK, Nandi S, Pradhan B, Sahoo S, Al-Ansari N (2024) A simplistic approach for monitoring meteorological drought over arid regions: A case study of Rajasthan. India. Appl Water Sci 14:36. https://doi.org/10.1007/s13201-023-02085-z

Swati S, Umesh S (2015) Nemerow's Pollution Index: For groundwater quality assessment. J Environ Sci Pollut Res 1:23–31

Szczerbińska N, Gałczyńska M (2015) Biological methods used to assess surface water quality. Arch Pol Fish 23:185–196. https://doi.org/10.1515/aopf-2015-0021

Tirkey P, Bhattacharya T, Chakraborty S (2013) Water quality indices—important tools for water quality assessment: A review. Int J Adv Chem 1:15–28

Tuan PV, Jiang Y, Stigter T, Zhou Y (2024) Understanding groundwater use and vulnerability of rural communities in the Mekong Delta: The case of Tra Vinh province. Vietnam. Groundw Sustain Dev 25:101095. https://doi.org/10.1016/j.gsd.2024.101095

Tyagi S, Sharma B, Singh P, Dobhal R (2013) Water quality assessment in terms of water quality index. Am J Water Resour 1:34–38. https://doi.org/10.12691/ajwr-1-3-3

Uddin MG, Nash S, Olbert AI (2021) A review of water quality index models and their use for assessing surface water quality. Ecol Indic 122:107218. https://doi.org/10.1016/j.ecolind.2020.107218

Vaiphei SP, Kurakalva RM (2021) Hydrochemical characteristics and nitrate health risk assessment of groundwater through seasonal variations from an intensive agricultural region of upper

Krishna River basin, Telangana. India. Ecotoxicol Environ Saf 213:112073. https://doi.org/10.1016/j.ecoenv.2021.112073

Wang Q, Su M (2020) Integrating blockchain technology into the energy sector - from theory of blockchain to research and application of energy blockchain. Comput Sci Rev 37:100275. https://doi.org/10.1016/j.cosrev.2020.100275

Wasay SA, Haron MdJ, Tokunaga S (1996) Adsorption of fluoride, phosphate, and arsenate ions on lanthanum-impregnated silica gel. Water Environ Res 68:295–300. https://doi.org/10.2175/106143096x127730

WHO (2022) Guidelines for drinking-water quality: Fourth edition incorporating the first and second addenda

WHO (2023) Drinking-water. https://www.who.int/news-room/fact-sheets/detail/drinking-water#:~:text=Inadequate management of urban%2C industrial, through exposure to infested water. Accessed 10 Jan 2025

WHO (2025) Primary health care. https://www.who.int/news-room/fact-sheets/detail/primary-health-care._Accessed_date:_28/10/2025

Xia Y, Zhang M, Tsang DCW, Geng N, Lu D, Zhu L, Igalavithana AD, Dissanayake PD, Rinklebe J, Yang X, Ok YS (2020) Recent advances in control technologies for non-point source pollution with nitrogen and phosphorous from agricultural runoff: Current practices and future prospects. Appl Biol Chem 63:8. https://doi.org/10.1186/s13765-020-0493-6

Yadav KK, Yadav JP, Bijarniya S (2019) Adoption of drip irrigation system by the farmers in Jhotwara panchayat samiti of district Jaipur, Rajasthan. Indian J Ext Educ 55:142–144

Yang L, Driscol J, Sarigai S, Wu Q, Lippitt CD, Morgan M (2022) Towards synoptic water monitoring systems: A review of AI methods for automating water body detection and water quality monitoring using remote sensing. Sensors 22:2416. https://doi.org/10.3390/s22062416

Techniques in Water Quality Modeling 13

Ankita Singh, Leelambar Singh, Nitika Mundetia,
and Ravindra Nath Tripathi

Abstract

In the context of water quality, modeling is a powerful and crucial technique for assessing the interaction of water bodies under different pressures and anthropogenic effects. The objective of this paper is, therefore, to present a brief discussion of the important techniques in water quality modeling, specifically addressing their applications and approaches. These techniques are employed to model and evaluate water quality parameters, including pollutants, dissolved oxygen, nutrient biology, and effects of land use change and climate change on the aquatic ecosystems. Methodologies used include deterministic models where the mathematically determined formulae are used to ascertain water quality, from the predictable processes, and stochastic models that challenge the regularity of processes involved in a water system with variation in the results obtained. Empirical models designed with historical data as the determinant provide basic strategies for data-rich scenarios, while mechanistic or process-based models give precise physical and chemical occurrence trends of the water basin. Other techniques, more complex, include hybrid models, artificial intelligence (AI), machine learning (ML), and professional Geographic Information Systems (GIS) that assimilate

A. Singh · L. Singh (✉)
Department of Civil Engineering, Malaviya National Institute of Technology, Jaipur, Rajasthan, India
e-mail: leelambar.ce@mnit.ac.in

N. Mundetia
Department of Basic and Applied Science, Sangam University, Bhilwara, Rajasthan, India
e-mail: nitika.mundetia@sangamuniversity.ac.in

R. N. Tripathi
Graphic Era University, Dehradun, Uttarakhand, India

S. Yadav et al. (eds.), *Water Resources in Delta Region in India*, Deltas of the World,
https://doi.org/10.1007/978-3-032-06500-1_13

various data and have high precision in the determination of water quality parameters. In the evaluation, we compare and contrast these models and specify unique contexts in which each model works best. The conclusion has revealed that the deterministic and mechanistic models are effective mainly for system understanding, and the AI, GIS-based, and hybrid models appear to have potential in water quality real-time monitoring/forecasting. The results underscore the need for further research and the application of complex modeling methodologies for efficient future water resource management. Moreover, the strategies presented in this paper are closely related to the United Nations Sustainable Development Goals (SDGs), and SDG 6 (Clean Water and Sanitation) by enhancing water quality management practices. These modeling methods also address climate-related and anthropogenic effects, supporting SDG 13 (Climate Action) and indirectly supporting SDG 14 (Life Below Water) and SDG 15 (Life on Land), as they enable the sustainable management of aquatic and terrestrial ecosystems.

13.1 Introduction

The use of water quality modeling is a fundamental component in current and actual environmental science as well as management since it allows for the effective description, anticipation, and conduct of changes within aquatic ecosystems (Wang et al. 2024). Demand for devices that help protect the water supply has increased due to the complexity of environmental conditions because of urbanization, industrialization, agricultural activities, and the enhanced rate of climate change (Jain et al. 2024). The importance is not only in measurements and controls but also in system models that help focus on how the water systems can behave under certain states. Such models help identify tendencies in water quality, evaluate the outcomes of management decisions, and develop long-term approaches to coping with water ecosystem health (Yan et al. 2022). Water quality models are vital for addressing the challenges posed by pollution, climate variability, and anthropogenic influences on aquatic systems (Ejigu et al. 2021). These models help the investigators and planners to establish the origin and distribution of pollutants and their impacts, as well as possible treatments through mimicking the physico-chemical as well as biochemical events that occur in water bodies (Hendren et al. 2013). The capability of these models to integrate diverse data sources from hydrological conditions to land-use patterns allows for comprehensive analyses that are crucial for informed decision-making.

The present established methods of modeling involve deterministic, empirical, and stochastic models, thus providing the basis for water quality evaluations (Ejigu et al. 2021). Some of the mathematical tools that have been used include equations that define certain processes, such as Streeter-Phelps models concerning the roles of DO and BOD (Kothandaraman 1968). These models are very beneficial for scenario appraisal and system comprehension based on their clear structure that defines the complex relations between pollutants and aquatic systems. In contrast, empirical models use past observations as the basis to define correlations between the quality characteristics of water and environmental conditions. Metrics such as regression analysis and water quality indices provide easily understandable and quantifiable comparison tools in data-intensive water quality analysis, irrespective of the actual kinetics and thermodynamics (Jadhav et al. 2023). While stochastic models contain probability factors to address fluctuations and the diversity of nature. This approach is illustrated by methods such as Monte Carlo simulation; it provides sound results for scenarios that include stochastic elements (Jamatut et al. 2024).

In coming up with complex methods for water quality models, the field of modeling has evolved logically from a methodological to a paradigm shift that is both computational and technical. Machine learning or Artificial intelligence-based approaches are now at the forefront of massive data analysis and nonlinear patterns in the water quality data (Lowe et al. 2022; Krishnan et al. 2022). Other AI techniques include decision trees, neural networks, and support vector machines (SVM); the use of which should improve both the accuracy and efficiency of water quality analysis (Hanoon et al. 2021). These methods fit in environments that are changing, where interactions between many variables can be challenging to capture with most models (Alnahit et al. 2022). Technological advances have also brought about dramatic changes in water quality modelling, especially through the use of Geographic Information Systems (GIS). Environmental models using GIS help in the depiction of concentrations of nutrients/pollutants, identification of areas that have high concentrations, and spatial analysis of the parameters or correlation of land uses and water quality (Wang and Xie 2018). The biggest advantage derived from the overlay of the various layers is that researchers are capable of taking a broad view of the issues with water quality, in a way directing efforts and resources in the right direction.

Climate change remains a notable threat to water quality management; it modifies the occurrence of extreme weather events and influences temperature regimes and precipitation (Singh and Saravanan 2020; Muyambo et al. 2023). Contemporary models of water quality take climate into account and use its predictions to estimate the constant impacts on aquatic ecosystems of these modifications. Among them, water quality models that depict the situation in which runoff patterns change, nutrient concentrations increase, and risks of algal blooms appear to help understand potential threats to water resources (Elhabashy et al. 2023). Incorporation of climate information not only improves the ability of a water quality model to predict future conditions but also provides management frameworks that consider future changes. Due to the complexity of water quality problems, the analysis and prediction frequently require the use of several models and data integrated into the research process (Rode et al. 2010). Watershed models, like SWAT (Soil and Water Assessment Tool) and HSPF (Hydrological Simulation Program–Fortran), illustrate the relationships between land use, hydrological, and water quality conditions and provide strategies to effectively control pollutants. These approaches provide a comprehensive view of water systems by bridging the gap between various scales and processes and sustaining the appropriate land-use management (Xie and Lian 2013; Keller et al. 2023). Likewise, models, including SWAT-MODFLOW, which combine surface and groundwater resources, provide clear information about relationships between these systems and provide a broader vision of water resource management (Mundetia et al. 2024).

Mathematical models are crucial in chemical and biological processes such as nutrient cycling, sediment interactions, and pathogen transport. Such models as the eutrophication model, for example, assume algal growth and dissolved oxygen as a function of nutrient concentrations, rating the impacts of agricultural runoff and urbanization (Jin et al. 2021). Sediment–water interaction models are concerned with sediment's function as a nutrient source or sink and contaminants, thus determining water quality through adsorption and desorption processes. Epidemiological models of pathogen transport quantify the propensity and rate of spread and decay of infective agents, which can be used for analysis of disease transmission and hazards associated with contamination (Alex et al. 2023).

Models of water quality have a vast array of applications, and there are many possibilities for immediate utilization: monitoring, dynamic modeling, and policy making. They are crucial in determining the effectiveness of the measures to control pollution, estimating the effects of changes in land use, and directing efforts in ecological restoration. For example, during the COVID-19-enforced lockdown, the Streeter-Phelps model was employed to comprehend the variations in the Ganga River's water quality metrics and identified the positive changes due to a decrease in anthropogenic activities (Singh et al. 2023).

In addition, the results of water quality models entail significant implications for sound water resource management for sustainable development. These models not only reduce deterministic approaches to people–place relations but also identify critical thresholds, predict future trends, and quantify social benefits of management interventions that lead to development-oriented policies while protecting the environment. The more complex models incorporate other techniques, including AI and GIS, which increase the model's ability to handle new scenarios, such as changes in weather conditions, as well as the growth and development of water systems within urban areas (Hanoon et al. 2021).

Water quality modeling is in an excellent position to drive the science that is needed to confront the complexity of water issues affecting aquatic systems. Analyzing the types of modeling methodologies that have been developed over the years, starting with the deterministic approach and ending with modern AI-based ones, it is noted that such a trend is associated with the increasing requirements for reliability, flexibility, and integration of the environmental management field (Wani et al. 2024). Water quality models play an increasingly important role in developing sustainable strategies as the effects of human activity and climate variability continue to worsen (Singh and Saravanan 2020). By bridging science, technology, and policy, these models provide the foundation for informed decision-making as well as the long-term management of water resources worldwide.

13.1.1 Building the Model

Model development is a series of procedures that turn a real-world system into one that can be used to analyze and predict. Every step makes the model physically viable, scientifically sound, and practically applicable in making decisions (Robinson). Most significant steps are as follows:

- Physical Conceptual Model

  This becomes possible by beginning with the creation of a physical conceptual model, which gives a simplified representation of the real-world system. It determines the key elements, processes, and interactions that control the behaviour of the system. To illustrate this, the conceptual model in a water quality setting can explain the inflow of pollutants, organic matter decomposition, and reaeration by the atmosphere. This is the step that is the blueprint of what processes should be incorporated in the model and how they are interrelated.
- Mathematical/Process Representation

  Governing equations are then used to convert the conceptual framework to a mathematical model. These equations are the physical, chemical, or biological processes governing the dynamics of the system (e.g., mass balance, decay kinetics, hydrodynamic transport). The representation can be empirical, deterministic, or stochastic, depending on the purpose. This step helps to make sure that the conceptual knowledge is translated into a quantitative form to compute it.
- Data Collection and Input Preparation

  The quality and completeness of input data determine the success of any model. Hydrological, meteorological, and water quality parameters, boundary conditions, and forcing factors are also key datasets. These data are processed, interpolated, and formatted in a way that they can be used in the modeling platform. The uncertainty of data is also considered and reduced by means of proper pre-processing.
- Model Setup and Implementation

  The model is set up in the modeling environment or software of choice using the available data. This consists of specifying the spatial and time domain, discretising the study area (grid or reaches), and setting initial and boundary conditions. At this stage, parameters like decay rates, dispersion coefficients, and reaction constants are added.
- Calibration

  Calibration is the procedure of fine-tuning model parameters so that the outputs of the model are consistent with actual conditions. As a step, this makes the model fine-tuned such that it replicates past measurements of the important variables (e.g., dissolved oxygen, BOD, and nutrient concentrations). Calibration gives the assurance that the model has the capability of reproducing the conditions of the study system at the base.

- Validation

  After the calibration, the model is then tested against independent data not used in the calibration. The validation tests the predictive power of the model and makes sure that the model is not overfit. The validation model shows that it is able to generate observed dynamics under different conditions, thus confirming its reliability.
- Assessment of Model Accuracy and Correction of Model Bias

  Upon validation, the model is evaluated based on the statistical values of Root Mean Square Error (RMSE), Nash-Sutcliffe Efficiency (NSE), correlation coefficient (R2), and percent bias (PBIAS). In the case of systematic errors (bias) found, corrections are made by either adjusting parameters, by modeling the process more accurately, or by handling the data more favorably. This continuous enhancement enhances the strength of the model.
- Scenario Applications

  Scenario analysis can then be applied to a fully calibrated and validated model. Various management and environmental conditions are modeled to evaluate the impacts they might have.

13.2 Mathematical and Numerical Modeling

Mathematical and numerical modeling is also important in water quality management because it offers a systematic view of the water system behaviour. These models use algebraic formulas for the biological, chemical, and physical processes that affect water quality, such as the nutrient cycle and pollution transport (Ziemińska-Stolarska and Skrzypski 2012; Kumar et al. 2024). Numerical models are able to solve an enormous number of equations simultaneously that can consider the entire database, as well as the temporal behavior of the systems related to water. These models simulate flow dynamics, weather, and land use to calculate the anticipated changes in water quality under certain conditions (Tayebzadeh Moghadam et al. 2021). They are useful for trend analysis, evaluating the effectiveness of the management actions, and decision-making on sustainable water management. Examples of such models include:

13.2.1 Deterministic Models

Forecasts that are accurate and precise and which utilize mathematical models that require exact inputs and coefficients are known as deterministic models. What makes these models deterministic is that, regardless of how often the model is applied, the results obtained will always be the same. These models are usually represented by mathematical equations, for example, differential equations, which describe how one variable relates to another in the system. For instance, in hydrology, deterministic models can be used to describe the flow of water as well as pollutants by the use of equations that describe physical laws. These models are more suited for figuring out how water quality metrics behave under certain conditions. For example, the Streeter-Phelps model is quite popular for studying the behaviour of DO and BOD in rivers to arrive at the degree of organic pollution (Rasheed et al. 2024).

The Streeter-Phelps equations were first introduced with linear decay kinetics in 1925 by Streeter and Phelps, who reported on the Ohio River for the U.S. Government (D'Agostino 2024). The Streeter-Phelps equations provide a mathematical framework for understanding the interactions between pollutants and dissolved oxygen in rivers.

Formulation of the streeter-phelps Equations

The equations indicated here give details on the fluctuation of the river's dissolved oxygen content during the degradation process of the pollutants over a distance as well as a time period. The basic assumptions underlying these equations include:

- Advection: BOD measured is transported along the river at the same rate as the DO since both are affected by the river flow.
- Degradation: BOD in the present concentration is in direct proportion to its degradation, while DO reduces as it is used in the degradation of BOD.
- Re-aeration: The re-aeration rate is directly related to the amount of oxygen that is required by the river, which is an estimation of the difference between the saturation concentration of oxygen of the river water and that of the present concentration of dissolved oxygen.
- External Pollution: An outside pollution source raises the BOD level.

The basic form of the Streeter-Phelps equations can be expressed as follows:

Singh (2024) conducted a study on the Ganga River valley, including its delta, by examining the DO using the Streeter-Phelps model to assess saturation deficit across 14 real-time monitoring stations along the Ganga River. The model's primary objective was to evaluate how the lockdown conditions affected the river's water quality, particularly focusing on the relationship between biochemical oxygen demand (BOD) and DO levels (Eq. 13.1 and 13.2).

- For Biological Oxygen Demand (BOD):

$$\frac{\partial b}{\partial t} + v\frac{\partial b}{\partial x} = -k_1 b + k_3 \tag{13.1}$$

where:

b BOD concentration (mg/L).
v flow velocity (m/s).
k_1 BOD decay rate (per day).
k_3 external pollution source term (mg/L).

- For Dissolved Oxygen (DO):

$$\frac{\partial c}{\partial t} + v\frac{\partial c}{\partial x} = -k_1 b + k_2(c_3 - c) \tag{13.2}$$

where:

c DO concentration (mg/L)
c_s oxygen saturation concentration (mg/L)
k_2 re-aeration rate (per day)

To support the present study, real-time water quality data were retrieved from CPCB for the Ganga River covering lockdown, post-lockdown, and pre-lockdown periods. The Streeter-Phelps model was developed based on differential equations that represent oxidizable organic and oxygen concentrations. The key equations (Eqs. 13.3 and 13.4) used in the model are:

$$\frac{dz}{dt} = -k_d \cdot Z \tag{13.3}$$

$$\frac{dX}{dt} = -k_d \cdot Z \cdot s + k_a(X_{sat} - X) \tag{13.4}$$

where

Z concentration of degradable organic matter (mg/L),
X DO level (mg/L),
k_d decay rate (1/Time),
k_a aeration rate (1/Time),
s stoichiometric factor,
X_{sat} saturation level of oxygen (mg/L) (Singh et al. 2022).

The findings, normalized by applying the Streeter-Phelps model, were found to provide useful information on the water quality of the Ganga River during lockdown. Through a comparison of the outcomes with those from earlier tests of gaseous DO levels, it was possible to observe a significant lift in DO levels at various monitoring stations, chiefly at Varanasi and other locations, by about 25–30%. A considerable decrease in BOD levels reflects an improvement in organic water quality because of the absence of industrial processes. The model showed a reasonably high R^2 value of 0.57, meaning that while the model gave useful insights, its ability to predict DO levels under dynamic conditions is not accurate (Singh et al. 2022).

13.2.2 Empirical Models

An empirical model is a mathematical representation that is derived from observed data rather than from theoretical principles (Andrews 2021). These models are built using the patterns experienced in that data and enable the researcher to predict something or understand phenomena within one context. Empirical models are those that are developed based on data collected from experimental observations or previous records. The primary source of model derivation is this data, so that actual conditions and behaviours are reconstructed. Some common empirical models include:

- *Quality Index (WQI)*: Contains more than one water quality characteristic in order to offer a single figure that symbolizes general water quality (Patel et al. 2023).
- *Regression Models*: Regression analysis, using either linear or non-linear models, to select the water parameters that correlate with the factors affecting them.
- *Artificial Neural Networks (ANNs):An* Algorithm designed by using past data and by learning new patterns for the purpose of forecasting the water quality (Kurani et al. 2023).
- *Fuzzy Logic Models*: Time series of pluviometric data must be used in conjunction with models for predicting the water quality conditions, and fuzzy sets must be applied to the real data, endowed with a stochastic nature (Trach et al. 2022).
- *Multivariate Statistical Models*: By getting into subpercentages, season, etc., use techniques such as PCA to discover the pattern and other variables affecting water quality.
- *Empirical Loading Models*: Forecast pollutants generally through land use, runoff, and precipitation parameters.
- *Box Models*: Gross frameworks that let the input–output flow, which is used to determine the water quality expectations.
- *Time Series Models*: Prescribe the general conditions of water quality in the future of some given history and tendencies together with the help of ARIMA models (Jain and Mallick 2017).

13.2.3 Stochastic Models

A stochastic model is a technique that predicts statistical characteristics of results by considering random variations in parameters across time, frequently using data from the past. A probabilistic method is necessary for stochastic models to provide system output that depends on input data in addition to other, typically random, inputs (Costa 2021). The Monte Carlo simulation is one example of a stochastic model. The Monte Carlo simulation was invented by mathematicians Stanislaw Ulam and John von Neumann during World War II, primarily while working on the Manhattan Project, and named after the famous casino in Monaco due to the method's reliance on random sampling, like a roulette wheel; Ulam initially conceived the idea while playing solitaire, wanting to calculate the probability of winning by simulating many games through random draws (Trang 2017). Numerous uncertain problems can be analyzed using the Monte Carlo method, which depends heavily on probability and statistics theory (Seifi et al. 2020). The study titled "A hybrid approach based on Monte Carlo simulation-VIKOR method for water quality assessment" presents a novel methodology for evaluating water quality, particularly in the context of the Songhua River tributary. It assumes that the probability distribution function of a random variable is known and uses multiple sampling to increase the output results' dependability (Cimorelli et al. 2020). The Monte Carlo (MC) approach is used in this study to overcome sampling uncertainty, and it is assumed that the water quality data follow a normal distribution. As a result, this work employs the MC approach, an uncertainty technique that is frequently used to address decision-making issues and which successfully mitigates the impact of human factors on data quality, to randomly replicate water quality monitoring data. The index weight is then determined using the CRITIC technique. As an objective weighting technique, it can guarantee that assessment index data can be used to its fullest potential and successfully steer clear of subjective issues. Lastly, the VIKOR method was used to assess the water quality (Yang and Chen 2023).

$$f\left(x_{ij}^{MC}\right) = \frac{1}{\theta_j\sqrt{2\pi}} e^{-\frac{\left(x_{ij}^{MC}-\mu_j\right)^2}{2\theta_j^2}} \tag{13.5}$$

$$x_{ij}^{MC} = \mu_j + \theta_j \cdot Z \tag{13.6}$$

where,

f()	represents the normal distribution function (Eq. 13.5),
x_{ij}^{MC}	denotes the simulation value (from MC) of the j_{th} index under the i_{th} evaluation object (Eq. 13.6),
μ_j and θ_j	are the mean and standard deviation of the distribution, respectively.
Z	represents a random variable drawn from a standard normal distribution (Yang and Chen 2023).

13.3 Chemical and Biological Process Simulation

Abstraction of chemicals and biological processes is a universally adopted modeling strategy for analyzing the complexities of water ecosystems, especially with regard to chemical quality and nutrients. Eutrophication, mainly resulting from nutrient enrichment due to agricultural and urban pollution, causes a massive and undesirable boost in algae, and it may have a negative effect on the aquatic ecosystem and seriously harm the quality of water (Withers et al. 2014). There are also other models like the CE-QUAL-W2 that should be useful in replicating the behaviour of other important nutrients like nitrogen and phosphorus that cause massive algal growth and water conditions in general (Zhang and Rao 2012).

Additional sediment–water interaction models enhance understanding regarding the ability of sediments to release or absorb nutrients depending on existing conditions (Chen et al. 2016). Moreover, pathogen transport models estimate the trends and rates of pathogenic microbes, facilitating contamination risk and water management plan estimation (Niazi et al. 2015; Weiskerger and Phanikumar 2020). Altogether, these simulations are critical in understanding the sustainable use of aquatic systems with the aim of predicting and managing the effects of nutrient loading, sedimentation, and pathogen translocation on water quality (Sabel et al. 2024).

13.3.1 Eutrophication Models

The main issue affecting the majority of surface waters nowadays is eutrophication, which refers to excessive plant growth brought on by nutrient enrichment by human activity (Smith et al. 2009). Inland water bodies that are naturally or artificially enriched typically experience eutrophication with typical algal blooms, which lower dissolved oxygen levels and degrade the water quality for human use, both of which have clear negative consequences on fisheries (Karul et al. 2000). Eutrophication models are essential tools for understanding and managing the nutrient dynamics in aquatic ecosystems, particularly in lakes and reservoirs. These models simulate the processes that lead to nutrient enrichment, primarily due to anthropogenic activities, and their subsequent effects on water quality and aquatic life. Both nitrogen and phosphorus are of concern in eutrophication, although nitrogen has garnered significantly more attention since it typically inhibits primary production in estuaries and coastal waterways, also because the worldwide utilization of nitrogen from synthetic fertilizers is far greater than that from phosphorus (Glibert et al. 2005). The models simulate the cycling of nitrogen and phosphorus through various forms, such as dissolved inorganic nitrogen (DIN), ammonium (NH_4^+), nitrate (NO_3^-), and total phosphorus (TP). The interactions between these forms are crucial for understanding how nutrient loading affects algal growth and water quality (Zhang and Rao 2012). CE-QUAL-W2, QUAL2K, EPA WASP, SWAT, EUTRO5, DYNHYDRO,

MIKE 11/3, PROTECH, BATHTUB, and HydroQual are examples of the eutrophication model. The U.S. Army Engineer Waterways Experiment Station (WES) Environmental and Hydraulics Laboratories created the 2-D, laterally-averaged, hydrodynamic water quality model CE-QUAL-W2 as a tool for managing water resources (Kim et al. 2006).

13.3.2 Sediment–Water Interaction Models

Soil erosion-derived sediment affects the usage of water resources in various chemical, physical, and biological ways. When sediment particles are present, the quality of the water for a given purpose is diminished. Sediments can act as both sources and sinks for nutrients such as nitrogen and phosphorus. During periods of low oxygen (anoxic conditions), sediments may release nutrients back into the water column, promoting algal blooms and eutrophication (Chen et al. 2016). Conversely, under aerobic conditions, sediments can absorb excess nutrients, thereby improving water quality. Sediments can also adsorb contaminants, including heavy metals and organic pollutants. The adsorption capacity of sediments for these contaminants can vary widely based on sediment characteristics and the nature of the contaminants (Yang et al. 2021). For example, the presence of iron and aluminium oxy-hydroxides in sediments can enhance the adsorption of lead, thereby reducing its bioavailability in the water column (Bakshe and Jugade 2024). However, under certain conditions, such as changes in pH or ionic strength, these contaminants can be desorbed and re-enter the water column, posing risks to aquatic life and human health (Wang et al. 2022).

Sediment–water interaction models are designed to simulate the dynamics of sediment transport, nutrient cycling, and contaminant behaviour in aquatic environments. These models help predict how changes in environmental conditions, such as flow velocity, sediment concentration, and water chemistry, affect the adsorption and desorption processes at the sediment–water interface (Tseng and Tinoco 2021). commonly used sediment–water interaction models are EUTRO5, CE-QUAL-W2, WASP, AQUATOX, MIKE 11/3, DYNHYDRO, HydroQual, SediNet, PROTECH, BATHTUB.

13.3.3 Pathogen Transport Models

In the water quality context, a pathogen transport model is a numeric model that defines and describes the transport and die-off of pathogens in the water body, considering influences such as water currents, sedimentation, solar radiation, and other factors to help evaluate contamination risks and water management scenarios. There are many watershed-scale models in which pathogen transport analysis can be included, of which two commonly used models are the Hydrological Simulation Program Fortran (HSPF) and the Soil and Water Assessment Tool (SWAT) (Niazi et al. 2015). The advection–dispersion-reaction equation for the fraction of inspired oxygen (FIO) concentration is the general microbial fate and transport equation. Among these terms, several have been included for advective transport, diffusion, or microbial breakdown and dispersion in the water column as shown in Eq. 13.7.

$$\frac{\partial C}{\partial t} + u\frac{\partial C}{\partial x} + v\frac{\partial C}{\partial y} + w\frac{\partial C}{\partial z} = \frac{\partial}{\partial x}\left(K_H\frac{\partial C}{\partial x}\right) + \frac{\partial}{\partial y}\left(K_H\frac{\partial C}{\partial y}\right) + \frac{\partial}{\partial z}\left(K_V\frac{\partial C}{\partial z}\right) - kC \tag{13.7}$$

where,

C	relates to the microbial concentration (mL^{-1}),
k	represents the overall microbial decay rate (d^{-1}),
K_V and K_H	are the vertical and horizontal mixing coefficients, respectively ($m^2\ s^{-1}$) (Weiskerger and Phanikumar 2020).

13.4 Integrated Modeling Approaches

Systems in water quality management use several models, tools, and data sources that solve different problems but are integrated into one system. These approaches take into account aspects like hydrology, land use change, climate change, and pollutant behavior in order to mimic the process occurring in watersheds, rivers, lakes, and estuaries (Talukdar et al. 2023). While combining different models, such approaches present higher results in the predictions and are entirely useful for allocating the management of water quality.

13.4.1 Watershed Models

The watershed models, as employed in water quality modeling, seek primarily to embed and forecast the behavior of water systems in a given watershed or catchment. These models are valuable in determining the effects of land use, climate, and such factors on the water quality of rivers, lakes, and streams (Cheng et al. 2022; Singh and Saravanan 2020). Typically, watershed models are employed for estimating the transport and fate of pollutants, sediment, nutrients, and contaminants in a watershed from their source to the receiving waters. First, it was possible to model the whole hydrological cycle, as proved by the release of the Stanford Watershed Model by Crawford and Linsley in 1966. Watershed models that are developed and used worldwide are SWAT (Soil and Water Assessment Tool) and HSPF (Hydrological Simulation Program–Fortran). SWAT is one of the process-based semi-distributed hydrologic models developed by the Agricultural Research Service, United States Department of Agriculture (Athira 2021). The SWAT model is an advanced eco-hydrological model that has been widely applied to watersheds across the globe (Arnold et al. 2018; Aloui et al. 2023).

The U.S. Environmental Protection Agency (EPA) sponsored the development of HSPF, a powerful, extremely intricate mathematical computer program designed to continually model water quantity and quality processes in both natural and artificial water systems (Skahill 2004).

Commonly used models include:

- SWAT (Soil and Water Assessment Tool): Models the hydrology, sediment transport, and chemical and physical quality of water concerning land use and climate change futures (Arnold et al. 2018).
- VSM (Vegetative Swale Model): Successively copies how pollutant removal is accomplished by vegetative swales of urban areas.
- HSPF (Hydrological Simulation Program—FORTRAN): Semi-distributed and Specific to model hydrology and water quality based on pollutants of the available land uses in the concerned area (Roostaee and Deng 2019).
- GWLF (Generalized Watershed Loading Function): Predicts the discharge of pollutants in a watershed, especially nutrients and sediment.
- MIKE SHE: Models and predicts depths of surface water, groundwater, and water quality in watersheds.
- WEPP (Water Erosion Prediction Project): Specializes in the erosion of soil and its effect on water quality.
- SLAMM (Stormwater Management Model): Simulation of events of stormwater runoff and its effect on water quality in the urban area (Haris et al. 2016).
- GLOBE (Global Land Use and Basin Environment Model): Simulates land use and climate change impacts on water quality at a global scale.

13.4.2 Groundwater-Surface Water Interaction Models

Groundwater and surface water, such as rivers, lakes, reservoirs, wetlands, and estuaries, interconnect with each other in nearly all parts of the globe. The relationship between the two water types is multilateral and a little bit more challenging to explain. For such interactions in regard to climate, landform, geology, and biotic variables, a clear hydrogeoecological framework should be established (Sophocleous 2002). The most often used model for GW-SW modeling is SWAT-MODFLOW (Ntona et al. 2022; Mundetia et al. 2024). The U.S. Geological Survey's MODFLOW modular finite difference groundwater flow model is essentially a computer program that solves the groundwater flow equation (Harbaugh et al. 1996). MODFLOW is a computer program that numerically solves the three-dimensional groundwater flow equation (Eq. 13.8) using a finite difference method

$$\frac{\partial}{\partial x}\left(K_{xx}\frac{\partial h}{\partial x}\right)+\frac{\partial}{\partial y}\left(K_{yy}\frac{\partial h}{\partial y}\right)+\frac{\partial}{\partial z}\left(K_{zz}\frac{\partial h}{\partial z}\right)-W=S_s\frac{\partial h}{\partial t} \tag{13.8}$$

where, K_{xx}, K_{yy}, and K_{zz} are the hydraulic conductivities along x, y, and z axes (m/day), respectively, h is the potential head (m), W is the volumetric flux per unit volume which indicates source and sink (m^3 /day), S_s is the specific storage (1/day), and finally, t is time (day) (Zeydalinejad 2022). Several studies have used MODFLOW to examine the SW-GW interaction with the River Package, where the volumetric flow is calculated using Darcy's law (Ntona et al. 2022). The paper titled "Enhancing SWAT + simulation of groundwater flow and groundwater-surface water interactions using MODFLOW routines" focuses on improving the simulation of groundwater and its interactions with surface water within watershed models. The study specifically applies the enhanced SWAT + model, which integrates MODFLOW routines, to the Middle Bosque River Watershed (MBRW) in Texas, USA (Zamani et al. 2023).

13.5 Advanced Techniques

The advanced techniques used in water quality modeling include.

13.5.1 Artificial Intelligence (AI) and Machine Learning (ML)

Water quality analysis and prediction of parameters is one of the departments that have benefited from advanced techniques in Artificial Intelligence (AI) and Machine Learning (ML). These technologies could help change the approach towards monitoring and protecting water resources, which is especially important, given the fact that global environmental problems are intensifying and new, more complex tasks are being set. AI and ML algorithms are capable of processing massive amounts of data, which traditional analysis would provide substantial results to extract (Sarker 2021). This capability is convenient when it comes to water quality management since this process includes the analysis of numerous values with a complex set of factors that may change over time due to natural and man-made impacts. Among the many AI approaches that can be used for water quality prediction, the best is a neural network. These models are good at identifying an exponential correlation between features in a data set, which is very useful when modeling water quality parameters because they are usually related in a fairly complicated way (Cheng et al. 2022). For example, neural networks can be trained to predict changes in parameters such as turbidity, pH levels, and contaminant concentrations by learning from past data. They are especially useful in detecting less significant and noticeable relations between different environmental factors, which will help in a more precise forecast of water quality at certain time intervals. This capacity for modeling non-linear relationships improves the power to predict potential threats to water bodies, such as an increase in pollution or the formation of harmful algae. Another quite popular AI algorithm is Support Vector Machines, abbreviated as SVMs, and they participate in classification and regression types of problems. Consequently, in the context of water quality analysis, SVM can be used to classify the quality of water samples in terms of safe, marginal, or unsafe by the chemical content or biological content of the sample (Shehab et al. 2023). They can also predict any real number data values as continuous variables, like amounts of pollution, and the dissolved oxygen content of water, by finding the regression hyperplane that most clearly divides data points in a high-volume space. The nonlinearity and extreme accuracy of SVMs make them ideal for sophisticated datasets, whereby basic models would prove unfit to handle, especially in conditions where the data includes noise or outlier values (Ding et al. 2023).

In addition to that, decision trees might be more transparent and interpretable in contrast to neural networks or SVM, yet they are highly effective when it comes to the prediction of water quality as well. These models disassemble the decision-making processes into easily manageable and easy-to-understand steps, which is why these are most suitable in the development of decision models that can easily be explained to a non-technical person. Some strengths of decision trees are that they can analyze large volumes of data, and factors such as temperature, industrial discharge, and rainfall, among others. They can create coherent decision trees that allow for understanding how each factor affects the water quality, which makes these models valuable for decision-makers who have to decide based on the calculations made by a model. This is because interpretability makes it possible for anyone, such as policymakers or environmental managers, to be confident in the results and intervene when called for. It is essential to understand that all these AI and ML techniques combine to allow solving tasks and analyzing large sets of heterogeneous data that may not be easy or even impossible for an analyst to process (Soori et al. 2023). As applied to data, these models always evolve and get better as new information is introduced into the model. This form of learning is dynamic and provides an iteratively correct value of the water quality predictions at any change in environmental conditions. Besides that, the availability of hidden patterns in the data could provide a better understanding of water quality management, and thus it would help prevent the loss of water resources rather than repairing them. Therefore, the application of AI and ML algorithms, such as neural networks, SVMs, and decision trees, yields valuable accuracy enhancement of water quality characterization and management (Nasir et al. 2022). So, within the scope of these technologies, collected and accumulated Super-Big data sets, with their systematic, pattern search, and learning from already available data history, could considerably help in enhancing the decision-making and thus improving the safety of water resources. As the amount of water data improves and the capabilities of AI and ML progress, the future of water quality management is set to be more data-led, predictive, and reactive.

13.5.2 Big Data Analytics

Big data analytics has now emerged as one of the key approaches to water quality management, a technique that helps in the analysis of large amounts of data from numerous resources (Gohil et al. 2021). These documents include remote camera streams, GIS imagery, sensor web, and archives that offer the current and potential status of water bodies. The constantly growing amount and complexity of data call for big data analytics since it would be impossible or highly inefficient to process all of this information individually and combine its results (Li et al. 2021). On this level, the sole value of big data analytics for water management is predictive performance. Clarified were such issues as patterns where there was a clear set of sequences, correlations whereby the cause and effects of the change in water quality were ascertained, and trends that made it possible to predict with a higher level of accuracy the changes in water quality. For example, big data models that use historical systems and even the systems of the present can have tendencies of PH, turbidity, DO, and pollutants to catch the mistake before it surfaces (Aliashrafi et al. 2021). This kind of prediction is very crucial when it comes to HIV and AIDS eradication, waterborne diseases, protection of our ecosystems, and sustainable water resources.

Furthermore, big data analytics also plays a key role in assessing the impact of climate change on water quality. By integrating long-term climate data, such as temperature, precipitation patterns, and extreme weather events, into water quality models, it becomes possible to project future scenarios and understand how changing environmental conditions might influence water bodies. This approach helps predict potential risks like increased pollution, algae blooms, and altered water availability, which are all linked to climate change. By incorporating these projections into water quality management strategies, decision-makers can prepare for future challenges and implement adaptive solutions to mitigate adverse impacts. The use of big data analytics also enables real-time analysis and decision-making; changes in the quality of the water can therefore be detected and addressed in real-time (Gohil et al. 2021). By receiving data flows from sensors and other monitoring systems, analysts can identify deviations or sharp changes in water conditions and report them to take appropriate measures. This responsiveness is essential to stop pollution of water by such agents as industrial wastes, farming effluents, or disaster-caused water contamination. Overall, big data analytics provides an invaluable framework for improving water quality prediction, management, and conservation. Processing and analyzing large, diverse datasets enhances our ability to monitor, predict, and address the challenges posed by both current and future water quality issues. In the face of growing environmental concerns and climate change, big data analytics is a critical tool for ensuring the sustainable use and protection of global water resources.

13.5.3 Climate Change Integration

The combination of climate models into water quality analyses has become essential within the climate change issue around the globe. Contemporary models of water quality now take into consideration climatic forecasts regarding water quality trends in the long run, influenced by effects like temperature change, precipitation, and others (Soomro et al. 2024). When these climate projections are incorporated, water quality models could provide better estimates of how water bodies might be influenced by climate change in the future. This enables the identification of additional exposure risks like higher pollution, changed water flows, or toxic algal blooms, which are conditioned by climate shifts. in addition, combining climate models with water quality prediction can also help in creating adaptive management. These strategies are to counteract or reduce the expected effects of climate change on water quality and to achieve more sustainability of water resources (Giri et al. 2021). When applied in decision-making, integrated models help decision-makers to come up with anticipatory strategies for tackling possible hitches that may arise in the future, such as modifying water treatment procedures, increasing pollution control or conservation activities among others. Subsequently, this approach leads to enhanced decision-making that will enable water quality management to respond appropriately to the future realities characterized by climate change.

13.5.4 Geographic Information Systems (GIS)

GIS is used in the assessment and protection of water quality as it facilitates the integration of spatial and temporal datasets (Uslu et al. 2024). With the help of GIS, measurement data, land use data, water quality data, and other environmental data can be combined into one. This enables one to get the full picture of how the various variables that affect water bodies are coupled spatially as well as temporally, which is useful in the identification of trends and threats to water bodies. Casting one dataset over another, GIS makes it easier to establish the correlation between water quality and factors such as urbanization or intensive agriculture, or the effects of industrialization on water quality, to uncover further drivers of water quality deterioration (Khouni et al. 2024). Besides data analysis, GIS is an application of visualization and mapping, making it easier to explain issues of water quality to stakeholders. By rendering spatial images, GIS can define problem sectors, areas to avoid contamination, areas prone to contamination, or areas sensitive to climate change (El Osta et al. 2022). Such information presentation helps to adapt the cognition of the location of water quality issues to the general idea, which can facilitate the identification of proper solutions by determining the priorities and distribution of resources. For either short-term or long-term trend analysis or evaluating the effectiveness of management actions, GIS holds analytic opportunities for sound and effective conclusions on the enhancement and maintenance of water quality.

13.6 Discussion

This study underscores the pivotal role of water quality modeling in tackling challenges linked to human activities, climate change, and aquatic ecosystem dynamics. Traditional models excel in system understanding and scenario analysis, while advanced techniques like AI, machine learning, and GIS enable real-time monitoring and precise predictions. The emphasis on integrated approaches highlights the importance of combining scales and processes to offer a holistic view of water systems, especially when integrating climate projections for adaptive management (Prosser et al. 2021). However, limitations exist, including data variability, computational demands, and reliance on assumptions. The uneven adoption of advanced methods due to resource constraints and expertise gaps is notable, particularly in low-income regions. Moreover, uncertainties in models, especially those incorporating climate data, point to the need for improved data collection and robust validation methods to enhance accuracy and applicability. Addressing these challenges through research and capacity building is essential for maximizing the potential of water quality modeling in sustainable resource management (Xiang et al. 2021). The approaches and findings in this chapter closely relate to the United Nations Sustainable Development Goals (SDGs). In particular, they directly support SDG 6 (Clean Water and Sanitation) by improving strategies for sustainable water quality management in rural and delta areas. They also address the effects of climate change and human activities on water systems. This contribution relates to SDG 13 (Climate Action) and also supports SDG 14 (Life Below Water) and SDG 15 (Life on Land) indirectly.

13.7 Conclusion

Water quality modeling represents a critical tool in understanding and managing aquatic ecosystems under the growing pressures of human activities and climate change. The chapter highlights a range of methodologies, from traditional deterministic, stochastic, and empirical models to advanced techniques leveraging artificial intelligence, machine learning, and Geographic Information Systems (GIS). These methods provide a foundation for predicting trends, assessing impacts, and formulating strategies to address water quality challenges. The integration of advanced techniques, such as AI and big data analytics, enhances the precision and adaptability of models, enabling real-time monitoring and predictive capabilities for complex environmental systems. Additionally, the inclusion of climate change projections into water quality models underscores the necessity for adaptive strategies that ensure long-term sustainability. While traditional models remain invaluable for understanding system dynamics and scenario analysis, the adoption of advanced and integrated approaches offers the potential to bridge knowledge gaps, especially in addressing the multifaceted nature of water quality issues. However, challenges persist, including the need for high-quality data, computational resources, and skilled expertise, particularly in resource-constrained regions. The evolution of water quality modeling reflects a growing commitment to precision, innovation, and sustainability in water resource management. As the field progresses, continued research, capacity building, and cross-disciplinary collaboration will be vital in addressing future water quality challenges and safeguarding the health of aquatic ecosystems globally. These advancements support the United Nations Sustainable Development Goals (SDGs), especially SDG 6 (Clean Water and Sanitation). They also contribute to SDG 13 (Climate Action) and aid SDG 14 (Life Below Water) and SDG 15 (Life on Land).

13.8 Recommendations

1. Strengthen Integrated Monitoring Systems: Set up high-resolution monitoring networks across river basins and delta areas to improve the accuracy of water quality assessments and model predictions.
2. Use modern computational tools: Combine machine learning, remote sensing, and GIS-based modeling techniques to allow for dynamic and large-scale assessment of water quality across different land use and climate conditions.
3. Model Calibration and Scenario Analysis: Use strong calibration and validation methods, along with scenario-based modeling, to evaluate the effects of human activities, seasonal changes, and extreme events on water quality.
4. Policy Integration and Decision Support: Translate model outputs into actionable recommendations for policymakers and water resource managers, facilitating data-driven decision-making for sustainable water management and pollution mitigation.
5. Focus on Deltaic and Vulnerable Ecosystems: Prioritize research and management efforts in deltaic environments, where hydrodynamics, sediment transport, and nutrient fluxes significantly influence water quality and ecosystem health.
6. Capacity Building and Knowledge Dissemination: Promote targeted training, stakeholder engagement, and community-based water management strategies to enhance the practical adoption of modeling approaches and sustainable practices.
7. Future Research Directions: Encourage interdisciplinary studies to explore the combined effects of climate change, urbanization, and agricultural intensification on water quality, ensuring adaptive and resilient management strategies.

Acknowledgements The authors would like to express their gratitude to Malaviya National Institute of Technology, Jaipur for their support during the preparation of this chapter.

State whether you used AI or not in preparing the chapter.

No AI was used in the preparation of this chapter.

Author Contribution Ankita Singh: Data collection, investigation, Formal analysis and Supervision; Leelambar Singh: Review and drafting; Nitika Mundetia: Review and supervision; Ravindra Nath Tripathi: Review and methodology.

Funding No.

Data Availability Not applicable.

Declarations

Ethics Approval and Consent to Participate Not applicable.

Consent for publication Not applicable.

Competing Interests The authors declare no competing interests.

References

Alex FJ, Tan G, Kyei SK, Ansah PO, Agyeman PK, Fayzullayevich JV, Olayode IO (2023) Transmission of viruses and other pathogenic microorganisms via road dust: emissions, characterization, health risks, and mitigation measures. Atmos Pollut Res 14(1):101642. https://doi.org/10.1016/j.apr.2022.101642

Aliashrafi A, Zhang Y, Groenewegen H, Peleato NM (2021) A review of data-driven modelling in drinking water treatment. Rev Environ Sci Bio/Technol 1–25. https://doi.org/10.1007/s11157-021-09592-y

Alnahit AO, Mishra AK, Khan AA (2022) Stream water quality prediction using boosted regression tree and random forest models. Stoch Env Res Risk Assess 36(9):2661–2680. https://doi.org/10.1007/s00477-021-02152-4

Aloui S, Mazzoni A, Elomri A, Aouissi J, Boufekane A, Zghibi A (2023) A review of Soil and Water Assessment Tool (SWAT) studies of Mediterranean catchments: applications, feasibility, and future directions. J Environ Manage 326:116799. https://doi.org/10.1016/j.jenvman.2022.116799

Andrews M (2021) The math is not the territory: navigating the free energy principle. Biol Philos 36(3):30. https://doi.org/10.1007/s10539-021-09807-0

Arnold JG, Moriasi DN, Gassman PW, Abbaspour KC, White MJ, Srinivasan R, Jha MK (2012) SWAT: model use, calibration, and validation. Trans ASABE 55(4):1491–1508. https://doi.org/10.13031/2013.42256

Athira P (2021) Calibration of hydrological models considering process interdependence: a case study of SWAT model. Environ Model Softw 144:105131. https://doi.org/10.1016/j.envsoft.2021.105131

Bakshe P, Jugade R (2024) Influence of Fe/Al oxyhydroxides and soil organic matter on the adsorption of Pb onto natural stream sediment. J Environ Health Sci Eng 1–9.https://doi.org/10.1007/s40201-024-00894-1

Brockwell PJ, Davis RA (eds) (2002) Introduction to time series and forecasting. Springer New York, New York, NY. https://doi.org/10.1007/b97391

Chapra SC (2008) Surface water-quality modeling. Waveland press. https://www.researchgate.net/publication/48447645
Chen X, Li H, Hou J, Cao X, Song C, Zhou Y (2016) Sediment–water interaction in phosphorus cycling as affected by trophic states in a Chinese shallow lake (Lake Donghu). Hydrobiologia 776:19–33. https://doi.org/10.1007/s10750-016-2811-x
Cheng C, Zhang F, Shi J, Kung HT (2022) What is the relationship between land use and surface water quality? A review and prospects from remote sensing perspective. Environ Sci Pollut Res 29(38):56887–56907. https://doi.org/10.1007/s11356-022-21348-x
Cimorelli L, Covelli C, De Vincenzo A, Pianese D, Molino B (2021) Sedimentation in reservoirs: evaluation of return periods related to operational failures of water supply reservoirs with Monte Carlo simulation. J Water Resour Plan Manag 147(1):04020096. https://doi.org/10.1061/(ASCE)WR.1943-5452.0001307
Costa CMDSB, Leite IR, Almeida AK, de Almeida IK (2021) Choosing an appropriate water quality model—a review. Environ Monit Assess 193:1–15. https://doi.org/10.1007/s10661-020-08786-1
D'Agostino G (2024) Uncertainty analysis of a river quality model using differential inequalities. Doctoral dissertation, University of Guelph. https://hdl.handle.net/10214/28610
Delpla I, Jung AV, Baures E, Clement M, Thomas O (2009) Impacts of climate change on surface water quality in relation to drinking water production. Environ Int 35(8):1225–1233. https://doi.org/10.1016/j.envint.2009.07.001
Ding C, Xia Y, Yuan Z, Yang H, Fu J, Chen Z (2023) Performance prediction for a fuel cell air compressor based on the combination of backpropagation neural network optimized by genetic algorithm (GA-BP) and support vector machine (SVM) algorithms. Therm Sci Eng Prog 44:102070. https://doi.org/10.1016/j.tsep.2023.102070
Ejigu MT (2021) Overview of water quality modeling. Cogent Eng 8(1):1891711. https://doi.org/10.1080/23311916.2021.1891711
Elhabashy A, Li J, Sokolova E (2023) Water quality modeling of a eutrophic drinking water source: impact of future climate on Cyanobacterial blooms. Ecol Model 477:110275. https://doi.org/10.1016/j.ecolmodel.2023.110275
Giri S (2021) Water quality prospective in Twenty First Century: Status of water quality in major river basins, contemporary strategies and impediments: a review. Environ Pollut 271:116332. https://doi.org/10.1016/j.envpol.2020.116332
Glibert PM, Seitzinger S, Heil CA, Burkholder JM, Parrow MW, Codispoti LA, Kelly V (2005) Eutrophication. Oceanography 18(2):198. https://www.academia.edu/download/99116794/18-2_glibert3.pdf
Gohil J, Patel J, Chopra J, Chhaya K, Taravia J, Shah M (2021) Advent of Big Data technology in environment and water management sector. Environ Sci Pollut Res 28(45):64084–64102. https://doi.org/10.1007/s11356-021-14017-y
Hanoon MS, Ahmed AN, Fai CM, Birima AH, Razzaq A, Sherif M, El-Shafie A (2021) Application of artificial intelligence models for modeling water quality in groundwater: comprehensive review, evaluation and future trends. Water Air Soil Pollut 232:1–41. https://doi.org/10.1007/s11270-021-05311-z
Harbaugh AW, McDonald MG (1996) Programmer's documentation for MODFLOW-96, an update to the US Geological Survey modular finite-difference groundwater flow model (No. 96–486). US Geological Survey; Branch of Information Services [distributor]. https://pubs.usgs.gov/of/1996/0486/report.pdf
Haris H, Chow MF, Usman F, Sidek LM, Roseli ZA, Norlida MD (2016) Urban stormwater management model and tools for designing stormwater management of green infrastructure practices. In: IOP conference series: earth and environmental science, vol 32, no 1. IOP Publishing, p 012022. https://doi.org/10.1088/1755-1315/32/1/012022

Hendren CO, Lowry M, Grieger KD, Money ES, Johnston JM, Wiesner MR, Beaulieu SM (2013) Modeling approaches for characterizing and evaluating environmental exposure to engineered nanomaterials in support of risk-based decision making. Environ Sci Technol 47(3):1190–1205. https://pubs.acs.org/doi/full/https://doi.org/10.1021/es302749u

Jadhav AR, Pathak PD, Raut RY (2023) Water and wastewater quality prediction: current trends and challenges in the implementation of artificial neural network. Environ Monit Assess 195(2):321. https://doi.org/10.1007/s10661-022-10904-0

Jain G, Mallick B (2017) A study of time series models ARIMA and ETS. Available at SSRN 2898968. https://papers.ssrn.com/sol3/papers.cfm?abstract_id=2898968

Jain S, Srivastava A, Khadke L, Chatterjee U, Elbeltagi A (2024) Global-scale water security and desertification management amidst climate change. Environ Sci Pollut Res 1–25.https://doi.org/10.1007/s11356-024-34916-0

Jamatutu SA, Abbass K, Gawusu S, Yeboah KE, Jamatutu IAM, Song H (2024) Quantifying future carbon emissions uncertainties under stochastic modeling and Monte Carlo simulation: insights for environmental policy consideration for the belt and road initiative region. J Environ Manage 370:122463. https://doi.org/10.1016/j.jenvman.2024.122463

Jin W, Spoerri L, Haass NK, Simpson MJ (2021) Mathematical model of tumour spheroid experiments with real-time cell cycle imaging. Bull Math Biol 83:1–23. https://doi.org/10.1007/s11538-021-00878-4

Karul C, Soyupak S, Çilesiz AF, Akbay N, Germen E (2000) Case studies on the use of neural networks in eutrophication modeling. Ecol Model 134(2–3):145–152. https://doi.org/10.1016/S0304-3800(00)00360-4

Keller AA, Garner K, Rao N, Knipping E, Thomas J (2023) Hydrological models for climate-based assessments at the watershed scale: a critical review of existing hydrologic and water quality models. Sci Total Environ 867:161209. https://doi.org/10.1016/j.scitotenv.2022.161209

Khouni I, Louhichi G, Ghrabi A (2021) Use of GIS based inverse distance weighted interpolation to assess surface water quality: case of Wadi El Bey, Tunisia. Environ Technol Innov 24:101892. https://doi.org/10.1016/j.eti.2021.101892

Kim Y, Kim B (2006) Application of a 2-dimensional water quality model (CE-QUAL-W2) to the turbidity interflow in a deep reservoir (Lake Soyang, Korea). Lake Reservoir Manage 22(3):213–222. https://doi.org/10.1080/07438140609353898

Kothandaraman V (1968) A probabilistic analysis of dissolved oxygen-biochemical oxygen demand relationship in streams. University of Illinois at Urbana-Champaign. https://search.proquest.com/openview/02b5a50747959113279d11bdb1d46c5a/1?pq-origsite=gscholar&cbl=18750&diss=y

Krishnan SR, Nallakaruppan MK, Chengoden R, Koppu S, Iyapparaja M, Sadhasivam J, Sethuraman S (2022) Smart water resource management using artificial intelligence—a review. Sustainability 14(20):13384. https://doi.org/10.3390/su142013384

Kumar D, Khurana P, Kumar D (2024) Mathematical models and computational approaches for assessing water quality in rural area of India. In: Integrated management of water resources in India: a computational approach: optimizing for sustainability and planning. Springer Nature Switzerland, Cham, pp 23–41. https://doi.org/10.1007/978-3-031-62079-9_2

Kurani A, Doshi P, Vakharia A, Shah M (2023) A comprehensive comparative study of artificial neural network (ANN) and support vector machines (SVM) on stock forecasting. Annal Data Sci 10(1):183–208. https://doi.org/10.1007/s40745-021-00344-x

Li W, Chai Y, Khan F, Jan SRU, Verma S, Menon VG, Li X (2021) A comprehensive survey on machine learning-based big data analytics for IoT-enabled smart healthcare system. Mobile Netw Appl 26:234–252. https://doi.org/10.1007/s11036-020-01700-6

Lowe M, Qin R, Mao X (2022) A review on machine learning, artificial intelligence, and smart technology in water treatment and monitoring. Water 14(9):1384. https://doi.org/10.3390/w14091384

Mundetia N, Sharma D, Sharma A (2024) Groundwater sustainability assessment under climate change scenarios using integrated modelling approach and multi-criteria decision method. Ecol Model 487:110544. https://doi.org/10.1016/j.ecolmodel.2023.110544

Muyambo F, Belle J, Nyam YS, Orimoloye IR (2023) Climate-change-induced weather events and implications for urban water resource management in the free state province of South Africa. Environ Manage 71(1):40–54. https://doi.org/10.1007/s00267-022-01726-4

Nasir N, Kansal A, Alshaltone O, Barneih F, Sameer M, Shanableh A, Al-Shamma'a A (2022) Water quality classification using machine learning algorithms. J Water Process Eng 48:102920. https://doi.org/10.1016/j.jwpe.2022.102920

Niazi M, Obropta C, Miskewitz R (2015) Pathogen transport and fate modeling in the Upper Salem River Watershed using SWAT model. J Environ Manage 151:167–177. https://doi.org/10.1016/j.jenvman.2014.12.042

Ntona MM, Busico G, Mastrocicco M, Kazakis N (2022) Modeling groundwater and surface water interaction: an overview of current status and future challenges. Sci Total Environ 846:157355. https://doi.org/10.1016/j.scitotenv.2022.157355

El Osta M, Niyazi B, Masoud M (2022) Groundwater evolution and vulnerability in semi-arid regions using modeling and GIS tools for sustainable development: case study of Wadi Fatimah, Saudi Arabia. Environ Earth Sci 81(9):248https://doi.org/10.1007/s12665-022-10374-0

Patel PS, Pandya DM, Shah M (2023) A systematic and comparative study of Water Quality Index (WQI) for groundwater quality analysis and assessment. Environ Sci Pollut Res 30(19):54303–54323. https://doi.org/10.1007/s11356-023-25936-3

Prosser IP, Chiew FH, Stafford Smith M (2021) Adapting water management to climate change in the Murray-Darling Basin, Australia. Water 13(18):2504. https://doi.org/10.3390/w13182504

Rasheed FA, Moradi M (2024). Analysis of organic pollution impact on purification capacity at the Tanjero River in Sulaimani City. Paddy Water Environ 1–10.https://doi.org/10.1007/s10333-024-00994-6

Robinson S (2014) Simulation: the practice of model development and use. Bloomsbury Publishing. https://lmsspada.kemdiktisaintek.go.id/pluginfile.php/123916/mod_label/intro/simulation-the-practice-of-model-development-and-use.9780470847725.21800.pdf

Rode M, Arhonditsis G, Balin D, Kebede T, Krysanova V, Van Griensven A, Van der Zee SE (2010) New challenges in integrated water quality modelling. Hydrol Process 24(24):3447–3461. https://doi.org/10.1002/hyp.7766

Sabel M, Yohannes E, Straile D, Rothhaupt KO (2024) Effects of oligotrophication and Dreissena littoral-pelagic coupling on littoral invertebrate and fish communities: insights from stable isotopes of archived samples. Hydrobiologia 1–13.https://doi.org/10.1007/s10750-024-05500-x

Sarker IH (2021) Machine learning: algorithms, real-world applications and research directions. SN Comput Sci 2(3):160. https://doi.org/10.1007/s42979-021-00592-x

Seifi A, Dehghani M, Singh VP (2020) Uncertainty analysis of water quality index (WQI) for groundwater quality evaluation: application of Monte-Carlo method for weight allocation. Ecol Ind 117:106653. https://doi.org/10.1016/j.ecolind.2020.106653

Shehab SA, Darwish A, Hassanien AE, Scientific Research Group in Egypt (2023) Water quality classification model with small features and class imbalance based on fuzzy rough sets. Environ Dev Sustain 1–19.https://doi.org/10.1007/s10668-023-03916-4

Singh VP (2018) Hydrologic modeling: progress and future directions. Geosci Lett 5(1):1–18. https://doi.org/10.1186/s40562-018-0113-z

Singh L, Saravanan S (2020) Impact of climate change on hydrology components using CORDEX South Asia climate model in Wunna, Bharathpuzha, and Mahanadi, India. Environ Monit Assess 192(11):678. https://doi.org/10.1007/s10661-020-08637-z

Singh J, Swaroop S, Sharma P, Mishra V (2023) Real-time assessment of the Ganga river during pandemic COVID-19 and predictive data modeling by machine learning. Int J Environ Sci Technol 20(7):7887–7910. https://doi.org/10.1007/s13762-022-04423-1

Singh VP, Frevert DK (eds) (2010) Watershed models. CRC press. https://doi.org/10.1201/9781420037432

Skahill BE (2004) Use of the hydrological simulation program: FORTRAN (HSPF) model for watershed studies. https://erdc-library.erdc.dren.mil/bitstream/11681/3641/1/TN-SMART-04-1.pdf

Smith VH, Schindler DW (2009) Eutrophication science: where do we go from here? Trends Ecol Evol 24(4):201–207. https://doi.org/10.1016/j.tree.2008.11.009

Soomro SEH, Soomro AR, Batool S, Guo J, Li Y, Bai Y, Wang Y (2024) How does the climate change effect on hydropower potential, freshwater fisheries, and hydrological response of snow on water availability? Appl Water Sci 14(4):65. https://doi.org/10.1007/s13201-023-02070-6

Soori M, Arezoo B, Dastres R (2023) Artificial intelligence, machine learning and deep learning in advanced robotics, a review. Cognit Robot 3:54–70. https://doi.org/10.1016/j.cogr.2023.04.001

Sophocleous M (2002) Interactions between groundwater and surface water: the state of the science. Hydrogeol J 10:52–67. https://doi.org/10.1007/s10040-001-0170-8

Talukdar P, Kumar B, Kulkarni VV (2023) A review of water quality models and monitoring methods for capabilities of pollutant source identification, classification, and transport simulation. Rev Environ Sci Bio/Technol 22(3):653–677. https://doi.org/10.1007/s11157-023-09658-z

Tayebzadeh Moghadam N, Abbaspour KC, Malekmohammadi B, Schirmer M, Yavari AR (2021) Spatiotemporal modelling of water balance components in response to climate and landuse changes in a heterogeneous mountainous catchment. Water Resour Manage 35:793–810. https://doi.org/10.1007/s11269-020-02735-w

Trach R, Trach Y, Kiersnowska A, Markiewicz A, Lendo-Siwicka M, Rusakov K (2022) A study of assessment and prediction of water quality index using fuzzy logic and ANN models. Sustainability 14(9):5656. https://doi.org/10.3390/su14095656

Trang M, Dudley MN, Bhavnani SM (2017) Use of Monte Carlo simulation and considerations for PK-PD targets to support antibacterial dose selection. Curr Opin Pharmacol 36:107–113. https://doi.org/10.1016/j.coph.2017.09.009

Tseng CY, Tinoco RO (2021) A two-layer turbulence-based model to predict suspended sediment concentration in flows with aquatic vegetation. Geophys Res Lett 48(3):e2020GL091255. https://doi.org/10.1029/2020GL091255

Uslu A, Dugan ST, El Hmaidi A, Muhammetoglu A (2024) Comparative evaluation of spatiotemporal variations of surface water quality using water quality indices and GIS. Earth Sci Inf 17(5):4197–4212. https://doi.org/10.1007/s12145-024-01389-1

Wang X, Xie H (2018) A review on applications of remote sensing and geographic information systems (GIS) in water resources and flood risk management. Water 10(5):608. https://doi.org/10.3390/w10050608

Wang S, Vogt RD, Carstensen J, Lin Y, Feng J, Lu X (2022) Riverine flux of dissolved phosphorus to the coastal sea may be overestimated, especially in estuaries of gated rivers: implications of phosphorus adsorption/desorption on suspended sediments. Chemosphere 287:132206. https://doi.org/10.1016/j.chemosphere.2021.132206

Wang B, Hua L, Mei H, Wu X, Kang Y, Zhao N (2024) Impact of climate change on the dynamic processes of marine environment and feedback mechanisms: an overview. Arch Comput Methods Eng 1–32.https://doi.org/10.1007/s11831-024-10072-z

Wani AK, Rahayu F, Ben Amor I, Quadir M, Murianingrum M, Parnidi P, Latifah E (2024) Environmental resilience through artificial intelligence: innovations in monitoring and management. Environ Sci Pollut Res 31(12):18379–18395. https://doi.org/10.1007/s11356-024-32404-z
Weiskerger CJ, Phanikumar MS (2020) Numerical modeling of microbial fate and transport in natural waters: review and implications for normal and extreme storm events. Water 12(7):1876. https://doi.org/10.3390/w12071876
Withers PJ, Neal C, Jarvie HP, Doody DG (2014) Agriculture and eutrophication: where do we go from here? Sustainability 6(9):5853–5875. https://doi.org/10.3390/su6095853
Xiang X, Li Q, Khan S, Khalaf OI (2021) Urban water resource management for sustainable environment planning using artificial intelligence techniques. Environ Impact Assess Rev 86:106515. https://doi.org/10.1016/j.eiar.2020.106515
Xie H, Lian Y (2013) Uncertainty-based evaluation and comparison of SWAT and HSPF applications to the Illinois River Basin. J Hydrol 481:119–131. https://doi.org/10.1016/j.jhydrol.2012.12.027
Yan T, Shen SL, Zhou A (2022) Indices and models of surface water quality assessment: review and perspectives. Environ Pollut 308:119611. https://doi.org/10.1016/j.envpol.2022.119611
Yang X, Chen Z (2023) A hybrid approach based on Monte Carlo simulation-VIKOR method for water quality assessment. Ecol Ind 150:110202. https://doi.org/10.1016/j.ecolind.2023.110202
Yang Y, Ye S, Zhang C, Zeng G, Tan X, Song B, Chen Q (2021) Application of biochar for the remediation of polluted sediments. J Hazard Mater 404:124052. https://doi.org/10.1016/j.jhazmat.2020.124052
Zamani N, Eslamian S, Koupai JA (2023) Hydroinformatics methods for groundwater simulation. In: Handbook of Hydroinformatics. Elsevier, pp 207–217. https://doi.org/10.1016/B978-0-12-821962-1.00023-4
Zeydalinejad N (2022) Artificial neural networks vis-à-vis MODFLOW in the simulation of groundwater: a review. Model Earth Syst Environ 8(3):2911–2932. https://doi.org/10.1007/s40808-022-01365-y
Zeydalinejad Moghadam N, Abbaspour KC, Malekmohammadi B, Schirmer M, Yavari AR (2021) Spatiotemporal modelling of water balance components in response to climate and landuse changes in a heterogeneous mountainous catchment. Water Resour Manage 35:793–810. https://doi.org/10.1007/s11269-020-02735-w
Ziemińska-Stolarska A, Skrzypski J (2012) Review of mathematical models of water quality. Ecol Chem Eng S 19(2):197–211. https://doi.org/10.2478/v10216-011-0015-x

Water Quality Assessment and Management 14

Leelambar Singh, Ankita Singh, Nitika Mundetia, and Ravindra Nath Tripathi

Abstract

Water quality assessment and management are critical for ensuring the long-term sustainability of water resources that support human consumption, industry, agriculture, and ecosystems. This study focuses on the Chambal River Basin, aiming to identify methodologies for monitoring water quality and evaluating approaches for maintaining water quality standards. The primary goal is to assess the effectiveness of monitoring techniques and management practices for protecting various ecosystems. A detailed analysis of key physicochemical parameters such as pH, EC, hardness, HCO_3, NO_3, SO_4, Ca, Mg, Cl, K, and F was carried out through systematic field sampling and laboratory investigations. Further, Geographic Information Systems (GIS) and remote sensing tools were integrated to perform spatial assessments of pollution sources and visualize water quality trends across the basin. The results were compared with World Health Organization (WHO) standards to identify non-compliant zones and high-risk areas. Findings reveal that while progress has been made in controlling point-source pollution, non-point sources like agricultural runoff, wastewater discharge, and urban stormwater remain significant threats. Climate change further exacerbates water quality degradation by intensifying extreme rainfall events that enhance pollutant transport

L. Singh (✉) · A. Singh
Department of Civil Engineering, Malaviya National Institute of Technology, Jaipur, India
e-mail: leelambar.ce@mnit.ac.in

N. Mundetia
Department of Basic and Applied Science, Sangam University, Bhilwara, India
e-mail: nitika.mundetia@sangamuniversity.ac.in

R. N. Tripathi
Graphic Era University, Dehradun, India

S. Yadav et al. (eds.), *Water Resources in Delta Region in India*, Deltas of the World,
https://doi.org/10.1007/978-3-032-06500-1_14

and groundwater vulnerability. The study emphasizes the necessity of integrated water management strategies combining advanced monitoring technologies, spatial tools, regulatory frameworks, and community participation to secure water resources for the future. Importantly, the work contributes to the achievement of Sustainable Development Goals (SDGs), particularly SDG 6 (Clean Water and Sanitation), SDG 3 (Good Health and Well-Being), SDG 13 (Climate Action), and SDG 15 (Life on Land), by promoting safe access to water, protecting human health, fostering climate resilience, and conserving ecosystems.

14.1 Introduction

An essential resource for maintaining ecological equilibrium, economic growth, and life is water. Environmental integrity, industrial operations, agricultural output, and public health are all directly impacted by its quality (Mishra et al. 2023). The suitability of groundwater depends upon the quality of water. Groundwater quality assessment is necessary for the suitability of daily use. Over 1.5 billion people worldwide are said to depend on groundwater for their necessities (Adimalla and Wu 2019; Zhang et al. 2020). Around 2.2 billion people did not have access to safe drinking water as of 2022 (UN 2024) and in some places, about half of the world's population facing severe water scarcity issue (Lee et al. 2023). Due to over-extraction, urban demand, domestic sewage, land use changes, climate change and agriculture pollution groundwater quality fluctuates (Sharma and Binda 2009; Kaushik and Ghosh 2015; Singh and Saravanan 2020; Ahmed et al. 2025). Physical, chemical and biological factors are analyzed as part of the water quality assessment whether the water is appropriate for drinking, irrigation, and industrial uses.

Indian river deltas, including the Godavari, Krishna, Mahanadi, and Ganges–Brahmaputra, are important in the fields of agriculture, the environment, the economy, and society. Due to their rich alluvial soils, these deltas are extremely productive agricultural areas that sustain millions of farmers and provide a substantial contribution to India's food security. In terms of ecology, Indian deltas are home to a variety of distinct ecosystems, such as vast mangrove forests like the Sundarbans in the Ganges Delta, which are recognized as UNESCO World Heritage Sites and the biggest mangrove forest in the world (Rahman et al. 2020; Bandyopadhyay 2007; Nath et al. 2021). These ecosystems serve as natural coastal buffers against floods and hurricane surges, offer vital habitat for a wide variety of animals, and aid in carbon sequestration, which lessens the consequences of climate change. However, pollution, rising salinity, climate change-induced severe events, and changed river flows brought on by dams and water diversion projects pose problems for these deltas. For Indian deltas to continue to benefit agriculture, biodiversity conservation, disaster risk reduction, and livelihood support, sustainable management through integrated water resource planning, habitat protection, and climate adaption techniques is essential.

Toxic metals, inorganic salts, cations (magnesium, calcium, sodium, and potassium), as well as anions (bicarbonate, carbonate, chloride and sulfate) are the principal contaminants of groundwater. Therefore it is widely studied groundwater quality analysis and risk assessment including in Egypt, China and India (Li et al. 2018; Adimalla and Wu 2019). In a study on groundwater quality in India, Kumar and Singh (2024) discovered that Mg^{2+}, F−, Cl−, and NO_3- were important factors in determining different water quality. Alemu et al. (2024) conducted a study on groundwater quality and found that due to anthropogenic activities groundwater quality was fully contaminated with fecal coliforms. Zhang et al. (2022) studied the groundwater quality in China and investigated that Ca–HCO_3 type are the general hydro-geochemical composition of groundwater. According to Thirumoorthy et al. (2024) assessment of the quality of the groundwater in southern India, industrial effluent increased the hardness and chloride levels of the water. There are different techniques to assess the groundwater quality, including Fuzzy (Zhao et al. 2025; Nagothuet al. 2025) and water quality index (WQI) (Pushkar 2025; Tan et al. 2025).The Chambal River basin falls under the semi-arid to sub-humid climatic regions and facing several groundwater-related challenges (Yadav et al. 2014; Reddy 2017; Kumar et al. 2017; Bagora and Narulkar 2024). Therefore it is essential to investigate the current condition of the Chambal river basin.

Water Quality Index (WQI) is a quantitative metric used to assess the overall water qualityfor drinking and other purposes (Banda and Kumarasamy 2020; Chabuk et al. 2020; Jayasankar et al. 2025). WQI is a unique and practical evaluation that aids in selecting the most effective treatment strategy by condensing the entire water quality condition into a single term. It was employed in a number of research to evaluate surface water quality (Dhanush et al. 2024; Akar et al. 2024), river bodies (Mishra et al. 2024; Adilakshmi and Venkatesan 2024), and groundwater (Kumar and Singh 2024; Gani et al. 2025).The Geographic Information System (GIS) is one of the best tools for managing water resources and comprehending nature (Stocks and Wise 2000; Lovett 2014). Spatial analysis in GIS is an extension of interpolating known values and using interpolation techniques to find unknown values (Childs 2004). According to Garnero and Godone (2014), ArcGIS offers a number of interpolation methods, including Thiessen Polygons, Kriging, Inverse Distance Weighting (IDW), Nearest Neighbor Interpolation, Arial interpolation, and kernel smoothing. Chabuk et al. (2020) evaluated a river's water quality by applying IDW techniques with GIS. Mardiana et al. (2022) used GIS to study groundwater pollution using the DRASTIC and water quality indices. Uslu et al. (2024) employ GIS to assess the condition of water quality utilizing a variety of water quality indices in order to obtain a thorough grasp of water health.

The sediment and water quality features that are essential for delta development and health in large river systems are reflected in the physical and chemical parameter evaluations. Techniques often employed to research and manage delta regions globally are mirrored in the use of GIS and remote sensing for spatial mapping of water quality.

Water quality assessments can benefit from the applications of advanced technologies such as remote sensing and GIS, which allow for a spatial study of pollution patterns and causes and facilitate improved decision-making. This study explores techniques for assessing water quality and considers efficient management strategies. The present study aims to (1) determine the location of inappropriate drinking groundwater and (2) evaluate the groundwater quality of the Chambal River basin using GIS and IDW interpolation techniques.

14.2 Materials and Methods

14.2.1 Study Area

The Chambal River Basin is considered as the study area. This river flows through Madhya Pradesh and Rajasthan and eventually meets the Yamuna River in Uttar Pradesh, and it is amongst the India's principal rainfed rivers. The study area lies between latitudes 22° 0′ N and 28° 0′ N and longitudes 74° 0′ to 79° 0′ E (see Fig. 14.1). Spanning a vast geographical area of approximately 141,987 km^2, the basin plays a crucial role in the region's hydrology and ecosystem. The elevation of the study area ranges from 76 to 1,313 m above mean sea level, providing diverse landscapes and ecological zones. The Chambal River Valley is an integral portion of the Vindhyan system, characterized by a unique geological composition of massive limestone, sand stone and slate that rests atop older rock formations (Crawford 1969). The study area is also acknowledged for its distinct topography, including ravines, gullies, and an undulating floodplain. These features contribute to the region's complex hydrological and environmental characteristics. The annual rainfall over the study area ranges from 600 to 700 mm, most of which is received during the monsoon season. This rainfall is vital for sustaining the river and its surrounding ecosystems.

14.2.2 Materials Used

A total of thirteen groundwater quality parameters (pH, TH, EC, HCO_3, NO_3, SO_4, Cl, Mg, Ca, Na, Kand F) were taken to analyse the spatial pattern of WQI in Chambal River Basin. Groundwater data taken from the Ground Water Year Book for 2021–22 of Central Ground Water Board of India (https://cgwb.gov.in/).Satellite-based precipitation, i.e., CHIRPS data from 2000 to 2021, is taken from the web portal (https://chc.ucsb.edu/data/chirps).

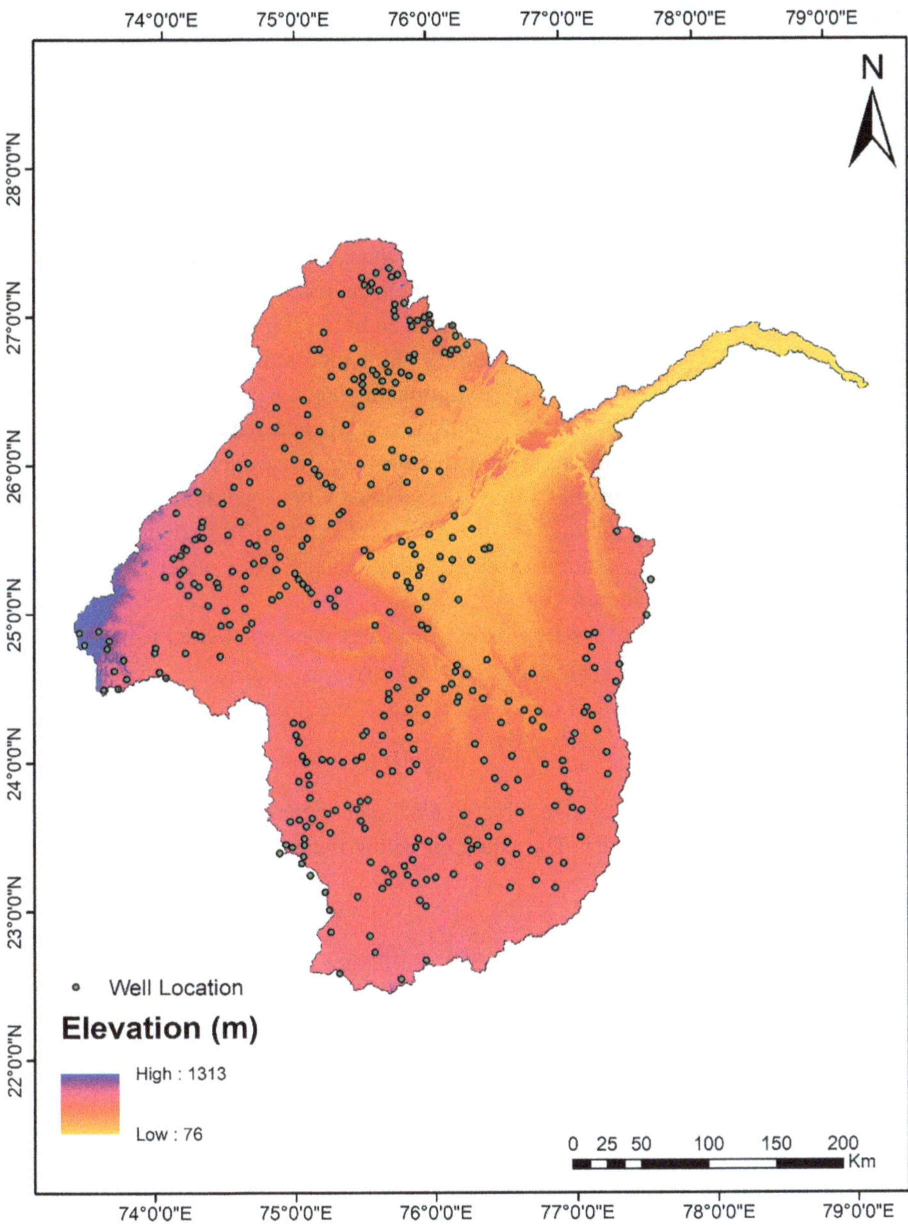

Fig. 14.1 Location of the Chambal River Basin study area

14.2.3 Water Quality Index (WQI)

WQI was developed in 1965 by the National Sanitation Foundation (NSF) in the United States (Horton 1965). This tool was developed by the NSF to streamline water quality monitoring and increase accessibility for both the general public and specialists. Since then, the WQI has been changed and adjusted by numerous nations and organizations to accommodate regional water quality requirements and environmental conditions. The index simplifies the comparison and evaluation of water bodies by relating several water quality characteristics into a particular value, which helps to improve pollution control and water management practices worldwide (Chidiac et al. 2023).

Based on several important factors, including pH, temperature, dissolved oxygen, turbidity, and the concentrations of different contaminants (such as nitrates, phosphates, and heavy metals), the Water Quality Index (WQI) is a statistical measure used to evaluate the overall waterquality. By combining information from these factors, the WQI generates a single, easily comprehensible score that indicates how suitable the water is for a range of uses, including drinking, irrigation, and recreation. A higher WQI number denotes safe and clean water, whereas a lower value denotes poor water quality (Banda and Kumarasamy 2020; Gupta and Gupta 2021).

The relative weights for the 13 water quality measures were calculated using Eq. (14.1) to evaluate their importance in evaluating water quality.

$$\text{Re}l\_Wt_i = \frac{AWt_i}{\sum AWt_i} \tag{14.1}$$

where Rwl_Wt_i represents the relative weight, AWt_i stands for the assigned weight, and n denotes the number of parameters. The equation below illustrates the quality rating scale for each parameter:

$$Q_i = \left(\frac{c_i}{s_i}\right) \times 100 \tag{14.2}$$

The concentration of each parameter (Ci, in milligrammes per litre) in the water sample and the acceptable standard (Si) established by the Bureau of Indian Standards (BIS) were used to compute the quality rating (Qi) for each individual parameter.

Equation (14.3) calculates Subindex (SIi) for each parameter:

$$SI_i = RWt_i \times Q_i \tag{14.3}$$

where SI_i is the subindex of the ith parameter, Qi rating is based on the ith parameter concentration.

Finally, WQI is proposed using the Eq. (14.4).

$$WQI = \sum SI_i \tag{14.4}$$

Table 14.1 Categorization of WQI

WQI value	Class
<50	Excellent
50–100	Good
101–200	Poor
201–300	Very poor
>300	Unsuitable

WQI was employed to comprehensively evaluate the groundwater suitability for drinking purpose. To calculate the WQI permissible limit of WHO (2004) was used.

Spatial analysis by Inverse Distance Weightage (IDW) techniques applied using ArcGIS environment. Characterization of WQI is presented in Table 14.1.

Overall adopted methodology flow chart is shown in Fig. 14.2.

14.3 Results

14.3.1 Statistics of Hydro-Chemistry

In Table 14.2, statistical values of different parameters of water quality are presented. The mean value of pH (7.92) suggests that the water is slightly alkaline, which is typical for many natural water sources. The pH range is narrow, from 7.04 (neutral) to 8.85 (moderately alkaline), with a low standard deviation of 0.31. This indicates consistent pH levels across samples, with minimal fluctuations. EC measures the ability of water to conduct electricity, which correlates with the concentration of dissolved ions. The mean EC of 2163.35 μS/cm suggests moderate salinity, but the wide range (from 320.00 to a striking 17,420.00 μS/cm) and high standard deviation of 2074.96 indicate substantial variability. This could be attributed to differing geological formations, industrial discharges, or agricultural runoff. With a mean value of 417.46 mg/L, bicarbonate levels are relatively high, reflecting its role as a buffering agent that stabilizes pH. The range spans from 73 to 1915 mg/L, and a standard deviation of 222.47 highlights average variability, likely influenced by natural weathering processes and human activities. Chloride is a key indicator of salinity. The mean concentration is 339.14 mg/L, with values ranging from 21 to an extreme 5176 mg/L. A high standard deviation of 502.62 reveals significant variation, which may result from domestic sewage, industrial effluents, or marine intrusion in coastal areas. Sulfate levels average at 217.92 mg/L, with a wide range from 3 to 2284 mg/L.

The standard deviation of 318.84 suggests considerable variability, potentially influenced by natural sources like mineral dissolution or anthropogenic activities such as agricultural runoff and industrial.

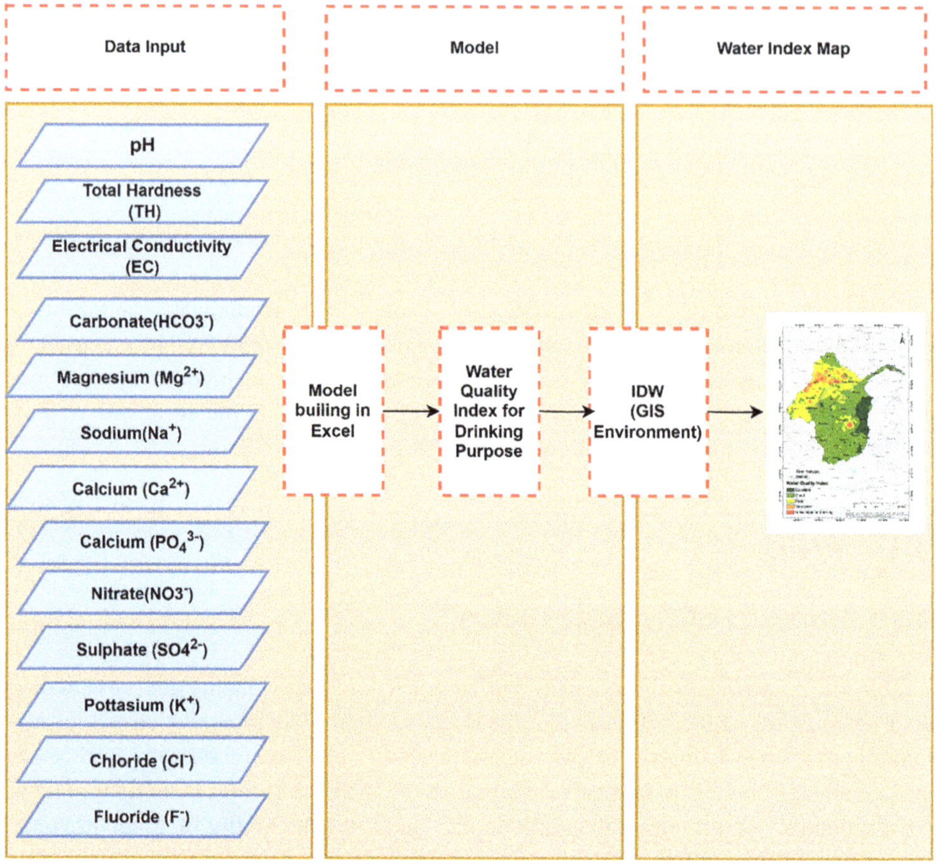

Fig. 14.2 Flow chart for adopted methodology

Nitrate concentrations have a mean of 63.13 mg/L, ranging from a negligible 0.02 to a concerning 1750.00 mg/L. A high standard deviation of 142.98 reflects significant differences between samples, possibly due to fertilizer application, septic systems, or organic matter decomposition. Phosphate levels are generally low, with a mean of 0.17 mg/L and values ranging from 0.00 to 8.85 mg/L. The standard deviation of 0.56 indicates minimal variability, which aligns with its typically low concentration in natural waters, unless influenced by agricultural runoff or wastewater discharge. The mean TH is 425.09 mg/L, indicating hard water, with a range from 90.00 to 4320.00 mg/L. A high standard deviation of 448.82 underscores considerable variability, likely due to variations in calcium and magnesium concentrations. Calcium levels average 74.10 mg/L, from 12.00 to 570.00 mg/L. The standard deviation of 64.26 reflects moderate variability, influenced by local geology, such as limestone dissolution, and anthropogenic factors like wastewater discharge. Magnesium shows average concentration of 58.33 mg/L varying from 2.00

Table 14.2 Summary of different parameters

Parameters	Mean	Maximum	Minimum	Standard deviation
pH	7.92	8.85	7.04	0.31
EC (μS/cm)	2163.35	17,420.00	320.00	2074.96
HCO_3 (mg/l)	417.46	1915.00	73.00	222.47
Cl (mg/l)	339.14	5176.00	21.00	502.62
SO_4 (mg/l)	217.92	2284.00	3.00	318.84
NO_3 (mg/l)	63.13	1750.00	0.02	142.98
PO_4 (mg/l)	0.17	8.85	0.00	0.56
TH (mg/l)	425.09	4320.00	90.00	448.82
Ca (mg/l)	74.10	570.00	12.00	64.26
Mg (mg/l)	58.33	807.00	2.00	76.04
Na (mg/l)	305.14	2500.00	12.00	339.32
F (mg/l)	0.93	9.80	0.01	1.10
K (mg/l)	18.69	843.00	0.40	68.48

to 807.00 mg/L. The high standard deviation of 76.04 suggests variability, likely caused by natural mineral dissolution and anthropogenic contributions. Sodium levels average 305.14 mg/L, with a wide range from 12.00 to 2500.00 mg/L. The standard deviation of 339.32 indicates substantial variation, which could stem from natural sources, irrigation return flows, or industrial pollution. Potassium has a mean concentration of 18.69 mg/L, from 0.40 to 843.00 mg/L. The standard deviation of 68.48 suggests some variability, which might arise from agricultural runoff or leaching of potassium-rich minerals. Fluoride levels average 0.93 mg/L varying from 0.01 to 9.80 mg/L. A standard deviation of 1.10 indicates relatively stable levels, although some samples exceed the permissible limit for drinking water, warranting closer monitoring. It was also found that, NO_3, EC, HCO_3, F, SO_4, Na contributed highly to high water quality pollution.

To categorize the water samples according to the quantities of important cations (K+, Na+, Mg^{2+}, Ca^{2+}) and anions (SO_4^{2-}, HCO_3-, Cl−), a Piper diagram was made, as shown in Fig. 14.3. The figure aids in classifying the water samples into different types by providing a visual depiction of their ionic composition. The findings show that the bicarbonate type of water predominates in the samples, indicating a high concentration of HCO3-ions in the water. Subsequently, the sodium and potassium type, in which sodium and potassium ions are more prevalent, becomes the second most prevalent categorization. Furthermore, a lesser percentage of samples are classified as sodium–potassium-chloride-sulfate, indicating that these ions are present in different amounts. This classification draws attention to the variety in water chemistry and offers important information about the geological or human-caused elements affecting the research area's water quality.

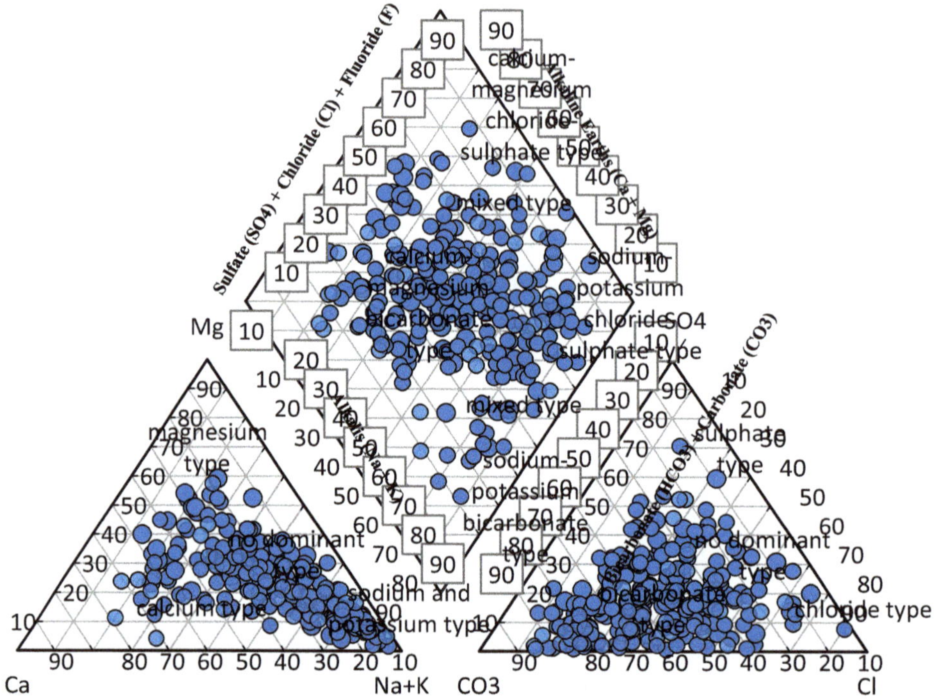

Fig. 14.3 Piper plot for groundwater samples displays the sort of water found in the study location

14.3.2 Spatial Analysis of WQI

The WQI spatial map is presented in Fig. 14.4, illustrating the varying levels of groundwater contamination across different regions. The analysis highlights that the northern region experienced significantly higher contamination compared to the southern region. Notably, the districts of Tonk and Ajmer were identified as having extremely polluted groundwater, rendering it unfit for drinking purposes. Additionally, the districts of Jaipur, Bhilwara, and Jhalawar were located in a highly polluted groundwater zone, further emphasizing the severity of contamination in these areas. On the other hand, some regions exhibited excellent drinking water quality. Specifically, the districts of Shivpuri, Guna, and Vidisha were categorized as having outstanding groundwater quality. Conversely, certain parts of Jhalawar, Chittaurgarh, and Udaipur were identified as areas with low groundwater quality, signifying localized contamination concerns. In contrast, several areas were noted for having good groundwater quality. These included districts such as Sikar, Morena, Kota, Mandsaur, Ratlam, Ujjain, Rajgarh, Shajapur, Dewas, Sehore, Bundi, Dholpur, and Banswara, which were characterized by relatively less polluted groundwater resources.

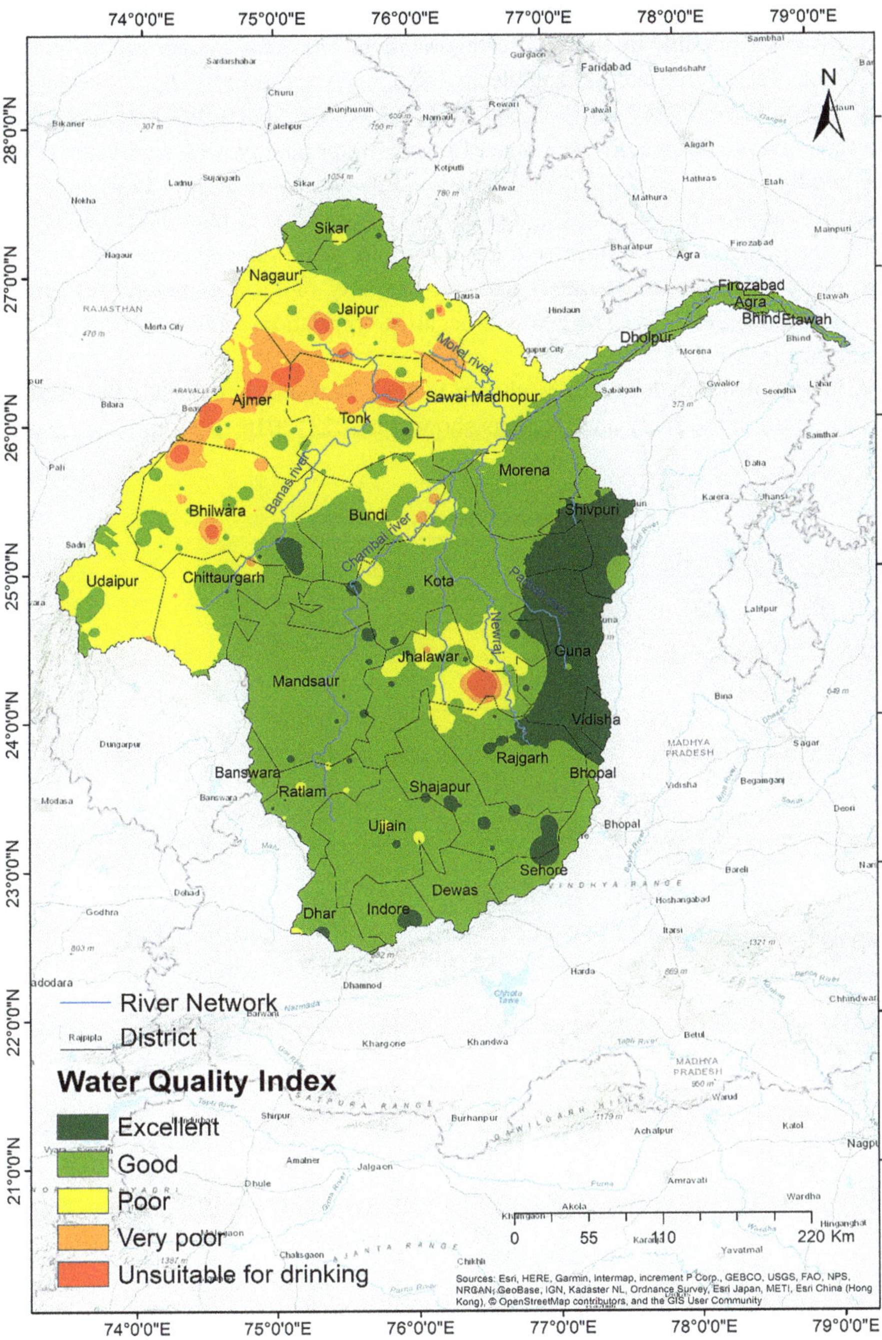

Fig. 14.4 Water quality index map of Chambal river basin

Additional understanding of the spatial distribution map of water quality throughout the region was made possible by the WQI categorization. The areas categorized as Excellent, Good, Poor, Very Poor, and Unsuitable for drinking were estimated to cover approximately 12,804 km^2, 77,207 km^2, 43,683 km^2, 6,303 km^2, and 1,989 km^2, respectively. These categories indicate a diverse range of groundwater quality, with significant portions falling into the "Good" and "Poor" classifications. The spatial map (Fig. 14.4) underscores the need for targeted interventions in regions with poor and very poor water quality while ensuring the preservation of areas with excellent and good water resources.

The most highly linked elements with the majority of the parameters were EC, Cl, SO4, TH, Ca, Mg, and Na. There was a negative correlation between pH and PO_4 and other metrics (see Fig. 14.5).

The long-term rainfall pattern is shown in Fig. 14.6 and it indicates the increasing trend of rainfall in the study area. Results show that 2013, 2016, 2019 and 2021 were the

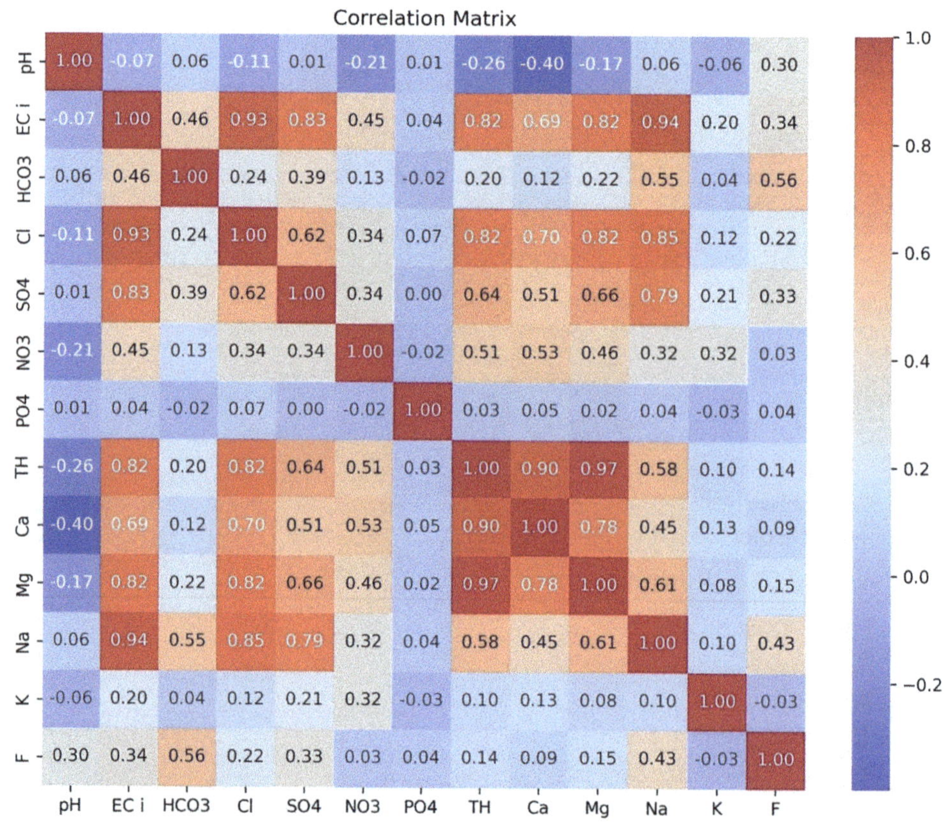

Fig. 14.5 Correlation matrix of water quality parameters

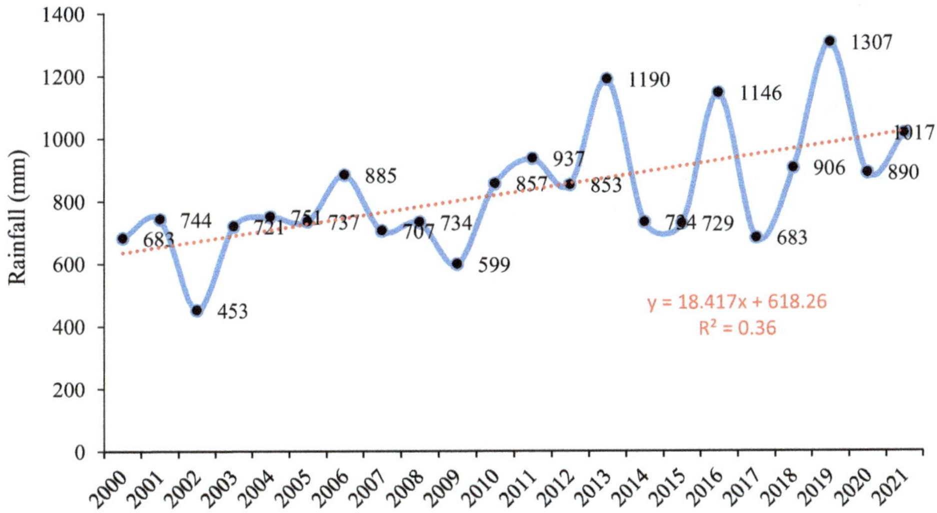

Fig. 14.6 Long term rainfall from 2000 to 2021

highest rainfall years and that's why agricultural runoff contributes to higher groundwater contamination.

14.4 Discussion

The study's conclusions provide important new information about Chambal River basin ground water quality dynamics and the difficulties in upholding water standards in an arid and semi-arid area. A thorough grasp of the main causes of water pollution was obtained by the examination of physical and chemical parameters, highlighting the necessity of focused interventions. Parameters like pH and phosphate exhibit low variability, suggesting stable environmental conditions. In contrast, parameters like EC, chloride, nitrate, and sodium show high variability, reflecting diverse sources and potential contamination hotspots (Saha and Paul 2019). Finding pollution hotspots and examining geographical trends were made possible by the incorporation of GIS and remote sensing techniques. High concentrations of pollutants including fluoride (F) and nitrate (NO3) were found in high-risk locations, especially those close to industrial districts and heavily farmed agricultural areas. This emphasizes how important it is to include spatial analysis in water quality monitoring programs in order to facilitate focused remedial operations. Elevated levels of chloride, sulfate, nitrate, and sodium in some samples point to human activities such as agricultural runoff, industrial effluents, and wastewater discharge. The high total hardness indicates that many samples fall into the "hard" or "very hard" water category, which can impact its use for domestic and industrial purposes (Oyewunmi et al. 2020).

The southern region of the study area has excellent groundwater quality for drinking. Due to climate change, the rainfall pattern may increase the agriculture runoff and leachate the industrial effluents and wastewater discharge to groundwater (Singh and Tiwari 2019; Sarker et al. 2021).

The analysis reveals that the northern region experiences higher contamination levels compared to the southern areas. Notably, the districts of Tonk and Ajmer exhibit extreme pollution, rendering their groundwater unfit for drinking purposes. Munoth et al. (2015) also discovered similar results in these districts. This distribution underscores the importance of targeted water management strategies. Efforts should prioritize areas classified as Very Poor or Unsuitable to mitigate health risks and ensure sustainable water use. Additionally, preserving regions with good and excellent water quality is crucial to maintaining their status and preventing future degradation.

By emphasizing significant obstacles to obtaining safe and accessible drinking water in vulnerable areas, the study closely relates these findings to SDG 6 (Clean Water and Sanitation). Due to the direct impact that poor groundwater quality has on public health, identifying pollution sources and hotspots also advances SDG 3 (Good Health and Well-Being). Additionally, the threats to groundwater quality posed by climate change are consistent with SDG 13 (Climate Action), which highlights the necessity of flexible approaches to managing water resources in the face of shifting rainfall patterns. In addition to supporting SDG 15 (Life on Land), protecting high-quality groundwater zones helps to maintain ecosystems that rely on freshwater supplies. Hence, the chapter's insights not only provide a scientific basis for groundwater management in the Chambal basin but also support the global agenda of safeguarding human health, ensuring equitable access to clean water, and promoting environmental sustainability.

14.5 Limitations and Future Research

Although the study offers a thorough evaluation of Rajasthan's water quality, there are a few things to keep in mind. First, because water quality parameters might change throughout the year, seasonal variability may have an impact on the data collected through field sampling and laboratory analysis. Furthermore, the study only looks at certain water quality parameters; a more thorough analysis might look at other aspects like microbiological contamination and newly discovered contaminants like medications and personal care items, which are raising concerns about water quality.

The integration of GIS and remote sensing tools, although valuable, relies on the availability and accuracy of spatial data, which might be limited in some areas. The use of these technologies for monitoring non-point source pollution, such as agricultural runoff, requires a more granular and localized approach to improve the precision of pollution source identification (Chatrabhuj et al. 2024). In terms of future research, a more comprehensive study on the long-term impacts of climate change on water quality and the role of extreme weather events would provide crucial insights into how changing environmental conditions influence water resources. Furthermore, developing and testing innovative, cost-effective water quality monitoring techniques, such as sensors and real-time data collection systems, could improve the efficiency and effectiveness of water quality management (Meng et al. 2017).

The social and economic effects of water quality on nearby populations, particularly in rural regions, also require further study. This could assist in customizing management plans that take into account the effects on the environment and human health. Last but not least, incorporating community-based water quality monitoring initiatives and raising public awareness through educational initiatives might be crucial to guaranteeing the region's water resources' long-term preservation.

14.6 Conclusion

The importance of comprehensive water quality management in Chambal river basin emphasized by this study. Pollution may be reduced and sustainable water supplies can be guaranteed by combining cutting-edge monitoring technologies, tools for spatial analysis, and proactive policy actions. The results offer a road map for resolving present issues and preserving water quality for coming generations. The spatial analysis of WQI was investigated through the Inverse Distance Weightage (IDW) techniques using the ArcGIS environment of the Chambal river basin and it was found that the northern region's groundwater quality parameters were more contaminated than the southern region. Notably, the districts of Tonk and Ajmer exhibit extreme pollution, rendering their groundwater unfit for drinking purposes. Additionally, it was discovered that NO_3, EC, HCO_3, F, SO_4, and Na all made significant contributions to the high levels of water quality impairment. In this study, WQI calculation may not include all relevant parameters, such as heavy metals, microbial contaminants, and emerging pollutants like pharmaceuticals and plastics. This could limit the comprehensiveness of the groundwater quality assessment. The study does not consider the socioeconomic implications of groundwater contamination, such as the cost of remediation, access to alternative water sources, or impacts on local communities. Addressing these limitations in future research will improve the robustness, applicability, and impact of groundwater quality assessments.

14.7 Recommendations

In the Chambal River watershed, proactive and coordinated groundwater quality management is desperately needed, according to the study's conclusions. The use of sophisticated monitoring technologies, geographical analysis, and predictive modeling have to be given top priority in order to reduce pollution and guarantee sustainability. In addition to supporting data-driven policy responses, field-based water quality measurement in conjunction with GIS and remote sensing can offer early warning systems for pollution hotspots. In regions like Tonk and Ajmer, where severe groundwater pollution has been found, targeted efforts are crucial to protecting public health and lowering reliance on contaminated water sources. It is also advised that future research use a wider range of water quality indicators. For a more thorough analysis, parameters including heavy metals, microbiological contaminants, and newly discovered pollutants like plastics and medications should be included in Water Quality Index (WQI) evaluations. The usefulness of the data for decision-making might also be improved by adding socioeconomic factors, such as the cost of remediation, the availability of substitute drinkable water, and effects on agricultural production. Local stakeholders and policymakers should work together to promote sustainable water use practices, tighten regulations on industrial discharge and agricultural runoff, and create community-based awareness campaigns. By working together, we can improve resilience, safeguard ecosystems, and eventually align with SDG 6's clean and safe water for all objectives.

Acknowledgements The author want to say thanks for Central Ground Water Board to provide the data (https://cgwb.gov.in/).

State whether you used AI or not in preparing the chapter.

We used artificial intelligence tools as an aid in researching, drafting, and refining the chapter content. AI assisted in gathering information, generating draft text, and improving clarity and coherence, but all content was reviewed, verified, and edited by me to ensure accuracy and alignment with the intended subject matter. The final intellectual responsibility and authorship of the chapter lie with us.

Author Contribution Leelambar Singh: Data collection, investigation, Formal analysis and supervision; Ankita Singh: Review and drafting; Nitika Mundetia: Review and supervision; Ravindra Nath Tripathi: Review and methodology.

Funding No.

Data Availability Not applicable.

Declarations

Ethics Approval and Consent to Participate Not applicable.

Consent for publication Not applicable.

Competing Interests The authors declare no competing interests.

References

Adilakshmi A, Venkatesan V (2024) Effective monitoring of Noyyal River surface water quality using remote sensing and machine learning and GIS techniques. Desalin Water Treat 320:100630
Adimalla N, Wu J (2019) Groundwater quality and associated health risks in a semi-arid region of south India: implication to sustainable groundwater management. Hum Ecol Risk Assess Int J 25(1–2):191–216
Ahmed MF, Ismail S, Khan MZ (2025) Aquifer characterization using geophysical borehole-logging and hydrochemical techniques—a case study from Lahore, Pakistan. Environ Earth Sci 84(1):33
Akar AU, Sisman S, Ulku H, Yel E, Yalpir S (2024) Evaluating lake water quality with a GIS-based MCDA integrated approach: a case in Konya/Karapınar. Environ Sci Pollut Res 31(13):19478–19499
Alemu CM, Aycheh YF, Angualie GS, Engidayehu SS (2024) Modeling on comprehensive evaluation of groundwater quality status using Geographic Information System (GIS) and Water Quality Index (WQI): a case study of Bahir Dar City, Amhara, Ethiopia. Water Pract Technol 19(4):1084–1098
Bagora P, Narulkar S (2024) Enhancing monsoon predictions for the upper chambal catchment through temporal and spatial downscaling of predicted future precipitation. J Inst Eng (India) Ser A 105(3):703–717
Banda TD, Kumarasamy MA (2020) Review of the existing water quality indices (WQIs). Pollut Res 39(2):487–512
Bandyopadhyay S (2007) Evolution of the Ganga Brahmaputra delta: a review. Geogr Rev India 69(3):235–268
Chabuk A, Al-Madhlom Q, Al-Maliki A, Al-Ansari N, Hussain HM, Laue J (2020) Water quality assessment along Tigris River (Iraq) using water quality index (WQI) and GIS software. Arab J Geosci 13:1–23
Chatrabhuj, Meshram K, Mishra U, Omar PJ (2024) Integration of remote sensing data and GIS technologies in river management system. Discov Geosci 2(1):67
Chidiac S, El Najjar P, Ouaini N, El Rayess Y, El Azzi D (2023) A comprehensive review of water quality indices (WQIs): history, models, attempts and perspectives. Rev Environ Sci Bio/Technol 22(2):349–395
Childs C (2004) Interpolating surfaces in ArcGIS spatial analyst. ArcUser, July-September 3235(569):32–35
Crawford AR (1969) India, Ceylon and Pakistan: new age data and comparisons with Australia. Nature 223(5204):380–384
Dhanush SK, Murthy M, Ayyappa S, Prabhuraj DK, Verma R (2024) Water quality assessment of Bheemasandra Lake, South India: a blend of water quality indices, multivariate data mining techniques, and GIS. Environ Sci Pollut Res 1–20
Gani A, Hussain A, Pathak S, Ahmed S, Omar PJ (2025) Impact of pollutants on groundwater quality and health risk assessment of quaternary aquifers in Northern India. J Hazard Toxic Radioact Waste 29(1):04024039
Garnero G, Godone D (2014) Comparisons between different interpolation techniques. Int Arch Photogramm Remote Sens Spat Inf Sci 40:139–144
Gupta S, Gupta SK (2021) A critical review on water quality index tool: genesis, evolution and future directions. Eco Inform 63:101299
Horton RK (1965) An index number system for rating water quality. p 300–305. Water quality criteria-stream vs. effluent standards. J Wat Pollut Contr Fed 37(3):292–315

Jayasankar P, Kumar KS, Vedavalli V, Zoysa SIDA, Meghana U (2025) Water quality index: measure of quality of drinking water for sustainable development. In: Sustainable environment and health. Apple Academic Press, pp 185–195

Kaushik P, Ghosh P (2015) Geomorphic evolution of Chambal river origin in Madhya Pradesh using remote sensing and GIS. Int J Adv Remote Sens GIS 4(1):1130–1141

Kumar A, Singh A (2024) Pollution source characterization and evaluation of groundwater quality utilizing an integrated approach of Water Quality Index, GIS and multivariate statistical analysis. Water Supply 24(10):3517–3539

Kumar A, Sharma MP, Rai SP (2017) A novel approach for river health assessment of Chambal using fuzzy modeling, India. Desalin Water Treat 58:72–79

Lee H, Calvin K, Dasgupta D, Krinner G, Mukherji A, Thorne P, Zommers Z (2023) Climate change 2023: synthesis report. Contribution of working groups I, II and III to the sixth assessment report of the intergovernmental panel on climate change. The Australian National University

Li P, He S, Yang N, Xiang G (2018) Groundwater quality assessment for domestic and agricultural purposes in Yan'an City, northwest China: implications to sustainable groundwater quality management on the Loess Plateau. Environ Earth Sci 77(23):775

Lovett A (2014) GIS and environmental management. In: Environmental science for environmental management. Routledge, pp 267–285

Mardiana S, Anzum R, Dwijendra NKA, Al Sarraf AAM, Timoshin A, Sergushina E, Tikhomirov E (2022) Assessment of groundwater quality and their vulnerability to pollution using GQI and DRASTIC indices. J Water Land Dev 138–142

Meng F, Fu G, Butler D (2017) Cost-effective river water quality management using integrated real-time control technology. Environ Sci Technol 51(17):9876–9886

Mishra RK (2023) Fresh water availability and its global challenge. Br J Multidiscip Adv Stud 4(3):1–78

Mishra AP, Singh S, Jani M, Singh KA, Pande CB, Varade AM (2024) Assessment of water quality index using Analytic Hierarchy Process (AHP) and GIS: a case study of a struggling Asan River. Int J Environ Anal Chem 104(5):1159–1171

Munoth P, Tiwari K, Goyal R (2015) Fluoride and nitrate groundwater contamination in Rajasthan, India: a review. In: 20th international conference on hydraulics

Nagothu SK, Bindu Sri P, Anitha G, Vincent S, Kumar OP (2025) Advancing aquaculture: fuzzy logic-based water quality monitoring and maintenance system for precision aquaculture. Aquacult Int 33(1):32

Nath A, Samanta S, Banerjee S, Danda AA, Hazra S (2021) Threat of arsenic contamination, salinity and water pollution in agricultural practices of Sundarban Delta, India, and mitigation strategies. SN Appl Sci 3(5):560

Oyewunmi OD, Safiabadi-Tali SH, Jahanshahi-Anbuhi S (2020) Dual-modal assay kit for the qualitative and quantitative determination of the total water hardness using a permanent marker fabricated microfluidic paper-based analytical device. Chemosensors 8(4):97

Pushkar P (2025) Ground-water quality assessment for three coastal-islands of India with the help of water quality index (WQI). Econ Polity Environ Int Peer-Rev J Soc Stud 6(2):11–36

Rahman MM, Ghosh T, Salehin M, Ghosh A, Haque A, Hossain MA, Hutton CW (2020) Ganges-Brahmaputra-Meghna delta, Bangladesh and India: a transnational mega-delta. Deltas Anthropocene 23–51

Reddy PB (2017) Productivity of chambal river in relation to water quality. WJ Pharm Pharm Sci 6(7):1466–1475

Saha P, Paul B (2019) Groundwater quality assessment in an industrial hotspot through interdisciplinary techniques. Environ Monit Assess 191:1–20

Sarker B, Keya KN, Mahir FI, Nahiun KM, Shahida S, Khan RA (2021) Surface and ground water pollution: causes and effects of urbanization and industrialization in South Asia. Sci Rev 7(3):32–41

Sharma H, Binda P (2009) Geo-hydrology of the Mej Basin, Rajasthan. Earth Syst Sci Felicitation Vol Honour Prof VK Verma 1:277

Singh S, Tiwari S (2019) Climate change, water and wastewater treatment: interrelationship and consequences. Water conservation, recycling and reuse: issues and challenges, 203–214

Singh L, Saravanan S (2020) Impact of climate change on hydrology components using CORDEX South Asia climate model in Wunna, Bharathpuzha, and Mahanadi, India. Environ Monitor Assess 192(11):678

Stocks CE, Wise S (2000) The role of GIS in environmental modelling. Geogr Environ Model 4(2):219–235

Tan JJ, Arumugasamy SK, Teo FY (2025) Water quality index prediction using artificial neural network: a case study of Selangor River, Malaysia. Int J Sustain Agric Manag Inform 11(1):48–71

Thirumoorthy P, Murugasamy MV, Dhanasekaran J, Sasikumar K, Periyasamy M, Selvam J (2024) Assessment of groundwater quality in Perundurai region of south India using water quality index and statistical modelling. Groundw Sustain Dev 25:101104

UN (2024) Water for prosperity and peace. Fatcs, Figures and Action Examples. The United Nations World Water Development report 2024, Prepared by WWAP, Engin Koncagül. Richard Connor and Valentina Abete, UNESCO

Uslu A, Dugan ST, El Hmaidi A, Muhammetoglu A (2024) Comparative evaluation of spatiotemporal variations of surface water quality using water quality indices and GIS. Earth Sci Inform 1–16

WHO (2004) Guidelines for drinking water quality: training pack. WHO, Geneva, Switzerland

Yadav NS, Kumar A, Sharma MP (2014) Ecological health assessment of Chambal River using water quality parameters. J Integr Sci Technol 2(2):52–56

Zhang Q, Xu P, Qian H (2020) Groundwater quality assessment using improved water quality index (WQI) and human health risk (HHR) evaluation in a semi-arid region of northwest China. Exp Health 12(3):487–500

Zhang Y, Jia R, Wu J, Wang H, Luo Z (2022) Uncertain in WQI-based groundwater quality assessment methods: a case study in east of Beijing, China. Environ Earth Sci 81(7):202

Zhao X, Ming D, Meng Y, Yang Z, Peng Q (2025) Analysis of the water quality of a typical industrial park on the Qinghai-Tibet plateau using a self-organizing map and interval fuzzy number-based set-pair analysis. Water 17(1):111

Water Quality Assessment and Management of Various Sources in Barak Valley, South Assam, India

15

Sadhana Rani Paul, Subhadeep Biswas, and Shanti Gopal Patra

Abstract

Water quality assessment represents evaluating the quality of a certain water sample based on several criteria depending on the purpose for which it will be utilized. On the other hand, the management process involves collecting, analyzing, and recording the physical, chemical, and biological parameters of the water sample. The present chapter describes the water quality of various sources in different parts of Barak Valley, Assam, India. Key findings of the recently conducted studies are presented in brief. Firstly, after the introduction section, a description of where the investigations are performed is provided. Traditional laboratory-based techniques and new analytical methods for measuring parameters are discussed. Apart from experimental methods, different statistical techniques are also practiced. The Water Quality Index of samples from different locations is highlighted as reported by different groups. Compilation of such reports will help identify which water from which location will be suitable for which purpose. Further, it will help the scientific community decide the treatment needed for purification purposes. It aims to contribute significantly to developing policies for overall water quality, management, and distribution in these parts of South Assam.

S. R. Paul · S. G. Patra
Chemistry Department, National Institute of Technology Silchar, Assam, India
e-mail: shanti@che.nits.ac.in

S. Biswas (✉)
Civil Engineering Department, National Institute of Technology Silchar, Assam, India
e-mail: subhadeep@civil.nits.ac.in; sdeep0819@gmail.com

S. Yadav et al. (eds.), *Water Resources in Delta Region in India*, Deltas of the World,
https://doi.org/10.1007/978-3-032-06500-1_15

15.1 Introduction

The term 'water quality' signifies the physical, chemical, thermal, and/or biological properties of water. It plays a vital role in the life span of any living species (Ritchie and Schiebe 2000). As per a report given by the World Health Organization (WHO) in 2002, due to unsound water, hygiene, and sanitation, 1.7 million deaths and 54.2 million Disability Adjusted Life Years (DALYs) lost in the entire world are reported annually. In India, waterborne diseases like diarrhea are taking the lives of more than a million children per year (Parikh et al. 1999). In establishing a community and making it survive, the necessary pre-requisite is the availability of safe and dependable sources of water (Mahapatra and Mishra 2005). Usually, we associate the term water quality with the health of human beings, but, in a broader sense, it applies to animals, birds, and all the aquatic creatures in the water system (Ritchie and Schiebe 2000). In India, in the state of Assam, the valley in the southern part and surrounded by the North Cachar Hills district is known as the Barak Valley, a region named after the river Barak. This valley consists of three districts called Cachar, Karimganj (officially, Sribhumi) and Hailakandi and it falls in the middle of 24°80' and 25°80' N, latitudinally and 92°15' and 93°15' E, longitudinally, possessing a geographical area of 6,921 km^2 (where, Cachar 3,786 km^2, Karimganj 1,809 km^2, and Hailakandi 1,326 km^2). It occupies nearly 9% of the whole land area of Assam (Barhai and Das 2024; Shah 2012).

The region is significant in various aspects, including sociological, ecological, cultural, and geographical aspects. The presence of different tribes, with their unique cultural features, tea gardens, the beauty of rivers, hills, and greenery, has made this valley a beautiful place to dwell in. On the one hand, there are so many good things to cherish; on the other hand, environmental problems are causing deep concern. Water pollutions and getting a supply of good water quality for various purposes are pressing issues in contemporary times.In the Barak Valley, the main reasons behind the water quality deterioration are mainly anthropogenic activities, population growth, and seasonal influence. The disposal of wastewater, agricultural pesticides from the farmland, sewage outlets, and automobile oil wastes are the prime anthropogenic cause (Hasan et al. 2007; Srivastava et al. 2003; Usha et al. 2006). The inhabitants of this valley who live near any surface water body, such as a river, pond, lake, etc., use this water source for cleaning and washing clothes, utensils, and vehicles. In rural areas, this pondwater is often used for drinking purposes. However, due to a lack of hygienic awareness and defecation in the open by humans and domestic animals, pathogenic contamination is obvious in these drinking water sources. On the other hand, sewage outlets in urban centers of this valley, improperly made septic tanks, and the absence of appropriate soak pits are creating the pollution of both surface and groundwater (Gupta 2004).

In a statistical survey, it was seen that the population in this valley was 1.116 million in 1951, which increased to more than 3.0 million in 2001, with a concomitant rise from 161 to about 436 persons per km^2 with respect to population density (Mazumdar et al.

1998). As per another source, i.e., data from the census of India of 2011, the population growth of this valley was 17.93% between 2001 and 2011 (Annayat et al. 2024). Due to the high rate of population growth, poor sanitation is noticed in rural and urban areas. In urban areas, leakage is experienced in ill-treated and old water distribution pipes, which, in turn, results in waterlogging and entry of contaminated water from the sewer in the drinking water pipelines (mostly during the months of monsoon) (Gupta 2004).

Most importantly, the climate is a very big factor in the Barak Valley. A subtropical monsoon climate is observed here. Due to this, high rainfall is experienced, 85% of which occurs from May to October. As a result of the high rainfall, the water availability is high, but this is constrained to the months of rain only, i.e., May and October. Water scarcity is noticed from December/January to March/April if the rainwater is not conserved properly. Though have been harvested traditionally in privately owned ponds, community-managed tanks, etc., for more than a century, now, due to the degradation of water quality and lack of proper management, they are on the brink of collapse (Gupta 2004). Also, floods mostly appear yearly in the rainy season on different scales. Due to the flood, the river Barak and its tributaries overflow and spread water throughout the valley. Though the flood brings lots of alluvial soil, it also drags many water pollutants like sewage, agricultural runoff, industrial chemicals, and contaminants. These pollutants significantly degrade the water quality of the water bodies in this valley. Therefore, getting good quality water for drinking and doing valuable work at that time becomes a tough job. Moreover, these aspects make the water quality assessment of Barak Valley very indispensable. Further, this need for assessment is often encountered in various research and surveys in different parts of the valley.

Additionally, drinking water scarcity is often encountered in the Barak Valley in any season. Besides accompanying all the aforementioned reasons, another notable cause is urbanization. Due to urbanization, several ponds have collapsed or been filled in already to build houses or shopping malls, resulting in decreased surface water. One such vivid example is the Goldighi shopping mall, which is present in the heart of Barak Valley, i.e., in the Silchar city of Cachar district. This mall was built by filling a big pond called 'Goldighi'.

This book chapter is going to be a blend of every aspect related to the water quality of Barak Valley, starting from the significant sources of water of this valley, the quality of water that can be obtained, different water quality assessment techniques comprising traditional laboratory and advanced analyticalmethodologies, variety of management techniques of available water sources to the hindrances which currently encountered and their possible remediations in future.

15.2 Major Sources and Quality of Water in Barak Valley

Barak Valley possesses several sources of water. Among them, surface water sources are rivers, ponds, rainfall, seasonal wetlands, floodplain lakes, marshes, tanks, waterlogged areas, streams and springs, etc. (Das and Das 2023). While mentioning rivers, the main river, as the name implies, the river Barak, which starts from the Manipur hills and flows through this region, enters Bangladesh as its tributaries Surma and Kushiyara rivers and ends up by joining the river Meghna (at Bhairab Bazar, Bangladesh) (Dutta and Sarma 2017). Some tributaries of the Barak River, which flow throughout the different parts of the valley, have become the essential sources of surface water for the residents of this valley, including Jiri, Chiri, Madhura, Larang, Harang, Kalainchera and Dolu-Jatinga, on the north bank and Ghagra, Sonai-Rukni, Katakhal, and Dhaleswari on the south bank. Apart from these, there are two branches of the river Barak, which are generated from the bisection that happened before the entrance to the Sylhet district of Bangladesh, and they are–Surma and Kushiyara (it's one of the major tributaries is Singla). Except for Barak, there is another river named Longai. This flows through the Karimganj district to move into a wetland in Bangladesh (Gupta et al. 2015). The water quality of the rivers and their tributaries is drinkable, but the presence of pathogens like coliform bacteria is found in different measures (Gupta 2004).

There are different lakes in Barak Valley. Among them, the river Barak itself has seven lakes of its own, specifically oxbow lakes, which represent the unique feature of this river. In the aqua-zone of India, this is the only river with seven ox-bow lakes in a row. In the local language 'Anua's, these ox-bow lakes all exist in the plain land of Barak Valley, and not a single ox-bow lake is present in the Manipur part of the river. These seven 'Anua's are 'Chiri' anua, 'Rupaibali' anua 'Banskandi' anua, 'Kaptanpur' anua, 'Gobindapur' anua, 'Ramnagar' anua and 'Shalchapra' anua (as marked in Fig. 15.1).

Among them, the largest and longest ox-bow lake, 'Rupaibalianua' in the Rupaibali village near to the Lakhipur subdivision of Cachar's water is urea polluted, as per an article published in Barak Bulletin (July 4, 2024). The issue came into the limelight after the tests were performed in the Assam University laboratory at the villagers' request. There were aquatic plants in this lake, and more than half of the lake's areas were reported to be covered by water hyacinth, specifically. After performing the test and analysis, it was revealed that the area stimulated the growth of this water hyacinth and other toxic phytoplankton in this lake. This test also revealed that the water of this lake had low dissolved oxygen and high BOD, the concentration of nitrates and phosphates was higher than the normal range, and the usual qualities of the water had been destroyed. Additionally, the reason behind the arrival of high urea content in lake water was the usage of urea as a catalyst to grow fish in high quantities by a fisherman community who was given the lease of this lake by the government ("Nature's Pride at Risk: River Barak's Pristine Ox-Bow Lakes Threatened by Urea Pollution–Barak Bulletin," n.d.). Coming to normal lakes, a prominent lake is there in this valley with a vast geographic significance, namely

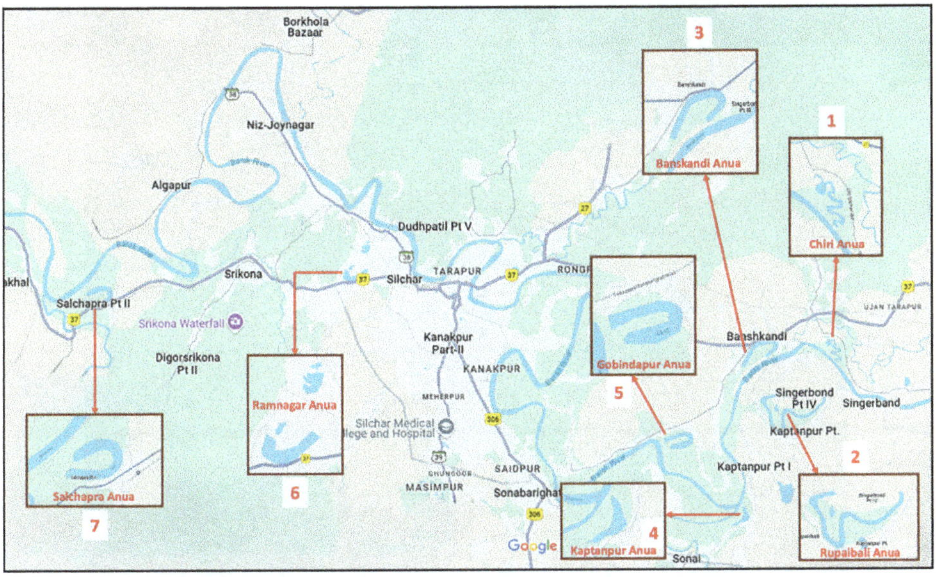

Fig. 15.1 Map of Barak River showing seven ox-bow lakes created by Google Maps

Son Beel (Fig. 15.2). This lake is particularly a freshwater lake, which is situated in the Karimganj (officially, Sribhumi) district, and it is Asia's second largest wetland. It is not permanent; it becomes a lake in the rainy season (March onwards), and in the winter, it dries up (Dutta et al. 2019). However, during the entire rainy season, water of this lake becomes a significant source of water for the people of that area.

It was earlier reported that for a sample of water of Son Beel collected in consecutive four months, January, February, March, and April, at different dates, on average, the pH obtained was near neutral, and the average values of BOD and COD found were around 3.65 mg/L and 11.70 mg/l. Apart from these, the presence of bacteria was also tested, and a high amount of coliform bacteria was observed (Action Plan for Son Beel- Priority V, 2019).

Ponds are another source of water in this region. However, there are limited ponds left in the urban areas of this valley; most ponds are mainly seen in the rural areas, and water quality is up to the mark. Various research groups surveyed the water quality of different ponds in this valley. In one such survey done by Bhuiyan and Gupta on nine ponds of Irongmara of Cachar district, it turns out that all nine ponds are suitable for drinking and usage in domestic works. According to the report, all of them are capable of fish production except one pond, in which the amount of nitrate is lower than the content required for the production of fish (Bhuiyan and Gupta 2007).

Tubewells, dig wells (in rural areas), hand pumps, borewells, aquifers, etc., are some of the significant groundwater sources of this region. There are mainly two types of aquifers

Fig. 15.2 Satellite picture of Son Beel made by Google Maps

found in this valley: a) unconsolidated aquifers (comprising alluvial deposits of Holocene to Recent age) in the valleys and b) semi-consolidated aquifers (by rocks of Tertiary to Pleistocene age), in the hilly areas (Shah 2012). The quality of groundwater varies from place to place within the valley, but one heavy metal that is often found in the water coming from tube wells in most places is arsenic (As). It has been highlighted in many research works and surveys done on the groundwater samples of this valley. One such research work was done by Barhai and Das, and as per the report of their investigation (done on the water samples gathered from 24 tubewells of this valley), groundwater is contaminated by different trace elements. Among them, As, Fe, and Cu were found to be present in high concentrations (Barhai and Das 2024).In recent times, the water quality index (WQI) has been used by various research groups as a standard method for describing the water quality of a particular source to be used for various purposes. In one of the works, Khangembam and Kshetrimayum reported the hydrogeochemical factors and WQI of water samples collected from different water resources of the Barak valley (Khangembam and Kshetrimayum 2019). A set of 33 samples was subjected to testing and analysis. The detailed investigation found that the water samples contain a high amount of HCO_3, and the WQI values recommended are unsuitable for drinking due to excessive concentrations of As, Fe, and Mn.

15.3 Assessment of Water Quality Parameters

15.3.1 The Vital Role of Water Analysis in Public and Environmental Well-Being

Considering environmental concerns, understanding water quality is essential to ensuring safe and clean water. Over the years, various methods of water quality have been developed. This section summarises the recent methodologies and technologies utilised in this field.

Water analysis is important for the environment and health, as well as maintaining the quality of water resources and long-term availability. This involves diverse techniques to detect, measure, and track various pollutants and characteristics within water bodies, encompassing water storage, lakes, rivers, groundwater, etc. The significance of the analysis of water lies in its capacity to precisely detect health-related issues, evaluate ecological dangers, and inform critical decisions concerning the management of water resources and safeguarding public health within our communities.

15.3.1.1 Maintaining the Quality of Drinking Water for Communities

One of the important aims of water analysis is to measure different water parameters to maintain the quality and safety measures for human consumption. Experts can pinpoint risks and implement effective solutions by examining factors such as bacteria, chemical pollutants, and physical attributes. Regular monitoring is crucial in preventing waterborne illnesses and securing all residents' access to clean drinking water.

15.3.1.2 Protecting the Health of Local Ecosystems

Analyzing water is indispensable for understanding how pollution impacts the delicate balance of aquatic ecosystems and their biodiversity. Contaminants like heavy metals, agricultural pesticides, excessive nutrients, and harmful microorganisms can severely affect fish, amphibians, and indigenous aquatic flora. By conducting thorough water quality assessments, we can accurately locate pollution sources, comprehend the ecological risks to our local environment, and formulate strategies to protect and restore these essential habitats.

15.3.1.3 Origin of Water Pollution

In water characterization, one important thing is identifying the origins of pollution within different water sources. Researchers can trace contamination pathways in the local environment by examining the unique chemical compositions, isotopic signatures, and microbial indicators. This knowledge is crucial for environmental authorities, policy creators, and regulatory organizations tasked with ensuring compliance with pollution controls and implementing clean-up strategies. Accurately identifying these sources is essential for preventing further contamination and ensuring accountability for those responsible.

15.3.1.4 Tracking Changes in Water Quality Over Time

Regularly tracking water quality characteristics allows researchers to identify gradual shifts and alterations in nearby water ecosystems. Examining past data helps us judge how well current pollution controls are working, understand the consequences of changes in land use, and predict upcoming water quality issues. This analysis of trends yields a vital understanding of the health of local aquatic life, supports flexible management approaches designed for particular situations in Silcoorie Grant, and provides the basis for informed choices about the long-term care of water resources in our region.

15.3.1.5 Supporting Public Health Initiatives

Analyzing water is essential for protecting the health of our community by pinpointing and reducing dangers from waterborne illnesses and chemical pollutants in our local water sources here in Silcoorie Grant. By identifying harmful germs, poisons, and new contaminants, our public health officials can put suitable treatments in place, issue important warnings, and stop outbreaks of water-related diseases within Assam. Prompt and precise monitoring of our water quality is key to avoiding health emergencies, lessening the strain on our healthcare system in Cachar, and protecting those most at risk, such as children and older adults in our area.

15.3.2 Different Adopted Methods

The water quality assessment can be categorized into two main methods: (i) Traditional laboratory methods and (ii) Advanced analytical methods (Skoog et al. 2022).

I. Traditional Laboratory Methods

Several laboratory methods are traditionally utilized to assess water quality by measuring various parameters in the water. Traditional laboratory methods include the analysis of water samples by physical, chemical, and biological to determine the presence of different contaminants and other essential substances, such as dissolved oxygen. Despite the inception of several modern techniques, the traditional methods provide reliable and accurate results.

15.3.2.1 Chemical Analysis

Titrimetric: Typically, water samples can be acidic or basic. The acid–base character of the water can be understood using acid–base titration. The alkalinity is generally determined by titrating water using a standard HCl solution with a phenolphthalein indicator. Further, hard water contains chloride, sulfate salts of Ca^{2+}, and Mg^{2+}. These hardness-causing cations are determined through complexometric titration using EDTA reagent and EBT indicator. Further, water contains dissolved oxygen, which is determined using redox titration. Here, the iodometric titration method is used.

Colorimetric: This method analyzes the change in intensity to measure the presence of different chemicals, such as dyes, surfactants, and other pollutants, such as arsenate. This method is the application of the Beer-Lambert law, which is expressed as:

$$\log\left(\frac{I_0}{I_s}\right) = \varepsilon C l$$

where I_0 and I_s are the intensity of incident light and after passing through the sample, respectively; ε is the molar extinction coefficient; C is the concentration, and l is the length of the cell.

Chromatography: Ion chromatography analyses anions such as nitrate, sulfate, and chloride. In a typical chromatographic column, the ions interact with the column differently, leading to different retention times. This difference provides the applicability of the process to separate ions present in water, which are then detected using a suitable detector.

15.3.2.2 Biological Analysis

The microbiological tests are performed to detect the presence of protozoa, viruses, bacteria, and other microorganisms. Typical methods include multiple tube fermentation, membrane filtration, polymerase chain reaction (PCR), etc.

The biological oxygen demand (BOD) is an important parameter to understand water quality. BOD is the amount of oxygen the microorganism requires to completely oxidize the organic matter in the water. Typically, the water sample is incubated for five days, and then the dissolved oxygen concentration is measured using redox titration. Differences in DO before and after incubation provide BOD.

15.3.2.3 Physical Analysis

The physical parameters routinely checked for water are pH, turbidity, light scattering, conductivity, etc. pH meter, nephelometer, dynamic light scattering (DLS), conductivity meters, etc., are used to determine water quality. Total dissolved solids (TDS) are determined using the gravimetric method. Here, the water is vaporized, and the remaining solid amount is measured by weighing.

II. Advanced Analytical Techniques

For advanced water analysis, techniques such as high-performance liquid chromatography (HPLC), mass spectrometry (MS), liquid chromatography-mass spectrometry (LC–MS), gas chromatography-mass spectrometry (GC–MS), atomic absorption spectroscopy (AAS), X-ray fluorescence spectroscopy, and electrochemical detection methods (amperometric, voltammetry, electrochemical impedance spectroscopy, potentiometric detection, etc.) are employed (Kadadou et al. 2024).

Sample preparation is crucial in analytical techniques, requiring many pre-concentration steps and clean-up procedures. Extraction techniques such as solid phase extraction (SPE) (Badawy et al. 2022) and liquid–liquid extraction (LLE) (Mahara et al. 1998) were used in the sample preparations. The LLE procedure separates one substance between immiscible organic and aqueous solvents (Mandal et al. 2023). The analyte is transferred using the SPE method, and the sample is isolated. Although a cheap separating funnel is used in the LLE procedure, the organic solvent makes it costly. As a result, the SPE becomes the most efficient method for sample preparation, which involves transferring the sample to the matrix, making the process selective, rapid, and a step of purification before the GC, TLC, and HPLC chromatography procedures (Bianchin et al. 2012). However, some drawbacks include the channelling phenomenon and blockage of cartridges. As a result, new methods are continuously being developed. Several advanced extraction techniques were utilized, including dispersive solid-phase extraction (DSPE), magnetic solid-phase extraction (MSPE), stir bar sorptive extraction (SBSE), dispersive liquid–liquid microextraction (DLLME), and solid-phase microextraction (SPME) (López-Paz et al. 2014; Nasiri et al. 2020).

Chromatography (Kucharska and Grabka 2010), a standard analytical method for separating organic compound mixtures based on differential passage through phases (gas chromatography [GC] uses inert gas, liquid chromatography [LC] uses liquid), provides qualitative and quantitative data. GC's advantage is its small sample size, but drawbacks include solvent use and hazardous reagents, leading to LC's increased use. Conventional LC uses C8/C18 stationary phases, 3–5 μm particles, and 10–25 cm columns (Kucharska and Grabka 2010). Newer miniaturized LC techniques with smaller particles or column diameters offer improved sensitivity, efficiency, resolution, and faster analysis times (Mazzeo et al. 2005).

The development of liquid chromatography (LC) techniques resulted in high-performance liquid chromatography (HPLC), a method that utilizes columns packed with fine particles (usually 3–10 μm) to improve separation. This enhanced efficiency necessitates high pressure to achieve optimal flow within the system. HPLC is now a versatile and standard chromatographic method. Ultra-pure liquid chromatography (UPLC), using 1.7 μm particles under high pressure, offers higher resolution and varied response times (Mijangos et al. 2018). Capillary liquid chromatography (CLC), using 500 μm internal diameter columns, lowers sample/solvent needs and increases sensitivity with microliter-per-minute flow rates. Recent research on fluorinated compounds in river water demonstrated that ultra-high-performance liquid chromatography (UHPLC) and pressurized liquid chromatography (PLC) offered better speed and resolution compared to high-performance liquid chromatography (HPLC). While UHPLC coupled with tandem mass spectrometry (UHPLC-MS/MS) provided superior method limits of detection (MLDs) (Tomai et al. 2018), it came with a higher cost. Techniques employing tandem mass spectrometry detectors (UHPLC-MS/MS and HPLC–MS/MS) exhibited enhanced selectivity over those using mass spectrometry (MS) alone, albeit at a more significant expense. Liquid chromatography coupled with mass spectrometry (LC–MS) or tandem mass spectrometry (LC–MS/MS) is frequently used for water analysis, often preceded by solid-phase extraction (SPE), though resolution can still be an issue. These methods detect persistent organic pollutants (POPs) and emerging organic pollutants (EOPs).Wang et al. reported the development of an LC–MS-based novel analytical technique for detection of drugs like sulfamethoxazole and trimethoprim in rat plasma (Wang et al. 2012). Different analytical techniques utilized to detect standard POPs are given in Table 1.

GC–MS was the main technique used in the 1990s for water pesticide analysis. Improved GC–MS with atmospheric pressure ionization now better analyzes environmental pesticide residues (μg/L range) (Campanale et al. 2021). However, polar, low-volatility pesticides require derivatization for efficient GC–MS. For these, LC–MS is used instead (Casado et al. 2018).

In inorganic analysis, AAS mainly quantifies metallic elements by measuring the absorption of specific wavelengths of light by atoms. Atomic emission spectrometry (AES) analyzes emitted light at characteristic wavelengths when heated atoms' electrons return to lower energy states. Calorimetry, a highly sensitive and adaptable method for quantitative analysis, quantifies compound concentration by assessing the amount of radiant energy absorbed by colored solutions at particular wavelengths. Table 2 presents a variety of analytical techniques employed for the detection of inorganic compounds.

Several analytical approaches exist for detecting and quantifying waterborne particulates, such as turbidity meters, particle counters, filtration and subsequent mass determination, microscopy, spectroscopic analysis, and sedimentation/centrifugation (illustrated in Fig. 15.3).

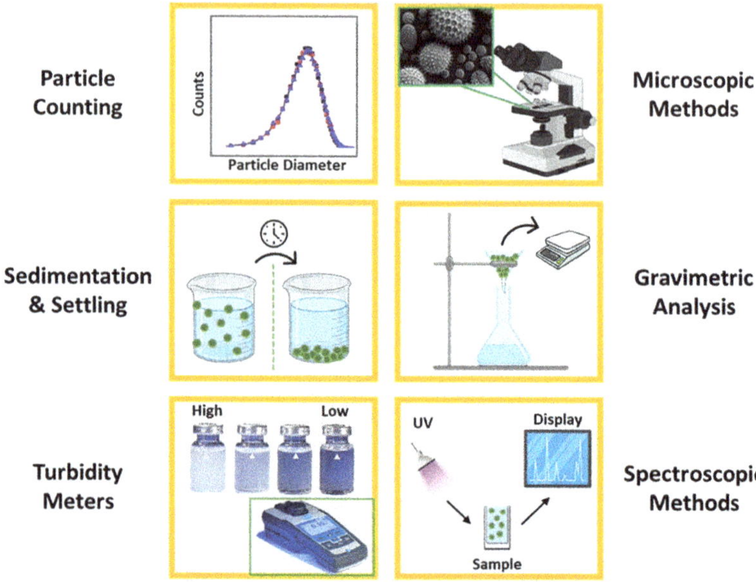

Fig. 15.3 Analytical methods used for the detection of particulates in water. Reproduced with permission from the Kadadou et al. 2024, © 2024 Elsevier. Kadadou et al. 2024

15.4 Different Management Techniques Available for the Conservation of Water Sources

Barak Valley is blessed with a vast quantity of natural water resources. As per the recent report, the total annual groundwater recharge for the whole state of Assam, including Barak Valley is 27.05 billion cubic meters (bcm), and out of which 21.97 bcm has been found extractable(Groundwater Management in India: A Multi-State Field Study of Availability, and Locally Appropriate Solutions for Sustainable, Equitable and Efficient Use of Groundwater, n.d.). At the present stage of groundwater extraction, it is 11.73%. Further, no region has been designated as saline in the entire state. However, due to the consistent increase in population at the current time, water scarcity and problems in uniform water distribution are often felt across the region. Hence, it is urgent to incorporate water management techniques to conserve existing water resources. Moreover, water resource management is one of the pressing environmental topics of contemporary times. It includes a set of practices including rainwater harvesting, groundwater recharge, desalination and reuse of saline water, irrigation, etc. Sustainable management techniques need to be practiced for proper water distribution and supply. Under the various provisions of the United Nations Sustainable Development Goals (UNSDGS), proper water resource management comes under SDG 6: Clean Water and Sanitation.

Firstly, regarding rainwater harvesting, it is essential to note that this practice is feasible in places where adequate rainfall is received, but no proper supply system is designed. Moreover, rainwater harvesting is necessary in places where water can be stored for agricultural purposes in the dry season. In this region of Assam (Barak valley), a good quantity of rainfall is obtained yearly. As per one of the recent reports, the region receives an annual average rainfall of nearly 3000 mm (Barhai and Das 2024). Secondly, rainwater harvesting can also be a good source of drinking water when groundwater is severely contaminated. In one of the studies, Shah reported that the groundwater reserves of Cachar and Karimganj districts of the Barak Valley are severely contaminated with As (Shah 2012). The maximum concentration of As has been found as 350 μg/L and 66% of tubewells from this region possess an As concentration above WHO guidelines of 10 μg/L. In this scenario, using harvested rainwater is undoubtedly a sustainable option. A recent study based on Khulna, a division of Bangladesh located in the coastal region, suggested a similar idea (Amos et al. 2020). Despite having a considerable number of tubewells and adequate groundwater storage, this region (Khulna) is highly contaminated by As pollution as well as enhanced salinity has made the groundwater unsuitable for drinking purpose.

Like rainwater harvesting, groundwater recharge is another potential water conservation technique. It forms an integral part of the water management program, replenishing the aquifers and sustainable water availability for a prolonged duration. In Assam, groundwater is mainly available in sandstone and alluvium aquifers. According to one of the recent articles, the Cachar district of Barak Valley contains mainly unconsolidated, semi-consolidated, and consolidated aquifers (Paul et al. 2024). As this region receives heavy rainfall,

Among different water management techniques, irrigation is one of the most prominent. The Barak Valley has a huge potential for irrigation, although utility has not been so attractive till recent times. However, the state of Assam has an age-old history of several irrigation projects. The Jamuna irrigation scheme was the first in Assam, completed in 1968–69. Other major irrigation projects include the Sukla irrigation project (covering parts of Kamrup and Baksa district), the Dhansiri irrigation project (in Udalguri district), etc. On the other hand, Bhumki FIS, Rupahi FIS, Dekadong FIS, etc., are some examples of completed medium irrigation projects in Assam. Recently, some initiatives, including the Sonai Irrigation Project, have been fascinating for the region. It is a proposed significant irrigation scheme under the non-National Irrigation Project.

15.5 Proposed Strategies for Effective Conservation and Distribution

Various water resource management techniques are available for proper storage and sustainable and equitable distribution. However, till now, only very few of them are being actively practiced in the Barak valley region. Besides, additional measures shall be adopted in various scenarios to judiciously protect the available water resources and have a rational supply. In various articles of the present era, the potential contamination of various water resources of this region is alarmingly highlighted. Choudhury et al. reported the aquifer contamination due to the landfill leachate. (Choudhury et al. 2022) Paul et al. also highlighted the open dumping sites located in the city and the groundwater pollution that occurs from them. (Paul et al. 2024)Moreover, groundwater reserves in some areas are As contaminated as reported in recent articles (Deb Kanungo 2015; Shah 2012). Gupta and Singh mentioned that the tubewells in the low-lying flood plains are more at risk of As pollution than those at higher elevation (Gupta and Singh 2019).Hence, stringent policies shall be recommended to protect the aquifers from environmental pollution. Provisions for engineered landfills may be adopted to prevent the leachate from reaching the groundwater reserves. Distinguishing features of the engineered landfill include liners, leachate collection systems, etc., which help in preventing groundwater contamination. Chowdhury et al., in one of the recent articles, described the potential of fly ash-based geopolymer as a suitable liner material for adsorbing heavy metals from landfill leachate to mitigate groundwater pollution (Chowdhury et al. 2022). Kumar and Paul also highlighted the sustainability of fly ash generated from the thermal power plants in constructing impermeable landfill liners (Kumar and Paul 2024). The authors further mentioned that adding clay materials like bentonite and kaolinite can provide various benefits in this regard. Regarding As remediation purposes various technologies such as coagulation-flocculation, oxidation, membrane separation, etc. are reported in the literature (Nicomel et al. 2015).

Although rainwater harvesting is an effective water resource management for the Barak Valley, developing and installing the proper setup for the purpose is lacking. Various studies of contemporary times explore several small-scale economic rainwater harvesting schemes often suitable for encountering water scarcity issues. Amos et al. reported the successful design of such a rural scheme for capturing natural water and meeting the demand of a coastal region of Bangladesh where many tube wells are contaminated with As and contain high salinity (Amos et al. 2020). Among various sizes tried out, the authors reported that a storage tank having dimensions of 3 m depth and 100 m × 100 m size would be sufficient for providing water for 85% of the days in a year at the rate of 100 lpcd.

The previous section highlighted that irrigation is one of the most effective water management options. Unfortunately, in Barak Valley, only a few small-scale irrigation schemes exist. Major irrigation projects are under consideration only. However, the region has several irrigation projects. Groundwater stored in the aquifers needs to be utilized to practice active irrigation. As per the study of Singh and Ghosh, there is abundant groundwater stored in untapped condition (Singh and Ghosh 2022). This water can be utilized for irrigation purposes and the growth of Rabi crops during the dry winter season.

15.6 Application in Case of Delta

The Barak River does not form any delta in India. In Bangladesh, its two bifurcations, the Surma River and the Khushiyara River (joins to become Meghna River), converges with the Ganges (called Padma, in Bangladesh) and Brahmaputra (called Jamuna, in Bangladesh) rivers before flowing into the Bay of Bengal (Fig. 15.4) to form the world's largest delta, namely, Bengal delta or Ganges delta or Ganges–Brahmaputra (GB) delta or, more precisely, Ganges–Brahmaputra-Meghna (GBM) delta. It is also known as the "Sundarban Delta", as the world's largest mangrove forest, called the 'Sundarbans mangrove forest', is situated on this deltaic region (Whitehead 2018; Whitehead et al. 2015).

As the three associated rivers of the GBM delta are indirectly connected to the river Barak, to some extent, the problems that arise in this delta are also responsible for the Barak River. There are a variety of problems reported. First problem is–**Sedimentation**. From a study done by Rashid et al. (2024), it is revealed that the sedimentation rate of

Fig. 15.4 Pictorial representation of delta formation by GBM as it merges to the Bay of Bengal (Reproduced with permission from Whitehead et al. 2015, © 2024 Royal Society of Chemistry) (Whitehead et al. 2015)

the Ganges–Brahmaputra (GB) tidal delta plain is about 2.12–4.59 mm yr^{-1} on average, which is high. This sedimentation leads to cloudy water; as a result, the sunlight penetration and oxygen level decrease. And, excessive sedimentation causes deposition of heavy metals and other pollutants, which eventually disrupts the natural aquatic migrations and damages the gills and other organs of the aquatic creatures. Also, the sediments are not distributed properly in the tidal and fluvio-tidal plains due to diverse human activities such as poldering, dams, and embankments, and because of this uneven distribution of sediments, the flow of the river gets disturbed (Rashid et al. 2024).

In addition to Sedimentation, **Organic Matter (OM)** is another problem. A study was performed on the sources and transport mechanisms of OM in the GBM river system in Bangladesh, and in that, they analysed the total organic carbon (TOC), total nitrogen (TN), their stable isotopes $\delta^{13}C$ and $\delta^{15}N$, and the grain size of the sediments. That analysis disclosed the presence of a heterogeneous mixture of OM derived from aquatic environments, terrestrial plants, and anthropogenic sources. In particular, the result demonstrated that the Brahmaputra River contained higher concentrations of TOC with $\delta^{13}C$ values and TN with δ15N values, which pointed to the fact that the source of the OM is primarily C_3 plants. On the other hand, for the Ganges River, the source of OM is the significant anthropogenic inputs (reflected from lower levels of TOC and higher values of the isotopes). And for the Lower Meghna, the values indicated a mixture of marine and terrestrial sources (Khan et al. 2024).

Furthermore, **Nutrient and Heavy Metal Risks**. From a study conducted on the aquacultured and wild mud crabs (*Scylla olivacea*) of GBM delta, the presence of heavy metals, mainly Ni, Cd, and Pb, beyond permissible limits is reported, and a difference in the nutritional quality between the two types of crabs is also explained (Ferdousi et al. 2025). So, this study, in a way, confirms the presence of heavy metals and the lack of nutrients in the surface water of this delta. On the other hand, since the Barak River is indirectly associated with this delta, this finding can be correlated with the investigation of trace elements (As, Fe, and Cu) conducted (Barhai and Das 2024) on the shallow aquifers of the Barak Valley, which revealed elevated concentrations of these elements in several groundwater samples.

At last, **Irrigation and Climatic Change**. Irrigation is a significant livelihood in the GBM delta. But due to the intensive irrigation (mainly in the dry season), groundwater depletion occurs in this delta, and this in turn decreases the depth of the surface water, causing a reduction of stream flow of the GBM basin flowing into the Bay of Bengal. As a result, salinity increases in the water, which can be witnessed in the elevation of the roots in the mangrove forest present in this delta.

Climatic condition is another factor that influences groundwater depletion. The increase in precipitation reduces the impact of groundwater depletion on the streamflow, and a decrease in precipitation and a decline in the snowmelt intensify the impact. In a study done on the Ganges and Brahmaputra River basins (flowing into the Bay of Bengal) by reanalysing land data, the effects of climate change and irrigation on the surface water have been quantified. And, between the two basins, it is found that the impact of these activities is mostly observed in the Ganges basin. The decrease in the flow of the freshwater is up to 1200 m^3/s/year, which is subject to increase or decrease depending on the climatic conditions of the specific area (Maina et al. 2024).

The groundwater depletion caused by irrigation and climate change affects the hydrology of the Ganges, Brahmaputra and Meghna rivers and shows a threat to the stability of the Bay of Bengal.

As it is understood that the Barak River is also a contributing factor in this stream, it is an urgent requirement to control and monitor the water resources present in the Barak Valley, so that the world's largest delta does not get negatively affected.

15.7 Discussions

Assessment of water quality and management of water resources is one of the prioritised tasks in the Barak Valley. It not only ensures a good water supply in the South Assam region but also helps in indirectly maintaining good water quality in the downstream delta region, i.e. in Bangladesh. Various quality monitoring techniques are available for scientists and engineers to use for monitoring purposes. Traditional titrimetric processes are often easy to handle. However, modern chromatographic techniques offer high accuracy and precision. UV–visible spectrophotometric methods are commonly adopted for determining the concentration of various pollutants in water resources. Regarding the management options, rainwater harvesting, groundwater recharge and irrigation are the predominant ones in the Barak valley. However, the current status of water conservation and management is not satisfactory from all perspectives.

Efficient water quality monitoring and management satisfy many sustainable development goals (UN SDGs). It directly supports SDG 6: clean water and sanitation, and SDG 3: good health and well-being. It also aligns with the objectives of SDG 14: life below water. Besides SDG 11: sustainable cities and communities, and SDG 12: responsible consumption and production, water quality monitoring and management tasks are also supported.

15.8 Conclusions

Barak Valley, located in southern Assam, has immense natural water resources. However, due to rapid population increase and unplanned urbanisation, scarcity and stress in the water supply are often felt. Besides, heavy yearly floods often cause deterioration in this region's existing surface and subsurface water resources. Moreover, several groundwater reserves in this region are contaminated with heavy metal pollution, such as As. Therefore, in contemporary times, it is extremely necessary to continuously assess several water quality parameters to find their suitability for diversified applications. In addition to age-old traditional methods, new analytical methods may also be adopted. Sustainable water management practice is highly recommended for judicious storage and supply of existing water resources. Rainwater harvesting, groundwater recharge, running of major irrigation projects, etc., techniques shall be rigorously carried out for efficient water resource management in this area.

Recommendations.

A few valuable recommendations can be put forward in this respect:

- Rigorous water quality monitoring programme shall be started in the Barak valley for the welfare of the region as well as those residing in the downstream delta region.
- Modern analytical techniques need to be integrated in the water management scheme.
- Efficient and inexpensive conservation strategies like rainwater harvesting, groundwater recharge need to be practiced more in number.
- Large scale irrigation projects must be promoted in this region.

Add Acknowledgement (if any).
Na

State whether you used AI or not in preparing the chapter.

No AI was used during the preparation of this chapter.

Table 15.1 Standard persistent organic pollutant classes, nomenclature, regulatory levels in water and effluent, and detection methods. Reproduced with permission from the reference number [21], © 2024 Elsevier

Class	Name	Drinking water EPA maximum contaminant level goal (mg/L)[a]	Surface water EPA maximum recommended level (µg/L)[b]	Analytical technique	Ref
Hydrocarbons	Benzene	0	16–58	Solid phase microextraction technique followed by GC–MS	(Bianchin et al. 2012)
	Toluene	1	520	Solid phase microextraction technique followed by GC–MS	(Bianchin et al. 2012)
	Xylenes	10		Solid phase microextraction technique followed by GC–MS	(Bianchin et al. 2012)
	Polycyclic aromatic Hydrocarbons	0		Solid phase microextraction technique followed by GC–MS	(Bianchin et al. 2012)
	Alachlor	0.002		GC–MS	(Hong and Lemley 1998)
	Atrazine	0.003		Oxidized buckypaper (BP) as a sorbent membrane of a stir-disc solid phase extraction module	(Mijangos et al. 2018; Tomai et al. 2018)
	Ethylbenzene	0.7		Solid phase microextraction technique followed by Gas Chromatography mass spectrometry	
				(GS-MS)	
Pesticides	Carbofuran	0.04		1- LLE and a TPI on-column GC/MS method	(Belmonte Vega et al. 2005; Huertas-Pérez and García-Campaña 2008; Tomai et al. 2018; Vandecasteele et al. 2000)
				2- LC	
				3- LC–MS with electrospray ionization (ESI)	

(continued)

Table 15.1 (continued)

Class	Name	Drinking water EPA maximum contaminant level goal (mg/L)[a]	Surface water EPA maximum recommended level (μg/L)[b]	Analytical technique	Ref
				4- Reversed phase liquid chromatography	
				(RP-HPLC)	
	2,4-D	0.07	12,000	1- HPLC	(Chen et al., 2018; James et al. 2022; Santilio et al. 2022; Vandecasteele et al. 2000; Velkoska-Markovska and Petanovska-Ilievska 2016)
				2- Reversed phase liquid chromatography	
				(RP-HPLC)	
	Dalapon	0.2		1- GC	(Hodgeson 2003; Wu et al. 2017; Xie 2001)
				2- IC-MS	
	Dinoseb	0.007		1- HPLC	(Arancibia et al. 2005; Espinosa-Mansilla et al. 1999; Morales and Cela 2000; Pedrero et al., 1994; Sirisha et al. 2007; Szeto and Price 1991; Vandecasteele et al. 2000)

(continued)

Table 15.1 (continued)

Class	Name	Drinking water EPA maximum contaminant level goal (mg/L)[a]	Surface water EPA maximum recommended level (μg/L)[b]	Analytical technique	Ref
				2- GC	
				3- Capillary electrophoresis	
				4- Electrochemical methods and kinetic-spectrophotometry	
	Diquat	0.02		1- Reversed phased liquid chromatography	(Hodgeson et al. 1997; Jing et al. 2024; Lee et al. 2004; Takino et al. 2000)
				(HPLC) with UV detection	
				2- Liquid chromatography–(electrospray Ionization) mass spectrometry [LC-(ESI)MS]	
				3- LC–MS	

(continued)

Table 15.1 (continued)

Class	Name	Drinking water EPA maximum contaminant level goal (mg/L)[a]	Surface water EPA maximum recommended level (μg/L)[b]	Analytical technique	Ref
				4- HPLC–MS/MS	
	Endothall	0.1		1- GC–MS 2- IC-MS	(Ameel et al. 1997; Barceló 1993; Hu and Rohrer 2020)
	Endrin	0.002	0.03	GC	(Hu and Rohrer 2020)
	Glyphosate	0.7		1- LC–MS/MS 2- HPLC with fluorescence detection 3- HPLC with UV detection 4- HPLC–MS/MS 5- GC–MS/MS 6- GC with detection by flame photometry (DFC)	(Caruso and Santoro 2014; Pires Rosa, 2018; Zhang et al. 2023)
	Lindane	0.0002	4.4	1- Solvent bar microextraction technique using GC 2- GC	(Caruso and Santoro 2014; Vergel et al. 2014)

(continued)

Table 15.1 (continued)

Class	Name	Drinking water EPA maximum contaminant level goal (mg/L)[a]	Surface water EPA maximum recommended level (μg/L)[b]	Analytical technique	Ref
	Methoxychlor	0.04	0.02	GC	(Caruso and Santoro 2014)
	Oxamyl	0.2		1- Photo-induced chemiluminescence (PICL) detection with FI methodology	(Belmonte Vega et al. 2005; Gu and Shamsi, 2010; López-Paz et al. 2014)
				2- LC–MS with electrospray ionization (ESI)	
				3- CEC-atmospheric pressure photoionization	
				(APPI)-MS system	
	Picloram	0.5		Reversed phase liquid chromatography	(Vandecasteele et al. 2000)
				(RP-HPLC)	
	Simazine	0.004		1- PES microextraction followed by LC–MS/MS	(Belmonte Vega et al. 2005; Mijangos et al. 2018; Vandecasteele et al. 2000)
				2- LC–MS with electrospray ionization (ESI)	
				3- Reversed phase liquid	
				Chromatography(RP-HPLC)	
	Toxaphene	0.003	7E-4	GC	(Caruso and Santoro 2014)

(continued)

Table 15.1 (continued)

Class	Name	Drinking water EPA maximum contaminant level goal (mg/L)[a]	Surface water EPA maximum recommended level (μg/L)[b]	Analytical technique	Ref
	2,4,5-TP (Silvex)	0.05	400	Reversed phase liquid chromatography	(Vandecasteele et al. 2000)
				(RP-HPLC)	
	Chlordane	0.002	3.2E-4	GC	(Caruso and Santoro 2014)
	Hexachlorobenzene	0.001	7.9E-5	GC	(Caruso and Santoro 2014)
Industrial	Polychlorinated biphenyls	0.0005	6.4E-5	1- Solid phase microextraction (SPE) and GC	(Ferrario et al. 1997; Ong and Hites 1995; Potter and Pawliszyn 1994; S. Zhang et al. 2007)
Chemicals				2- Solid phase extraction followed by GC/MS	
				3- Gas chromatographic electron capture mass spectroscopy (GC-ECMS)	
				4- SPE and GC–MS	
				5- High resolution gas chromatography/high resolution mass spectrometry (HRGC/HRMS)	
	Carbon tetrachloride	0.005	5		
	o-Dichlorobenzene	0.6	3000		
	p-Dichlorobenzene	0.075	900	Capillary GC–MS	(Fielding et al. 1981)

(continued)

Table 15.1 (continued)

Class	Name	Drinking water EPA maximum contaminant level goal (mg/L)[a]	Surface water EPA maximum recommended level (μg/L)[b]	Analytical technique	Ref
	1,2-Dichloroethane	0	650		
	1,1-Dichloroethylene	0.007	20,000		
	cis-1,2-Dichloroethylene	0.07			
	trans-1,2-Dichloroethylene	0.1	4000		
	1,2-Dichloropropane	0	31		
	Epichlorohydrin	0			
	1,1,2-Trichloroethane	0	8.9	Capillary GC–MS	(Fielding et al. 1981)
	Dioxin (2,3,7,8-TCDD)	0.00000003	5.1E-9		

[a]Environmental Protection Agency (EPA). National primary drinking water regulations: Long Term 1 Enhanced Surface Water Treatment Rule. Final rule. Fed Regist. 2002 Jan 14;67(9):1811–44. PMID: 11,800,007

[b]US Environmental Protection Agency ("EPA"), State-specific water quality standards effective under the Clean Water Act (CWA), (n.d.)

Table 15.2 Allowable levels of common inorganic pollutants in drinking and surface water, and the analytical techniques used to detect them. Reproduced with permission from the reference number 21, © 2024 Elsevier

Name	EPA maximum contaminant level goal (mg/L)[a]	Surface water maximum recommended level (μg/L)[b]	Analytical technique	Ref
Antimony	0.006	640	Atomic absorption spectrometric	(Fishman and Friedman 1989)
Arsenic	0	0.14	1- Inductively coupled plasma-mass spectrometry (ICP-MS) 2- Mass spectroscopy 3- X-ray fluorescence spectroscopy 4- Atomic absorption spectroscopy (AAS) 5- Calorimetric 6- Electrochemical detection: amperometric, voltammetry 7- Electrochemical impedance spectroscopy 8- Potentiometric detection	(Sharma and Negi 2020)
Asbestos (fiber > 10micrometers)	7 million fibers per liter (MFL)			
Barium	2		1- Atomic absorption spectrometric	(Fishman and Friedman 1989)
Beryllium	0.004	641	2- Atomic emission spectrometric 1- Atomic absorption spectrometric 2- Atomic emission spectrometric (ICP)	(Fishman and Friedman 1989)

(continued)

Table 15.2 (continued)

Name	EPA maximum contaminant level goal (mg/L)[a]	Surface water maximum recommended level (μg/L)[b]	Analytical technique	Ref
Cadmium	0.005		1- Inductively coupled plasma-mass spectrometry(ICP-MS)	(Cobben et al. 1992; Fishman and Friedman 1989; Sharma and Negi 2020)
			2- Mass spectroscopy	
			3- X-ray fluorescence spectroscopy	
			4- Atomic absorption spectroscopy (AAS)	
			5- Atomic emission spectrometric	
			6- Electrochemical detection: amperometric, voltammetry,	
			7- Electrochemical impedance spectroscopy	
			8- Potentiometric detection	
Chromium (total)	0.1		1- Inductively coupled plasma-mass spectrometry(ICP-MS)	(Fishman and Friedman 1989; Sharma and Negi 2020)
			2- Mass spectroscopy	
			3- X-ray fluorescence spectroscopy	
			4- Atomic absorption spectroscopy (AAS),	
			5- Calorimetric	

(continued)

Table 15.2 (continued)

Name	EPA maximum contaminant level goal (mg/L)[a]	Surface water maximum recommended level (μg/L)[b]	Analytical technique	Ref
			6- Electrochemical detection: amperometric, voltammetry	
			7- Electrochemical impedance spectroscopy	
			8- Potentiometric detection	
Copper	1.3		1- Inductively coupled plasma-mass spectrometry(ICP-MS)	(Cobben et al. 1992; Fishman and Friedman 1989; Sharma and Negi 2020)
			2- Mass spectroscopy	
			3- X-ray fluorescence spectroscopy	
			4- Atomic absorption spectroscopy (AAS),	
			5- Atomic emission spectrometric	
			6- Electrochemical detection: amperometric, voltammetry	
			7- Electrochemical impedance spectroscopy	
			8- Potentiometric detection	
			9- CHEMFET	
Cyanide (as free cyanide)	0.2	400	Calorimetric	(Fishman and Friedman 1989)

(continued)

Table 15.2 (continued)

Name	EPA maximum contaminant level goal (mg/L)[a]	Surface water maximum recommended level (μg/L)[b]	Analytical technique	Ref
Fluoride	4.0		Ion exchange chromatographic	
Lead	0		1- Inductively coupled plasma-mass spectrometry(ICP-MS)	(Cobben et al. 1992; Sharma and Negi 2020)
			2- Mass spectroscopy	
			3- X-ray fluorescence spectroscopy	
			4- Atomic absorption spectroscopy (AAS)	
			5- Atomic emission spectrometric	
			6- Electrochemical detection: amperometric, voltammetry	
			7- Electrochemical impedance spectroscopy	
			8- Potentiometric detection	
			9- CHEMFET	
Mercury (inorganic)	0.002		1- Inductively coupled plasma-mass spectrometry(ICP-MS)	(Sharma and Negi, 2020)
			2- Mass spectroscopy	
			3- X-ray fluorescence spectroscopy	
			4- Atomic absorption spectroscopy (AAS)	
			5- Electrochemical detection: amperometric, voltammetry	

(continued)

Table 15.2 (continued)

Name	EPA maximum contaminant level goal (mg/L)[a]	Surface water maximum recommended level (µg/L)[b]	Analytical technique	Ref
			6- Electrochemical impedance spectroscopy	
			7- Potentiometric detection	
Nitrate (measured as Nitrogen) Nitrite (measured as Nitrogen)	10		Ion exchange chromatographic	(Fishman and Friedman 1989)
	1		1- Calorimetric	(Fishman and Friedman 1989)
			2- Ion exchange chromatographic	
Selenium	0.05	4200	Atomic absorption spectrometric	(Fishman and Friedman 1989)
Thallium	0.0005	0.47	Atomic absorption spectrometric	(Fishman and Friedman 1989)

[a]Environmental Protection Agency (EPA). National primary drinking water regulations: Long Term 1 Enhanced Surface Water Treatment Rule. Final rule. Fed Regist. 2002 Jan 14;67(9):1811–44. PMID: 11,800,007

[b]US Environmental Protection Agency ("EPA"), State-specific water quality standards effective under the Clean Water Act (CWA), (n.d.)

References

Action Plan for Son Beel-Priority V (Issue April) (2019)

Ameel JJ, Axler RP, Owen CJ, Johnson KE, Guyton B (1997) The determination of hydrothol 191 (amine salt of endothall) in eutrophic wastewater ponds. Chemosphere 34(3):641–654. https://doi.org/10.1016/S0045-6535(96)00359-1

Amos CC, Ahmed A, Rahman A (2020) Sustainability in water provision in rural communities: the feasibility of a village scale rainwater harvesting scheme. Water Resour Manage 34(15):4633–4647. https://doi.org/10.1007/s11269-020-02679-1

Annayat W, Ashwini K, Sil BS (2024) Relative prioritizing of sub-watersheds of Barak River built on, morphometric parameters, LULC and SCS-CN model. Int J Energy Water Resour 2012. https://doi.org/10.1007/s42108-024-00291-2

Arancibia JA, Delfa GM, Boschetti CE, Escandar GM, Olivieri AC (2005) Application of partial least-squares spectrophotometric-multivariate calibration to the determination of 2-sec-butyl-4,6-dinitrophenol (dinoseb) and 2,6-dinitro-p-cresol in industrial and water samples containing hydrocarbons. Anal Chim Acta 553(1–2):141–147. https://doi.org/10.1016/j.aca.2005.08.017

Badawy MEI, El-Nouby MAM, Kimani PK, Lim LW, Rabea EI (2022) A review of the modern principles and applications of solid-phase extraction techniques in chromatographic analysis. Anal Sci 38(12):1457–1487. https://doi.org/10.1007/s44211-022-00190-8

Barceló D (1993) Chapter 5 official methods of analysis of priority pesticides in water using gas chromatographic techniques, pp 149–180. https://doi.org/10.1016/S0167-9244(08)70125-2

Barhai A, Das S (2024) A multivariate investigation of groundwater chemistry data in Barak River Valley, South Assam India. Environ Qual Manag 34(2):1–10. https://doi.org/10.1002/tqem.22316

Belmonte Vega A, Garrido Frenich A, Martínez Vidal JL (2005) Monitoring of pesticides in agricultural water and soil samples from Andalusia by liquid chromatography coupled to mass spectrometry. Anal Chim Acta 538(1–2):117–127. https://doi.org/10.1016/j.aca.2005.02.003

Bhuiyan JR, Gupta S (2007) A comparative hydrobiological study of a few ponds of Barak Valley, Assam and their role as sustainable water resources. J Environ Biol 28(4):799–802

Bianchin JN, Nardini G, Merib J, Dias AN, Martendal E, Carasek E (2012) Simultaneous determination of polycyclic aromatic hydrocarbons and benzene, toluene, ethylbenzene and xylene in water samples using a new sampling strategy combining different extraction modes and temperatures in a single extraction solid-phase microextra. J Chromatogr A 1233:22–29. https://doi.org/10.1016/j.chroma.2012.02.022

Campanale C, Massarelli C, Losacco D, Bisaccia D, Triozzi M, Uricchio VF (2021) The monitoring of pesticides in water matrices and the analytical criticalities: a review. TrAC, Trends Anal Chem 144:116423. https://doi.org/10.1016/j.trac.2021.116423

Caruso A, Santoro M (2014) Detection of Organochlorine Pesticides by GC-ECD Following U.S. EPA Method 8081

Casado J, Santillo D, Johnston P (2018) Multi-residue analysis of pesticides in surface water by liquid chromatography quadrupole-Orbitrap high resolution tandem mass spectrometry. Anal Chim Acta 1024:1–17. https://doi.org/10.1016/j.aca.2018.04.026

Chen X, Zhang H, Wan Y, Chen X, Li Y (2018) Determination of 2,4-Dichlorophenoxyacetic acid (2,4-D) in rat serum for pharmacokinetic studies with a simple HPLC method. PLoS ONE 13(1):e0191149. https://doi.org/10.1371/journal.pone.0191149

Chouhury P (2024) Nature's pride at risk: river Barak's Pristine Ox-Bow lakes threatened by urea pollution. Barak Bulletin, 4 July. https://Barakbulletin.Com/En_US/Natures-Pride-at-Risk-River-Baraks-Pristine-Ox-Bow-Lakes-Threatened-by-Urea-Pollution/

Choudhury M, Jyethi DS, Dutta J, Purkayastha SP, Deb D, Das R, Roy G, Sen T, Bhattacharyya KG (2022) Investigation of groundwater and soil quality near to a municipal waste disposal site in Silchar, Assam, India. Int J Energy Water Resour 6(1):37–47. https://doi.org/10.1007/s42108-021-00117-5

Chowdhury A, Naz A, Chowdhury A (2022) Waste to resource: Applicability of fly ash as landfill geoliner to control ground water pollution. Materials Today: Proceedings 60:8–13. https://doi.org/10.1016/j.matpr.2021.10.367

Cobben PLHM, Egberink RJM, Bomer JG, Bergveld P, Verboom W, Reinhoudt DN (1992) Transduction of selective recognition of heavy metal ions by chemically modified field effect transistors (CHEMFETs). J Am Chem Soc 114(26):10573–10582. https://doi.org/10.1021/ja00052a063

Das S, Das T (2023) Flood, livelihood, and community resilience: a study from Barak Valley region of Assam in Northeast India. In: International handbook of disaster research. Springer Nature Singapore, pp 2189–2202. https://doi.org/10.1007/978-981-19-8388-7_63

Deb Kanungo T (2015) Arsenic mitigation processes on trial and tested in Barak Valley, Assam, India. Int J Pharm Drug Anal 3(1), 12–18. www.ijpda.com

Dutta N, Sarma JN (2017) River channel change of Barak River and its seven tributaries, Barak Valley, North-East India. J Appl Geochem 19(4):457–463. https://doi.org/10.13140/RG.2.2.31501.23524

Dutta N, Dutta Choudhury K, Dutta Choudhury M, Nath SS (2019) Hydro power generation: a case study of Sonbeel. Int J Innov Eng Technol (IJIET) 13(4):45–50. https://doi.org/10.21172/ijiet.134.07

Espinosa-Mansilla A, Madera AZ, Salinas F (1999) Use of a stopped-flow pneumatic mixing module to analyze dinitrophenol pesticides. Simultaneous determination of dinoseb and dinobuton. J Agric Food Chem 47(5):1976–1980. https://doi.org/10.1021/jf980821b

Ferdousi L, Gain AK, Bithi UH, Begum M, Yeasmin MS, Siddique MAB, Tasnim N, Al Noman Z, Khatun MR, Shaikh MEA, Sharmin S, Ahmed S, Goshwami A, Rodgers EM (2025) Nutritional prospects and heavy metal risks in fattened versus wild mud crabs of the Ganges-Brahmaputra delta: Implications for sustainable management. Aquaculture 595:741619. https://doi.org/10.1016/j.aquaculture.2024.741619

Ferrario J, Byrne C, Dupuy AE (1997) Background contamination by coplanar polychlorinated biphenyls (PCBs) in trace level high resolution gas chromatography/high resolution mass spectrometry (HRGC/HRMS) analytical procedures. Chemosphere 34(11):2451–2465. https://doi.org/10.1016/S0045-6535(97)00083-0

Fielding M, Gibson TM, James HA (1981) Levels of trichloroethylene, tetrachloroethylene and p-dichlorobenzene in ground waters. Environ Technol Lett 2(12):545–550. https://doi.org/10.1080/09593338009384086

Fishman MJ, Friedman LC (1989) Methods for determination of inorganic substances in water and fluvial sediments. US Geological Survey

Groundwater management in India: A multi-state field study of availability, and locally appropriate solutions for sustainable, equitable and efficient use of groundwater (n.d.) In Assam State Report, 2022

Gu C, Shamsi SA (2010) CEC-atmospheric pressure ionization MS of pesticides using a surfactant-bound monolithic column. Electrophoresis 31(7):1162–1174. https://doi.org/10.1002/elps.200900739

Gupta A, Bhattacharjee D, Borah P, Debkanungo T, Paulchoudhury C (2015) Arsenic contamination of groundwater in Barak Valley, Assam, India: topography-based analysis and risk assessment. In: safe and sustainable use of arsenic-contaminated aquifers in the Gangetic plain. Springer International Publishing, pp 81–96. https://doi.org/10.1007/978-3-319-16124-2_6

Gupta A, Singh EJ (2019) Arsenic-Iron relationships in aquifers of North East India: implications for public health and the environment. Environ Manage 63(4):437–443. https://doi.org/10.1007/s00267-018-0999-4

Gupta A (2004) Drinking water quality in Barak Valley, Assam, North East India: prioritizing management options. In: Proceedings of the 2nd APHW Conference, January 2004, pp 1–7

Hasan GO, Paul PM, Don P (2007) Distribution of heavy metals in vegetation surrounding the Blackstone River, USA: considerations regarding sediment contamination and long term metals transport in freshwater riverine ecosystems. J Environ Biol 28:493–502

Hodgeson JW (2003) Determination of Haloacetic acids and Dalapon in drinking water by liquid-liquid extraction, derivatization and gas chromatography with electron capture detection. In EPA Method 552:2

Hodgeson JW, Bashe WJ, Eichelberger JW, Munch JW (1997) Method 549.2: determination of diquat and paraquat in drinking water by liquid-solid extraction and high performance liquid chromatography with ultraviolet detection. In: EPA Rep

Hong S, Lemley AT (1998) Gas chromatographic–mass spectrometric determination of alachlor and its degradation products by direct aqueous injection. J Chromatogr A 822(2):253–261. https://doi.org/10.1016/S0021-9673(98)00625-6

Hu J, Rohrer J (2020) Determination of four polar pesticides in drinking water by IC-MS. Thermo Fish Sci Inc

Huertas-Pérez JF, García-Campaña AM (2008) Determination of N-methylcarbamate pesticides in water and vegetable samples by HPLC with post-column chemiluminescence detection using the luminol reaction. Anal Chim Acta 630(2):194–204. https://doi.org/10.1016/j.aca.2008.09.047

James S, Ele Yaya E, Singh Chandravanshi B, Zewge F (2022) Determination of 2,4-dichlorophenoxyacetic acid in water, sediment and soil using high performance liquid chromatography. Bull Chem Soc Ethiop 37(2):251–263. https://doi.org/10.4314/bcse.v37i2.1

Jing C, Qun X, Rohrer V (2024) Sensitive and rapid determination of paraquat and diquat in tap and environmental waters. Thermo Fish Sci Appl Note 1051

Kadadou D, Tizani L, Alsafar H, Hasan SW (2024) Analytical methods for determining environmental contaminants of concern in water and wastewater. MethodsX 12:102582. https://doi.org/10.1016/j.mex.2024.102582

Khan MHR, Liu J, Rahman A, Huang Y, Tan L, Liu S, Chen Z (2024) Sources and transport of organic matter in the Ganges-Brahmaputra-Meghna River system of Bengal Basin. Bangladesh. Marine Pollution Bulletin 208:116973. https://doi.org/10.1016/j.marpolbul.2024.116973

Khangembam S, Kshetrimayum KS (2019) Evaluation of hydrogeochemical controlling factors and water quality index of water resources of the Barak valley of Assam, Northeast India. Groundw Sustain Dev 8:541–553. https://doi.org/10.1016/j.gsd.2019.02.001

Kucharska M, Grabka J (2010) A review of chromatographic methods for determination of synthetic food dyes. Talanta 80(3):1045–1051. https://doi.org/10.1016/j.talanta.2009.09.032

Kumar K, Paul B (2024) Towards a sustainable geoliner construction in landfills by potential blending of fly ash with kaolin clay alternative: a review with an insight to Indian scenario. Waste Dispos Sustain Energy 6(2):243–258. https://doi.org/10.1007/s42768-023-00178-8

Lee X, Kumazawa T, Fujishiro M, Hasegawa C, Arinobu T, Seno H, Ishii A, Sato K (2004) Determination of paraquat and diquat in human body fluids by high-performance liquid chromatography/tandem mass spectrometry. J Mass Spectrom 39(10):1147–1152. https://doi.org/10.1002/jms.695

López-Paz JL, Catalá-Icardo M, Langa-Sánchez A (2014) Determination of N -methylcarbamate pesticides using flow injection with photoinduced chemiluminescence detection. Int J Environ Anal Chem 94(6):606–617. https://doi.org/10.1080/03067319.2013.879295

Mahapatra MK, Mishra HS (2005) Ground water pollution in Subarnapur and Nuapada districts of Orissa. India. Poll. Res. 24:863–865

Mahara BM, Borossay J, Torkos K (1998) Liquid-liquid extraction for sample preparation prior to gas chromatography and gas chromatography-mass spectrometry determination of herbicide and pesticide compounds. Microchem J 58(1):31–38. https://doi.org/10.1006/mchj.1997.1517

Maina FZ, Getirana A, Kumar SV, Saharia M, Biswas NK, McLarty S, Appana R (2024) Irrigation-driven groundwater depletion in the Ganges-Brahmaputra basin decreases the streamflow in the Bay of Bengal. Commun Earth and Environ 5(1):169. https://doi.org/10.1038/s43247-024-01348-0

Mandal S, Poi R, Hazra DK, Ansary I, Bhattacharyya S, Karmakar R (2023) Review of extraction and detection techniques for the analysis of pesticide residues in fruits to evaluate food safety and make legislative decisions: Challenges and anticipations. J Chromatogr B 1215:123587. https://doi.org/10.1016/j.jchromb.2022.123587

Mazumdar A, Acharjee PR, Bhattacharya J (1998) Statistical Profile of Barak Valley 1998. North Eastern Centre for Advanced Studies, Silchar, Assam, India

Mazzeo JR, D.Neue U, Kele M, Plumb RS (2005) Advancing LC performance with smaller particles and higher pressure. Anal Chem 77(23), 460 A-467 A. https://doi.org/10.1021/ac053516f
Mijangos L, Ziarrusta H, Olivares M, Zuloaga O, Möder M, Etxebarria N, Prieto A (2018) Simultaneous determination of 41 multiclass organic pollutants in environmental waters by means of polyethersulfone microextraction followed by liquid chromatography–tandem mass spectrometry. Anal Bioanal Chem 410(2):615–632. https://doi.org/10.1007/s00216-017-0763-2
Morales S, Cela R (2000) Highly selective and efficient determination of US Environmental Protection Agency priority phenols employing solid-phase extraction and non-aqueous capillary electrophoresis. J Chromatogr A 896(1–2):95–104. https://doi.org/10.1016/S0021-9673(00)00581-1
Nasiri M, Ahmadzadeh H, Amiri A (2020) Sample preparation and extraction methods for pesticides in aquatic environments: A review. TrAC, Trends Anal Chem 123:115772. https://doi.org/10.1016/j.trac.2019.115772
Nicomel N, Leus K, Folens K, Van Der Voort P, Du Laing G (2015) Technologies for arsenic removal from water: current status and future perspectives. Int J Environ Res Public Health 13(1):62. https://doi.org/10.3390/ijerph13010062
Ong VS, Hites RA (1995) Determination of pesticides and polychlorinated biphenyls in water: a low-solvent method. Environ Sci Technol 29(5):1259–1266. https://doi.org/10.1021/es00005a018
Parikh KS, Parikh J, Ram TLR (1999) Air and water quality management: new initiatives needed. India Development Report, 2000, pp 85–96
Paul MK, Dey M, Sharma C (2024) Impact of open dumping site on groundwater quality in Silchar City, Assam, India. Water Air Soil Pollut 235(10):659. https://doi.org/10.1007/s11270-024-07434-5
Pedrero M, Casado B, de Villena FJM, Pingarrón JM (1994) Determination of dinoseb by adsorptive stripping voltammetry using a mercury film electrode. Fresenius' J Anal Chem 349(7):546–551. https://doi.org/10.1007/BF00323992
Pires Rosa PC (2018) Brief review analytical methods for the determination of glyphosate. MOJ Toxicology 4(1). https://doi.org/10.15406/mojt.2018.04.00088
Potter DW, Pawliszyn J (1994) Rapid determination of polyaromatic hydrocarbons and polychlorinated biphenyls in water using solid-phase microextraction and GC/MS. Environ Sci Technol 28(2):298–305. https://doi.org/10.1021/es00051a017
Rashid MB, Ahsan K, Majlis ABK, Ahsan MK, Mahmud A (2024) Sedimentation and coastal area management in the human-modified Ganges-Brahmaputra tidal delta plain of Bangladesh. Int J River Basin Manag 22(1):31–43. https://doi.org/10.1080/15715124.2022.2092489
Ritchie JC, Schiebe FR (2000) Water quality. Remote Sens Hydrol Water Manag 287–303. https://doi.org/10.1007/978-3-642-59583-7_13
Santilio A, Girolimetti S, Picardo V (2022) Simple and rapid high-performance liquid chromatography method for simultaneous determination of picloram and 2,4-D in pesticide formulations. Analytica 3(4):430–438. https://doi.org/10.3390/analytica3040029
Shah BA (2012) Role of Quaternary stratigraphy on arsenic-contaminated groundwater from parts of Barak Valley, Assam. North-East India. Environ Earth Sci 66(8):2491–2501. https://doi.org/10.1007/s12665-011-1472-3
Sharma C, Negi YS (2020) Methods of inorganic pollutants detection in water. In: Inorganic Pollutants in Water. Elsevier, pp 115–134. https://doi.org/10.1016/B978-0-12-818965-8.00007-X
Singh MK, Ghosh S (2022) River aquifer interaction modeling for the waterlogged aquifer and sustainable groundwater irrigation along river flood plain. J Geol Soc India 98(9):1271–1282. https://doi.org/10.1007/s12594-022-2162-3

Sirisha K, Mallipattu S, Jayarama Reddy SR (2007) Differential pulse adsorptive stripping voltammetric determination of Chlorpyrifos at a Sepiolite Modified Carbon Paste Electrode. Anal Lett 40(10):1939–1950. https://doi.org/10.1080/00032710701384766

Skoog DA, West DM, Holler FJ, Crouch SR (2022) Fundamental of Analytical Chemistry (10th ed.). Cengage Learning India Pvt. Ltd

Srivastava N, Agrawal M, Tyagi A (2003) Study of physico-chemical characteristics of water bodies around Jaipur. J Environ Biol 24:177–180

Szeto SY, Price PM (1991) High-performance liquid chromatography method for the determination of dinoseb: application to the analysis of residues in raspberries. J Agric Food Chem 39(9):1614–1617. https://doi.org/10.1021/jf00009a015

Takino M, Daishima S, Yamaguchi K (2000) Determination of diquat and paraquat in water by liquid chromatography/electrospray–mass spectrometry using volatile ion-pairing reagents. Anal Sci 16(7):707–711. https://doi.org/10.2116/analsci.16.707

Tomai P, Martinelli A, Morosetti S, Curini R, Fanali S, Gentili A (2018) Oxidized Buckypaper for stir-disc solid phase extraction: evaluation of several classes of environmental pollutants recovered from surface water samples. Anal Chem 90(11):6827–6834. https://doi.org/10.1021/acs.analchem.8b00927

Usha R, Ramalingam K, Bharathi Rajan UD (2006) Freshwater lakes–a potential source for aquaculture activities–a model study on Perumal Lake, Cuddalore, Tamil Nadu. J Environ Biol 27:713–722

Vandecasteele K, Gaus I, Debreuck W, Walraevens K (2000) Identification and quantification of 77 pesticides in groundwater using solid phase coupled to liquid—liquid Microextraction and reversed-phase liquid chromatography. Anal Chem 72(14):3093–3101. https://doi.org/10.1021/ac991359c

Velkoska-Markovska L, Petanovska-Ilievska B (2016) Quantitative Determination of 2, 4-D in Pesticides Monosan Herbi and Dma-6. J Agric, Food Environ Sci, JAFES 67(1985):62–67

Vergel C, Guerrero EJ, Mendiguchía C, Moreno C (2014) Determination of Organochloride and Triazine Pesticides in Natural Waters by Solvent Bar Microextraction. Anal Lett 47(13):2209–2220. https://doi.org/10.1080/00032719.2014.902462

Wang X, Zhang Y, Qin K, Nian S, Xiao T, Zhang Q, Wang X, Ma J (2012) Simultaneous determination of sulfamethoxazole and trimethoprim in rat plasma by LC-ESI-MS and its application to a pharmacokinetic study. J Liq Chromatogr Relat Technol 35(7):951–962. https://doi.org/10.1080/10826076.2011.615087

Whitehead PG, Barbour E, Futter MN, Sarkar S, Rodda H, Caesar J, Butterfield D, Jin L, Sinha R, Nicholls R, Salehin M (2015) Impacts of climate change and socio-economic scenarios on flow and water quality of the Ganges, Brahmaputra and Meghna (GBM) river systems: low flow and flood statistics. Environ Sci Process Impacts 17(6):1057–1069. https://doi.org/10.1039/C4EM00619D

Whitehead PG (2018) Biophysical modelling of the Ganges, Brahmaputra, and Meghna Catchment. In: Ecosystem services for well-being in deltas. Springer International Publishing, pp 249–262. https://doi.org/10.1007/978-3-319-71093-8_13

Wu S, Anumol T, Gandhi J, Snyder SA (2017) Analysis of haloacetic acids, bromate, and dalapon in natural waters by ion chromatography–tandem mass spectrometry. J Chromatogr A 1487:100–107. https://doi.org/10.1016/j.chroma.2017.01.006

Xie Y (2001) Analyzing Haloacetic Acids Using Gas Chromatography/Mass Spectrometry. Water Res 35(6):1599–1602. https://doi.org/10.1016/S0043-1354(00)00397-3

Zhang S, Zhang Q, Darisaw S, Ehie O, Wang G (2007) Simultaneous quantification of polycyclic aromatic hydrocarbons (PAHs), polychlorinated biphenyls (PCBs), and pharmaceuticals and

personal care products (PPCPs) in Mississippi river water, in New Orleans, Louisiana, USA. Chemosphere 66(6):1057–1069. https://doi.org/10.1016/j.chemosphere.2006.06.067
Zhang H, Dou J, Miao R, Hu J, Huo Z, Zhang F, Ji W (2023) An analytical method for the determination of glyphosate and aminomethylphosphoric acid using an anionic polar pesticide column and the application in urine and serum from glyphosate poisoning patients. Anal Methods 15(13):1668–1673. https://doi.org/10.1039/D3AY00039G

www.ingramcontent.com/pod-product-compliance
Ingram Content Group UK Ltd.
Pitfield, Milton Keynes, MK11 3LW, UK
UKHW050301220926
475829UK00006B/110
9783032064998